T0141686

METEOROLOGICAL MONOGRAPHS

VOLUME 24 | DECEMBER 1993 | NUMBER 46

THE REPRESENTATION OF CUMULUS CONVECTION IN NUMERICAL MODELS

Edited by

Kerry A. Emanuel
David J. Raymond

American Meteorological Society
45 Beacon Street, Boston, Massachusetts 02108

ISBN 1-878220-13-6
ISSN 0065-9401

Published by the American Meteorological Society
45 Beacon St., Boston, MA 02108

Printed in the United States of America
by Lancaster Press, Lancaster, Pennsylvania

TABLE OF CONTENTS

PREFACE

Of the many subgrid-scale processes that must be represented in numerical models of the atmosphere, cumulus convection is perhaps the most complex and perplexing. It is by now well known that the simulation of many individual phenomena, ranging from tropical and extratropical cyclones to the Madden–Julian oscillation, is sensitive to the way convection is represented. It has also been recognized that the water vapor content of large parts of the atmosphere is strongly controlled by cloud microphysical processes, including those operating within cumulus clouds, yet scant attention has been paid to this problem in formulating most existing convection schemes. Given that water vapor variability is the strongest feedback in climate simulations, it would seem timely to reconsider this and other aspects of the cumulus parameterization problem.

As a step in this direction, a workshop on cumulus parameterization was held on 3–5 May 1991, at Key Biscayne, Florida, bringing together many of the leading specialists in convection and convective parameterization. The main objectives of the workshop were to promote a vigorous discussion of the major issues in representing cumulus convection in numerical models and to produce a monograph suitable both for describing the state of the art of the field and for use in graduate education. This volume is the fruit of the labors of most of the individuals involved in the workshop. Each chapter has been reviewed by one or more of the authors of other chapters, and an attempt has been made to make the material accessible to nonspecialists.

The monograph is divided into six parts. Part I provides an overview of the problem, including descriptions of cumulus clouds and the effects of ensembles of cumulus clouds on mass, momentum, and vorticity distributions. A review of closure assumptions is also provided. A review of "classical" convection schemes in widespread use is provided in Part II. These schemes include the convective adjustment scheme developed by Manabe, the Kuo parameterization, the Betts–Miller scheme, and the Arakawa–Schubert parameterization. The special problems associated with the representation of convection in mesoscale models are discussed in Part III, along with descriptions of some of the commonly used mesoscale schemes. Part IV covers some of the problems associated with the representation of convection in climate models, while the parameterization of slantwise convection is the subject of Part V. The monograph concludes with a single paper describing some recent and very promising efforts to use explicit numerical simulations of ensembles of convective clouds to test cumulus representations.

No single chapter is devoted to the issue of validation of cumulus parameterizations. In the opinion of the editors, this is an issue that needs far more attention and will, we trust, be the subject of future publications.

Many organizations and individuals contributed to this volume. In particular, we would like to thank Mr. Joel Sloman of MIT, who was responsible for a major part of the editing, and Dr. Ronald Taylor of the National Science Foundation, which in part subsidized both the workshop and this monograph.

Kerry A. Emanuel
David J. Raymond
Editors

Part I

General Considerations

Chapter 1

Closure Assumptions in the Cumulus Parameterization Problem

AKIO ARAKAWA

Department of Atmospheric Sciences, University of California, Los Angeles, Los Angeles, California

1.1. Introduction

Physical processes associated with condensation of water vapor are inherently nonlinear and, therefore, their collective effects can directly interact with larger-scale circulations. But most individual clouds, in which condensation takes place, are subgrid-scale for the conventional grid size of general circulation and numerical weather prediction models. Then, for a set of model equations to be closed, we must *formulate the collective effects of subgrid-scale clouds in terms of the prognostic variables of grid scale.* This is the problem of cumulus parameterization in numerical modeling of the atmosphere.

It should be emphasized that the need for cumulus parameterization is not limited to numerical models. Understanding the interaction between moist-convective and large-scale processes is one of the most fundamental issues in dynamics of the atmosphere, and cumulus parameterization is needed for a closed formulation of that interaction regardless of whether we are using numerical, theoretical, or conceptual models. Even if we had a numerical model that resolved all scales, understanding its results inevitably requires simplifications that involve various levels of "parameterization." In Fig. 1.1, the upper half of the loop represents the effect of large-scale processes on moist-convective processes, while the lower half represents that of moist-convective processes on large-scale processes. For the loop to be closed, it must include the segments shown by heavy curves, a formulation of which is precisely the objective of a cumulus parameterization. (We refer to the upper half as "control" and the lower half as "feedback," although there is no need to identify which is the first in the loop. These terminologies are convenient only when we are concentrating on the parameterization problem represented by the right half of the loop.)

There are a number of cumulus clouds in virtually all tropical and most extratropical disturbances. Some of these clouds may be developing, some may be fully developed, and some may be decaying. In the cumulus parameterization problem we are concerned with the statistical behavior of such cloud ensembles under different large-scale conditions. The problem of cumulus parameterization, therefore, is analogous to that of climate dynamics. In the latter problem, we are concerned with time and space means and their statistical significance, identification of external and internal parameters for different temporal and spatial scales, free fluctuations under given external conditions, interactions between low- and high-frequency variations, and possible multiple equilibria of the overall regime. All of these may have their counterparts in the cumulus parameterization problem. Difficulties in a cumulus parameterization can then be compared with those in parameterizing transient processes in the climate system.

Since cumulus parameterization is an attempt to formulate the collective effect of cumulus clouds without predicting individual clouds, it is a *closure problem* in which we seek a limited number of equations that govern the statistics of a system with huge dimensions. The core of the parameterization problem is, therefore, in the choice of appropriate closure assumptions. When we have global models with comprehensive physics in mind, rather than idealized models with a more limited scope, closure assumptions must meet the following requirements:

(i) *Closure assumptions must not lose the predictability of large-scale fields.* This is an obvious requirement since we need to parameterize clouds for predicting the time evolution of large-scale fields. If we wish to assume that a certain variable is in an equilibrium, the variable must be one whose prediction is not intended by the model.

(ii) *Closure assumptions must be valid quasi-universally.* This is also an obvious requirement because comprehensive global models must be valid for a variety of synoptic and surface conditions.

One may then ask, Can we really find closure assumptions satisfying these requirements? In other words, To what extent is it possible to parameterize cumulus clouds? These are difficult questions to answer. The difficulty is amplified by the existence of intermediate scales in cloud organization, which are gen-

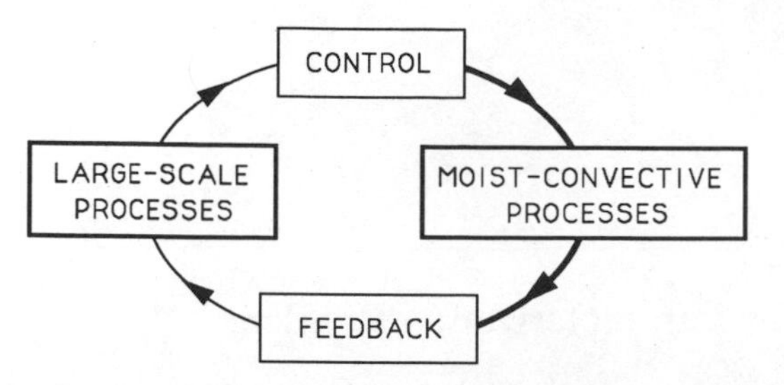

FIG. 1.1. A schematic figure showing the interaction between large-scale and moist-convective processes.

erally termed "mesoscale." We will discuss problems associated with this situation in a later section. Here we emphasize that, even when there was no mesoscale organization, or even when the model grid resolves such organization, the answer to the question of *parameterizability* is by no means obvious.

The purpose of this chapter is to review the conceptual basis for cumulus parameterization in view of closure assumptions. We will concentrate on the thermodynamical aspect of parameterizing deep cumulus clouds, for which most of the conceptual difficulties exist. In section 1.2, we define the thermodynamical aspect of the cumulus parameterization problem. Section 1.3 presents a classification of closure assumptions. Sections 1.4–1.6 review some basic examples of closure assumptions and their logical consequences. Section 1.7 reviews some of the current studies on parameterizability of cumulus convection. Finally, section 1.8 presents summary and further discussions.

1.2. The thermodynamical aspect of the cumulus parameterization problem

In this section we define the objective of cumulus parameterization more specifically, concentrating on its thermodynamical aspect. With the pressure coordinate, the large-scale potential temperature and water vapor budget equations may be written as

$$c_p\left[\frac{\partial\bar{\theta}}{\partial t} + \bar{\mathbf{v}}\cdot\nabla\bar{\theta} + \bar{\omega}\frac{\partial\bar{\theta}}{\partial p}\right] = \left(\frac{p_0}{p}\right)^{R/c_p} Q_1 \quad (1.1)$$

and

$$L\left[\frac{\partial\bar{q}}{\partial t} + \bar{\mathbf{v}}\cdot\nabla\bar{q} + \bar{\omega}\frac{\partial\bar{q}}{\partial p}\right] = -Q_2, \quad (1.2)$$

where an overbar denotes the Reynolds average with respect to a large-scale horizontal area. The quantities Q_1 and Q_2 are the *apparent heat source* and *apparent moisture sink* (Yanai et al. 1973), respectively. All other symbols are standard except that subscript p for time and horizontal derivatives is omitted. The area-averaged continuity equation,

$$\nabla\cdot\bar{\mathbf{v}} + \frac{\partial\bar{\omega}}{\partial p} = 0, \quad (1.3)$$

determines $\bar{\omega}$ from $\bar{\mathbf{v}}$.

Obviously, Q_1 and Q_2 are different from the true heat source and moisture sink, because (1.1) and (1.2) are applied to averaged fields. When the large-scale domain is a grid box of a model, Q_1 and Q_2 include the collective effects of subgrid-scale processes. Using the usual assumptions in Reynolds averaging and those in the anelastic approximation, and neglecting the horizontal transports due to subgrid-scale processes, we may write

$$Q_{1C} \equiv Q_1 - Q_R = L\bar{C} - \frac{\partial\overline{\omega' s'}}{\partial p}, \quad (1.4)$$

$$-Q_2 = -L\bar{C} - \frac{\partial\overline{\omega' Lq'}}{\partial p}. \quad (1.5)$$

Here Q_R is the radiation heating, Q_{1C} is the part of Q_1 due to condensation and associated transport processes, $\bar{C}$ is the rate of *net* condensation (per unit mass of dry air), s is the dry static energy, $c_pT + gz$, and a prime denotes the deviation from the horizontal area average. As (1.4) and (1.5) show, the difference of Q_{1C} and Q_2 from the net heat of condensation, $L\bar{C}$, is due to the transport of sensible and latent heat, respectively, by convective (and turbulent) motions. Eliminating $\bar{C}$ from (1.4) and (1.5), we obtain

$$Q_{1C} - Q_2 = -\frac{\partial\overline{\omega' h'}}{\partial p}, \quad (1.6)$$

where h is the moist static energy, $s + Lq$. For further discussion of (1.4), (1.5), and (1.6), see the chapters by Yanai and Johnson (chapter 4) and by Arakawa and Cheng (chapter 10) in this monograph.

Terms Q_{1C} and Q_2 represent the "feedback" shown in Fig. 1.1. If we interpret observed and subsequently spatially smoothed values of $\mathbf{v}$, θ, and q as $\bar{\mathbf{v}}$, $\bar{\theta}$, and $\bar{q}$, and if we have a time sequence of observed $\bar{\theta}$ and $\bar{q}$, then all terms on the left-hand sides of (1.1) and (1.2), including the time derivative terms, are known. Then we can calculate Q_1 and Q_2 from (1.1) and (1.2) as residuals. This procedure, which is standard in diagnostic studies of cumulus ensembles, follows the thin segment of the loop in Fig. 1.1 *backward* from "large-scale processes." Since the path does not follow the heavy segments shown in the figure, the values of Q_1 and Q_2 obtained in this way are independent of cumulus parameterization. Figure 1.2 shows an example of such calculations for the GATE [GARP (Global Atmospheric Research Program) Atlantic Tropical Experiment] phase III period. Here and in all subsequent figures for Q_1, Q_{1C}, and Q_2, the unit we have chosen is equivalent to degrees Celsius per day when divided by c_p.

From Fig. 1.2 we see that

(i) both Q_{1C} and Q_2 highly fluctuate in time;

(ii) the fluctuations of Q_{1C} and Q_2 are strongly coupled;

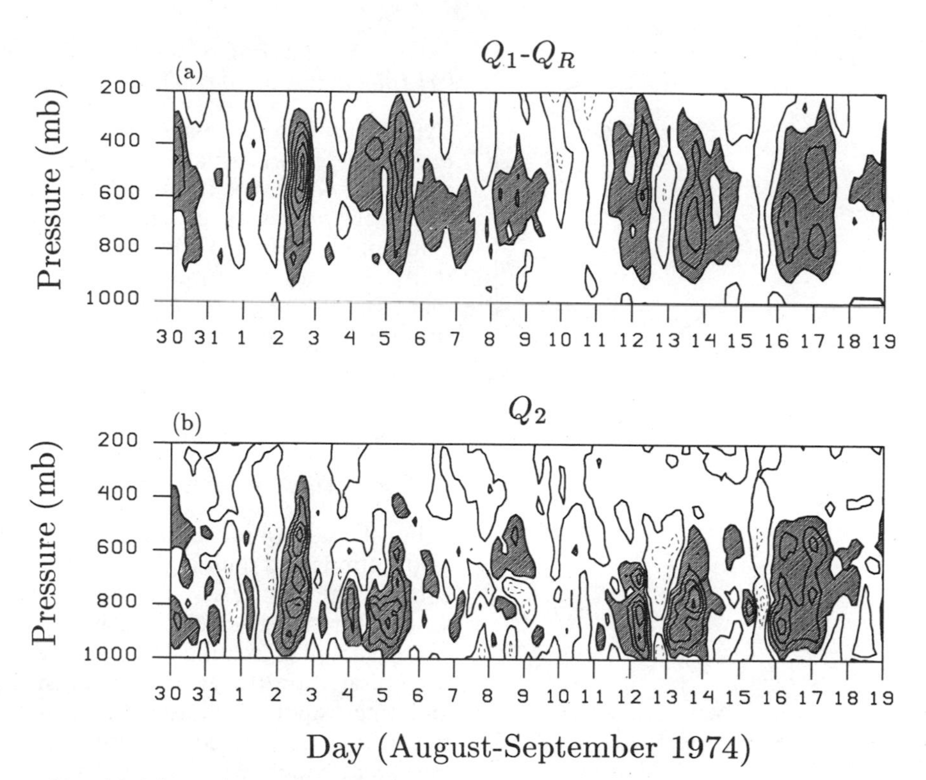

FIG. 1.2. Diagnostically calculated (a) Q_{1C} ($\equiv Q_1 - Q_R$) and (b) Q_2 for the GATE phase III period averaged over the $3° \times 3°$ grid box centered at 8.5°N, 23.5°W. Contour intervals are equivalent to 5°C day^{-1} when divided by c_p. Negative values are shown by dashed lines, and areas representing values larger than 5°C day^{-1} are shaded; redrawn from Arakawa and Chen (1987).

(iii) the level of maximum Q_{1C} is generally higher than the level of maximum Q_2; and

(iv) the separation of these maximum levels varies from case to case.

From (ii) we can infer that the process responsible for the residuals involves condensation of water vapor. From (iii) we can infer that the process includes convective transports. From (iv) we can also infer that the relative importance of convective transports varies from one case to another.

It is clear that any useful parameterization of cumulus convection must be able to predict the large fluctuations of Q_{1C} and Q_2. Figure 1.3 shows the mean vertical profiles of (a) temperature lapse rate Γ, and moist-adiabatic temperature lapse rate Γ_m, and (b) relative humidity RH for the GATE dataset. The error bars show corresponding standard deviations. (The standard deviation of Γ_m is not shown since it is negligibly small.) From the mean profiles and standard deviations of Γ and Γ_m, we see that conditional instability (as measured by $\Gamma - \Gamma_m$) almost steadily exists below 550 mb. Thus, the fluctuations of Q_{1C} and Q_2 cannot be attributed to fluctuations in conditional instability.

1.3. A classification of closure assumptions

In this section we classify closure assumptions into basic types. For convenience, we rewrite (1.1) and (1.2) as

$$\frac{\partial T}{\partial t} = \left(\frac{\partial T}{\partial t}\right)_{\text{LS}} + \frac{1}{c_p} Q_1, \tag{1.7}$$

$$\frac{\partial q}{\partial t} = \left(\frac{\partial q}{\partial t}\right)_{\text{LS}} - \frac{1}{L} Q_2. \tag{1.8}$$

Here and throughout the rest of this chapter, overbars for area average are omitted. The subscript LS denotes the contribution to the time derivative by large-scale advective processes: that is,

$$\left(\frac{\partial T}{\partial t}\right)_{\text{LS}} \equiv -\left(\frac{p}{p_0}\right)^{R/c_p}\left(\bar{\mathbf{v}} \cdot \nabla\bar{\theta} + \bar{\omega}\,\frac{\partial\bar{\theta}}{\partial p}\right), \tag{1.9}$$

$$\left(\frac{\partial q}{\partial t}\right)_{\text{LS}} \equiv -\left(\bar{\mathbf{v}} \cdot \nabla\bar{q} + \bar{\omega}\,\frac{\partial\bar{q}}{\partial p}\right). \tag{1.10}$$

In the diagnosis of Q_1 and Q_2 from observations shown in section 2, we treated the tendency terms as known quantities. In the parameterization problem,

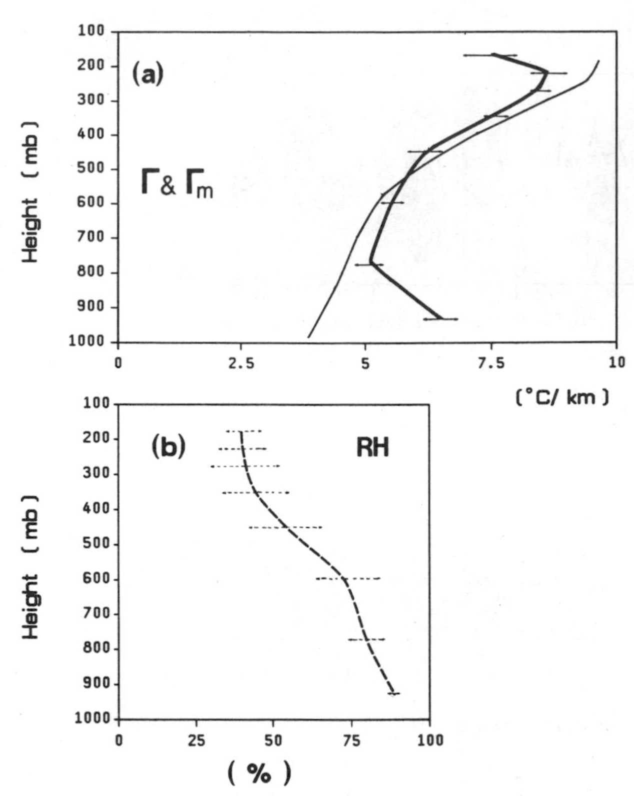

FIG. 1.3. Time-averaged vertical profiles of (a) Γ (heavy line) and Γ_m (thin line) and (b) RH for the GATE phase III as in Fig. 1.2. Error bars show standard deviations, from Chen (1989).

however, these two terms, as well as Q_1 and Q_2, are unknowns. It is then obvious that we need *at least* two types of closure assumptions since at this stage we have only two equations for four unknowns. (Closure assumptions may introduce new unknowns requiring additional closure assumptions such as those for parameterizing cloud microphysical processes.) Since closure assumptions are for reducing the number of degrees of freedom, they can be valid only when different degrees of freedom are coupled.

For the rest of this and the next three sections, we concentrate on parameterization of condensation and associated convective transports, assuming that Q_1 and Q_2 are entirely due to those processes. Thus, the difference between Q_{1C} and Q_1 is ignored. [If we stay with this definition of Q_1 and Q_2 in a practical application, (1.9) and (1.10) should be modified to include parameterized radiation and turbulent effects.]

Arakawa and Chen (1987) classified closure assumptions used in typical cumulus parameterization schemes into four basic types, three of which are given below:

type I: coupling of the $\partial T/\partial t$ term in (1.7) (*net* warming) and the $\partial q/\partial t$ term in (1.8) (*net* moistening);

type II: coupling of the Q_1 term in (1.7) and the Q_2 term in (1.8);

type IV: coupling of the Q_1 and Q_2 terms with the $(\partial T/\partial t)_{LS}$ and/or $(\partial q/\partial t)_{LS}$ terms.

Type I closure is a constraint on (the time changes of) *large-scale states,* usually through an equilibrium assumption, while type II closure is a constraint on *moist-convective processes,* usually through the choice of a cloud model. Type IV closure, on the other hand, directly assumes a coupling between *large-scale and moist-convective processes.* The next two sections give simple examples of these types of closure assumptions.

1.4. Simple examples of type I and type II closure assumptions

a. Large-scale condensation

For the sake of illustrating type I and type II closure assumptions and their logical consequences, let us first consider the closure assumptions for a commonly used scheme for "large-scale" condensation. In this scheme, condensation is assumed to occur when the air is supersaturated on the grid scale. Then the temperature and water vapor are adjusted to a saturated state. When this scheme is applied continuously in time, the air follows a sequence of saturated states after the initial adjustment, as long as the combined effect of $(\partial T/\partial t)_{LS}$ and $(\partial q/\partial t)_{LS}$ *tends* to generate a supersaturated state. Thus, as long as

$$\left(\frac{\partial q}{\partial t}\right)_{LS} > \left(\frac{\partial q^*}{\partial t}\right)_{LS}\left[=\gamma\frac{c_p}{L}\left(\frac{\partial T}{\partial t}\right)_{LS}\right] \quad (1.11)$$

holds,

$$q - q^* = 0 \quad (1.12)$$

for all t. Here $q^*(T, p)$ is the saturation value of q and $\gamma \equiv (L/c_p)(\partial q^*/\partial T)_p$. The time derivative of this equation gives type I closure for this scheme. In addition, the scheme neglects convective transports associated with condensation. Then, from (1.4) and (1.5),

$$Q_1 - Q_2 = 0. \quad (1.13)$$

This is type II closure for this scheme. [Recall that Q_1 in (1.13) corresponds to Q_{1C} in (1.4).]

Differentiating (1.12) with respect to time and using (1.7) and (1.8), we obtain

$$\gamma Q_1 + Q_2 = L\left(\frac{\partial q}{\partial t}\right)_{LS} - \gamma c_p\left(\frac{\partial T}{\partial t}\right)_{LS}. \quad (1.14)$$

Note that the right-hand side is positive from (1.11). Equation (1.14) is a consequence of type I closure, but the dependence of a combined effect of Q_1 and Q_2 on large-scale processes is now explicit. Using type II closure given by (1.13) in (1.14), we can express Q_1 and Q_2 separately in terms of the large-scale processes as

$$Q_1 = Q_2 = \frac{1}{1+\gamma}\left[L\left(\frac{\partial q}{\partial t}\right)_{\mathrm{LS}} - \gamma c_p\left(\frac{\partial T}{\partial t}\right)_{\mathrm{LS}}\right]. \quad (1.15)$$

Note that (1.15) has the form of type IV closure. Thus, type I and type II closures combined are equivalent to type IV closure in effect, at least for this example, although the actual implementation of the scheme is through the adjustment of T and q to an equilibrium state given by (1.12) under the constraint of (1.13).

Using (1.15) in (1.7) and (1.8), we can express net warming and net moistening in terms of the large-scale processes alone. The results are

$$\frac{\partial T}{\partial t} = \frac{1}{1+\gamma}\frac{1}{c_p}\left(\frac{\partial h}{\partial t}\right)_{\mathrm{LS}}, \quad (1.16)$$

$$\frac{\partial q}{\partial t} = \frac{\gamma}{1+\gamma}\frac{1}{L}\left(\frac{\partial h}{\partial t}\right)_{\mathrm{LS}}, \quad (1.17)$$

where, by definition,

$$\left(\frac{\partial h}{\partial t}\right)_{\mathrm{LS}} \equiv c_p\left(\frac{\partial T}{\partial t}\right)_{\mathrm{LS}} + L\left(\frac{\partial q}{\partial t}\right)_{\mathrm{LS}}. \quad (1.18)$$

b. The moist-convective adjustment scheme

This scheme, originally proposed by Manabe et al. (1965), is perhaps the simplest scheme among those currently being used in general circulation models. In the original version of the scheme, moist convection is assumed to occur when and where the air is conditionally unstable ($\Gamma > \Gamma_m$) and supersaturated on the grid scale (RH > 100%). Then the temperature and water vapor mixing ratio are adjusted to a moist-adiabatic ($\Gamma = \Gamma_m$) and saturated (RH = 100%) state. The only constraint on Q_1 and Q_2 imposed by the scheme is conservation of energy expressed as

$$\int_{p_T}^{p_B} (Q_1 - Q_2)dp = 0, \quad (1.19)$$

where p_T and p_B are the pressures at the top and bottom of the convective layer, respectively.

This scheme can easily be criticized because it requires *grid-scale* saturation for *subgrid-scale* moist convection. However, the scheme can be considered as a prototype for a family of schemes called *adjustment schemes* [e.g., Kurihara (1973); Arakawa and Schubert (1974), as implemented by Lord (1982) and Lord et al. (1982); Betts (1986), and Betts and Miller (1986)].

When the scheme is applied continuously in time, the air follows a sequence of moist-adiabatic saturated states after the initial adjustment, as long as the combined effect of $(\partial T/\partial t)_{\mathrm{LS}}$ and $(\partial q/\partial t)_{\mathrm{LS}}$ tends to generate a conditionally unstable, supersaturated state. Thus, as long as

$$\frac{\partial}{\partial p}\left(\frac{\partial h^*}{\partial t}\right)_{\mathrm{LS}}\left\{= \frac{\partial}{\partial p}\left[(1+\gamma)c_p\left(\frac{\partial T}{\partial t}\right)_{\mathrm{LS}}\right]\right\} > 0 \quad (1.20)$$

and (1.11) hold,

$$q - q^* = 0 \quad (1.21)$$

and

$$\frac{\partial h^*}{\partial p} = 0 \quad (1.22)$$

for all t. Here $\gamma \equiv (L/c_p)(\partial q^*/\partial T)_p$ as previously defined and h^* is the saturation moist static energy given by $c_pT + gz + Lq^*$. In deriving (1.20) and (1.22), the relation

$$\frac{\partial h^*}{\partial p} \gtreqless 0 \quad \text{as} \quad \Gamma \gtreqless \Gamma_m \quad (1.23)$$

has been used. The time derivatives of (1.21) and (1.22) give type I closure for this scheme.

In Fig. 1.4 the heavy line and the solid dot show the equilibrium states for the large-scale condensation scheme given by (1.12) and for the moist-convective adjustment scheme given by (1.21) and (1.22), respectively. Note that type I closure for the moist-convective adjustment scheme is a stronger constraint than that in the large-scale condensation scheme. In fact, after type I closure, there is only one degree of freedom left, which is a constraint on (the time change of) the vertical mean value of h. Since the vertical integration of $c_pT + Lq$ with respect to mass should be conserved during the adjustment, the scheme requires (1.19), which gives

$$\langle Q_1\rangle - \langle Q_2\rangle = 0, \quad (1.24)$$

where the angle brackets denote the vertical mean with respect to mass over the convective layer. This is type II closure for this scheme. The parameterization has

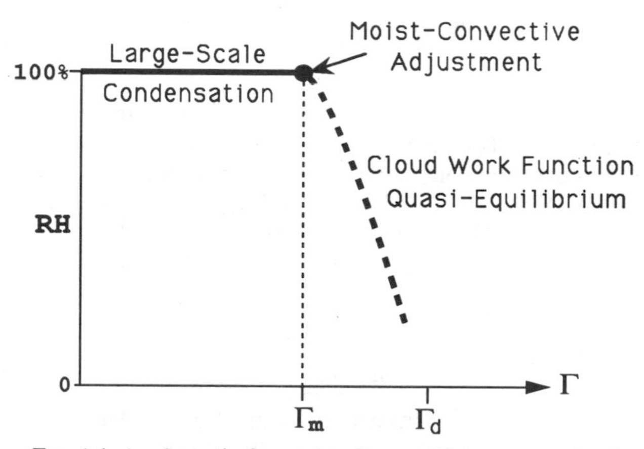

FIG. 1.4. A schematic figure showing equilibrium states for the large-scale condensation and moist-convective adjustment schemes, and the cloud work function equilibrium for deepest clouds.

now been completed and thus there is no room for a cloud model beyond (1.24).

Since (1.21) is identical to (1.12), (1.14) still holds. Using (1.24) in the vertical mean of (1.14), we obtain

$$\langle (1+\gamma)Q_1 \rangle = \left\langle L\left(\frac{\partial q}{\partial t}\right)_{\rm LS} - \gamma c_p\left(\frac{\partial T}{\partial t}\right)_{\rm LS} \right\rangle. \quad (1.25)$$

From (1.22), on the other hand, $\partial h^*/\partial t$ is constant in height so that $\partial h^*/\partial t = \langle \partial h^*/\partial t \rangle$. Then, $(1+\gamma)\partial T/\partial t = \langle (1+\gamma)\partial T/\partial t \rangle$. Using (1.7) in this relation,

$$(1+\gamma)\left[c_p\left(\frac{\partial T}{\partial t}\right)_{\rm LS} + Q_1\right] = \left\langle (1+\gamma)\left[c_p\left(\frac{\partial T}{\partial t}\right)_{\rm LS} + Q_1\right]\right\rangle. \quad (1.26)$$

Substituting (1.25) into the rhs of (1.26), we can express Q_1 in terms of the advective process as

$$Q_1 = \frac{1}{1+\gamma}\left\langle \left(\frac{\partial h}{\partial t}\right)_{\rm LS} \right\rangle - c_p\left(\frac{\partial T}{\partial t}\right)_{\rm LS}. \quad (1.27)$$

Then, from (1.14),

$$Q_2 = L\left(\frac{\partial q}{\partial t}\right)_{\rm LS} - \frac{\gamma}{1+\gamma}\left\langle \left(\frac{\partial h}{\partial t}\right)_{\rm LS} \right\rangle. \quad (1.28)$$

Here again we see that type I and type II closures combined are equivalent to type IV closure in effect, although the actual implementation of the scheme is through the adjustment of T and q to an equilibrium state given by (1.21) and (1.22) under the constraint of (1.24).

Using these results in (1.7) and (1.8) and then using (1.18) we obtain expressions of net warming and net moistening as

$$\frac{\partial T}{\partial t} = \frac{1}{1+\gamma}\frac{1}{c_p}\left\langle \left(\frac{\partial h}{\partial t}\right)_{\rm LS} \right\rangle, \quad (1.29)$$

$$\frac{\partial q}{\partial t} = \frac{\gamma}{1+\gamma}\frac{1}{L}\left\langle \left(\frac{\partial h}{\partial t}\right)_{\rm LS} \right\rangle. \quad (1.30)$$

These expressions are identical to (1.16) and (1.17) except that $(\partial h/\partial t)_{\rm LS}$ is now replaced by its vertical mean over the convective layer. *If* $(\partial h/\partial t)_{\rm LS}$ *is primarily due to the vertical advection,* which is approximately true in most tropical situations, (1.22) with $h = h^*$ indicates $(\partial h/\partial t)_{\rm LS} \approx 0$ as long as the adjustment is taking place. Then, from (1.29) and (1.30),

$$\frac{\partial T}{\partial t}, \frac{\partial q}{\partial t} \approx 0. \quad (1.31)$$

Thus, the moist-convective adjustment scheme effectively "locks" the temperature and moisture fields.

The conditions (1.20) and (1.11) are important; they are reproduced below:

$$\frac{\partial}{\partial p}\left[(1+\gamma)c_p\left(\frac{\partial T}{\partial t}\right)_{\rm LS}\right] > 0, \quad (1.32)$$

$$\left(\frac{\partial q}{\partial t}\right)_{\rm LS} - \gamma\frac{c_p}{L}\left(\frac{\partial T}{\partial t}\right)_{\rm LS} > 0. \quad (1.33)$$

With this scheme, moist convection exists only when these conditions are satisfied. This leads to the concept of *large-scale forcing,* which is implicit in adjustment schemes. Further discussion of large-scale forcing will be given in section 1.6.

1.5. An example of schemes based on type IV closure

The scheme discussed in the last section is based on type I and type II closures. As we have shown in (1.29) and (1.30), its effect can be expressed in terms of large-scale processes. But these expressions are *consequences* of the scheme rather than the scheme itself. On the other hand, some other schemes directly hypothesize such expressions, leading to type IV closure defined in section 1.3. An example of this type is Kuo's scheme.

The scheme proposed by Kuo (1974) is a prototype of a family of schemes (e.g., Krishnamurti et al. 1980; Krishnamurti et al. 1983; Krishnamurti and Bedi 1988; Anthes 1977; Kuo and Anthes 1984a), which are widely used especially in numerical weather prediction (NWP) models. In the original scheme,

$$Q_1 = (1-b)\frac{T_c - T}{\langle T_c - T \rangle} L\left\langle \left(\frac{\partial q}{\partial t}\right)_{\rm LS} \right\rangle, \quad (1.34)$$

$$L\left(\frac{\partial q}{\partial t}\right)_{\rm LS} - Q_2 = b\frac{q_c - q}{\langle q_c - q \rangle} L\left\langle \left(\frac{\partial q}{\partial t}\right)_{\rm LS} \right\rangle. \quad (1.35)$$

Here T_c and q_c are T and q of a model cloud, b is a constant called "moistening parameter" yet to be specified, and the angle brackets denote the vertical mean over the convective layer as previously defined. The vertical mean of $(\partial q/\partial t)_{\rm LS}$, or the vertical mean moisture convergence, plays the key role in this scheme.

We can derive (1.34) and (1.35) from the following assumptions:

$$\frac{Q_1}{c_p} = \alpha_T(T_c - T), \quad (1.36)$$

$$\left(\frac{\partial q}{\partial t}\right)_{\rm LS} - \frac{Q_2}{L} = \alpha_q(q_c - q), \quad (1.37)$$

where the coefficients α_T and α_q are assumed to be independent of height. Substitution of (1.36) and (1.37) into (1.24) yields

$$\alpha_T c_p\langle T_c - T \rangle + \alpha_q L\langle q_c - q \rangle = L\left\langle \left(\frac{\partial q}{\partial t}\right)_{\rm LS} \right\rangle, \quad (1.38)$$

which is automatically satisfied by

$$\alpha_T = (1-b)\frac{L}{c_p}\frac{1}{\langle T_c - T\rangle}\left\langle\left(\frac{\partial q}{\partial t}\right)_{\mathrm{LS}}\right\rangle, \quad (1.39)$$

$$\alpha_q = b\frac{1}{\langle q_c - q\rangle}\left\langle\left(\frac{\partial q}{\partial t}\right)_{\mathrm{LS}}\right\rangle. \quad (1.40)$$

Here b is an arbitrary parameter. Since α_T and α_q are independent of height by assumption, the same must be assumed for b. The use of these expressions in (1.36) and (1.37) gives (1.34) and (1.35).

The above derivation of Kuo's scheme suggests that its essence must be in (1.36) and (1.37). One may interpret these equations as adjustments of T and q to T_c and q_c with finite time scales given by $(\alpha_T)^{-1}$ and $(\alpha_q)^{-1}$. In (1.37), however, the *net* $\partial q/\partial t$ including $(\partial q/\partial t)_{\mathrm{LS}}$ is parameterized, while in (1.36) Q_1 is parameterized. Consequently, $\langle(\partial q/\partial t)_{\mathrm{LS}}\rangle$ becomes a key factor in determining the time scale, as shown in (1.39) and (1.40).

More common adjustment schemes, on the other hand, are based on

$$\frac{Q_1}{c_p} = \alpha_T(T_c - T), \quad (1.41)$$

$$-\frac{Q_2}{L} = \alpha_q(q_c - q). \quad (1.42)$$

In the limit as α_T and α_q approach infinity with finite Q_1 and Q_2, $T = T_c$ and $q = q_c$, which are T and q of an equilibrium state. By modifying T_c and q_c and/or by relaxing the time scale to finite values, we can derive a family of generalized adjustment schemes (e.g., Betts and Miller 1986). It is difficult to view Kuo's scheme as a member of this family due to the qualitative difference between (1.37) and (1.42). In particular, (1.34) and (1.35) do not involve $(\partial T/\partial t)_{\mathrm{LS}}$. Then, for example, the scheme cannot be applied to the radiative–convective equilibrium problem in which moist convection is forced by vertical differential heating.

1.6. Equilibrium states, large-scale forcing, and adjustment

We now go back to parameterization schemes that use type I closure, in which the existence of equilibrium states is explicitly assumed. There are good reasons to believe that these states are essentially neutral (or marginally unstable) for moist convection (Arakawa and Schubert 1974; Xu and Emanuel 1989). If we interpret the equilibrium states in this way, we can introduce (or clarify) various other concepts.

As we pointed out earlier, the equilibrium state for the moist-convective adjustment scheme is $\Gamma - \Gamma_m = 0$, which is the neutral state for saturated air. If the requirement of *grid-scale* saturation for *subgrid-scale* moist convection is relaxed, the instability criterion $\Gamma - \Gamma_m > 0$ must be generalized to include a humidity variable. Let $F(\mathbf{T}, \mathbf{q}) > 0$ be an instability criterion generalized in this sense. Here $\mathbf{T}$ and $\mathbf{q}$ are the *vertical profiles* of T and q, and F is an operator involving vertical differentiation of T.

Assume that initially the air is neutral so that

$$F(\mathbf{T}, \mathbf{q}) = 0. \quad (1.43)$$

Then, the total effect of $(d\mathbf{T}/dt)_{\mathrm{LS}}$ and $(d\mathbf{q}/dt)_{\mathrm{LS}}$ on the change of F gives *large-scale forcing* if

$$\left(\frac{dF}{dt}\right)_{\mathrm{LS}} \equiv \left(\frac{\partial F}{\partial \mathbf{T}}\right)\left(\frac{d\mathbf{T}}{dt}\right)_{\mathrm{LS}} + \left(\frac{\partial F}{\partial \mathbf{q}}\right)\left(\frac{d\mathbf{q}}{dt}\right)_{\mathrm{LS}} > 0 \quad (1.44)$$

is satisfied (destabilization). On the other hand, the total effect of $(d\mathbf{T}/dt)_C$ and $(d\mathbf{q}/dt)_C$ on the change of F, where the subscript C denotes the cloud effect, gives *adjustment* if

$$\left(\frac{dF}{dt}\right)_C \equiv \left(\frac{\partial F}{\partial \mathbf{T}}\right)\left(\frac{d\mathbf{T}}{dt}\right)_C + \left(\frac{\partial F}{\partial \mathbf{q}}\right)\left(\frac{d\mathbf{q}}{dt}\right)_C < 0 \quad (1.45)$$

is satisfied (stabilization). If the adjustment takes place in such a way that (1.43) is maintained (*equilibrium* of F), as in the moist-convective adjustment scheme,

$$\frac{dF}{dt} \equiv \left(\frac{dF}{dt}\right)_{\mathrm{LS}} + \left(\frac{dF}{dt}\right)_C = 0 \quad (1.46)$$

(type I closure). Then, we have

$$\left(\frac{\partial F}{\partial \mathbf{T}}\right)\left(\frac{d\mathbf{T}}{dt}\right)_C + \left(\frac{\partial F}{\partial \mathbf{q}}\right)\left(\frac{d\mathbf{q}}{dt}\right)_C = -\left(\frac{dF}{dt}\right)_{\mathrm{LS}}. \quad (1.47)$$

Let

$$G\left[\left(\frac{d\mathbf{T}}{dt}\right)_C, \left(\frac{d\mathbf{q}}{dt}\right)_C\right] = 0 \quad (1.48)$$

be a constraint on the relation between $(d\mathbf{T}/dt)_C$ and $(d\mathbf{q}/dt)_C$ through a cloud model (type II closure). Then we can find both $(d\mathbf{T}/dt)_C$ and $(d\mathbf{q}/dt)_C$ from (1.47) and (1.48) for given $(dF/dt)_{\mathrm{LS}}$ so that the parameterization is closed. [Recall that $(dT/dt)_C = Q_1/c_p$ and $(dq/dt)_C = -Q_2/L$.]

An example of $F(\mathbf{T}, \mathbf{q})$ is the cloud work function $A(\lambda)$ introduced by Arakawa and Schubert (1974), which is the work done by the buoyancy force per unit mass flux at cloud base. Here λ is the fractional rate of entrainment used to identify cloud type. For deepest clouds with $\lambda = 0$, we can show

$$A(0) = \int_{z_B}^{z_T} \beta(z)\left[h_M - h^*_{B+} - \int_{z_{B+}}^{z} \frac{\partial h^*(z')}{\partial z'}\,dz'\right]dz. \quad (1.49)$$

[See Eq. (10.31) of chapter 10.] Here M and $B+$ denote the subcloud mixed layer and a level slightly above the mixed layer, z_T is the height of the deepest cloud top, and $\beta \equiv g/c_pT(1+\gamma)$. Since $\partial h^*/\partial z = -(1+\gamma)(\Gamma - \Gamma_m)$, we may rewrite (1.49) as

$$A(0) = (\mathrm{RH} - 1)Lq_S^* \int_{z_B}^{z_T} \beta(z)dz + \int_{z_B}^{z_T} \beta(z) \times \int_{z_{B+}}^{z} c_p[1+\gamma(z')](\Gamma - \Gamma_m)dz'dz, \quad (1.50)$$

where S denotes a level in the mixed layer and RH is the relative humidity at level S [see Eq. (10.32) of chapter 10]. Since $A(0) \gtreqless 0$ as $\Gamma - \Gamma_m \gtreqless 0$ when RH $= 1$, $A(0)$ can be interpreted as a generalized bulk measure of moist convective instability for cloud type $\lambda = 0$. Since $A(0)$ increases as either RH or $\Gamma - \Gamma_m$ increases, the curve $A(0) = 0$ in the Γ–RH space has a slope shown in Fig. 1.4 by the heavy dashed line.

Type II closure is usually implicit in the cloud model used, especially in the formulation of the "feedback" branch of the loop shown in Fig. 1.1. In the Arakawa–Schubert parameterization, for example, *both* Q_1 and Q_2 are expressed in terms of a *single* one-dimensional variable $m_B(\lambda)$, which represents the spectral distribution of cloud-base mass flux. Elimination of $m_B(\lambda)$ between these two expressions gives an integral relationship between the vertical distributions of Q_1 and Q_2. This relationship can be greatly simplified for a layer in which the effect of detrainment from clouds is small. For such a layer, Q_1 and Q_2 in the Arakawa–Schubert parameterization can be approximated as

$$\rho Q_1 = M_c \frac{\partial s}{\partial z}, \quad (1.51)$$

$$-\rho Q_2 = M_c L \frac{\partial q}{\partial z}, \quad (1.52)$$

where s is the dry static energy as previously defined and M_c is the total cloud mass flux. The right-hand sides of these expressions represent warming and drying of the environment through *cloud-induced* subsidence. (This is a hypothetical subsidence and should be distinguished from the *net* subsidence, which does not necessarily appear in the region of active cumulus convection.) Eliminating M_c between (1.51) and (1.52), we obtain

$$\frac{Q_1}{\partial s/\partial z} = \frac{Q_2}{-L\partial q/\partial z}. \quad (1.53)$$

In many parameterization schemes, type I and type II closures are not as explicit as those in the Arakawa–Schubert parameterization. Yet these concepts are useful to understand why and to what extent cumulus convection can be parameterized. Figure 1.5 shows a hypothetical equilibrium curve in the Γ–RH space. The curve is basically a combination of the equilibrium

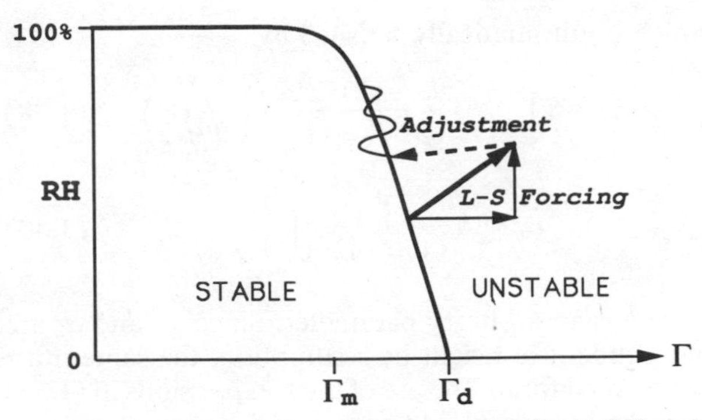

FIG. 1.5. A hypothetical equilibrium curve in the Γ–RH space that separates stable and unstable regions and an illustration of large-scale forcing (heavy solid arrow), its components (solid arrows), and the corresponding adjustment (heavy dashed arrow). See text for further explanation.

states shown in Fig. 1.4. The figure also shows an example of large-scale forcing by the heavy solid arrow and its two components (thin arrows), one representing destabilization of the temperature lapse rate and the other representing moistening. The corresponding adjustment is shown by the dashed arrow. Type II closure determines its direction, while type I closure determines its destination.

In nature, however, the exact balance given by (1.47) is not likely to exist. The equilibrium states should then be interpreted as *quasi*-equilibrium states with fluctuations such as those schematically shown in Fig. 1.5. These fluctuations may simply be a consequence of delayed adjustment, either due to a finite adjustment time scale or due to the absence of triggering. They may also represent an inherent fluctuation of the moist-convective system. Finally, they may be a consequence of some stochastic processes.

If the existence of such fluctuations is neglected, the parameterization is *diagnostic* in the sense that no additional prognostic equations are introduced and *deterministic* in the sense that no stochastic processes are involved. In a diagnostic parameterization, cumulus effects do not have their own history, recognizing the past only through the grid-scale variables of the model. All existing parameterization schemes are deterministic and most of them are diagnostic. The latter situation is analogous to the first-order turbulence closure models (or the "K theory"), in which there is no prognostic equation for turbulence, and the energy balance climate models, in which there is no prognostic equation for synoptic-scale eddies.

To better understand fluctuations of cumulus activity and the possible existence of quasi-equilibrium states, a simple mathematical model is constructed. [For a similar but slightly different formulation, see Xu (1991); Arakawa and Xu (1992); and chapter 11 by Randall and Pan in this monograph.] The cloud-scale kinetic energy equation may be written as

$$\frac{dK}{dt} = M(A - \delta). \tag{1.54}$$

Here K is the kinetic energy of cloud-scale circulation, M is the cumulus mass flux (at cloud base), A is the cloud work function (CWF) defined earlier, and δ is the dissipation of K per unit M. The time derivative of the CWF may be symbolically written as

$$\frac{dA}{dt} = -kM + F, \tag{1.55}$$

where the first and second terms on the rhs represent the adjustment and the large-scale forcing, respectively. For a given large-scale forcing F, we have a system of two equations at this stage, (1.54) and (1.55), for three unknowns, K, M, and A. The system becomes closed if we assume

$$K = \alpha M^2. \tag{1.56}$$

There is some observational evidence that δ is an approximate constant (see Lord and Arakawa 1980). If we further assume that α is also an approximate constant, (1.54)–(1.56) yield

$$(\tau_{\rm ADJ})^2 \frac{d^2M}{dt^2} + M = \frac{F}{k}, \tag{1.57}$$

where

$$\tau_{\rm ADJ} \equiv \left(\frac{2\alpha}{k}\right)^{1/2}. \tag{1.58}$$

Equation (1.57) is a forced oscillation equation for M.

The CWF equilibrium used in the Arakawa–Schubert parameterization neglects the lhs of (1.55) so that the mass flux M is given by

$$M = \frac{F}{k}. \tag{1.59}$$

In view of (1.57), this equilibrium can be interpreted as the result of *filtering* free oscillations of M by dropping the first term on the lhs. More specifically, the equilibrium assumption can be justified if both of the following assumptions are valid:

(i) M is forced by F so that the time scale of M is the same as that of F, which is the time scale of the large-scale processes, $\tau_{\rm LS}$;

(ii) for that time scale, the first term is negligible compared to the second term on the lhs of (1.57).

The assumption (i) is inevitable in any parameterization, at least implicitly, since free fluctuations are in principle not parameterizable *in terms of the large-scale processes.* The assumption (ii) can then be justified if

$$\tau_{\rm ADJ} \ll \tau_{\rm LS}. \tag{1.60}$$

There are great uncertainties in formulating processes that effectively determine $\tau_{\rm ADJ}$. This is especially because $\tau_{\rm ADJ}$ depends on α defined by (1.56). Here K is the sum of the vertical kinetic energy and horizontal kinetic energy, and, when the vertical scale is fixed, the relative importance of the latter increases as the characteristic horizontal scale increases. Consequently, α is large when subgrid-scale motions are dominated by mesoscale circulations. Then, from (1.59), $\tau_{\rm ADJ}$ is large and (1.60) may not be satisfied.

Recently there is a trend to relax (or avoid the explicit use of) the quasi-equilibrium assumption [see, for example, Betts and Miller (1986), Emanuel (1991), Pan and Randall (1991), and Moorthi and Suarez (1992); see also chapter 11 by Randall and Pan in this monograph]. In general, schemes developed along this line are more convenient and computationally efficient than the schemes that strictly enforce the quasi-equilibrium assumption. Transient solutions of such "relaxed" or "soft-adjustment" schemes are, however, not necessarily realistic due to the uncertainty of $\tau_{\rm ADJ}$ and, in the case of NWP, due to uncertainties in the initial condition. After the transient period, however, these schemes are expected to produce the quasi-equilibrium solution. It is important to note that the quasi-equilibrium solution is independent of the actual value of $\tau_{\rm ADJ}$ if (1.60) is satisfied.

Since the two time scales that appear in (1.60) are only loosely defined, the following question still remains: After all, to what extent can we parameterize cumulus convection diagnostically and deterministically? The answer to this question very likely depends on the grid size of the large-scale model and many other factors. The subject of the next section is related to this basic question.

1.7. Verification of closure assumptions: Examples

The conceptual framework for cumulus parameterization is still in its developing stage. In particular, great uncertainties exist in choosing appropriate closure assumptions. What is needed at this stage for verification of a scheme is more than testing of its overall products. Observations and other research tools should be used for directly verifying closure assumptions, for discovering improved closure assumptions, and even for assessing the limit of parameterizability. The purpose of this section is to give examples of such studies by reproducing some of the results presented by Chen and Arakawa (Arakawa and Chen 1987; Chen 1989) and to summarize more recent research at the University of California, Los Angeles (UCLA) on the macroscopic behavior of moist convection and its parameterizability.

a. "Quick-look" analysis of observed type I and type II coupling

Chen and Arakawa performed statistical analysis of observed data under a number of conditionally unstable large-scale conditions. The analysis can be divided into two categories: one is analysis of vertical profiles of temperature and humidity to find statistical constraints on their coupling under the existence of moist convection (type I coupling); the other is analysis of vertical profiles of Q_1 and Q_2, obtained from the residuals in the observed budgets, to find statistical constraints on their coupling (type II coupling). The analysis is based on two gridded datasets, the GATE dataset and Asian dataset. For description of these datasets, see Arakawa and Chen (1987). This subsection shows the results from the preliminary or "quick-look" phase of the analysis and those from similar analysis by Yuezhen Liu at UCLA using the gridded SESAME (Severe Environmental Storms and Mesoscale Experiment) data (Grell et al. 1991).

Figure 1.6 presents plots of the surface relative humidity RH_S versus the normalized lapse rate Γ_N, defined by

$$\Gamma_N \equiv \frac{\Gamma - \Gamma_{mS}}{\Gamma_d - \Gamma_{mS}}, \qquad (1.61)$$

for (a) the GATE dataset and (b) the cases satisfying $\langle Q_2 \rangle / c_p > 2°\text{C day}^{-1}$ in the Asian dataset. Here Γ is the mean lapse rate between the surface and 500 mb, Γ_{mS} is the moist-adiabatic lapse rate at the surface, and Γ_d is the dry-adiabatic lapse rate; $\langle Q_2 \rangle / c_p$ is used as a measure of the precipitation rate. (When the surface evaporation is neglected, $\langle Q_2 \rangle / c_p = 2°\text{C day}^{-1}$ is equivalent to approximately 8 mm day^{-1} of precipitation.) For the GATE dataset, the plots are confined to a small segment in the high-humidity, conditionally unstable region of the parameter space. The plots for the Asian dataset, on the other hand, show that 1) the (large-scale) relative humidity can be much lower than 100% even for precipitating cases and 2) the plots cluster around a line similar to the CWF equilibrium line shown in Fig. 1.4. These tendencies can be found for convective storms over middle-latitude land, as shown in Fig. 1.7 for SESAME IV and V.

The plots for the Asian and SESAME datasets, however, show considerable fluctuations around the equilibrium line. Also, there seem to be some systematic differences between the two datasets. These fluctuations and systematic differences can be due to a number of reasons, most importantly the use of a two-parameter space to represent the coupling of temperature and humidity *profiles*. It is still encouraging to see in these plots an indication for the existence of type I coupling, at least in a statistical sense.

As an example of type II coupling, Chen and Arakawa analyzed Q_1 and Q_2 of the GATE and Asian datasets from the point of view of (1.53). The results from the GATE dataset are shown in Fig. 1.8. Figure 1.8a presents plots of Q_{1C} ($\equiv Q_1 - Q_R$) versus Q_2 for each of the six layers in the vertical. The figure shows that the coupling between Q_{1C} and Q_2 is reasonably strong. However, the regression line deviates from the diagonal line and rotates clockwise with height. This is another way of seeing the effect of convective transports on the difference of Q_{1C} and Q_2. Figure 1.8b presents similar plots but for $\hat{Q}_{1C}$ versus $\hat{Q}_2$ defined by

$$\hat{Q}_{1C} \equiv Q_{1C} \frac{-L\partial q/\partial z}{(\partial \bar{s}/\partial z)(-L\partial \bar{q}/\partial z)}, \qquad (1.62)$$

$$\hat{Q}_2 \equiv Q_2 \frac{\partial s/\partial z}{(\partial \bar{s}/\partial z)(-L\partial \bar{q}/\partial z)}, \qquad (1.63)$$

where the overbar denotes the time average. If $Q_{1C}(\partial s/\partial z)^{-1} = Q_2(-L\partial q/\partial z)^{-1}$ holds, as in (1.53), we have

$$\hat{Q}_{1C} = \hat{Q}_2. \qquad (1.64)$$

From the plots of $\hat{Q}_{1C}$ and $\hat{Q}_2$ shown in Fig. 1.8b, we see that the regression line is now almost diagonal for all layers. In view of (1.53), this suggests that the subsidence induced by clouds is primarily responsible for Q_{1C} and Q_2. There are, however, considerable fluctuations around the diagonal line, especially for the upper layers, indicating that processes other than the subsidence, such as the detrainment from clouds, are also operating.

b. Canonical correlation analysis

Encouraged by the results of the quick-look analysis, Chen and Arakawa studied the coupling between two variables in more detail. The method chosen for this purpose is the canonical correlation analysis (see, for example, Cooley and Lohnes 1971). In our application of the analysis, we consider a pair of two variables, X and Y, and treat their vertical profiles as vectors $\mathbf{X} = (X_1, X_2, \cdots, X_k, \cdots)$ and $\mathbf{Y} = (Y_1, Y_2, \cdots, Y_k, \cdots)$. Here k is the index for levels or layers. Given a statistical sample of observed $\mathbf{X}$ and $\mathbf{Y}$, we first remove the respective sample mean from each vector. We then consider linear transformations from $\mathbf{X}$ to $\mathbf{X}' = (X'_1, X'_2, \cdots, X'_i, \cdots)$ and $\mathbf{Y}$ to $\mathbf{Y}' = (Y'_1, Y'_2, \cdots, Y'_i, \cdots)$. Here i is the index for the transformed (orthogonal) components. The transformations are chosen in such a way that each pair of the transformed components, $(X'_1, Y'_1), (X'_2, Y'_2), \ldots, (X'_i, Y'_i), \ldots$, has a (locally) maximum correlation for the given sample. The transformed components obtained in this way are the *canonical components* of $\mathbf{X}$ and $\mathbf{Y}$. By examining the correlation and associated variances for each pair of the canonical components, we determine how strongly the vertical profiles of X and Y are coupled.

Here we show the results from the GATE dataset obtained by Chen (1989) for type II coupling, in which Q_{1C} and Q_2 are chosen as X and Y. [Similar results with a lower vertical resolution are shown in Arakawa and Chen (1987).] The correlation coefficients obtained from the analysis are 0.925 for the first component, 0.721 for the second component, and 0.497

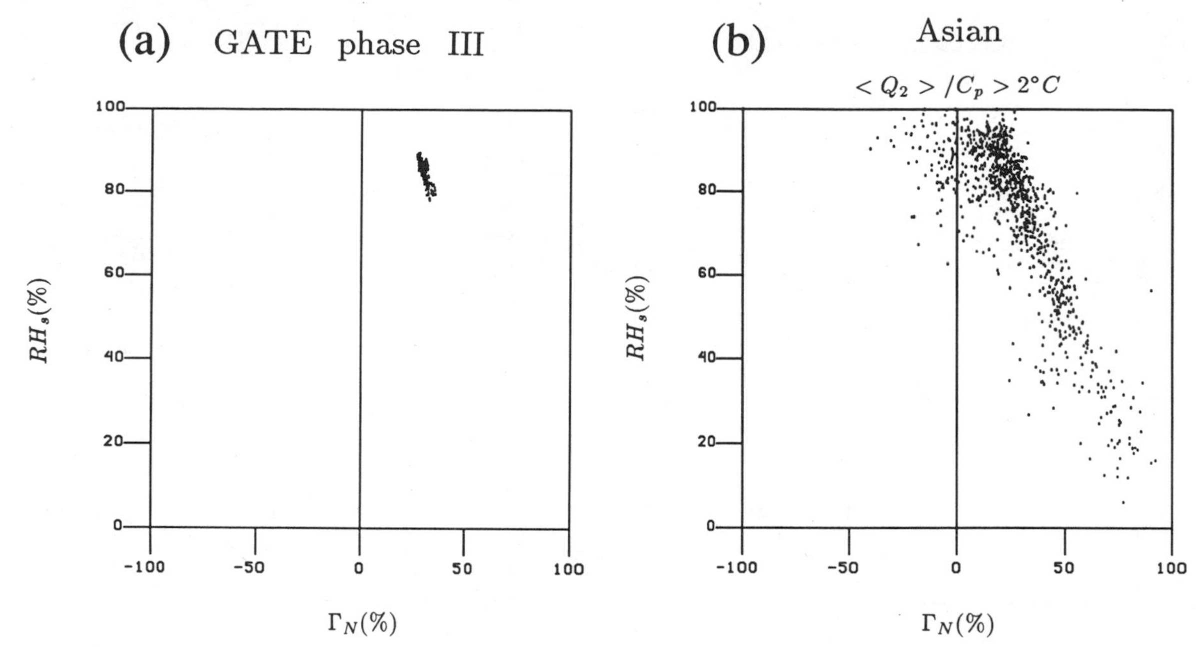

FIG. 1.6. Plots of the surface relative humidity RH_s vs the normalized lapse rate Γ_N, defined by (1.61). (a) For the GATE dataset; (b) for cases satisfying $\langle Q_2 \rangle / c_p > 2°C\ day^{-1}$ in the Asian dataset; replotted from Arakawa and Chen (1987).

for the third component. From the accumulated variances shown in Fig. 1.9, we can see that the total variances are mainly due to the highly correlated components, indicating that type II coupling is very strong for this dataset.

Chen (1989) extended the analysis to the Asian dataset and found that type II coupling is quasi-universally strong. Including type I coupling in the analysis, he concluded that cumulus convection is basically parameterizable. This conclusion is important since it is derived purely empirically without using any models or any physical hypotheses.

c. Rotated EOF analysis

The canonical correlation analysis described above is useful in finding the strength of coupling between

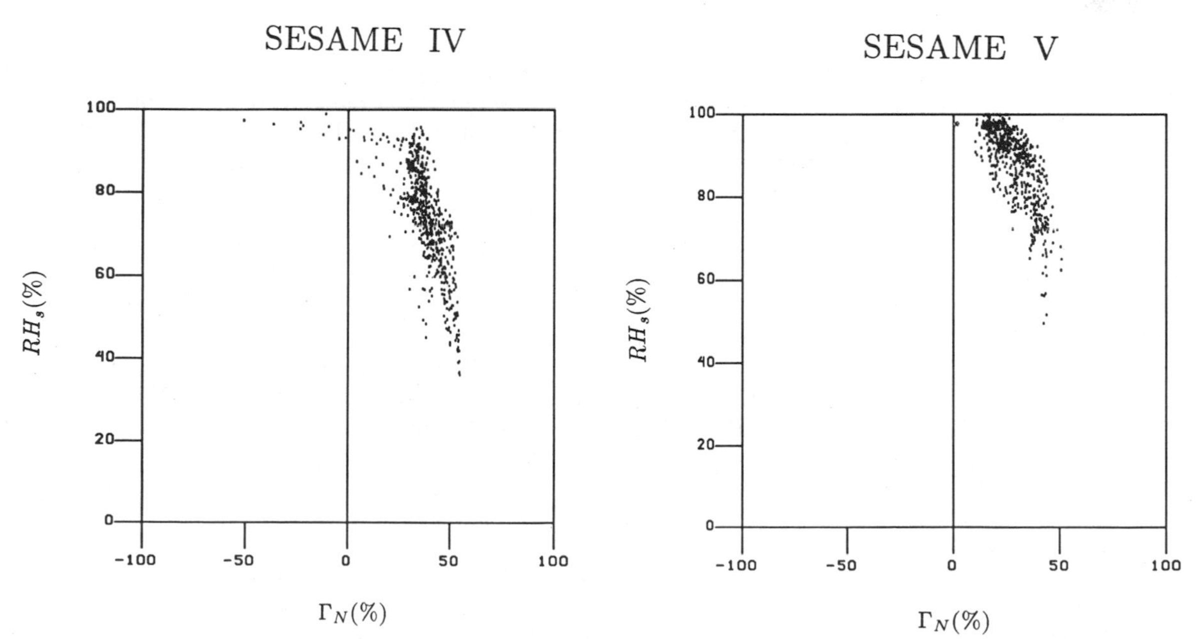

FIG. 1.7. Same as Fig. 1.6 but for the SESAME IV and V datasets.

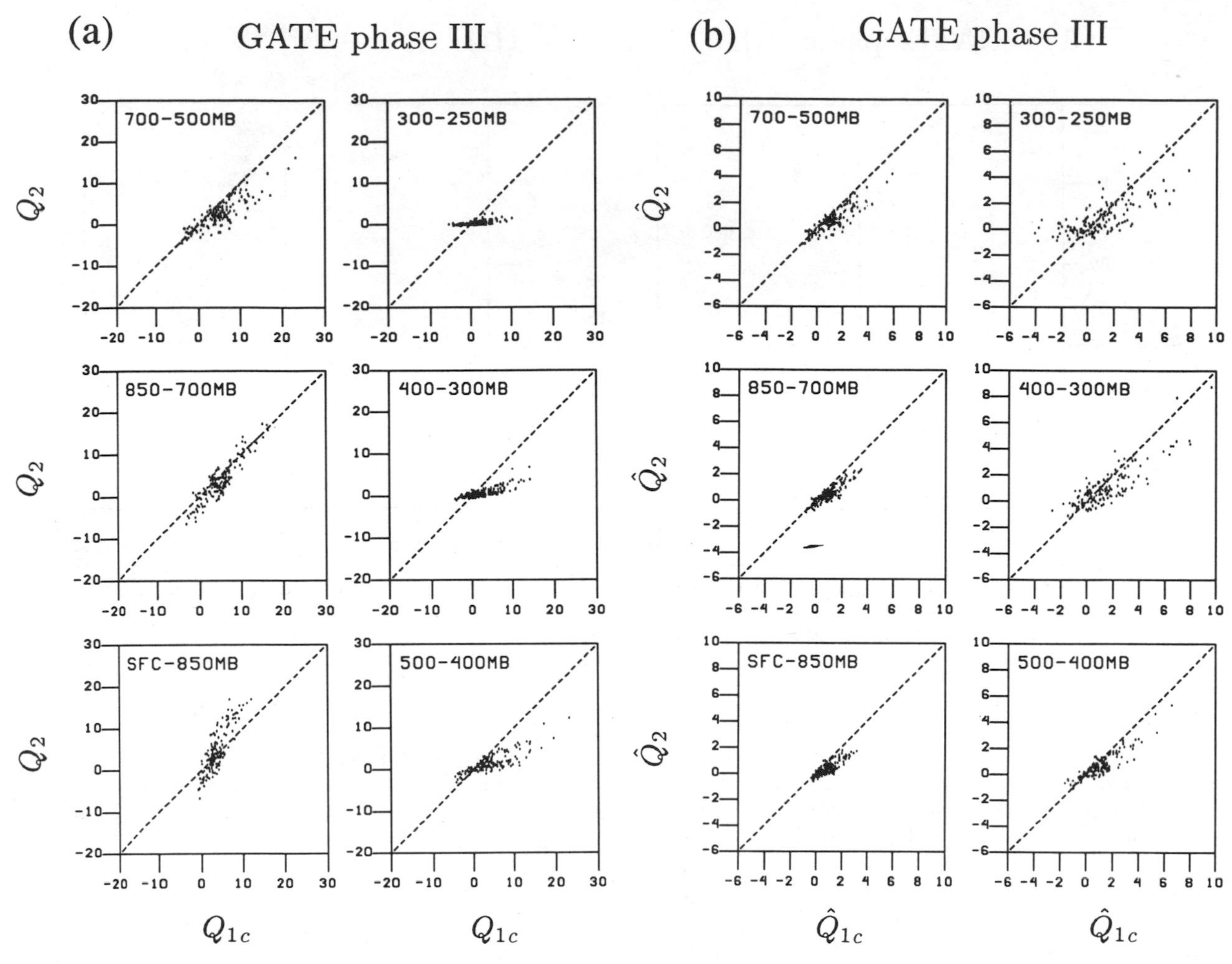

FIG. 1.8. (a) Plots of Q_{1C} vs Q_2 for the GATE dataset for each of the six layers. (b) Same as (a) but for $\hat{Q}_{1C}$ vs $\hat{Q}_2$ defined by Eqs. (1.62) and (1.63).

two variables. The structure associated with each component, however, does not necessarily have its own physical meaning, as in the case of the standard empirical orthogonal function (EOF) analysis (see Richman and Lamb 1985). The rotated EOF (REOF) analysis can better identify individual modes of variation, and, when applied to $\mathbf{Z} = (X_1, X_2, \cdots, X_k, \cdots, Y_1, Y_2, \cdots, Y_k, \cdots)$, where $\mathbf{X} = (X_1, X_2, \cdots, X_k, \cdots)$ and $\mathbf{Y} = (Y_1, Y_2, \cdots, Y_k, \cdots)$, it can be a useful tool for understanding the nature of the coupling between $\mathbf{X}$ and $\mathbf{Y}$.

Yuezhen Liu is performing such analysis using the GATE and Asian datasets. To analyze type II coupling, first Q_{1C} and Q_2 are individually normalized using the vertical mean of each with respect to mass. The resulting normalized vertical profiles are then combined to form a vector. Given a sample of that vector, REOF analysis is performed following the Harris–Kaiser II oblique rotation method (Harris and Kaiser 1964). The resulting REOFs are shown in Fig. 1.10 for the GATE dataset. Here the magnitude is chosen to be one standard deviation of the corresponding (rotated) principal component, and then Q_{1C} and Q_2 are individually denormalized. From this figure, we can see three dominant "modes" of coupling. REOF3 represents deep convection cases with Q_2 (dashed line) concentrated near 900 mb and Q_{1C} (solid line) almost evenly distributed between 900 and 300 mb. REOF2, on the other hand, essentially represents middle-tropospheric stratiform cloud cases, with the Q_{1C} and Q_2 maxima both around 500 mb. The difference $Q_{1C} - Q_2$, however, is not quite negligible even for this component. REOF1 represents cases in which the effects of shallow convection (or low-level stratiform clouds) and deep convection are mixed.

d. Experiments with a cumulus ensemble model

As we pointed out in section 1, the problem of cumulus parameterization is analogous to that of climate dynamics since both deal with the statistical behavior of complex systems. In climate dynamics, general circulation models (GCMs) are indispensable for verifying various hypotheses for climate change. Similarly,

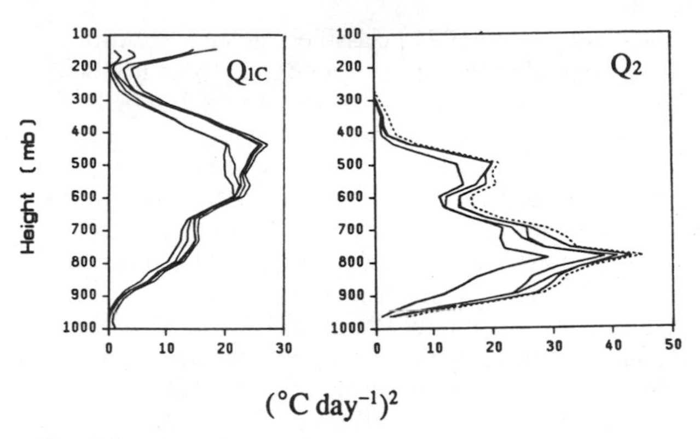

FIG. 1.9. Accumulated variances of Q_{1C} and Q_2 obtained by the canonical correlation analysis applied to the vertical profiles of Q_{1C} and Q_2 for the GATE dataset. In each panel, the dashed line represents the total variance, the leftmost line represents the contribution from the component 1, the next line to the right represents the contributions from the components 1 and 2, etc.

cumulus ensemble models (CEMs), which can simulate individual clouds while covering a large-scale domain, should be used for verifying closure assumptions in the cumulus parameterization problem. At UCLA, we are using a CEM developed by Krueger (Krueger 1988; see also Xu and Krueger 1991) to study the modulation of cumulus activity by large-scale processes and to verify the Arakawa–Schubert parameterization following the semiprognostic approach applied to the simulated data (Xu 1991; Arakawa and Xu 1992; Xu et al. 1992; Xu and Arakawa 1992). Some of the conclusions from these studies are listed below. For more details, see chapter 21 by Xu in this monograph.

1) Cumulus activity is rather strongly modulated by large-scale processes.

2) There are, however, some unmodulated fluctuations of cumulus activity.

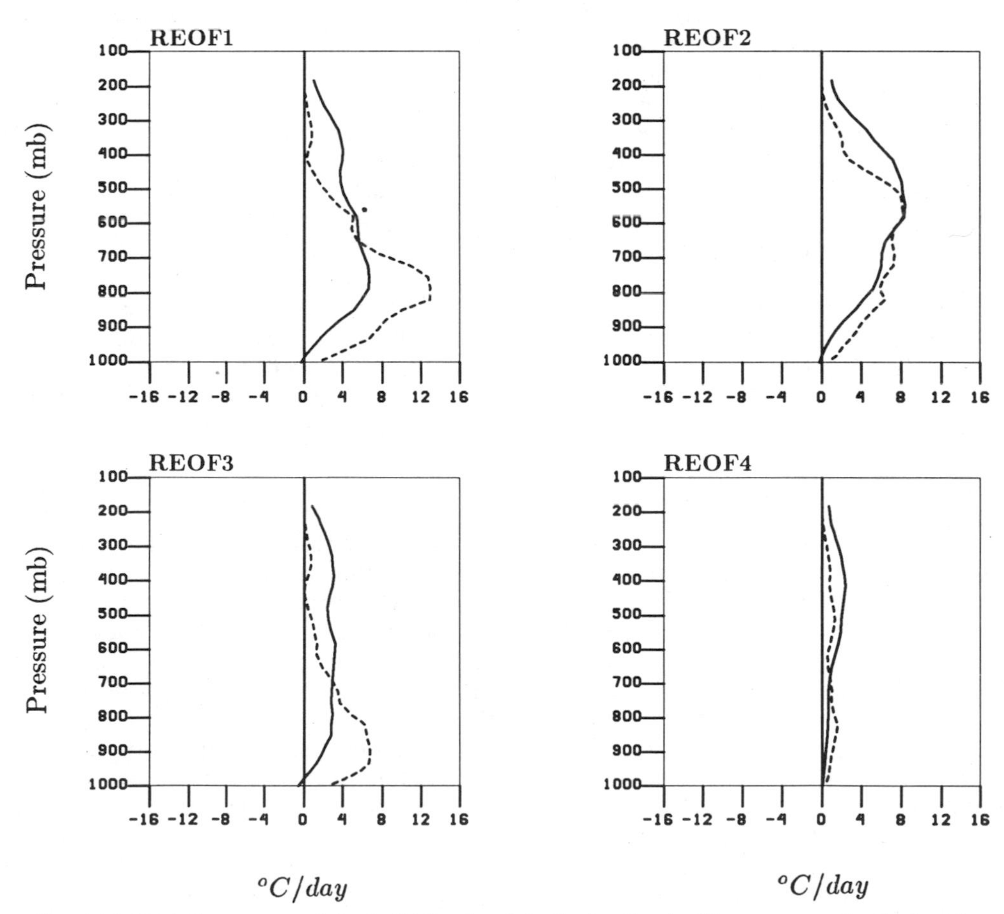

FIG. 1.10. Vertical structures of Q_{1C} (solid lines) and Q_2 (broken lines) of four components obtained by the REOF analysis applied to the combined vertical profiles of Q_{1C} and Q_2 for the GATE dataset.

3) Under the existence of pronounced unresolved mesoscale organization, there are some phase delays (approximately a few hours) in the modulation.

4) The CWF quasi-equilibrium assumption is valid for the modulated part of cumulus activity.

5) The CWF quasi-equilibrium assumption is better when it is applied to averages over smaller intervals so that mesoscale processes are included in the "large-scale" processes.

6) Errors of the parameterization when applied to averages over smaller intervals seem to be mainly due to the nondeterministic nature of the problem.

In conclusion, cumulus activity is basically parameterizable in terms of large-scale processes. The parameterizability increases as the mesoscale organization of cumulus activity is resolved. When the grid size becomes closer to the size of individual clouds, however, the grid-scale average tends to lose its statistical significance so that it cannot be parameterized in a deterministic sense. (This does not mean, however, that cumulus activity averaged over a *fixed* large-scale interval becomes less deterministic as the grid size becomes smaller.)

1.8. Summary and further discussions

Cumulus parameterization in numerical modeling is the problem of formulating the collective effects of subgrid-scale clouds in terms of the prognostic variables of grid scale. It is emphasized, however, that cumulus parameterization is needed in any formulation of scale interactions in the moist atmosphere, regardless of whether we are using numerical, theoretical, or conceptual models. Cumulus parameterization is, therefore, a scientific problem rather than a technical problem, as it may sound.

Given large-scale and moist-convective processes, one may argue about which is the cause and which is the effect. Such an argument is meaningless as far as the macroscopic behavior of moist convection is concerned. The important point here is that these two processes can *interact,* and cumulus parameterization is needed for a closed formulation of the interaction. When we are concentrating on the parameterization problem, however, we may use the terminologies "control" and "feedback" for convenience as we did in Fig. 1.1.

It is also emphasized that cumulus parameterization is a closure problem, in which the choice of appropriate closures is crucial. The conceptual framework for cumulus parameterization, however, is still in its developing stage, and there exist great uncertainties in choosing appropriate closures. Correspondingly, a number of parameterization schemes with different closures have been proposed.

What is important at this stage is identifying, analyzing, and verifying closures, which is more than testing the overall products of individual schemes. Arakawa and Chen (1987) classified closure assumptions in existing schemes, used either explicitly or implicitly, into four basic types, among which type I, type II, and type IV closures are discussed here with examples. (See section 1.3 for the definitions of these types.) In short, type I and type II refer to closures used in the adjustment schemes. Type II closure determines the direction of adjustment in a parameter space, while type I closure determines its destination. Type IV closure, on the other hand, formulates the cumulus effect (or the net change of state variables under the existence of cumulus convection) directly in terms of large-scale processes. It is shown that type I and type II closures combined are equivalent to type IV closure in effect. However, the converse is not true in general, making the physical basis for type IV closure more obscure.

In principle, the cloud model on which cumulus parameterization is based should be made as realistic as possible. It should be noted, however, that the inclusion of new effects increases the degrees of freedom so that additional closures are required. It should also be noted that there is no point in refining details of cloud model *unless* it is under the umbrella of reasonably formulated "control" shown in Fig. 1.1. The basic closures, such as those we have discussed here, are still of fundamental importance in effectively determining the "control."

It is emphasized that observations and other research tools should be used for directly verifying closure assumptions, for discovering improved closure assumptions, and even for assessing the limit of parameterizability. Examples of such studies are given in section 1.7.

In this article, the term "large scale" is used rather loosely without really defining it. In a parameterization of cloud-scale processes, "large scale" should be interpreted as the next larger scale, which is mesoscale. In this sense the problem of cumulus parameterization is best defined for mesoscale models. Moreover, as we pointed out in section 1.7, the assumption of cloud work function quasi equilibrium is better for mesoscale. This is because the magnitude of "large-scale" forcing, which is primarily due to advective processes, becomes more rapidly dominant than the cloud work function change does as the grid size decreases. However, if the grid size becomes too small, the cumulus effect averaged over the grid interval loses its statistical significance and parameterization becomes less deterministic for the grid scale. In addition, grid-scale and subgrid-scale processes become nonseparable even approximately, so that there is a danger of "double counting" the same effect.

For models that do not resolve mesoscale, the collective effect of mesoscale organization of clouds should be considered, but some components of that effect are not likely to be parameterizable at least in the "diagnostic" sense. (For the definition of this terminology, see section 1.6.) In spite of this, the results of the studies reviewed in section 1.7 seem to suggest that moist-convective processes (including mesoscale) are *basically* (though not fully) parameterizable. Further research is obviously needed in this area.

Finally, it should be remembered that closure assumptions must be valid quasi-universally. Different closure assumptions may appear to be valid for different cloud regimes. Large-scale models, however, must predict spatial and temporal distributions of cloud regimes (Arakawa 1975) *including* their transitions. Then, what we eventually need in these models is a unified cloud parameterization, covering deep, shallow, high, low, cumuliform, and stratiform clouds with and without mesoscale organizations. Constructing such a parameterization is obviously a huge task. Without the capability of doing so, however, our understanding of the moist atmosphere will not be complete.

Acknowledgments. The author would like to thank Ms. Yuezhen Liu for her help in preparing figures, and Ms. Kristin Mah for typing the manuscript. This material is based on the research supported jointly by NSF under Grant ATM-8910564 and NASA under Grant NAG 5-789.

Chapter 2

Observational Constraints on Cumulus Parameterizations

DAVID J. RAYMOND

Physics Department and Geophysical Research Center, New Mexico Institute of Mining and Technology, Socorro, New Mexico

2.1. Introduction

Observational tests of cumulus parameterizations are difficult. Since cumulus convection is generally embedded in a larger-scale flow pattern in which many feedbacks are acting, it is hard to tell what is causing what to happen. For those aspects of convection tightly embedded in feedback loops, the *change* in convective behavior caused by environmental changes needs to be correctly represented in parameterizations. However, it is precisely these changes that are difficult to observe, since negative large-scale feedback often minimizes the range of naturally occurring variations. For instance, suppose the convective mass flux were very sensitive to environmental instability, such that a slight increase in instability resulted in a large increase in convective mass flux. This in turn would generate large amounts of convective heating, which in turn would quickly stabilize the environment, thus reducing the vertical mass flux. In normal circumstances, large deviations from neutral stability would rarely occur, and observations of convective behavior in highly unstable conditions would be difficult to obtain. However, it is *this abnormal behavior* that needs to be studied in order to understand the feedback mechanisms that prevent it from occurring more frequently.

Fortunately, the embedding of convection in a multitude of feedbacks has a positive side as well. In order to obtain an accurate representation of the effects of convection on the larger scale, it is not necessary to have a parameterization that is highly accurate. It is sufficient that *changes* in large-scale forcing result in changes in the parameterized convection that are correct in sign and of the right order of magnitude. Under the influence of negative feedback, this behavior will often force the convective parameterization to do the right thing.

We also assert that the development of a cumulus parameterization completely empirically, without appeal to any theoretical basis, is impractical. The parameter space that needs to be filled by observation in this case is huge and is sparsely populated by real clouds. Furthermore, as we shall see, the effort to obtain adequate observations in any given situation is nontrivial. Thus, any realizable parameterization of convection must be based as solidly as possible on reliable physical theory, so as to reduce the required number of observations.

To summarize, a satisfactory parameterization of moist convection (a) will be based as closely as possible on physical theory, (b) need not yield highly accurate results in a situation without feedback, but (c) must be uniformly valid over the broadest possible range of conditions, and (d) respond appropriately to changes in natural forcing.

Any moist convective parameterization needs to make predictions in four general areas. First, for given environmental conditions, the vertical profile of convective mass flux needs to be predicted. As will be shown, the vertical mass flux is closely related to the vertical profile of heating. Second, the vertical profile of detrained water vapor flux (or equivalent potential temperature flux) is required. This may be the single most important prediction that a convective parameterization can make, given the role of moisture in radiative processes. Third, estimates of precipitation production and ice-particle detrainment are needed. These are closely correlated with the vapor detrainment, as all form a part of the water budget. Finally, the convective transport of momentum needs to be understood.

This paper is organized as follows. Section 2.2 describes a theoretical framework in which to view observations. Observational tools are described in section 2.3. Section 2.4 outlines the types of observations needed to constrain cumulus parameterizations and indicates which observations already exist. Section 2.5 proposes an observational program to fill the gaps in our knowledge.

2.2. Theoretical framework

In this section the phenomenology of the cumulus parameterization problem is set forth, with the purpose of developing a framework into which observations can be placed. This work is derived in part from the analysis of Raymond and Wilkening (1985), though the basic ideas extend back to Ooyama (1971), Yanai et al.

(1973), Arakawa and Schubert (1974), etc. The special problems with particular fields are then individually discussed. Care taken at this stage pays off in eliminating ambiguities and incompatibilities between parameterizations and large-scale equations later on. Finally, a simple graphical scheme for representing the flow of air through cumulus clouds is introduced.

a. General phenomenology

The main problem is to define a consistent separation between convective and large-scale flow. We focus on some intensive quantity χ that obeys the governing equation

$$\frac{d\chi}{dt} = \frac{\partial\chi}{\partial t} + \mathbf{v}\cdot\nabla\chi + \omega\frac{\partial\chi}{\partial p} = X_c + X_e, \quad (2.1)$$

where X_c and X_e are, respectively, the source terms for χ in cloud and in the surrounding environment.

Dividing χ into an averaged part and a fluctuation,

$$\chi = \langle\chi\rangle + \chi', \quad (2.2)$$

and using the mass continuity equation in pressure coordinates,

$$\nabla\cdot\mathbf{v} + \frac{\partial\omega}{\partial p} = 0, \quad (2.3)$$

where $\mathbf{v}$ is the horizontal velocity and ω is the vertical pressure velocity, (2.1) can be written

$$\frac{\partial\chi}{\partial t} + \mathbf{v}\cdot\nabla\langle\chi\rangle + \nabla\cdot(\mathbf{v}\chi') + \omega\frac{\partial\langle\chi\rangle}{\partial p} - X_e = X_c - \frac{\partial(\omega\chi')}{\partial p} \equiv S_\chi. \quad (2.4)$$

The idea is to arrange the terms so that S_χ is only nonzero inside of cumulus clouds. Term X_c is zero outside of cloud by definition, and $\partial(\omega\chi')/\partial p$ is very small outside of cloud when the fractional cloud coverage ϵ is small. This is because both ω and χ' in the environment scale as ϵ times mean cloud values, so that the product scales as ϵ^2. In an area-integrated sense, the ratio of environment to cloud contribution to $\omega\chi'$ therefore scales as ϵ.

Notice that this is not a complete Reynolds decomposition, because the velocities have not been divided into averaged and fluctuating parts. The details of the averaging process will be left undefined but could be either a spatial low-pass filter in the horizontal or a temporal low-pass filter. However, we insist that the averaging process be linear, so that it commutes with partial derivatives in space and time. The notion is to construct it so that $\langle\chi\rangle$ does not vary much over the scale of a single cumulus cloud but reflects the structure of the large-scale disturbance that is forcing the cloud. This *scale separation* assumption is dubious in some instances but is a necessary part of cumulus parameterization. Implicit in this decomposition is that $\langle\chi'\rangle = 0$.

Applying the averaging process to (2.4) leads to the large-scale governing equation for χ:

$$\frac{\partial\langle\chi\rangle}{\partial t} + \langle\mathbf{v}\rangle\cdot\nabla\langle\chi\rangle + \langle\omega\rangle\frac{\partial\langle\chi\rangle}{\partial p} - \langle X_e\rangle = \langle S_\chi\rangle. \quad (2.5)$$

In deriving (2.5), we have assumed that $\langle\nabla\cdot(\mathbf{v}\chi')\rangle$ is small enough to be ignored. This is a consequence of the scale separation assumption, since this term represents the average of a quantity with zero mean and mainly high spatial frequency fluctuations. It is also assumed that $\langle\mathbf{v}\cdot\nabla\langle\chi\rangle\rangle = \langle\mathbf{v}\rangle\cdot\nabla\langle\chi\rangle$. Thus, only averaged quantities remain on the left side of (2.5), and the scale separation is formally complete.

Suppose that (as is generally observed) cumulus convection is sparse, with individual cumulus or cumulonimbus clouds separated by regions of stable air. For a particular cloud an areal integration can be made over just the cloud. Representing this integration by a caret, (2.4) becomes

$$\frac{\partial\hat{\chi}}{\partial t} + \int\nabla\cdot[(\mathbf{v}-\mathbf{v}_t)\chi']dA + \hat{\omega}\frac{\partial\langle\chi\rangle}{\partial p} = \hat{S}_\chi. \quad (2.6)$$

In the integration over the cloudy area, $\langle\chi\rangle$ has been assumed to be constant. The term $\hat{\mathbf{v}}\cdot\nabla\langle\chi\rangle$ has been ignored for this reason. Since the cloud, and hence the area of integration, may be moving, the areal integration applied to the first term of (2.4) leads to the cloud translation velocity $\mathbf{v}_t$ being subtracted from the wind speed in the second term of (2.6). The term $\hat{X}_e$ has been ignored since the integration is over a cloudy region and not the environment.

Using the divergence theorem, the second term in (2.6) can be written

$$F_\chi \equiv \int\nabla\cdot[(\mathbf{v}-\mathbf{v}_t)\chi']dA = \int_{\text{periphery}}\chi'(\mathbf{v}-\mathbf{v}_t)\cdot\mathbf{n}\,ds, \quad (2.7)$$

where $\mathbf{n}$ is the horizontal, outward-pointing unit normal around the cloudy area, and ds is a line element along the periphery of the area. Here F_χ is the *detrained flux* of χ. This quantity can be determined by measurements on the periphery of the cloud in terms of the above line integral.

The third term in (2.6) can also be determined without penetrating the cloud. Integrating the mass continuity equation (2.3) over the cloud area and again applying the divergence theorem, we find

$$\frac{\partial\hat{\omega}}{\partial p} = -\int\nabla\cdot\mathbf{v}\,dA = -\int_{\text{periphery}}\mathbf{v}\cdot\mathbf{n}\,ds \equiv -F, \quad (2.8)$$

where F is the detrained mass flux. Integrating this in pressure and assuming that $\omega = 0$ at the surface where the pressure is p_s,

$$\hat{\omega}(p) = -\int_{p_s}^{p} F(p')dp'; \quad (2.9)$$

$\hat{S}_\chi$ can then be written

$$\hat{S}_\chi = \hat{\omega}\frac{\partial\langle\chi\rangle}{\partial p} + F_\chi + \frac{\partial\hat{\chi}}{\partial t}. \quad (2.10)$$

In steady situations $\partial\hat{\chi}/\partial t$ (sometimes called the *storage term* since it represents the time rate of change of χ stored in the cloud) is ignored and $\hat{S}_\chi$ is expressed totally in terms of quantities measured outside the cloud. In other instances some physical constraint bounds the storage term, allowing it to be ignored.

An alternative representation of $\hat{S}_\chi$ may be obtained by integrating the right side of (2.4) over the cloud area:

$$\hat{S}_\chi = \int\left[X_c - \frac{\partial(\omega\chi')}{\partial p}\right]dA$$

$$= \hat{X}_c - \frac{\partial}{\partial p}\int \omega\chi' dA. \quad (2.11)$$

Whether (2.10) or (2.11) is used to determine $\hat{S}_\chi$ depends on the situation. The advantages of (2.10) are that all measurements except that of the storage term can be made outside of the cloud and that the in-cloud source term X_c does not explicitly enter. Equation (2.11) is more useful when the storage term is important and the source term can be neglected or is easily accounted for. We refer to (2.10) as the *budget equation* and (2.11) as the *direct equation* for the convective source of χ.

We now relate $\langle S_\chi\rangle$ to $\hat{S}_\chi$. If $n(x, y, t)$ clouds are found per unit area, then $\langle S_\chi\rangle = n\hat{S}_\chi$. This follows from the hypothesis that S_χ is nonzero only inside cloud. If several different cloud types are found in the same area with areal number density n_i for type i, then the more general formula

$$\langle S_\chi\rangle = \sum_i n_i\hat{S}_{\chi i} \quad (2.12)$$

holds, where $\hat{S}_{\chi i}$ is the profile of $\hat{S}_\chi$ for clouds of type i.

b. *Virtual potential temperature*

The virtual potential temperature θ_v earns special treatment by virtue of the fact that air parcels seek their levels of neutral buoyancy. In other words, they tend to exit clouds at the level at which their virtual potential temperature equals that of the environment. This condition implies

$$F_v = \int_{\text{periphery}} \theta_v'(\mathbf{v} - \mathbf{v}_t)\cdot\mathbf{n}\,ds \approx 0. \quad (2.13)$$

In addition, buoyancy effects make it unlikely that a significant virtual temperature excess will build up in a cloud, thus making the storage term $\partial\theta_v/\partial t \approx 0$ there, even in unsteady situations. Under these conditions a simple relationship exists between the vertical mass current and the virtual temperature source term:

$$\hat{S}_v = \hat{\omega}\frac{\partial\langle\theta_v\rangle}{\partial p}. \quad (2.14)$$

Defining the vertical mass flux in cloud as $\langle\omega_c\rangle = \sum n_i\hat{\omega}_i$ in analogy with (2.12), then

$$\langle S_v\rangle = \langle\omega_c\rangle\frac{\partial\langle\theta_v\rangle}{\partial p}; \quad (2.15)$$

$\langle S_v\rangle$ is related to the parameter Q_1 (Yanai et al. 1973). If we further define the "environmental" vertical velocity as $\langle\omega_e\rangle = \langle\omega\rangle - \langle\omega_c\rangle$, then the governing equation for $\langle\theta_v\rangle$ takes the simple form

$$\frac{\partial\langle\theta_v\rangle}{\partial t} + \langle\mathbf{v}\rangle\cdot\nabla\langle\theta_v\rangle + \langle\omega_e\rangle\frac{\partial\langle\theta_v\rangle}{\partial p} = 0. \quad (2.16)$$

This analysis breaks down at the surface where rain-cooled air may be unable to descend to its level of neutral buoyancy and can therefore cause F_v to differ significantly from zero.

c. *Water vapor and equivalent potential temperature*

The storage term for total water can be significant, but so can the source term. Storage of water vapor is limited on the high end by saturation values, while the storage of condensed water is reduced by its tendency to fall out of the cloud. Given the complexity of the source terms for various moisture components and the difficulty of measuring moisture content within clouds, it is probably most productive to adopt the budget approach to the moisture parameterization problem and either make a crude estimate of the storage of moisture in clouds or ignore it altogether.

If, as assumed above, parcels exit clouds at or near their level of neutral buoyancy, then there is a one-to-one relationship between the detrained water vapor and equivalent potential temperature fluxes. Since equivalent potential temperature is nearly conserved during condensation and evaporation, it is perhaps simplest to deal directly with the equivalent potential temperature budget. The budget equation for the source of equivalent potential temperature becomes

$$\hat{S}_e = \hat{\omega}\frac{\partial\langle\theta_e\rangle}{\partial p} + F_e, \quad (2.17)$$

where F_e is the detrained equivalent potential temperature flux, and where the storage term has been neglected. If equivalent potential temperature is conserved inside clouds, then the source term for equivalent potential temperature is zero and integration of the direct equation in pressure yields

$$\int_{\text{cloud top}}^{\text{cloud base}} \hat{S}_e dp = -\int (\omega\theta_e')|_{\text{cloud base}} dA, \quad (2.18)$$

which says that the pressure-integrated source of equivalent potential temperature is simply the influx from below cloud base. The flux in or out of cloud top is assumed zero. This provides a potential way to check the correctness of the measured equivalent potential temperature source.

d. Condensed water

A simplified characterization of cloud microphysical processes in moist convection places condensed water into three classes. *Cloud water* consists of small liquid droplets that have grown primarily by condensation and are carried along by the flow. *Snow* consists of pristine ice crystals and aggregates of such crystals. These particles fall relative to the air at about 1 m s^{-1}, which means that they are essentially carried along by convective updrafts, but fall out of mesoscale circulations. *Convective precipitation* consists of particles that have grown by accretion, and therefore are of high density, with terminal fall speeds of order 5–20 m s^{-1}, and are able to fall out of many convective updrafts. This category includes graupel, hail, and raindrops.

Convective precipitation tends to fall out in the immediate vicinity of its point of formation, and induces strong, convective-scale downdrafts as it descends. Snow is typically ejected from the upper regions of convective clouds, and in highly sheared situations is carried 100 km or more from its point of origin before it falls out. There is evidence as well that further growth of snow can occur in the weak updrafts found in the upper regions of mesoscale circulations. When snow falls through the melting level, it turns into *anvil rain.* Evaporation of anvil rain is responsible for the generation of mesoscale downdrafts.

In very weak updrafts vapor deposition on snow is sufficient to deplete the supply of cloud water that is being renewed by condensation. Thus, only snow can be formed in such situations. When updrafts are stronger, vapor deposition is unable to keep up with condensation, and cloud water is available for accretion onto ice-crystal aggregates or larger cloud droplets. Thus, convective precipitation indicates the presence of vigorous updrafts.

Downdrafts produced by convective precipitation have the same lateral scale as convective updrafts, and therefore need to be parameterized. However, since snow can drift long distances before falling through the freezing level, it may be most convenient to treat snow and anvil rain explicitly in the larger-scale model. In this case a snow source term is needed from the parameterization.

The fraction of the available cloud water that is turned into convective precipitation and snow is an important parameter. Also important is the fraction of convective precipitation that evaporates before reaching the ground and the vertical profile of this evaporation. The latter is likely to be a function of how tilted the updraft is and its strength. For weak, vertically oriented updrafts, precipitation is likely to fall out through the updraft, thus causing evaporation only below cloud base. For updraft speeds exceeding the terminal fall velocity of precipitation, the precipitation will be ejected from the cloud at high levels and fall a great distance through the unsaturated environment, thus causing deep downdrafts. If the updraft is slanted, convective precipitation will likewise tend to fall into clear air above cloud base.

More important than the precipitation characteristics of a particular cloud are the changes in these characteristics as environmental conditions change. Environmental parameters of particular importance are wind shear, relative humidity, degree of conditional instability, and concentrations of cloud condensation and ice nuclei.

e. Momentum

Partitioning the velocity in the Euler equation in a manner similar to (4) yields

$$\frac{\partial \mathbf{v}}{\partial t} + \mathbf{v}\cdot\nabla\langle\mathbf{v}\rangle + \nabla\cdot(\mathbf{v}\mathbf{v}') + \omega\frac{\partial\langle\mathbf{v}\rangle}{\partial p} + \nabla\phi + f\mathbf{k}\times\mathbf{v} = -\frac{\partial(\omega\mathbf{v}')}{\partial p} \equiv \mathbf{S}_m, \quad (2.19)$$

where f is the Coriolis parameter and ϕ is the geopotential. Unlike the situation for virtual potential temperature and equivalent potential temperature, obtaining the momentum source term $\mathbf{S}_m$ from observations using the budget equation is difficult because the geopotential must be measured on the periphery of the cloud of interest to great precision. On the other hand, recourse to the direct equation is simpler than in the previous cases, because only the vertical flux of horizontal momentum occurs there:

$$\hat{\mathbf{S}}_m = -\frac{\partial}{\partial p}\int(\omega\mathbf{v}')dA. \quad (2.20)$$

Averaging (19) yields the large-scale momentum equation,

$$\frac{\partial\langle\mathbf{v}\rangle}{\partial t} + \langle\mathbf{v}\rangle\cdot\nabla\langle\mathbf{v}\rangle + \langle\omega\rangle\frac{\partial\langle\mathbf{v}\rangle}{\partial p} + \nabla\langle\phi\rangle + f\mathbf{k}\times\langle\mathbf{v}\rangle = \langle\mathbf{S}_m\rangle \equiv \sum_i n_i\hat{\mathbf{S}}_{mi}, \quad (2.21)$$

where $\langle\nabla\cdot(\mathbf{v}\mathbf{v}')\rangle$ is dropped as before.

f. Vorticity and potential vorticity

It is tempting to try to parameterize vorticity transport. One might imagine, for instance, that an excess of the vertical component of low-level vorticity could be carried aloft by convection, making cumulus clouds a net source of vorticity at high levels. However, a simple application of the circulation theorem for circulation loops embedded in low- and high-level isentropic surfaces shows that cumulonimbus clouds are net sources of vorticity at low levels and net sinks at high levels due to their net transfer of mass upward, independent of the vertical distribution of ambient vorticity (R. Rotunno 1980, personal communication; Esbensen et al. 1987; Haynes and McIntyre 1987).

Expressing this conclusion in terms of the direct equation (2.11), the vorticity source term, which contains the effects of the tilting and stretching of vortex lines, totally compensates the eddy flux divergence of vorticity. Furthermore, the storage of vorticity within the cloud is unbounded by any physical constraint, which makes the budget equation, (2.10), useless. Thus, the simplifications that make (2.10) and (2.11) useful tools in the study of cumulus parameterization for many variables are absent for the vorticity field.

Similar considerations apply to any attempt to parameterize the potential vorticity source using conventional techniques. However, as Haynes and McIntyre (1987) showed, the potential vorticity governing equation can be written

$$\frac{dq}{dt} = \frac{1}{\langle \rho \rangle} \nabla \cdot (\langle S_v \rangle \langle \zeta \rangle - \nabla \langle \theta_v \rangle \times \langle \mathbf{S}_m \rangle), \quad (2.22)$$

where $\langle S_v \rangle$ and $\langle \mathbf{S}_m \rangle$ have their previous meanings as virtual temperature and momentum sources, $\langle \rho \rangle$ is the air density, $\langle \zeta \rangle = \nabla \times \langle \mathbf{v} \rangle + f\mathbf{k}$ is the absolute vorticity vector, and $\langle \theta_v \rangle$ is the virtual potential temperature. In this formulation the potential vorticity is defined in terms of averaged quantities:

$$q = \frac{1}{\langle \rho \rangle} \nabla \langle \theta_v \rangle \cdot \langle \zeta \rangle. \quad (2.23)$$

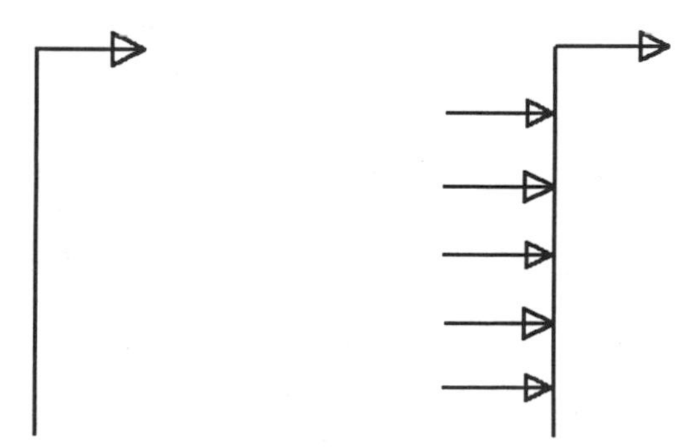

FIG. 2.1. Wiring diagrams for the nonentraining parcel model and the entraining plume. Air enters at cloud base and exits at cloud top for both of the models, as indicated by vertical lines topped with a vector to the right. In addition, the entrainment in the entraining plume is symbolized by the vectors entering from the left.

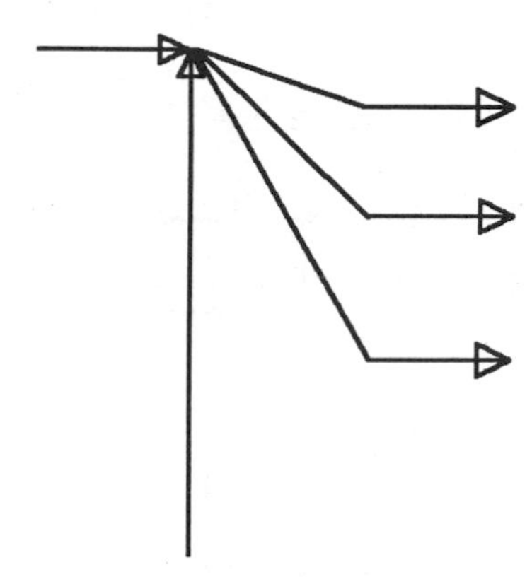

FIG. 2.2. Air is entrained at just one level in the cloud-top entrainment model. The multiple exiting arrows indicate that different mixtures of cloud-top and cloud-base air have different buoyancies and will therefore exit the cloud at different levels.

Thus, if the virtual temperature and momentum source terms are known, the potential vorticity source term can be immediately inferred.

g. Wiring diagrams and the connected current model

With the exception of storage effects, which in many cases can be neglected, the above analysis shows that the effects of cumulus clouds on the environment can often be characterized by observations of what is flowing in and out of the cloudy region. Thus, a conceptual model of clouds that focuses on the flows through clouds and the interactions of these flows would be useful to the parameterization problem. We therefore introduce the *connected current model,* which obeys the following postulates:

1) Air flowing into and out of the top, sides, and bottom of clouds is treated as a finite set of discrete currents, each of which is homogeneous.

2) These currents may merge with each other, in which case the combined current is assumed to be well mixed. In addition, currents may split, each subcurrent experiencing a different ultimate fate.

3) Currents exit the cloud at their level of neutral buoyancy.

This is a very general model of a cumulus ensemble, the only serious constraint being that the ensemble is assumed to be in a statistically steady state, so that changes in storage are not represented.

A graphical representation of this model is given by *wiring diagrams,* which show the various currents active in a given situation. The generality of this model is shown in Figs. 2.1–2.4, in which various specific models of convection are represented by different wiring diagrams.

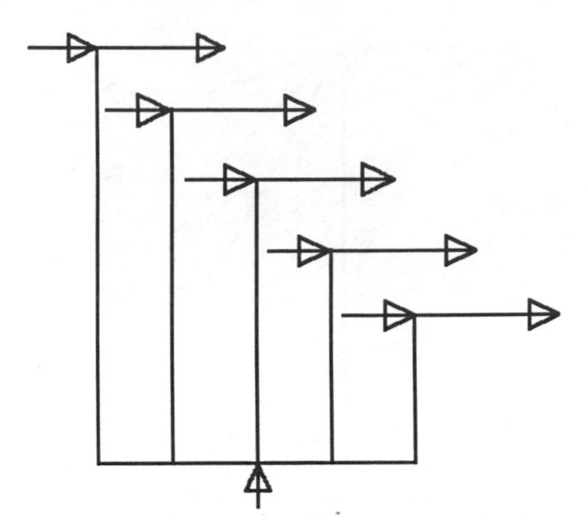

FIG. 2.3. In the detraining plume, subparcels ascending from cloud base mix with environmental air at various heights and exit the cloud.

2.3. Experimental tools

We now describe the types of results that can be obtained from three experimental tools that are commonly used to study convective clouds, namely, Doppler radar, instrumented aircraft, and sounding arrays. Each of these systems by itself provides only partial information about cumulus clouds. Used together, they may be able to significantly constrain cumulus parameterization models. It is essential to understand the advantages and limitations of these tools, so that intelligent use can be made of them.

a. Doppler radar

Doppler radar provides the most detailed information about the inner workings of cumulonimbus clouds. In regions where there is precipitation, each radar returns the component of mean precipitation particle velocity along the line of sight of the radar. The average is weighted by the sixth power of particle diameter, so Doppler radar essentially sees only the largest particles present. If three or more radars focus on a particular region of cloud, all three components of particle velocity can be recovered, assuming that the region of interest is not in or near the plane containing the three radars. For ground-based radars, this means that the radars must be situated so that the bulk of the cloud being studied has a high elevation angle relative to at least one of the radars. The simplest way to accomplish this is to place the radars relatively close together, say, within a typical cloud depth of each other. This creates the optimal spatial resolution for the radars but introduces time resolution problems, as the cloud can significantly change in the time it takes a typical radar to scan a large solid angle.

If there are only two radars, or if the radars are widely separated, the vertical component of particle velocity cannot be directly obtained, but it is still possible to obtain the horizontal components, subject to some limitations. For three or more radars, the horizontal velocities can be obtained irrespective of elevation angle. For only two radars, the vertical component of particle velocity significantly contributes to the measured line-of-sight velocity to the extent that elevation angles are large. This makes the interpretation of the observations difficult. The strategy for the two-radar configuration is thus to operate at relatively low elevation angles, so that the beams can be considered nearly horizontal. The horizontal components of the particle velocities can thus be determined to reasonable accuracy without serious recourse to assumptions about the magnitude of the vertical component of particle velocity.

It is reasonable to assume that the horizontal components of precipitation velocity are the same as the horizontal air velocity. However, neither dual nor triple Doppler radar observations directly yield vertical air velocity. This must be obtained from the pressure-integrated continuity equation. To relate pressure to geometric coordinates in this case, it is sufficient to use the hydrostatic law on a nearby sounding or, lacking that, a climatological sounding.

Sometimes radar observations are not available at low enough levels, and the alternative procedure of integrating down from the radar-observed cloud top may be attempted:

$$\omega(p) = \omega(p_t) - \int_{p_t}^{p} \nabla \cdot \mathbf{v}\, dp', \qquad (2.24)$$

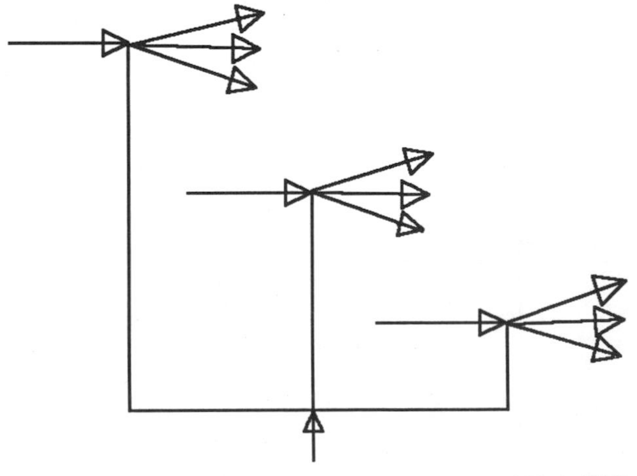

FIG. 2.4. The stochastic mixing model (Raymond and Blyth 1986) differs from the detraining plume model in that different mixing fractions are assumed for different subparcels mixing with environmental air at each level. This results in multiple exit levels for each mixing level.

where $p_t = p_t(x, y)$ is the pressure at cloud top. However, this leaves one with the task of determining $\omega(p_t)$. It is tempting to assume that $\omega(p_t) = 0$, but this assumption is unwarranted in situations with rapidly rising tops or when attenuation prevents the radar beam from penetrating to cloud top. An alternative assumption is that the vertical air velocity at the radar-observed cloud top equals the vertical precipitation velocity there plus an assumed particle terminal velocity. In some but not all cases only small particles reach cloud top, which means that a small terminal velocity may be assumed. This method depends on having high-elevation-angle triple-Doppler data and has the minor disadvantage of depending on assumptions about the microphysical character of the cloud top.

There are other related methods to determine vertical air motion from Doppler radar data, but they all share a dependency on cloud microphysical assumptions. The message is that horizontal air motions are relatively easy to determine using Doppler radar but that vertical air motions are difficult.

Two additional problems plague Doppler radar measurements. First, Doppler radar sees only that part of the cloud that contains precipitation-sized particles. In many thunderstorms the updraft is free of precipitation, thus making the radar-observed vertical motions incomplete. In addition, tropical oceanic systems often have very weak reflectivity at high altitudes because all large particles have fallen out. Thus, high-altitude motions may be missed. Second, for radar observations to be useful in the parameterization problem, the complete life cycle of a convective system needs to be observed. This is difficult because convection previous to the development of precipitation cannot be seen at all by conventional radar, and because many convective systems are migratory, so that a ground-based radar array sees the system only as it passes through the array.

The final difficulty with Doppler radar is that only velocities are directly measured. There are ways to estimate thermodynamic and microphysical quantities by using the dynamic and cloud microphysical governing equations in conjunction with radar observations, but these studies, while interesting, are fraught with uncertainty at this time.

In summary, relating Doppler radar measurements of clouds to the cumulus parameterization problem is difficult. However, no other tool has the potential of exploring the interior flows of cumulonimbus clouds to the extent of Doppler radar.

b. Aircraft

Aircraft have been used to study clouds in two ways: by direct penetration and by exploring the immediate environment of a cloud. In addition, airborne remote sensing tools are now becoming available, such as airborne Doppler radar and lidar.

With direct penetration, many factors of interest to cumulus parameterization can in principle be explored. However, numerous problems plague even the best observational efforts. With inertial platforms and gust probes, it is relatively easy to document horizontal and vertical motions in clouds. Thermodynamic measurements are more difficult. Accurate radiometric measurements of temperature in cloud are only now becoming available. Immersion thermometers often fail because temperature elements get wet to an unknown degree, and thus act partially like wet-bulb thermometers. This is a problem even in saturated regions of cloud where wet- and dry-bulb temperatures are essentially equal, because the adiabatic compression that occurs when air approaches the thermometer element makes the air in the immediate vicinity of the element warmer than ambient air and therefore unsaturated.

Humidity measurements are also difficult in clouds and rain. Whereas updrafts are generally quite close to saturation, downdraft regions are often unsaturated. They nevertheless contain condensed water in the form of large drops, ice crystals, or graupel. The rate of evaporation of these hydrometeors depends on the degree of subsaturation, which in turn depends on both the temperature and the dewpoint. The commonly used cooled mirror dewpoint devices respond slowly and will yield incorrect results if wetted by cloud water or rainwater. A combination of a wet-bulb thermometer for humidity and a radiometric thermometer for air temperature may be the best and simplest way to make these measurements. Unfortunately, airborne wet-bulb thermometers have fallen out of common use and are also subject to subtle errors in rapidly moving airstreams.

Thermodynamic fields are needed if one desires to make direct measurements of the vertical fluxes of heat and moisture in cumulus clouds. The above-noted problems with thermodynamic quantities make such measurements difficult with currently available instrumentation.

Also of interest are the fluxes of various categories of condensed water. Instruments exist to measure the concentrations of cloud particles of all sizes, with some caveats. Low concentrations of very large particles are missed, and no instrument in common use is able to characterize the population of very small ice crystals. Given the irregular geometry and uncertain densities of ice particles, it is difficult to compute the mass concentration of in-cloud ice, leaving large uncertainties in an important part of the total water budget.

An additional problem of aircraft measurements is representativeness. An aircraft traces a single line through a cloud, and it is rarely clear what has been missed. In addition, the minimum possible time between aircraft passes through a small cloud is comparable to the evolution time scale of the cloud. Thus, the sampling done by aircraft is very sparse indeed. Penetrating aircraft are probably best used in conjunction with radar, which can provide a context for the aircraft measurements.

The two maneuvers external to cloud that are most useful in the study of cumulus clouds are the sounding and the convergence box. The problems of making thermodynamic measurements are less severe here, since such maneuvers are generally flown in clear air around the cloud being studied. When interpreting soundings, it is important to remember that aircraft move primarily in the horizontal direction even when climbing or descending. Thus, care must be taken to ensure that changes in observed quantities are due to vertical rather than horizontal structure in the atmosphere.

A more subtle problem has to do with wind measurements. The aircraft velocity provided by an inertial navigation system (INS) is subject to a semiregular oscillation with an 84-min period and an amplitude of a few meters per second. This error contaminates the wind measurement from aircraft. A recently available solution to this problem is the satellite-based Global Positioning System (GPS). This system provides aircraft speeds and positions that are very accurate when averaged over a few minutes. GPS is therefore complementary to inertial navigation, since INS errors are in the low-frequency component. Combining the high-frequency part of the INS signal with the low-frequency part of the GPS signal provides very accurate aircraft velocities, and hence winds.

GPS correction of INS winds is particularly important in the convergence box, in which a closed horizontal path is flown around the cloud of interest. In this maneuver, measurements of detrained fluxes of various quantities as defined by (2.7) and (2.8) may be made. Since detrained fluxes of mass and of thermodynamic quantities are sufficient to characterize the effect of a cloud on the environment in many cases [see (2.9) and (2.10)], such aircraft measurements are very valuable. Unfortunately, these measurements are subject to the usual problem, namely, sparse sampling, and compositing techniques are generally required to make sense out of them.

Remote-sensing tools from aircraft have only recently become available, and we are just now learning how to use them. Due to antenna size limitations, airborne radars typically have somewhat broader beams than ground-based radars, and thus have less effective range. However, the ability to move the aircraft to the immediate vicinity of the cloud being studied potentially more than overcomes this limitation. Airborne radars also typically work in X band, which is more subject to attenuation effects than ground-based radars working at longer wavelengths.

c. *Sounding arrays*

Radio sounding arrays can in principle measure detrained fluxes, and a great deal of work has been done with such arrays. The strength of such arrays is that repeated soundings can be made, yielding a time sequence of the development of a convective system, as long as the system does not move out of the array area. However, the time resolution (determined by the time taken by a radiosonde to ascend) is usually of order 1 h, so such arrays are best used to study aggregates of clouds or large, long-lived convective systems.

Problems encountered in using sounding arrays are typically those of accuracy and representativeness of the observations. Wind observations, especially in strong wind situations, can be significantly in error, and subject to noise due to tracking oscillations. Newer devices based on the loran navigational system have the potential for eliminating this problem. However, an array typically has at most only five or six sounding sites surrounding the region of interest. This may not be enough to characterize the flow with sufficient accuracy to allow a good estimate of detrained mass flux.

Another difficulty is that the region defined by the sounding array will typically encompass much more clear air than, say, the region defined by an aircraft convergence box. Clear-air processes thus are more likely to contaminate the signal from the clouds. This can happen in several ways. Information about the occurrence of vertical mass flux in the atmosphere is transmitted outward by gravity waves with a dominant vertical wavelength equal to about twice the vertical scale Z of the mass flux pattern. Hydrostatic waves with this vertical scale travel at a speed of about $c = NZ/\pi$, where N is the Brunt frequency of the atmosphere. For deep convection, c is typically 30–50 m s^{-1}, so a sounding station 100 km from a thunderstorm will typically experience a lag of 2–3 h before it sees a local response. Detrained mass flux measured by an array of sounding stations at nonuniform distances from a storm will therefore contain a confusing transient response and will be useful only for observing relatively steady systems.

An additional problem is that gravity waves cannot propagate much farther than a Rossby radius in the atmosphere, that is, that distance that takes a gravity wave the inverse of the Coriolis parameter f to travel. For the dominant gravity waves emitted by convection, this is equal to $L = c/f = NZ/\pi f$. Typical values for deep convection are 300–500 km. Even for these distances significant detail is lost in the vertical structure of the observed mass flux, as smaller vertical scales have smaller Rossby radii. Thus, closely spaced arrays will yield the finest detail in vertical structure.

The water vapor emitted by convection travels at material rather than wave speeds, which typically means even greater lags in the measurement of detrained moisture fluxes, especially in weak wind situations. Observations from sounding arrays must be interpreted with these rather significant limitations in mind.

2.4. Existing observations

In this section we summarize selected observations made on cumulus and cumulonimbus clouds and attempt to place these observations in the context of the framework outlined in section 2.2. We first look at observations of convective mass flux. We then pass to studies pertinent to the convective source of equivalent potential temperature. Ice and precipitation particle fluxes are investigated next. Finally, we turn to momentum fluxes.

a. Mass fluxes and heating profiles

Vertical mass flux may be directly measured by Doppler radar and aircraft observations or may be inferred from the detrained mass flux $F(p)$ using the mass continuity equation, (2.9). A close relationship exists between convective heating and vertical mass flux, that is, that expressed by (2.14). This relationship should be approximately valid everywhere except the planetary boundary layer, where the storage term associated with cold-air outflows can be important. Thus, once the vertical mass flux is known above the boundary layer, the vertical heating profile is known as well.

Many observations of cumulus mass fluxes have been reported, though not always with enough supporting data to make the observations useful in testing cumulus parameterizations. Perhaps the first observations of cumulonimbus mass flux were made during the Thunderstorm Project (Byers and Braham 1949) and were reported by Byers and Hull (1949). By using swarms of pilot balloons, they were able to deduce the detrained mass flux profile around clouds before and after rain appeared at the surface. In both cases, net convergence was found at all observed levels (up to 6 km before rain started, up to 8 km afterward), with the exception that divergence was found below 1 km and above 7 km after the onset of rain.

With the use of networks of radiosondes (Fankhauser 1969, 1974; Betts 1973a; McNab and Betts 1978), radar observations (Miller et al. 1982; Frank and Foote 1982; Raymond et al. 1991), or instrumented aircraft (Telford and Wagner 1974; LeMone and Zipser 1980; Zipser and LeMone 1980; Raymond and Wilkening 1982, 1985; Raymond and Blyth 1986), various estimates have been made of vertical and detrained mass fluxes, and a wide variety of patterns has been seen.

Curiously, there is no definitive answer to a very simple question about cumulus clouds, namely, whether or not they typically ascend to their level of undilute neutral buoyancy. The question is complicated by the fact that cumulus clouds often occur in regimes of near-neutral conditional instability, so that cloud top can be very sensitive to buoyancy in the cloud. Furthermore, cumulus clouds generally feed from the planetary boundary layer, which is typically not as well mixed in equivalent potential temperature as it is in virtual potential temperature. Thus, very careful measurements at cloud base must be made to determine the buoyancy available to parcels entering the cloud. LeMone and Pennell (1976) found that trade-wind cumuli fed from near the top of the boundary layer but deep convection almost certainly taps more unstable air from lower levels.

Indirect evidence that cumulus clouds tend to reach their level of undilute neutral buoyancy comes from observations that cumulus congestus clouds contain unmixed parcels in their upper regions (Heymsfield et al. 1978). This observation also supports the notion that mixing in clouds is a random, impulsive process, with some fraction of cloud parcels managing to avoid mixing altogether. It is thus in conflict with the assumption of the entraining plume model (e.g., Squires and Turner 1962; Simpson et al. 1965) that mixing occurs instantaneously after entrainment of environmental air.

The fact that updraft mass fluxes can decrease with height also means that a single entraining plume is inappropriate as a model for convective ensembles. It may be, as suggested by Arakawa and Schubert (1974), that cumulus ensembles can be adequately represented by many entraining plumes with varying entrainment rates. However, downdrafts induced by evaporation of cloud water as proposed by Squires (1958), Telford (1975), etc., may be important to the parameterization problem. These downdrafts have been documented by Paluch (1979), Raymond and Wilkening (1982, 1985), Raymond and Blyth (1986), Blyth et al. (1988), Raga et al. (1990), and others, and can cause the net upward mass flux to decrease with height even if the updraft part of the mass flux increases with height.

Without question, downdrafts induced by precipitation are important in determining cumulus mass flux profiles. The downdraft mass current in the lower parts of cumulonimbi may equal or exceed the updraft mass current there (see, e.g., Knight and Squires 1982; Houze and Betts 1981). Just how much of the cloud liquid water evaporates in the production of these downdrafts is a question with no easy answers, as it depends on the precipitation efficiency, the degree to which rain shafts are ventilated by ambient air, the thermodynamic characteristics of this air, and the evolving particle size distributions in the rain shaft. Generally speaking, environments with higher shear and drier air are thought to promote stronger downdrafts, but even this is hard to quantify.

Figures 2.5 and 2.6 illustrate the major contributions to vertical mass fluxes in cumulus and cumulonimbus clouds, as observed by Raymond and Wilkening (1985). In the cumulus cloud, downdrafts caused by the evaporation of cloud droplets are comparable in magnitude to the updrafts, resulting in near-zero mass flux at high levels and significant detrainment at low levels. The maximum vertical mass flux occurs near cloud base. In the cumulonimbus, precipitation-induced downdrafts roughly balance updraft mass fluxes at low levels but are confined to the lower half of the cloud. Maximum vertical mass flux thus occurs at high levels.

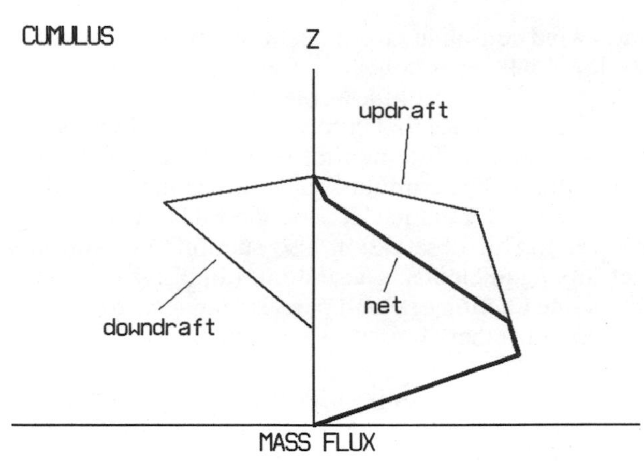

FIG. 2.5. Example of possible vertical mass flux profiles for a non-precipitating cumulus cloud. In the upper regions evaporative cooling of cloud material results in downdrafts that are comparable in strength to the updrafts. The net vertical mass flux is thus close to zero there.

b. Equivalent potential temperature

As was shown in section 2.2, the water vapor source term is closely related to the source term of equivalent potential temperature. Since this variable is nearly conserved within clouds, it behaves in a simple fashion. Recall, however, that equivalent potential temperature is not conserved in freezing and melting processes or when visible or infrared radiation is emitted or absorbed.

As (2.17) shows, the source term for equivalent potential temperature may be estimated from the vertical mass flux, the ambient lapse rate of equivalent potential temperature, and the detrained flux of equivalent potential temperature at each level. Measurements of this type have been made by large-scale sounding arrays (Betts 1973a; Yanai et al. 1973; etc.) and aircraft box maneuvers (Raymond and Wilkening 1985; Raymond and Blyth 1986).

At high levels, outflows are generally saturated. This means that the equivalent potential temperature source there is indeed specified once the vertical mass flux is known. Uncertainties in this quantity are therefore confined primarily to low and middle levels.

If no mixing occurs between streams passing through a cloud, the distribution of equivalent potential temperature in the detrained air will be the same as that in the air entering the cloud. If mixing does occur, then the detrained distribution will be narrower. In the former case, the value of equivalent potential temperature detrained at each level gives a strong clue as to the level at which that air entered the cloud. Betts (1973a) used this technique to infer composite flow patterns within Venezuelan squall lines.

Unfortunately, measurements of the completeness required for evaluations of this type are very hard to obtain. Sounding arrays are potentially useful in this regard but are typically subject to inaccuracies induced by large spacings and inadequate sampling. Aircraft box maneuvers can obtain accurate measurements at a given level but are unable to sample many levels during the lifetime of a cloud. Compositing is probably the best hope for obtaining adequate sampling for use with this technique.

It has long been known that organized convective systems transport air from near the midlevel equivalent potential temperature minimum downward to the surface (Newton 1950; Zipser 1969). Thus, the convective source of equivalent potential temperature at low levels is intimately tied in with the dynamics of precipitation-induced downdrafts.

Another region in which the equivalent potential temperature source term is uncertain is in the midlevel outflows documented by Raymond and Wilkening (1982, 1985) and Raymond and Blyth (1986). These are often, but not always, saturated, and probably result from mixing of cloud-base air with air from middle levels. They also often contain snow (A. Blyth 1993, personal communication).

c. Condensed water

Cloud microphysical observations are difficult to make. There is also a dearth of simple theories to explain such processes as the partitioning of cloud water into snow and convective precipitation. Observations of tropical mesoscale convective systems show that approximately 40% of the total precipitation is in the form of anvil rain; that is, it is derived from snow (Houze 1977; Houze and Rappaport 1984; Churchill and Houze 1984). It is uncertain what fraction of this snow is produced in the convective towers and how much is generated in mesoscale updrafts. It is also not clear how this fraction would change if environmental conditions changed.

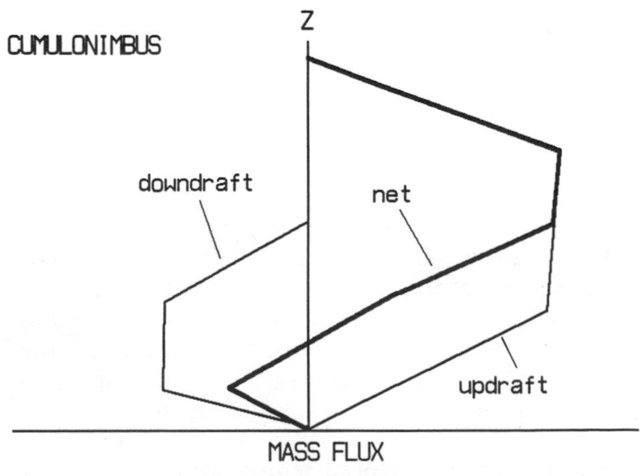

FIG. 2.6. Example of possible vertical mass flux profiles for a cumulonimbus cloud. Significant precipitation-induced downdrafts are confined to below the midlevel equivalent potential temperature minimum, causing the peak vertical mass flux to be above this level.

One difference between weak oceanic and strong continental updrafts is that the former tend to produce large quantities of ice particles just above the freezing level, while the latter must typically reach temperatures of $-15°C$ before significant ice is produced (Jorgensen et al. 1991; Raymond and Blyth 1989). The reasons for this are not understood but may indicate that elapsed time as well as temperature is important in ice formation (Vonnegut 1987). A possible alternative explanation is that the large droplets that are common at low levels in moist tropical systems readily freeze when lifted above the 0°C level, while the small droplets of continental clouds do not.

Whatever the reason, the above results suggest that virtually all the water substance carried above the freezing level in tropical oceanic clouds is converted into snow, while the fraction is much less for continental clouds.

Surprisingly few observations have been made on convective-scale downdrafts. This is partly because such currents form a very hostile environment for aircraft, with heavy precipitation and strong downward motions near the ground. Reliable radar observations are scarce as well, due to the fact that they tend to occur low in the cloud where ground clutter, terrain, and the curvature of the earth typically interfere. As discussed in section 2.2, many complex factors enter into the determination of convective downdraft profiles, and it is therefore difficult to construct a simple but accurate parameterization.

d. Momentum fluxes

As discussed in section 2.2, the momentum source term is best obtained observationally by direct measurement of the vertical flux of horizontal momentum. This is accessible using aircraft and Doppler radar measurements.

Aircraft observations of momentum transport were pioneered by LeMone (1983) for a weak oceanic convective line. LeMone et al. (1984) made a systematic study of tropical, oceanic squall lines in the eastern Atlantic using aircraft measurements and found that the vertical transfer of horizontal momentum was (a) downshear in the direction along the line and (b) negative in the cross-line direction (i.e., enhanced front-to-rear motions were correlated with updrafts), irrespective of the cross-line shear. Isolated cumulonimbi and less-well-organized lines tended to transfer momentum down the gradient in this study.

Results similar to those above were found for a Doppler radar study of a West African squall line (LaFore et al. 1988). In addition, this study showed that most of the momentum transport was due to the mean mesoscale circulation, at least at middle levels. Convective eddies tended to extend the momentum flux to higher and lower levels.

LeMone and Jorgensen (1991) found a more complex picture in a study of a convective line near the island of Taiwan. Cross-line momentum transfer was negative below 5 km, as in the eastern Atlantic, but positive above, and along-line transfer was not all downgradient.

The main lesson from these observations seems to be that momentum transfer is primarily a mesoscale process except in situations with isolated or disorganized cumuli. In the former case, the momentum transfer is a complex function of the morphology and orientation of the mesoscale system, while in the latter case downgradient transfer of momentum takes place. If this is true, then care must be taken to avoid "double counting" momentum transfer in mesoscale models—it would be incorrect to attempt to parameterize the process if the model were explicitly simulating it. A goal of further observational studies should be to separate the convective from the mesoscale part of the momentum transfer, if indeed such a separation is possible. It is conceivable that the convective part of the momentum transfer is always downgradient relative to the local shear, even though it is upgradient relative to ambient conditions. If this were so, it would simplify the convective part of the parameterization.

2.5. Proposed observations

Much work remains to be done before we understand cumulus clouds well enough to develop a reliable parameterization. The deficiencies lie mainly in our understanding of cloud microphysical processes and momentum fluxes, but some mysteries remain in mass and equivalent potential temperature fluxes as well. As indicated in the Introduction, understanding the response of convection to *changes* in forcing is more important than obtaining a highly accurate picture of convection in any given situation. Probably the only way this can be accomplished is by testing simple semi-empirical models of cumulus clouds against nature. Examples of such models (correct or incorrect) are the entraining plume, Telford's (1975) cloud-top mixing model, the stochastic mixing model of Raymond and Blyth (1986), the entraining–detraining plume model of Kain and Fritsch (1990), Emanuel's (1991) model, and the steady, two-dimensional models of Moncrieff (1981). Particularly lacking at this point are simple cloud microphysical models.

A wish list of observations follows:

1) It is important to extend current work on determining the distribution and characteristics of mass currents through clouds, including the location and degree of mixing events, the frequency with which the level of undilute neutral buoyancy is reached, and the region of the boundary layer from which various types of clouds feed. These studies would provide tests of current theories of the cumulus circulation, leading to reliable models of cumulus mass flux and equivalent potential temperature source. A key ingredient in this type of study is the development of accurate in-cloud

temperature measurements, probably of the radiometric type, for use on instrumented aircraft. This project would require intensive use of aircraft for cloud penetrations, cloud-base monitoring, and box maneuvers to obtain detrained mass and equivalent potential temperature fluxes. This program must be carried out on a large variety of clouds so that candidate cloud models are sufficiently exercised.

2) Cloud microphysical measurements designed to learn more about the production of ice and precipitation in convective clouds need to be made. In particular, the way in which cloud water is partitioned into convective precipitation and snow needs to be clarified. This will likely require the development of new instruments, particularly devices to measure the concentration and size distribution of very small ice particles as well as low concentrations of large precipitation particles. In addition, this project would benefit from theoretical, laboratory, and modeling work on ice nucleation and the formation of precipitation embryos.

3) A better understanding of convective downdrafts is important to the cumulus parameterization problem, since these flows are central contributors to low-level mass and equivalent potential temperature fluxes. Instruments that measure the concentrations of small raindrops in the presence of large drops are needed, as well as devices to measure temperature and humidity in the presence of rain. Since this question is so tied up with the dynamics of downdrafts, it would best be done in conjunction with numerical modeling efforts to put the measurements in context.

4) More studies of convective momentum transfer are needed, first to build up our sample of cases, and second to determine whether momentum transfer is in essence a convective or mesoscale process. These studies are probably best done using Doppler radar with help from aircraft observations at low levels. The development of better ground-clutter suppression on meteorological radars would improve low-level observations.

Most if not all of the above studies could be part of the same project since they use many of the same tools. They also share the need for new instrument development, with measurement of thermodynamic properties in cloud and rain having highest priority and with improved hydrometeor measurements not far behind. The primary tools in all cases are advanced meteorological radars and well-instrumented aircraft that can reach all levels of the clouds under study and penetrate them. Numerous past projects have been hampered by the lack of the proper tools and instruments.

Some recent observational programs have been greatly enhanced by concurrent exploratory numerical and theoretical work, and these proposed projects would benefit from this as well. The tightening of the feedback between observation and theory engendered by this approach will surely lead to more rapid progress in solving these problems.

Acknowledgments. The author is grateful for the input of Alan Blyth, especially in the area of cloud microphysics. Marcia Baker's review of the manuscript was helpful. This work was supported by National Science Foundation Grant ATM-8914116.

Chapter 3

Trade Cumulus Observations

MARCIA BAKER

University of Washington, Seattle, Washington

3.1. Introduction

a. Data sources

The first systematic series of observations in trade cumulus was carried out in the 1950s and 1960s by the Woods Hole Group (Riehl et al. 1951), followed by the National Hurricane Research Project (Brown 1959; Simpson, 1983a), who were attempting to determine the basic dynamical structure of the clouds and to document the evolution of warm rain. They were followed by large-scale boundary-layer experiments combining satellite, ship, and aircraft observations {the Atlantic Tradewind Experiment (ATEX) (Augstein and Ostapoff 1974), the Barbados Oceanographic and Meteorological Experiment (BOMEX) (Holland and Rasmussen 1973), and two experiments focused on the Hawaiian rainbands [the Joint Hawaii Warm Rain Project (JHWRP) (1985) and the Hawaiian Rainband Project (HaRP) (1991)]}. In addition, several individual observers (Takahashi 1977, 1981; Takahashi et al. 1989; Warner 1955, 1969, 1970b, 1971, 1973b, 1977) contributed very detailed and systematic aircraft studies of trade-wind cumulus. Many of these measurements are summarized and discussed in the reviews by Simpson (1983a,b,c). In this chapter we synthesize the data collected on the structure of the trade-wind sounding and the spatial and temporal distributions of dynamic, thermodynamic, and microphysical properties of the individual clouds and cloud clusters.

b. Fractional cloudiness and cloud organization

The mean fractional cloudiness in the trade cumulus layer is on average only about 11%–19% (Warren et al. 1988), with active, growing turrets occupying only 2% or 3% of the layer (Augstein et al. 1974; Riehl 1979). However, fractional cloudiness varies with height within the layer and depends not only on cloud layer properties (Albrecht 1981; Augstein et al. 1974) but on the subcloud moisture content and organization, as discussed below. The typical aspect ratio (width to height) of an isolated cloud (averaged over its lifetime) is about 2/3:1, and only the deeper clouds precipitate. While aircraft missions limited to the boundary layer (roughly the first 2 km above the ocean) gave the impression that the spatial distribution of the clouds was random, or at least highly irregular, flights above the boundary layer and satellite pictures reveal that the clouds are often highly organized, either in two-dimensional streets along the wind or in so-called mesoscale cellular convection. The streets are typically up to 100 km long; they are formed by clouds whose individual aspect ratios are about 2:1, and are separated by 2–8 km (Kuettner and Soules 1966). The cells are usually tens of kilometers across, and thus have aspect ratios on the order of 10:1, tend to be "open" (rings of clouds surrounding clear air) over warm water to the east of major continents, and "closed" (cloudy centers surrounded by rings of clear air) over cooler waters to the west of continents (Agee 1982).

In another form of organization, or semiorganization, cloud clusters and bands form during disturbed conditions at sea. Bands also form as a result of interaction of the trade winds with landmasses, such as the band clouds off the eastern shore of the island of Hawaii. The individual cells in larger cloud masses tend to be somewhat wetter, more vigorous, and longer lasting than isolated clouds (i.e., those separated by 5 km or more), so we will distinguish the two situations in describing the trade-wind cumulus.

3.2. Characteristics of the trade-wind atmosphere

a. General

Figure 3.1 shows a typical trade-wind sounding in clear air within a patch of cumulus clouds. The subcloud layer is fairly well mixed and is separated from the cloud layer by a slight temperature inversion at about 400–500 m from the sea surface (about 860–940 mb). The cloud layer is conditionally unstable (Brunt–Väisälä frequency $N_{BV} \approx 0.01$) and somewhat drier than the subcloud layer. The trade inversion caps the layer at about 750–850 mb. The same sounding is shown in Fig. 3.2 in terms of the two conservative thermodynamic tracers, Q (total water mixing ratio) and

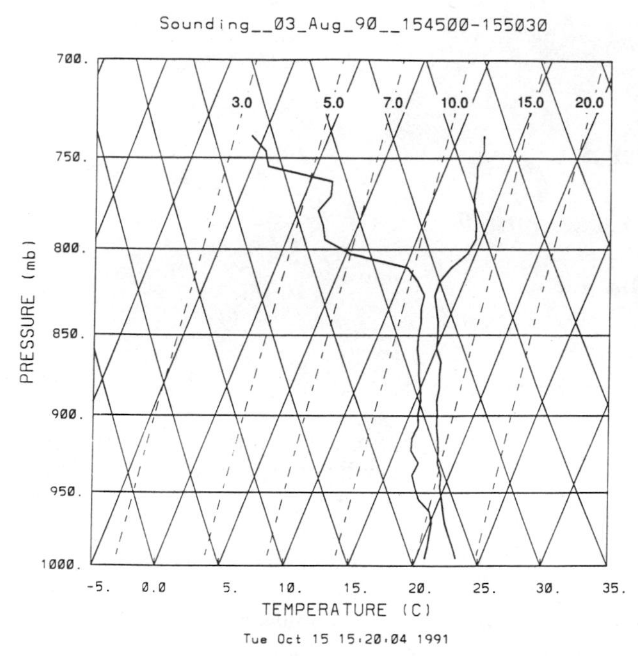

FIG. 3.1. Trade-wind sounding as measured by aircraft: 3 August 1990 (HaRP).

θ_q, the total wet equivalent potential temperature. The cloud layer properties in this case map onto a mixing line between the points characteristic of air close to cloud base and of air above the inversion. The mixing line in the cloud layer crosses the isopleths of virtual potential temperature, whereas the subcloud layer is neutrally buoyant. In some cases (Betts and Albrecht 1987; Raga 1989) the structure is more complicated, showing kinks in the middle of the cloud layer. A fairly common feature of the soundings is the extremely dry layer above the inversion, whose origin(s) is not yet fully explained.

Betts (1973b) has parameterized the thermodynamic structure of cloudy layers in terms of parameters $p^*(p)$, the pressure level at which air lying at pressure p is exactly saturated, and the derived quantities $\Delta(p) \equiv p - p^*$ and $\beta(p) \equiv dp^*/dp$; $\beta(p)$ ranges from near zero to a few tenths in the subcloud layer, between 1 and 2 in the cloud layer, and rises to 5–10 in the inversion, while Δ is fairly constant at around 20–50 mb below the inversion (Raga et al. 1990).

Figure 3.3 shows a profile passing through cloud of cloud condensation nucleus (CCN) concentrations in a cloudy region near the island of Hawaii. This figure shows that just above and below the clouds the CCN concentrations fall to close to zero as all particles are scavenged, thus possibly providing a break on further cloud formation (J. Hudson 1990, personal communication). The CCN composition is dominated by $(NH_4)_2SO_4$, with sea salt playing a very minor role (Pruppacher and Klett 1978).

b. Subcloud layer

LeMone and Pennell (1976) have shown that the positions of the trade-wind cumulus clouds are linked to the structure of the subcloud layer. They examined three cloudy patches roughly 40 km in scale, in two of which conditions were suppressed and in one with slightly enhanced but nonprecipitating clouds. In the suppressed case they found the clouds are passive tracers marking the positions of the updraft segments of subcloud roll vortices; the correlations between vertical fluxes of temperature, buoyancy, and horizontal momentum were maintained down to almost 300 m below cloud base. In the enhanced cases the clouds were deeper, more numerous, and more random, and stemmed from updrafts that could be traced down 200 m below cloud base. Velocities were higher in the subcloud layer below the enhanced convection ($\bar{w}$ was about 10–40 cm s^{-1} 30 m below cloud base). In the suppressed cases vertical fluxes in the subcloud layer were more or less evenly carried by up- and downdrafts; there was less vertical momentum flux carried by downdrafts in the subcloud layer in the enhanced convection case. Subcloud convergence organizes the Hawaiian band clouds (Raga 1989), where parcels of negatively buoyant air are lifted to the condensation level by the mechanical forcing.

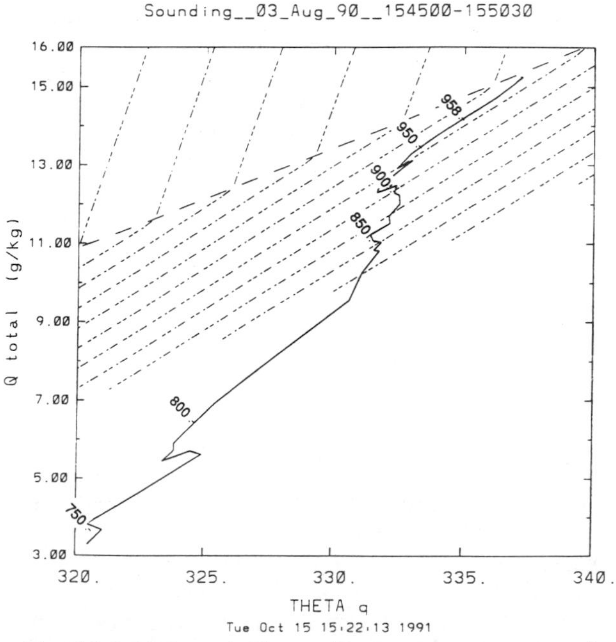

FIG. 3.2. Same data as in Fig. 3.1 displayed in terms of Q (g kg^{-1}), the total water content, and θ_q (K) (solid line). Dashed line shows Clausius–Clapeyron relation between Q and θ_q at the level of cloud base, 958 mb. All points above this line correspond to properties of air that is saturated at this pressure level. Dash–dot lines are isopleths of virtual potential temperature for this pressure level. Contour spacing: 2.5 K.

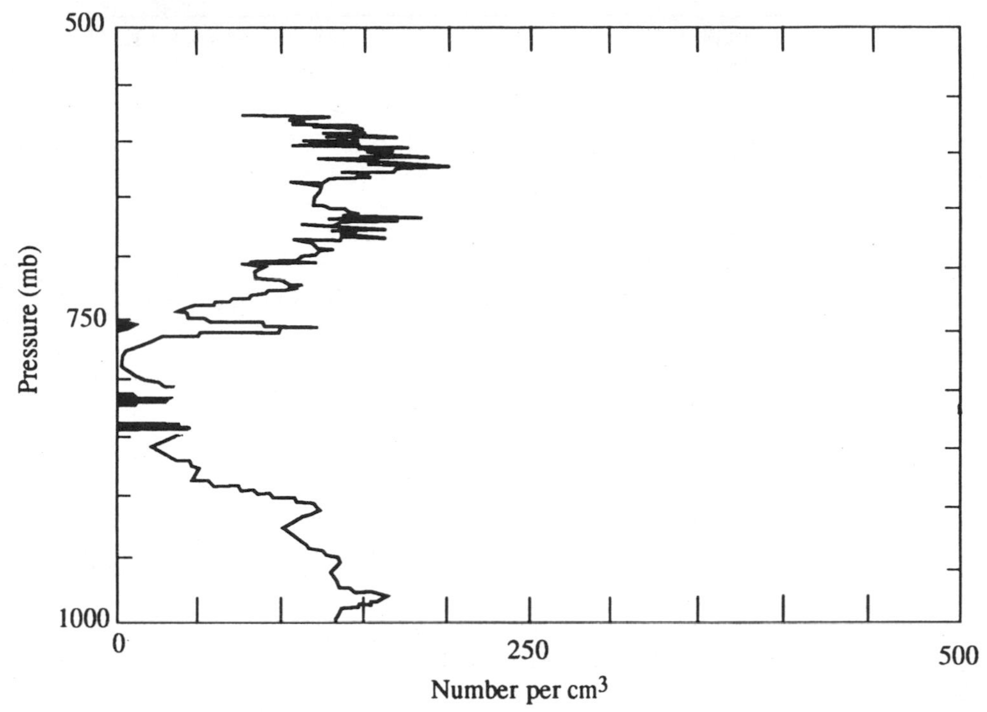

FIG. 3.3. Profile passing through cloud of CCN concentration (solid line); cloud-droplet concentration (blackened areas). Aircraft measurements: 1 August 1990 (by kind permission from J. Hudson).

c. Cloud layer

The bases of trade cumulus clouds are often quite flat and quite uniform, reflecting the nearly well-mixed nature of the subcloud layer, although this is not always the case. The outlines of isolated clouds, or cells within cloud clusters, tend to be oval in shape, with mean horizontal radii approximately ½–1 km in the growing phase. The environment around an active cloud tends to be disturbed in the sense that turbulent velocities are higher, relative humidity and temperature are higher, and CCN count is often lower than that in the environment far from clouds, suggested by Fig. 3.3. The disturbed area is roughly one cloud diameter in horizontal scale and is largely confined to the downshear shadow of the cloud (Warner 1977). These near cloud effects have not been sufficiently quantified to make numerical estimates of their magnitudes worthwhile. However, care must be taken in interpreting "trade-wind soundings" to distinguish those measured far from active clouds from those taken in a cloudy region.

If there is shear in the cloud-layer horizontal wind, the motion of the cloud outlines is usually along the mean wind at cloud base, rather than the mean wind in the cloud layer, as discussed by Wagner and Telford (1976). The clouds move more slowly than the surrounding air when the shear is positive (mean wind increasing with height in the along-wind direction) and more rapidly when it is negative (mean wind decreasing with height). In the former case the outflow from the cloud interferes with the inflow and destroys the supply of buoyant air needed to maintain the cloud, whereas in the latter case the outflow falls behind the moving cloud and cloud life is extended. Thus, locally fractional cloudiness and the time evolution of individual clouds are determined in part by the vertical variation of horizontal winds within the cloud layer.

d. Trade inversion

The inversion that caps the trade cumulus layer in undisturbed conditions tends to inhibit the rise of all but the most undilute, vigorous cumulus turrets. The inversion layer itself is a result of large-scale subsidence and varies somewhat in thickness and altitude depending on synoptic conditions. Vigorous convection can raise or even destroy the inversion. The reader is referred to Riehl (1979) for a review of many observations of inversion structure. Betts and Albrecht (1987) showed that the air mass descending into the inversion is often, but not always, thermodynamically distinct from that in the convective boundary layer. Typically, the air descending into the boundary layer sinks at a rate of 40–60 mb day^{-1} (Augstein et al. 1974; Holland and Rasmussen 1973); it is this sinking that is more or less balanced by the tendency of convection to raise the inversion.

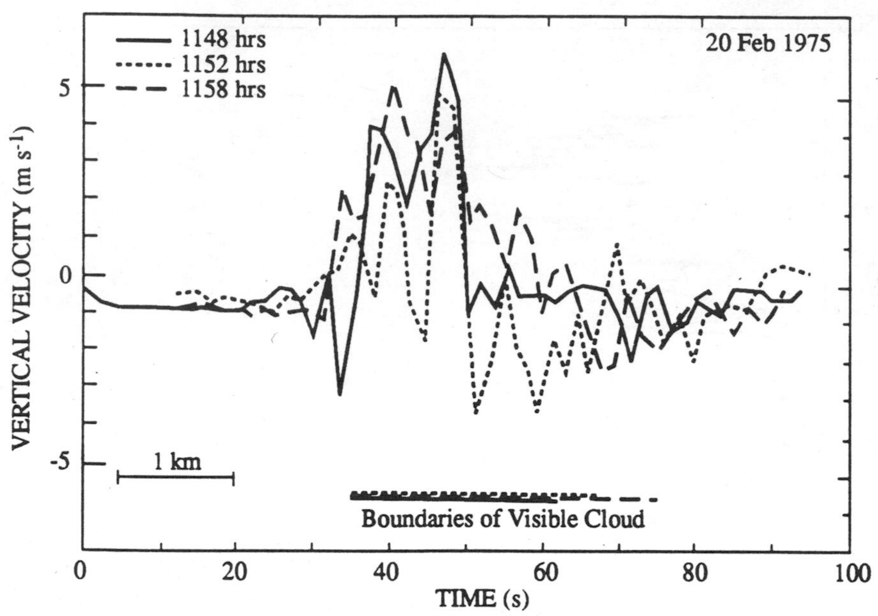

FIG. 3.4. Cross section of updraft structure from successive aircraft penetrations of a cumulus cloud (from Warner 1970).

3.3 Cloud properties

a. Dynamics

Since it is difficult to predict when and where a cumulus cloud will form, most of the aircraft measurements of in-cloud properties over the oceans are limited to measurements starting after the first 5 min or so of cloud formation, and most measurements have been made during the growing and mature stages of the cloud development, with very few in the later stages of dissipation. The most complete set of measurements of dynamical structure in clouds as a function of cloud age were made by Warner (1977) in isolated cumuli. He found that near cloud base the updraft is not a single coherent entity but rather a collection of smaller upward-moving turbules, a finding similar to that of Raga (1989) in the band clouds off Hawaii. A coherent updraft forms on the upshear side of the cloud about 100 m above cloud base, and it extends up to about 100 m below cloud top. In the updraft the liquid water content q_l [kg H_2O(kg air)$^{-1}$] and the updraft velocity w (m s^{-1}) are correlated. Near cloud top, in the inversion region, there is a highly turbulent region in which q_l and w are again uncorrelated and in which there may be fairly strong downdrafts. Only almost undilute cloudy parcels have sufficient buoyancy to penetrate the inversion, so that observations show that the tops of clouds found at high levels tend to be narrow, well-defined turrets with thermodynamic properties very close to those at cloud base (Raga 1989).

Figure 3.4 shows a cross section of the velocity structure in the middle section of the cloud. The updraft is surrounded by downdrafts that are in part cloudy throughout the cloud layer. This complex horizontal variation suggests that the average vertical velocity may not be a physically useful parameter, and in general it varies little with height. There appears to be little dependence of either maximum or mean vertical velocity on cloud width. The root-mean-square vertical velocity [$w_{rms} \equiv (w^2)^{1/2}$] can be taken as a measure of the turbulence in the cloud and therefore is related to the entrainment rate; w_{rms} varies linearly with height above cloud base and decreases with increasing stability (N_{BV}) in the cloud layer, suggesting that as the cloud becomes more buoyant it mixes more actively with the surrounding clear air.

The foregoing description of the updraft structure in trade cumulus pertains to the 10 or 15 min or so that characterize the growth and mature stages. Downdrafts become progressively more prominent and the updraft weaker as the cloud life continues, and final dissipation takes place as the cloudy air mixes in place with its environment.

b. Thermodynamic structure

In general, in-cloud composition at any level consists of those mixtures of cloud base and low-level environmental air that are positively buoyant, plus some negatively buoyant mixtures of environmental and upper-level cloudy air. The former are generally ascending, while the latter are in general descending. In growing and mature turrets the downdrafts represent a small fraction of the cloud mass.

Figures 3.5a–d show the profiles of the mean values of several important thermodynamic parameters inside

Hawaiian band clouds (Raga et al. 1990). These measurements are composites of those from 17 growing clouds in the bands. The figures clearly show that cloud parcels do not, in general, rise adiabatically (although some parcels with nearly the same thermodynamic properties as those at cloud base are found at all levels within the clouds). The mean thermodynamic properties at a fixed height above cloud base remain fairly constant throughout the growth and mature stages of the cloud (Warner 1977). The dotted lines represent estimates of these variations from a laterally entraining plume model, discussed below. The variability in these quantities is very high, and these mean values hide the fact that at each level each quantity takes on values varying over the entire possible range. (See, for example, Fig. 3.6.)

c. *Entrainment*

The fact that clouds entrain clear air from their surroundings above cloud base was evident to the early Woods Hole scientists, who found, as implied by Fig. 3.5b, that the temperature profile inside trade cumulus clouds was considerably steeper than that predicted on the basis of moist-adiabatic ascent. Stommel (1947) concluded on the basis of aircraft data that roughly half the air inside small cumulus is entrained and half enters through cloud base. Early attempts to draw analogies between clouds and laboratory tank similarity flows were moderately successful in predicting cloud-top height. In the similarity flows, the entrainment velocity is assumed proportional to the mean updraft velocity, and environmental air enters at all levels with this entrainment velocity. It is then instantaneously and uniformly mixed across the turbulent fluid (the cloud), yielding a "top-hat" structure in all conserved variables. The dotted lines in Figs. 3.5a–d result from application of a similarity plume model to the composited Hawaiian band cloud data. The results are fairly consistent with the measured mean profiles, largely because the height variation in the mean thermodynamic properties is small. As Warner (1970a) showed, however, these models cannot simultaneously reproduce liquid water content profiles and cloud-top height. He

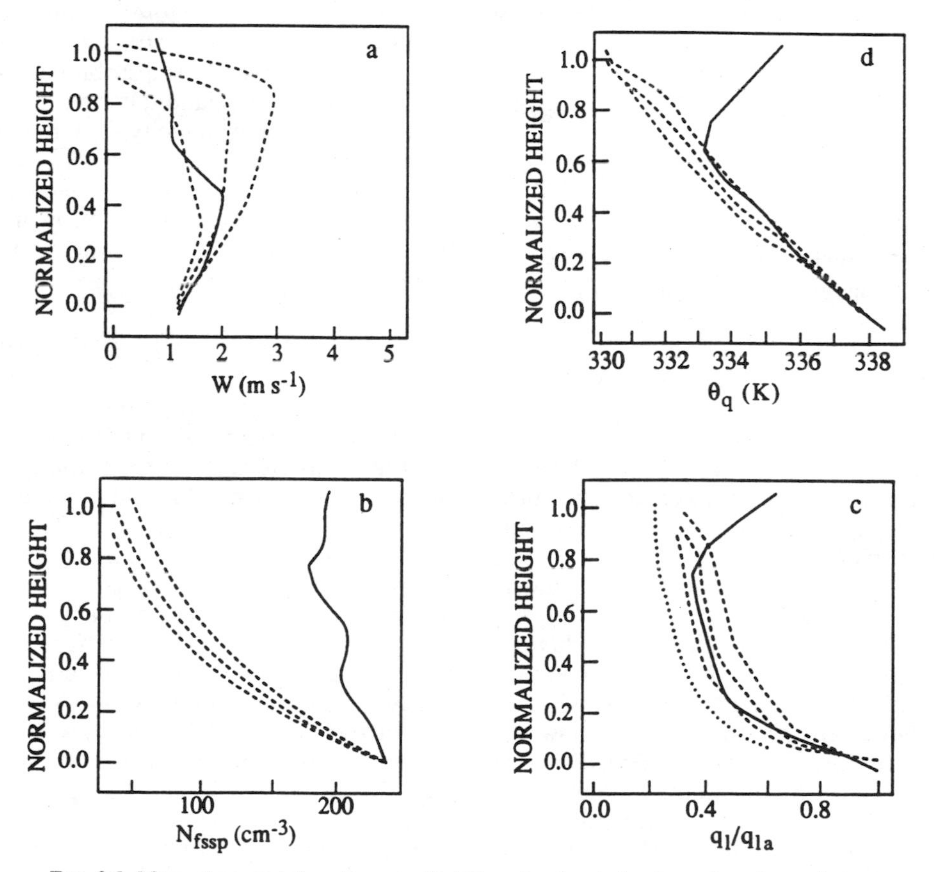

FIG. 3.5. Mean characteristics of composited Hawaiian band clouds, as functions of z^*, the height above cloud base, divided by the total cloud depth. (a) Vertical velocity; (b) cloud-droplet concentration; (c) ratio of observed liquid water content to the liquid water content in air rising adiabatically from cloud base; (d) θ_q. The dashed lines correspond to results from a laterally entraining plume model, assuming values of the entrainment parameter ranging from 0.5 to 1.5 km^{-1} (from Raga et al. 1990).

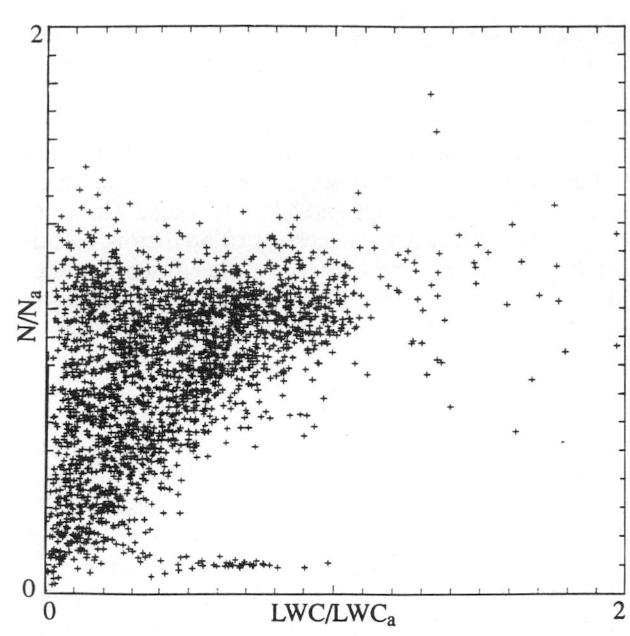

FIG. 3.6. Ratio of observed to undiluted values of total droplet number and liquid water content, as measured by aircraft in 19 clouds, showing the high degree of observed variability in both parameters. All 1-s, precipitation-free samples are included (by kind permission from J. Jensen).

pointed out, furthermore, that the physical basis for the analogy between small cumulus clouds and laboratory tank similarity plumes and thermals is weak. (For a review of the arguments on this issue, see Blyth et al. 1988.) Later work (Telford 1975; Paluch 1979) suggested that the actively entraining region of a convective cloud is its top, and that air, once entrained at cloud top, moves up or down depending on its buoyancy. The extent to which entrainment occurs at cloud top is difficult to establish using thermodynamic variables as tracers in trade cumulus because the clouds are so shallow and the environment in the cloud layer is so moist. Betts (1973b) showed that parcels mixed into the top of a trade-wind cumulus cloud can in principle become sufficiently negatively buoyant that they might descend to cloud base; however, Raga et al. (1990) did not find evidence of downdrafts in growing turrets within the cloud layer.

Figures 3.7a–e show the in-cloud vertical fluxes of mass, horizontal momentum, vertical momentum, heat, and water inside growing turrets, as measured during aircraft penetrations, and Fig. 3.8 shows the profile of mean in-cloud buoyancy. The fluxes derived from the aircraft measurements increase in the lower half of the cloud, where the buoyancy is increasing, and then either remain fairly constant or decrease as the buoyancy decreases toward the inversion.

From these and similar data, assuming lateral entrainment, we estimate that the equivalent entrainment rate is approximately 1 km^{-1} in the lower half of the clouds—a value that appears to apply to a large family of clouds, from trade-wind cumulus to Montana cumulonimbus.

d. Cloud microphysics

The early aircraft observations of cloud microphysics relied on impressions made by hydrometeors on sooted or coated slides for microphysical studies, while more sophisticated optical and electronic devices were used in later flights. While the data resolution and recording frequency characteristics of the modern instruments are far superior to those of the earlier instruments, the accuracy and reliability of the more indirect measurements have frequently proven disappointing. Therefore, we refer to both kinds of studies in this summary.

Warner (1969) was one of the first to examine the microphysical properties of trade-wind cumulus. His sooted slide measurements each represented a sampling of 10 cm^3, at 100-m intervals. He found both positively and negatively skewed droplet distributions, with some bimodal distributions, where "droplet distribution" here means the function $N(d, p)dd$, the number of particles per volume with diameters in the interval $(d, d + dd)$ at pressure p. Figure 3.9 shows typical droplet spectra in both precipitating and nonprecipitating clouds. The breadth of the distribution is clearly related to the onset of precipitation; large drops, with relatively high sedimentation velocity, fall relative to smaller drops, colliding and coalescing with a certain fraction of the latter. This figure shows the danger of parameterizations in which precipitation is predicted on the basis of liquid water content alone. Typically, the total droplet concentration

$$N_{\mathrm{TOT}}(p) \equiv \int_0^\infty N(d, p)dd \qquad (3.1)$$

ranges from around one hundred to a few hundred per cubic centimeter. Bimodality, which results from mixing of different air parcels, increases with cloud buoyancy and is not limited to cloud edges, again suggesting a nonuniform, nondiffusive, not strictly lateral entrainment process. By comparison of droplet spectra at different heights with those calculated assuming purely condensational growth, one can estimate an effective supersaturation, which is usually on the order of a few tenths of a percent. The normalized spectral width $R(d) \equiv \sigma(d)/\bar{d}$ remains within the bounds $0.2 < R(d) < 0.8$ and either is constant or increases with height. Warner (1973b) later showed that the integral properties of the droplet distribution in nonprecipitating clouds remain constant over approximately 15 min and that $N_{\mathrm{TOT}}(p)$ first rises slightly and then falls during the dissipation phase. Figure 3.5b shows the vertical profile of mean $N(p)$ for the composited Hawaiian band clouds. The mean droplet number does not decrease appreciably with height, suggesting that activation of new droplets

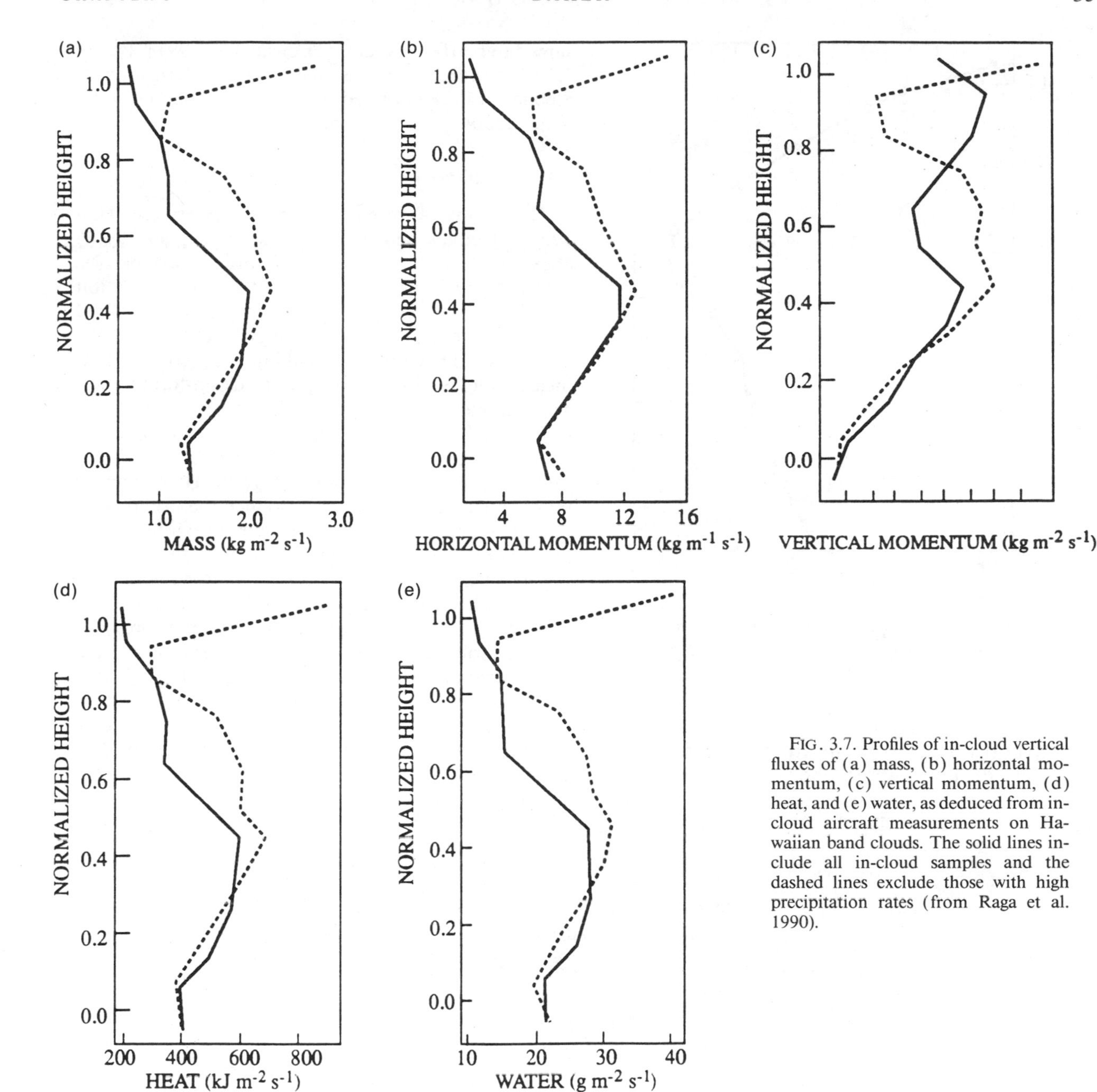

FIG. 3.7. Profiles of in-cloud vertical fluxes of (a) mass, (b) horizontal momentum, (c) vertical momentum, (d) heat, and (e) water, as deduced from in-cloud aircraft measurements on Hawaiian band clouds. The solid lines include all in-cloud samples and the dashed lines exclude those with high precipitation rates (from Raga et al. 1990).

approximately balances precipitation loss plus evaporation of drops due to entrainment of dry air, although Squires (1958a) showed earlier that inside an individual cloud N can decrease appreciably with height. Squires did not have reliable means by which to estimate rainfall production inside the clouds he was examining.

As interest in the cloud–climate interaction increases, it becomes important to characterize cloud-droplet spectra in ways useful for radiative transfer calculations. A parameter often used in this regard is the effective radius r_{eff}, defined as

$$r_{\text{eff}} \equiv \frac{\int_0^\infty N(d, p)d^3 dd}{\int_0^\infty N(d, p)d^2 dd.} \tag{3.2}$$

To a very good approximation, the value of $r_{\text{eff}}(p)$ is equal to that obtained in adiabatic ascent of a monodisperse collection of drops (Bower et al. 1991); that is

$$r_{\text{eff}}(p) \approx \left[\frac{3\rho_{\text{air}} q_{l\text{ad}}(p)}{4\pi\rho_L N_{\text{TOT}}(p)}\right]^{1/3}, \tag{3.3}$$

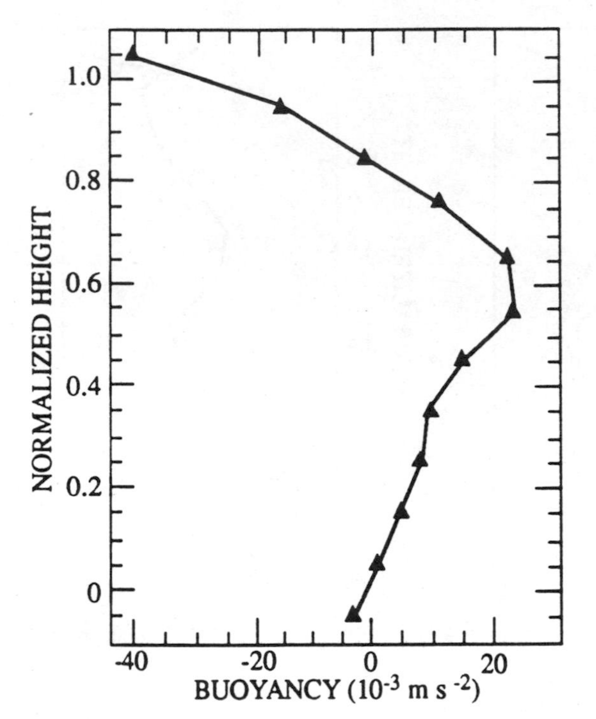

FIG. 3.8. Vertical profile of mean in-cloud buoyancy for clouds in composite Hawaiian band clouds (from Raga et al. 1990).

where $q_{lad}(p)$ is the liquid water content at pressure p in a parcel rising adiabatically from cloud base, ρ_L is the density of liquid water, and ρ_{air} is the density of cloudy air. Here N_{TOT} varies relatively little in the trade cumulus layer and, as stated above, can usually be assumed to be 100–200 cm^{-3} (Bower et al. 1991). Figure 3.10 shows Eq. (3.3), as well as parameterization of Stephens (1978) applied to data from the Hawaiian band clouds.

3.4. Precipitation

The results of an early "census" (Byers and Hall 1954) of cloud height (measured by a plane flying above the clouds) versus precipitation occurrence (as measured by 3-cm radar) revealed a clear correlation between the occurrence of precipitation and cloud height. However, it was clear from the census that the occurrence of precipitation could not be predicted from measurements of cloud height (or cloud-top temperature) alone.

Takahashi (1981) combined 300 h of aircraft flight data to examine the evolution of warm rain inside cumulus clouds off the island of Hawaii. He concludes that condensation alone can produce the first precipitation embryos at the top of the cloud updraft (where the diminished upward velocity permits the drops to begin to fall). Figure 3.11 shows his measured drizzle drop sizes at various heights within the clouds. He suggests that precipitation depends in part on the presence of large drops at low levels. The concentration of such drops varies depending on the concentrations of large nuclei, the level of cloud base itself, and the circulation patterns within the cloud. Under favorable conditions, vertical shear in the horizontal wind causes the first drizzle drops to cycle up and down within saturated columns of air (Takahashi et al. 1989; Rauber et al. 1991). The large drizzle droplets tend to appear first

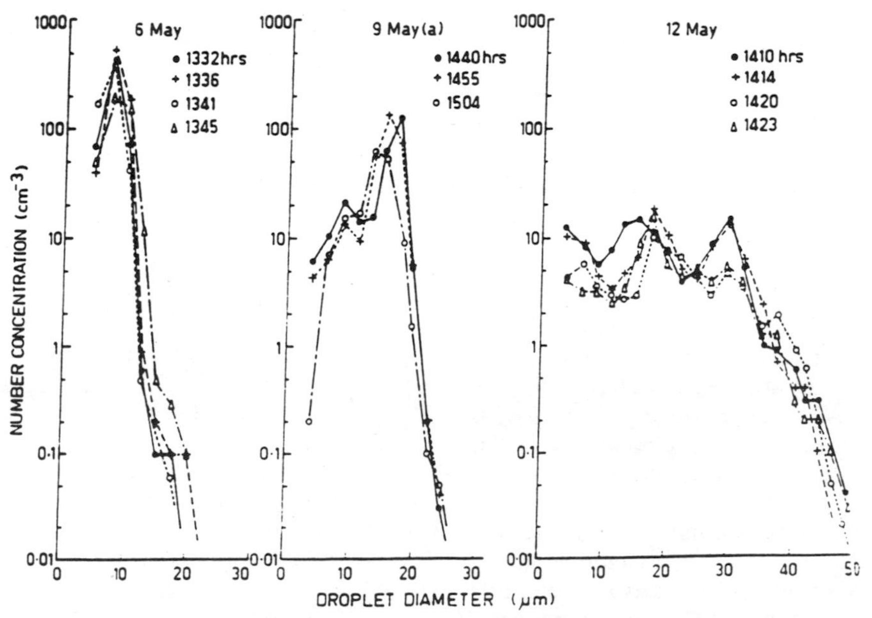

FIG. 3.9. Droplet spectrum as a function of cloud age. The cloud on the right exhibited high precipitation during the measurement period (from Warner 1973b).

in the region between up- and downdrafts. For sufficiently high drizzle water content, which Takahashi estimates to be on the order of 0.05 g m^{-3}, raindrops develop, and precipitation falls through the cloud-base level, due to the low magnitudes of the vertical velocity in the lower part of the clouds (see Fig. 3.5a). Isolated rain showers are characterized by shorter lifetimes and lower rainwater content, and they require deeper clouds than rain showers from cells in cloud groups or bands (Takahashi 1977). In general, shower lifetimes are 10–30 min, with intensity decreasing with time. Rainwater content can be up to 1 g m^{-3} in isolated showers, and up to three times that in line showers.

Thus, several factors determine the precipitation efficiency in these clouds, and these factors operate on very different scales. First is the CCN spectrum, which determines the droplet spectrum at low levels. Then the updraft characteristics, including vertical shear, determine the circulation patterns of the embryonic raindrops. The extent of growth during recycling of drops is also a function of cloud width and height, environmental moisture content, and inversion shape.

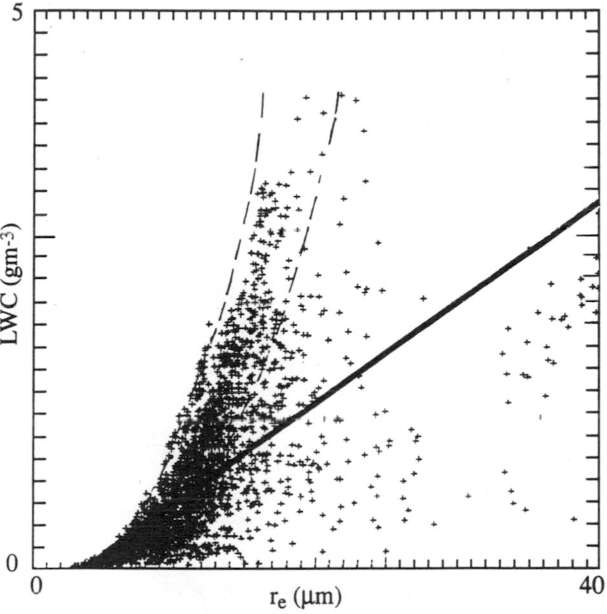

FIG. 3.10. Effective radius defined in Eq. (3.2) as a function of liquid water content, for all 1-s precipitation-free observations made during JHWRP. The thick solid line shows the parameterization of Stephens (1978), while the two dashed lines show the parameterization discussed in the text for N_{TOT} = 100 and 200 cm^{-3} (data by kind permission from J. Jensen).

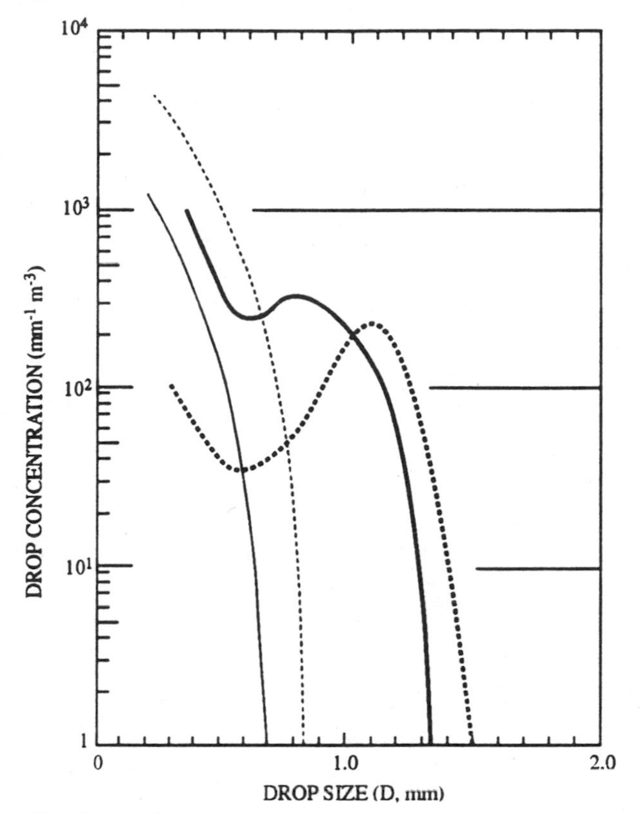

FIG. 3.11. Drizzle drop size distribution as a function of height. Thin solid line: cloud top, q_d = 0.04 g m^{-3}; thin dashed line: cloud top, q_d = 0.4 g m^{-3}; thick solid line: cloud top, q_r = 0.4 g m^{-3}; thick dashed line: beneath cloud base, q_r = 0.4 g m^{-3}, where q_d is the drizzle water content (200 μm $\lessapprox d \lessapprox$ 800 μm) and q_r is the raindrop water content ($d >$ 800 μm) (from Takahashi 1981).

Acknowledgments. This work was supported by NSF Grant ATM 89079 and by the Laboratoire de Metéorologie Dynamique, École Normale Supérieure, Paris, France.

Chapter 4

Impacts of Cumulus Convection on Thermodynamic Fields

MICHIO YANAI

Department of Atmospheric Sciences, University of California, Los Angeles, Los Angeles, California

RICHARD H. JOHNSON

Department of Atmospheric Science, Colorado State University, Fort Collins, Colorado

4.1. Heat and moisture budgets: Background and basic concepts

The important role played by deep cumulus clouds in the heat balance of the tropical atmosphere was first pointed out by Riehl and Malkus (1958). They studied the heat budget over the equatorial trough zone and postulated that concentrated upward heat transport due to undilute ("hot") cumulonimbus towers is required to maintain the heat balance of the upper troposphere against the loss due to radiation and poleward export. The role of penetrative cumulus towers in the tropical cyclone was discussed by Riehl and Malkus (1961) and Yanai (1961, 1964). They proposed that the role of cumulus convection as a heating agent must be adequately parameterized in the framework of large-scale motions. In the last three decades our understanding of the interaction of organized cumulus convection with its environment made significant progress through many observational and parameterization studies.

By the release of latent heat and the vertical transport of sensible heat and water vapor, cumulus clouds modify the temperature and moisture structure of the environment. The modification occurs primarily through the subsidence of environmental air that compensates the convective mass flux and through the detrainment of water substance from clouds (Ooyama 1971; Arakawa 1972; Gray 1973; Yanai et al. 1973; Arakawa and Schubert 1974). These two processes tend to oppose each other. Deep cumulonimbi tend to warm and dry the environment because they force the environmental air to sink. On the other hand, shallow cumuli tend to moisten and cool the environment by the detrainment of water vapor and cloud water that evaporates. Therefore, it is important to consider a spectrum of cumulus clouds with various sizes in the study of cumulus–environment interaction. Because cumulus convection occurs on scales much smaller than those normally resolvable from data gathered by conventional networks, its effects on the larger-scale circulation cannot be directly measured. Instead, effects are inferred indirectly from heat and moisture budgets of the larger-scale circulation system.

a. Heat and moisture budget equations

We consider an ensemble of cumulus clouds that is embedded in a large-scale motion system. We write the equations of mass continuity, thermodynamic energy, and moisture continuity, averaged over a large-scale horizontal area:

$$\nabla \cdot \bar{\mathbf{v}} + \frac{\partial \bar{\omega}}{\partial p} = 0, \tag{4.1}$$

$$\frac{\partial \bar{s}}{\partial t} + \nabla \cdot \overline{s\mathbf{v}} + \frac{\partial \overline{s\omega}}{\partial p} = Q_R + L(\bar{c} - \bar{e}), \tag{4.2}$$

$$\frac{\partial \bar{q}}{\partial t} + \nabla \cdot \overline{q\mathbf{v}} + \frac{\partial \overline{q\omega}}{\partial p} = \bar{e} - \bar{c}. \tag{4.3}$$

In the above, $s \equiv c_pT + gz$ is the dry static energy,[1] q the mixing ratio of water vapor, $\mathbf{v}$ the horizontal velocity, ω the vertical p velocity, Q_R the heating rate due to radiation, c the rate of condensation[2] per unit mass of air, e the rate of evaporation of cloud water, ∇ the isobaric del operator, and L the latent heat of vaporization. The overbar denotes the running horizontal average.

---

[1] Equation (4.2) is not exact. When s is used, production of kinetic energy must be included. Dry static energy s is used here for convenience in relating (4.2) to the budget equations for a cumulus ensemble. A more exact form of the thermodynamic energy equation is

$$c_p\left(\frac{p}{p_0}\right)^{R/c_p}\left(\frac{\partial \bar{\theta}}{\partial t} + \nabla \cdot \overline{\theta \mathbf{v}} + \frac{\partial}{\partial p}\overline{\theta\omega}\right) = Q_R + L(\bar{c} - \bar{e}), \tag{4.2a}$$

where θ is the potential temperature.

[2] The phase changes involving ice are neglected here but can be included in a straightforward manner (e.g., Johnson and Young 1983; Gallus and Johnson 1991).

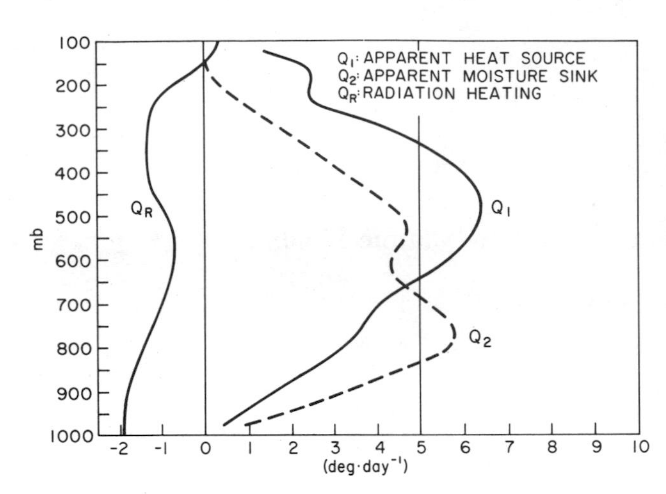

FIG. 4.1. The mean apparent heat source Q_1 (solid) and moisture sink Q_2 (dashed) over the Marshall Islands. On the left is the radiational heating rate given by Dopplick (1972) (from Yanai et al. 1973).

With the aid of (4.1), Eqs. (4.2) and (4.3) may be rearranged to give

$$Q_1 \equiv \frac{\partial \bar{s}}{\partial t} + \bar{\mathbf{v}} \cdot \nabla \bar{s} + \bar{\omega} \frac{\partial \bar{s}}{\partial p}$$

$$= Q_R + L(\bar{c} - \bar{e}) - \nabla \cdot \overline{s'\mathbf{v}'} - \frac{\partial}{\partial p} \overline{s'\omega'}, \quad (4.4)$$

$$Q_2 \equiv -L\left(\frac{\partial \bar{q}}{\partial t} + \bar{\mathbf{v}} \cdot \nabla \bar{q} + \bar{\omega} \frac{\partial \bar{q}}{\partial p}\right)$$

$$= L(\bar{c} - \bar{e}) + L\nabla \cdot \overline{q'\mathbf{v}'} + L \frac{\partial}{\partial p} \overline{q'\omega'}, \quad (4.5)$$

where deviations from the horizontal averages are denoted by primes. Terms Q_1 and Q_2 are the residuals of heat and moisture budgets of the "resolvable" motion. They are called the "apparent" heat source and moisture sink, respectively, because they include true sources and sinks as well as the correlation terms resulting from unresolved eddies.

In deriving (4.4) and (4.5) we have assumed that the *Reynolds conditions* and their consequences such as

$$\overline{\bar{A}\bar{B}} = \bar{A}\bar{B}, \quad \overline{\bar{A}B'} = 0 \quad (4.6)$$

hold with sufficient accuracy (e.g., Kampé de Fériet 1951; Monin and Yaglom 1971, p. 207). This is considered plausible when we imagine a horizontal area that is large enough to contain the ensemble of clouds but small enough to be regarded as a fraction of the large-scale system (Yanai et al. 1973). However, the presence of eddies of intermediate scale (see chapter 3 by M. Baker) may invalidate (4.6) and cause ambiguity in the interpretation of Q_1 and Q_2.

Traditionally, the eddy horizontal transport terms $-\nabla \cdot \overline{s'\mathbf{v}'}$ and $-\nabla \cdot \overline{q'\mathbf{v}'}$ have been ignored. The justification for this omission has been that the net lateral transports across the boundary of the fixed area by cumulus convection are negligible compared to the horizontal transports by the large-scale motion (Arakawa and Schubert 1974). However, we cannot rule out possible contributions from mesoscale eddies (Wu 1993b).

The eddy vertical flux terms may have significant contributions to Q_1 and Q_2 in a highly convective situation. If we can ignore the horizontal transports by eddies, we find

$$Q_1 - Q_2 - Q_R = -\frac{\partial}{\partial p} \overline{h'\omega'}, \quad (4.7)$$

where $h = s + Lq$ is the moist static energy. Equation (4.7) has been widely used to measure the activity of cumulus convection (e.g., Ninomiya 1968, 1971; Nitta 1972, 1975, 1977; Yanai et al. 1973; Yanai et al. 1976; Johnson 1976; Gallus and Johnson 1991).

When the horizontal eddy flux terms are ignored, integration of (4.4) and (4.5) from the tropopause pressure p_T to the surface pressure p_s yields

$$\langle Q_1 \rangle = \langle Q_R \rangle + L\langle \bar{c} - \bar{e} \rangle - \frac{1}{g} (\overline{s'\omega'})_{p=p_s}$$

$$\approx \langle Q_R \rangle + LP + \rho_s c_p (\overline{T'w'})_{p=p_s}$$

$$= \langle Q_R \rangle + LP + S, \quad (4.8)$$

$$\langle Q_2 \rangle = L\langle \bar{c} - \bar{e} \rangle + \frac{L}{g} (\overline{q'\omega'})_{p=p_s}$$

$$\approx LP - \rho_s L (\overline{q'w'})_{p=p_s}$$

$$= L(P - E), \quad (4.9)$$

where

$$\langle \quad \rangle \equiv \frac{1}{g} \int_{p_T}^{p_s} (\quad) dp,$$

and P, S, and E are, respectively, the precipitation rate, the sensible heat flux, and the evaporation rate per unit area at the surface. We note that

$$\langle Q_1 \rangle - \langle Q_2 \rangle = \langle Q_R \rangle + S + LE. \quad (4.10)$$

Equations (4.8), (4.9), and (4.10) can be used to check the accuracy of the estimates of Q_1, Q_2, and Q_R with the surface observations of P, S, and E (e.g., Yanai et al. 1973; Thompson et al. 1979; Song and Frank 1983). Equations (4.4), (4.5), (4.7) and (4.8), (4.9), (4.10) are useful expressions for the interpretation of the budget results. Careful comparison between the vertical distributions of Q_1 and Q_2 as well as comparison between the horizontal distributions of $\langle Q_1 \rangle$ and $\langle Q_2 \rangle$ yield valuable information on the nature of heating processes (e.g., Yanai et al. 1973; Nitta and Esbensen 1974; Luo and Yanai 1984; Gallus and Johnson 1991).

Figure 4.1 shows the time-averaged vertical distributions of Q_1 and Q_2 obtained over the Marshall Islands in the western Pacific using the data from 15 April to 22 July 1956 (390 samples) (Yanai et al. 1973). Similar profiles of Q_1 had been obtained by Reed and Recker (1971) and Nitta (1972) for this region. The climatological vertical profile of Q_R for this latitude, taken from Dopplick (1972), is shown on the left side of the figure. Note that there is a considerable difference between the Q_1 and Q_2 profiles. The average Q_1 has a maximum (6.4 K day^{-1}) at 475 mb. On the other hand, the average Q_2 has its principal maximum at a much lower level (775 mb) and a secondary peak at 525 mb. As shown by (4.7) this difference suggests the presence of eddy vertical transport of moist static energy associated with cumulus convection.

Extensive analyses of heat and moisture budgets have been made using the three-hourly data taken during phase III (30 August–19 September 1974) of GATE [the GARP (Global Atmospheric Research Program) Atlantic Tropical Experiment]. Time-averaged vertical distributions of Q_1 and Q_2 over the GATE A/B-scale ship network were presented by Nitta (1978), Thompson et al. (1979), and Sui et al. (1989). Figure 4.2 shows phase III–mean vertical profiles of Q_1 and Q_2 averaged over a 3° × 3° area at the center of the network. The phase III–mean vertical profile of Q_R estimated by Cox and Griffith (1979) is also shown. There is a systematic difference between the $Q_1 - Q_R$ and Q_2 profiles, indicating large contributions from the vertical eddy flux terms in (4.4) and (4.5).

Figure 4.3 shows the time–height sections of the observed $Q_1 - Q_R$, Q_2, and $Q_1 - Q_2 - Q_R$ averaged over the same area, from 0000 UTC 31 August (index 9) to 0000 UTC 18 September (index 153) (Cheng and Yanai 1989). The dark and open arrows indicate the periods when squall and nonsquall clusters are observed within the GATE A/B array, respectively. In Fig. 4.3 we see that both $Q_1 - Q_R$ and Q_2 are highly variable in time. The magnitudes of $Q_1 - Q_R$ and Q_2 associated with the nonsquall clusters are larger than those associated with the squall clusters (Figs. 4.3a,b). However, there are no significant differences in the vertical distributions and magnitudes of $Q_1 - Q_2 - Q_R$ between the squall and nonsquall cluster cases. For all cloud cluster cases $Q_1 - Q_2 - Q_R$ is positive in the upper troposphere and negative in the lower troposphere (Fig. 4.3c).

Other examples of the Q_1 and Q_2 analyses include those for 1) the Atlantic trades (Nitta and Esbensen 1974), 2) tropical mesoscale anvil clouds (Johnson 1980; Johnson and Young 1983), 3) the tropical South Pacific (Miller and Vincent 1987; Pedigo and Vincent 1990), 4) the Tibetan Plateau and adjacent monsoon regions (Nitta 1983; Luo and Yanai 1984; He et al. 1987; Yanai et al. 1992), 5) the Australian monsoon (Frank and McBride 1989; McBride et al. 1989), and 6) midlatitude mesoscale convective systems (Lewis

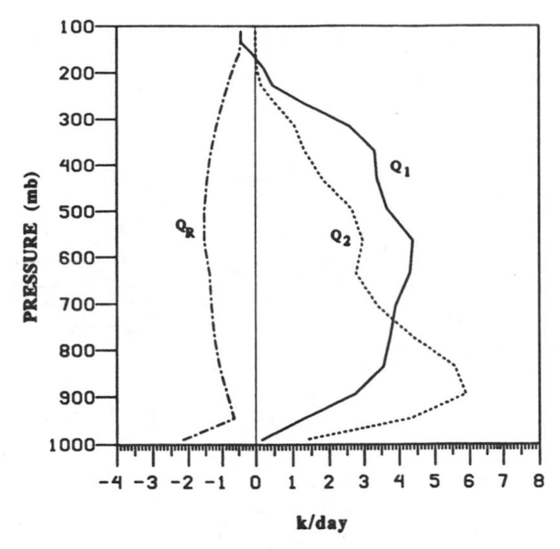

FIG. 4.2. Phase III–mean profiles of Q_1, Q_2, Q_R over the 3° × 3° area at the center of the GATE ship network.

1975; Ogura and Chen 1977; Kuo and Anthes 1984b,c; Grell et al. 1991; Gallus and Johnson 1991; Wu 1993a).

b. Representation of cumulus and mesoscale effects

The expressions for the eddy fluxes $-\overline{s'\omega'}$ and $-\overline{q'\omega'}$ may be considered in terms of cumulus properties such as fractional area coverage, mass flux, and temperature and moisture excesses over the environment (e.g., Yanai 1964; Ninomiya 1968, 1971; Yanai et al. 1973; Arakawa and Schubert 1974; Houze et al. 1980). However, here we follow a more general formulation given by Cheng and Yanai (1989).

Let us consider that the observed $Q_1 - Q_R$ and Q_2 are contributed from the effects of cumulus convection (Q_{1c} and Q_{2c}) and mesoscale circulations (Q_{1m} and Q_{2m}); that is,

$$Q_1 - Q_R = Q_{1c} + Q_{1m}, \tag{4.11}$$

$$Q_2 = Q_{2c} + Q_{2m}. \tag{4.12}$$

Similarly, the mean condensation and evaporation in (4.4) and (4.5) are decomposed into

$$\bar{c} - \bar{e} = \sum_i (c_i - e_i) + c_m - e_m, \tag{4.13}$$

where c_i and e_i are the condensation and evaporation rates associated with the ith cumulus cloud and c_m and e_m are those associated with either mesoscale updrafts or downdrafts.

To see how the eddy flux terms in (4.4), (4.5), and (4.7) are related to the cumulus and mesoscale processes, we divide the mean vertical mass flux[3] into the

[3] In this chapter, the mass flux is defined by $M = -\omega \approx g\rho w$.

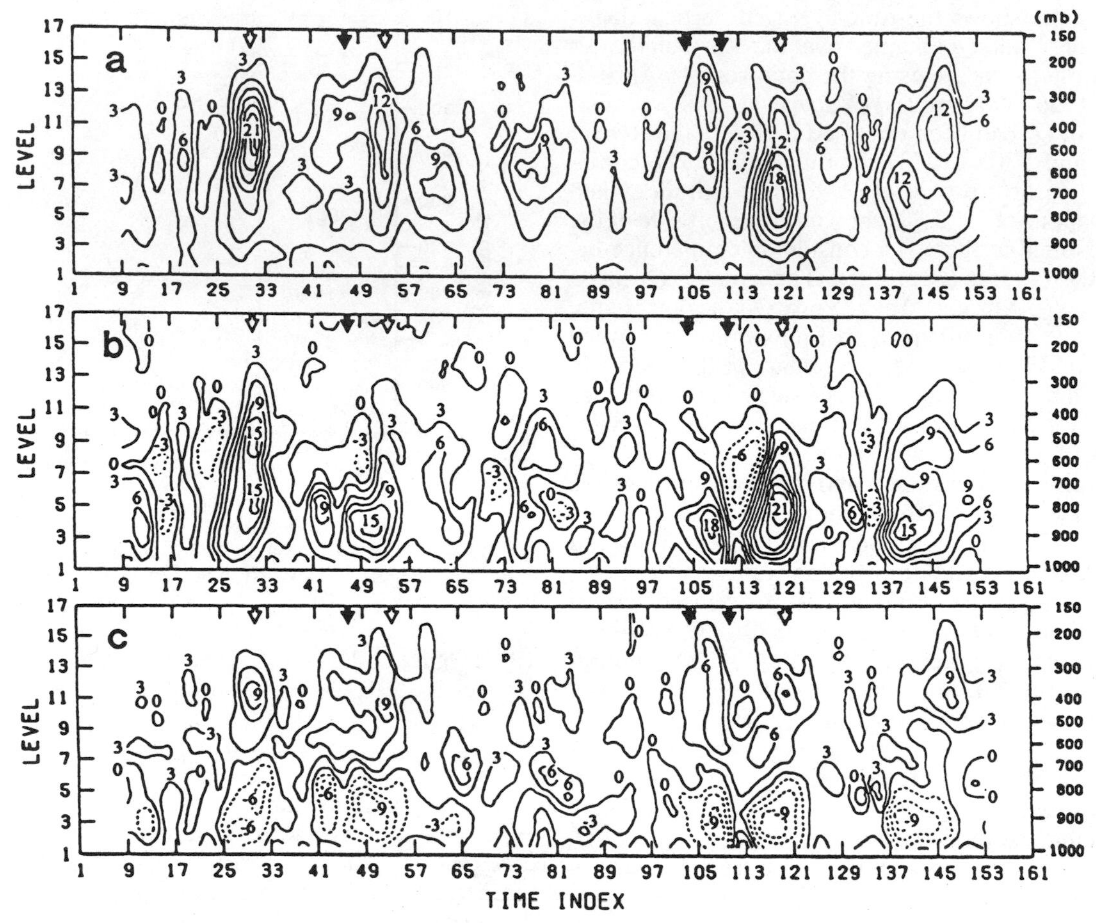

FIG. 4.3. Time–height sections of the observed (a) $Q_1 - Q_R$, (b) Q_2, and (c) $Q_1 - Q_2 - Q_R$, averaged over the 3° × 3° area at the center of the GATE network during a period from 0000 UTC 31 August (index 9) to 0000 UTC 18 September 1974 (index 153). Units are kelvins per day (from Cheng and Yanai 1989).

cumulus mass flux M_i, mesoscale mass flux M_m, and the residual mass flux in the environment $\tilde{M}$; that is,

$$\bar{M} = -\bar{\omega} = -\sum_i \sigma_i\omega_i - \sigma_m\omega_m - (1 - \sum_i \sigma_i - \sigma_m)\tilde{\omega} \tag{4.14a}$$

$$= \sum_i M_i + M_m + \tilde{M}, \tag{4.14b}$$

where σ is the fractional area coverage of updrafts and downdrafts. Here $(\ )_i$ denotes the quantities of the ith cumulus cloud, $(\ )_m$ the quantities of the mesoscale updraft or downdraft, and $(\tilde{\ })$ the quantities of the environment that is assumed to be shared by cumulus clouds and by a mesoscale updraft or downdraft.

Similarly, the area mean of a quantity A and its upward flux are written

$$\bar{A} = \sum_i \sigma_i A_i + \sigma_m A_m + (1 - \sum_i \sigma_i - \sigma_m)\tilde{A}, \tag{4.15}$$

and

$$-\overline{\omega A} = -\sum_i \sigma_i\omega_i A_i - \sigma_m\omega_m A_m - (1 - \sum_i \sigma_i - \sigma_m)\tilde{\omega}\tilde{A} \tag{4.16a}$$

$$= \sum_i M_i A_i + M_m A_m + \tilde{M}\tilde{A}, \tag{4.16b}$$

respectively.

Using $A' = A - \bar{A}$ and $\omega' = \omega - \bar{\omega}$, the eddy vertical flux of A is given by

$$\overline{\omega' A'} = \overline{\omega A} - \bar{\omega}\bar{A}. \tag{4.17}$$

Substituting (4.14), (4.15), and (4.16) into (4.17), we obtain the expression for the upward eddy flux

$$-\overline{\omega' A'} = \sum_i [(M_i - \sigma_i\bar{M})(A_i - \tilde{A})] + (M_m - \sigma_m\bar{M})(A_m - \tilde{A}). \tag{4.18}$$

If we equate A in (4.18) to s, q, and h, respectively, the eddy flux terms in (4.4), (4.5), and (4.7) can be written in terms of the vertical transports by cumulus and mesoscale mass fluxes. With the aid of (4.13) and (4.18), we can divide the contributions to $Q_1 - Q_R$ and Q_2 into cumulus and mesoscale components; that is,

$$Q_{1c} = \sum_i \left\{ L(c_i - e_i) + \frac{\partial}{\partial p}[(M_i - \sigma_i \bar{M})(s_i - \tilde{s})] \right\}, \tag{4.19}$$

$$Q_{2c} = \sum_i \left\{ L(c_i - e_i) - L\frac{\partial}{\partial p}[(M_i - \sigma_i \bar{M})(q_i - \tilde{q})] \right\}, \tag{4.20}$$

and

$$Q_{1m} = L(c_m - e_m) + \frac{\partial}{\partial p}[(M_m - \sigma_m \bar{M})(s_m - \tilde{s})], \tag{4.21}$$

$$Q_{2m} = L(c_m - e_m) - L\frac{\partial}{\partial p}[(M_m - \sigma_m \bar{M})(q_m - \tilde{q})]. \tag{4.22}$$

Because the fractional coverage of active cumulus clouds is very small (e.g., Malkus et al. 1961; Warner 1981), the assumption $M_i \gg \sigma_i \bar{M}$ is well justified. Also, we can approximate $\tilde{A}$ by $\bar{A}$ and use $\bar{A}$ to estimate the quantities of the environment of a cumulus ensemble (e.g., Yanai et al. 1973; Arakawa and Schubert 1974). Therefore, for the cumulus components we have

$$Q_{1c} = L\sum_i (c_i - e_i) + \frac{\partial}{\partial p}\sum_i M_i(s_i - \bar{s}), \tag{4.23}$$

$$Q_{2c} = L\sum_i (c_i - e_i) - L\frac{\partial}{\partial p}\sum_i M_i(q_i - \bar{q}). \tag{4.24}$$

We now make use of the stratiform character of the mesoscale cloud systems; namely, the eddy fluxes in (4.21) and (4.22) are assumed to be small compared to the condensation and evaporation terms:

$$Q_{1m} \approx Q_{2m} \approx L(c_m - e_m). \tag{4.25}$$

This assumption has some observational support in the study of tropical mesoscale anvils by Johnson and Young (1983), wherein they found the profiles of Q_{1m} and Q_{2m} to closely resemble each other and the amplitudes, on the average, to agree within 20% to 30%. However, further research is needed to verify (4.25). Methods to isolate the effects of mesoscale stratiform precipitation from the observed Q_1 and Q_2 will be discussed in section 4.3. With (4.25) we now find

$$Q_1 - Q_2 - Q_R \approx Q_{1c} - Q_{2c}, \tag{4.26}$$

which provides a basis for diagnosing cumulus ensemble properties using the observed $Q_1 - Q_2 - Q_R$.

The differences in the vertical profiles of $Q_1 - Q_R$ (Fig. 4.3a) and Q_2 (Fig. 4.3b) and the similarity of the $Q_1 - Q_2 - Q_R$ profiles (Fig. 4.3c) between the squall and nonsquall cluster cases observed during GATE phase III are qualitatively explainable as the results of different magnitudes of Q_{1m} and Q_{2m} as interpreted by (4.25). Because these terms contribute equally to the observed $Q_1 - Q_R$ and Q_2, they have no effect on the $Q_1 - Q_2 - Q_R$ profiles.

c. *Effects of cumulus-induced subsidence and detrainment*

A large amount of latent heat of condensation is liberated in cumulus clouds and the released heat is transported upward, but how this heat is utilized in warming the environment is not so obvious. To understand the mechanisms of cumulus heating (and moistening), we shall consider the exchange of mass, heat, and water substance between clouds and their environment and its consequence.

We consider the budgets of mass, heat, moisture, and liquid water of the ith cumulus cloud,

$$\frac{\partial \sigma_i}{\partial t} + \delta_i - \epsilon_i - \frac{\partial M_i}{\partial p} = 0 \quad \text{(mass)} \tag{4.27}$$

$$\frac{\partial}{\partial t}(\sigma_i s_i) + \delta_i s_{Di} - \epsilon_i \bar{s} - \frac{\partial M_i s_i}{\partial p} = Lc_i, \quad \text{(heat)} \tag{4.28}$$

$$\frac{\partial}{\partial t}(\sigma_i q_i) + \delta_i q_{Di} - \epsilon_i \bar{q} - \frac{\partial M_i q_i}{\partial p} = -c_i, \quad \text{(water vapor)} \tag{4.29}$$

$$\frac{\partial}{\partial t}(\sigma_i l_i) + \delta_i l_{Di} - \frac{\partial M_i l_i}{\partial p} = c_i - r_i, \quad \text{(liquid water)}, \tag{4.30}$$

where δ and ϵ are the rates of mass detrainment and entrainment per unit pressure interval, l the mixing ratio of liquid water, and r the rate of generation of rainwater. The subscript $(\ )_D$ signifies the value in the detraining air.

Eliminating c_i from (4.28) and (4.29) and from (4.29) and (4.30), we obtain

$$\frac{\partial}{\partial t}(\sigma_i h_i) + \delta_i h_{Di} - \epsilon_i \bar{h} - \frac{\partial M_i h_i}{\partial p} = 0, \tag{4.31}$$

$$\frac{\partial}{\partial t}[\sigma_i (q + l)_i] + \delta_i (q + l)_{Di} - \epsilon_i \bar{q} - \frac{\partial}{\partial p}[M_i (q + l)_i] = -r_i. \tag{4.32}$$

Equations (4.31) and (4.32) describe the conservation of moist static energy and water substance of the cloud.

Because the air is saturated in clouds, s_i and $q_i = q^*(T_i, p)$ are related to h_i by

$$s_i - \bar{s} = \frac{1}{1+\gamma}(h_i - \bar{h}^*), \quad (4.33)$$

$$L(q_i - \bar{q}^*) = \frac{\gamma}{1+\gamma}(h_i - \bar{h}^*), \quad (4.34)$$

where $\gamma \equiv (L/c_p)(\partial \bar{q}^*/\partial \bar{T})_p$, q^* is the saturation mixing ratio, and $h^* \equiv s + Lq^*$ is the saturation moist static energy (Arakawa 1969).

Even though individual clouds undergo various phases of growth and decay, we assume that the cloud ensemble maintains an approximate balance of mass, heat, and moisture with the slowly varying large-scale environment. We may ignore the storage terms after summing up (4.27)–(4.30) for the cumulus ensemble; that is,

$$\sum_i \left(\delta_i - \epsilon_i - \frac{\partial M_i}{\partial p}\right) = 0, \quad (4.35)$$

$$\sum_i \left(\delta_i s_{Di} - \epsilon_i \bar{s} - \frac{\partial M_i s_i}{\partial p}\right) = L \sum_i c_i, \quad (4.36)$$

$$\sum_i \left(\delta_i q_{Di} - \epsilon_i \bar{q} - \frac{\partial M_i q_i}{\partial p}\right) = -\sum_i c_i, \quad (4.37)$$

$$\sum_i \left(\delta_i l_{Di} - \frac{\partial M_i l_i}{\partial p}\right) = \sum_i (c_i - r_i). \quad (4.38)$$

Eliminating $L\sum_i c_i$ and $\sum_i \epsilon_i$ between (4.23), (4.35), and (4.36), we find

$$Q_{1c} = \sum_i \delta_i(s_{Di} - \bar{s}) - L\sum_i e_i - M_c \frac{\partial \bar{s}}{\partial p}. \quad (4.39)$$

Similarly, we eliminate $\sum_i c_i$ and $\sum_i \epsilon_i$ between (4.24), (4.35), and (4.37) to obtain

$$Q_{2c} = -L\sum_i \delta_i(q_{Di} - \bar{q}) - L\sum_i e_i + LM_c \frac{\partial \bar{q}}{\partial p}, \quad (4.40)$$

where

$$M_c = \sum_i M_i \quad (4.41)$$

is the total cumulus mass flux.

Assuming further that the rate of evaporation e_i equals that of detrainment of liquid water,[4]

$$e_i = \delta_i l_{Di}, \quad (4.42)$$

we obtain

$$Q_{1c} = -M_c \frac{\partial \bar{s}}{\partial p} + \sum_i \delta_i(s_{Di} - \bar{s} - Ll_{Di}), \quad (4.43)$$

$$Q_{2c} = LM_c \frac{\partial \bar{q}}{\partial p} - L\sum_i \delta_i(q_{Di} - \bar{q} + l_{Di}). \quad (4.44)$$

The first terms on the right-hand sides of (4.43) and (4.44) represent the vertical advection of heat and moisture by the part of environmental vertical motion that compensates the convective mass flux.[5] The second terms express the effects of heat and moisture detrained from cumulus clouds and evaporation of detrained cloud water. Equations (4.43) and (4.44) were originally derived by Ooyama (1971) and Arakawa (1972) and have been used extensively in diagnostic and parameterization studies of cumulus convection (e.g., Yanai et al. 1973; Arakawa and Schubert 1974). Equations (4.43) and (4.44) show that, to evaluate the cumulus heating and drying, we need to know not only the total cumulus mass flux M_c but also s, q, and l of the air detraining from all clouds. In the following sections, we introduce diagnostic methods to obtain M_c, Q_{1c}, and Q_{2c}.

4.2. Diagnosis of cumulus effects

a. The bulk method

Yanai et al. (1973) considered a bulk model of a cumulus cloud ensemble for use in a diagnostic study of the cumulus effect on the heat and moisture budgets of the Marshall Islands region (see Fig. 4.1). In this model cumulus clouds are classified according to the height of their tops and all clouds are assumed to have a common cloud-base height. The values of s_i, q_i (thus h_i), and l_i, at a given height, are assumed to be the same for every cloud that has the same top height. We may then use the suffix i as the index of a classified cloud type, instead of as the index of an individual cloud. This model did not consider the effects of convective-scale downdrafts or mesoscale stratiform clouds. Thus,

$$Q_1 - Q_R = Q_{1c}, \quad Q_2 = Q_{2c} \quad (4.45)$$

were assumed.

Noting that

$$\left.\begin{aligned} M_c &= \sum_i M_i, & \epsilon &= \sum_i \epsilon_i, & \delta &= \sum_i \delta_i, \\ c &= \sum_i c_i, & r &= \sum_i r_i, & e &= \sum_i e_i, \end{aligned}\right\} \quad (4.46)$$

[4] Thus, the evaporation of falling raindrops is ignored here.

[5] This does not necessarily imply that $\bar{M} - M_c$ is downward (see Cheng and Yanai 1989).

and defining weighted averages

$$\left.\begin{aligned}\bar{\bar{s}}_c &= \frac{\sum_i M_i s_i}{M_c}, & \bar{\bar{q}}_c &= \frac{\sum_i M_i q_i}{M_c},\\ \bar{\bar{h}}_c &= \frac{\sum_i M_i h_i}{M_c}, & \bar{\bar{l}} &= \frac{\sum_i M_i l_i}{M_c},\end{aligned}\right\}, \quad (4.47)$$

Yanai et al. simplified Eqs. (4.35)–(4.38) and further introduced the following assumptions.

(i) Cumulus clouds detrain at the level where they lose buoyancy (e.g., Simpson et al. 1965). Thus,

$$h_{Di} = \bar{h}^*. \quad (4.48)$$

Then $s_{Di} = \bar{s}$ and $q_{Di} = \bar{q}^*$ according to (4.33) and (4.34).

(ii) The liquid water mixing ratio of the detrained air, l_{Di}, is the same as $\bar{\bar{l}}$; that is,

$$l_{Di} = \bar{\bar{l}}. \quad (4.49)$$

(iii) The conversion of cloud droplets into rainwater is proportional to the average liquid water mixing ratio. The rate of precipitation is parameterized by

$$r = K(p)\bar{\bar{l}}, \quad (4.50)$$

where $K(p)$ is an empirical function that varies with pressure.

With these assumptions, they obtained a closed set of ten equations for the ten unknowns given by (4.46) and (4.47). The equations are

$$Q_1 - Q_R = -M_c \frac{\partial \bar{s}}{\partial p} - Le, \quad (4.51)$$

$$Q_2 = LM_c \frac{\partial \bar{q}}{\partial p} - L\delta(\bar{q}^* - \bar{q}) - Le \quad (4.52)$$

$$\epsilon - \delta + \frac{\partial M_c}{\partial p} = 0, \quad (4.53)$$

$$(\epsilon - \delta)\bar{s} + \frac{\partial M_c \bar{\bar{s}}_c}{\partial p} + Lc = 0, \quad (4.54)$$

$$\epsilon\bar{q} - \delta\bar{q}^* + \frac{\partial M_c \bar{\bar{q}}_c}{\partial p} - c = 0, \quad (4.55)$$

$$-\delta\bar{\bar{l}} + \frac{\partial M_c \bar{\bar{l}}}{\partial p} + c - r = 0, \quad (4.56)$$

$$r = K\bar{\bar{l}}, \quad (4.57)$$

$$e = \delta\bar{\bar{l}}, \quad (4.58)$$

and (4.33), (4.34) applied to $\bar{\bar{s}}_c$, $\bar{\bar{q}}_c$, and $\bar{\bar{h}}_c$. The values of M_c and $\bar{\bar{h}}_c$ at the cloud base (assumed at 950 mb) were determined from the heat and moisture budgets of the subcloud layer.

Yanai et al. (1973) applied this system of equations to the four times daily data of $\bar{s}$, $\bar{q}$, $Q_1 - Q_R$, and Q_2 over the Marshall Islands in 1956 (386 cases) and obtained solutions for 366 cases. In Fig. 4.4 the average vertical distributions of the cumulus mass flux M_c obtained by the bulk method (dots) and the large-scale mean mass flux $\bar{M} = -\bar{\omega}$ are shown (from Yanai et al. 1976). A notable feature seen in this figure is that M_c exceeds $\bar{M}$ except near the tropopause; that is, the upward mass flux in active cumulus clouds is larger than the mass flux required from the large-scale horizontal convergence. As a result of the large M_c, the residual mass flux $\tilde{M} = \bar{M} - M_c$ is downward. Another significant feature is that M_c is very large near the cloud base. This was thought to suggest that a large number of shallow cumulus clouds coexist in the region with taller cumulonimbi. The average vertical distribution of mass detrainment obtained by the bulk method is shown by dots in Fig. 4.5. The average detrainment has two strong maxima, that is, in the lowest layer and near the 200-mb level. The lower maximum is the result of large number of shallow clouds that detrain immediately above the cloud base.

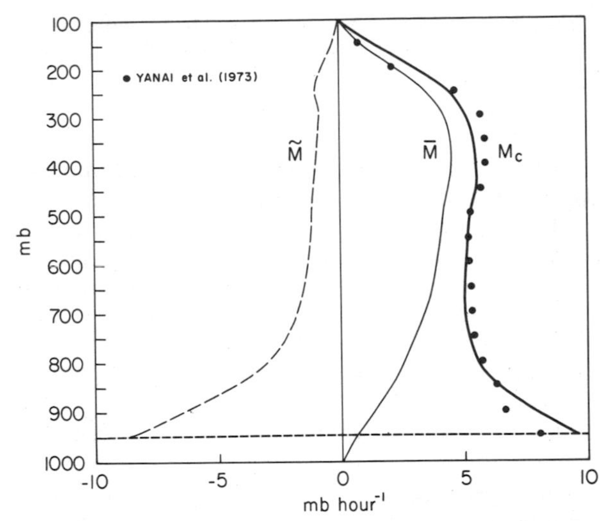

FIG. 4.4. The average total cloud mass flux M_c over the Marshall Islands obtained by the spectral method (thick solid), large-scale mass flux $\bar{M}$ (thin solid), and residual mass flux $\tilde{M}$ (dashed). The estimates of M_c obtained by the bulk method are shown by dots (from Yanai et al. 1976).

Having obtained all the variables that characterize the cumulus ensemble, we can interpret the observed apparent heat source in terms of the interaction between the clouds and the environment. Each term of (4.51) is shown in Fig. 4.6, that is, the observed $Q_1 - Q_R$, the adiabatic heating due to the compensating downward motion in the environment, and the cooling due to evaporation of cloud droplets. Figure 4.6 shows that the clouds act as a heating agent primarily through their induction of the compensating descent ($-M_c$). This interpretation does not necessarily mean that the actual mass flux in the environment $\tilde{M} = \bar{M} - M_c$ is downward. However, nearly all the cases show that $\tilde{M}$ is negative. The second important contribution to the apparent heat source is the cooling due to reevaporation of the detrained cloud droplets. This effect is pronounced in the lower troposphere.

The average moisture balance of the environment is shown in Fig. 4.7. The apparent moisture sink (Q_2) is primarily due to the induced downward motion ($-M_c$). This process alone would dry out the environmental air. It is the detrainment of water vapor and of liquid water from the clouds that makes up the balance needed to maintain the moisture of the environmental air. This would imply that the shallower, nonprecipitating cumulus clouds are needed to help the growth of the deep, precipitating cumulus towers by supplying the moisture.

In this method, no explicit model is introduced for cumulus clouds except that the use of (4.48) classifies the cloud types according to the level of detrainment. A weak point of this method is a crude parameterization of the conversion process of cloud droplets to raindrops. Furthermore, the solutions of the set of nonlinear equations are obtainable only by an iterative method. When $Q_1 - Q_R$ is negative (a frequent occurrence in the trades), the iteration will not converge. The bulk method does not give explicit information of the mass spectrum of various cloud types, other than the vertical distribution of the mass detrainment (Fig. 4.5). By demonstrating the possibility of obtaining the cumulus properties such as M_c, the bulk method has served a useful purpose. However, the results shown in Figs. 4.4–4.7 must be treated with caution. The most serious deficiencies of this method are 1) the neglect of possible contributions from mesoscale stratiform precipitation to the observed Q_1 and Q_2, and 2) the omission of the convective-scale downdrafts. This omission led to the overestimation of cumulus mass flux in the lower layer (Johnson 1976; Nitta 1977; Cheng 1989b). These problems will be addressed in sections 4.2c and 4.3.

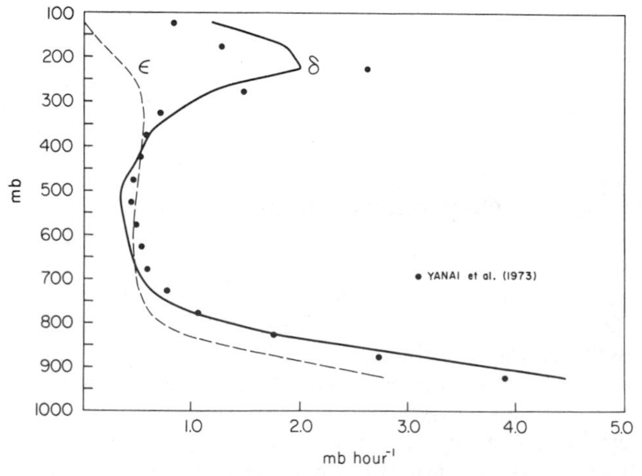

FIG. 4.5. The average mass detrainment $\delta\Delta p$ (solid) and entrainment $\epsilon\Delta p$ (dashed) over the Marshall Islands obtained by the spectral method. The estimates of $\delta\Delta p$ obtained by the bulk method are shown by dots (from Yanai et al. 1976).

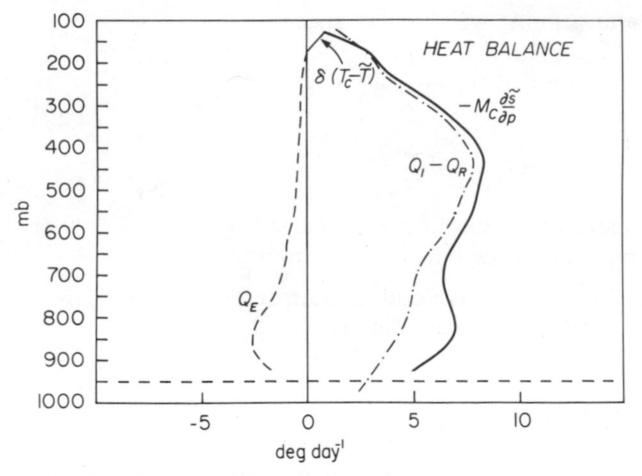

FIG. 4.6. The observed heat source $Q_1 - Q_R$ (dash–dot), adiabatic heating by compensating downward mass flux $-M_c(\partial\bar{s}/\partial p)$ (thick solid), evaporative cooling $Q_E = -Le$ (dashed), and detrainment of heat $\delta(T_c - \bar{T})$ (thin solid) (from Yanai et al. 1973).

b. The spectral method

1) BASIC EQUATIONS AND SOLUTIONS

Substitution of (4.43) and (4.44) into (4.26) yields

$$Q_1 - Q_2 - Q_R = -M_c \frac{\partial \bar{h}}{\partial p} + \sum_i \delta_i (h_{Di} - \bar{h}). \quad (4.59)$$

The spectral diagnostic method, developed by Ogura and Cho (1973) and Nitta (1975), uses (4.59) in conjunction with a spectral representation of the cumulus ensemble, which was introduced in a theory of cumulus parameterization proposed by Arakawa and Schubert (1974). The spectral method classifies clouds into subensembles according to their level of detrainment, and explicit assumptions on the subensemble mass and moist static energy budget equations based on a plume-type cloud model are used to uniquely relate the height reached by the clouds to their constant fractional rate of entrainment λ. In this cloud model, the mass flux increases exponentially with height and all the mass detrains at the top of clouds.[6] The top of the clouds, that is, the height of detrainment level $z_D(\lambda)$, will decrease as λ increases because the larger entrainment reduces the buoyancy. In the following, we use the pressure at the detrainment level $p_D(\lambda)$ instead of $z_D(\lambda)$.

The total mass flux in the clouds is expressed by

[6] We can include detrainment from the side of clouds (lateral detrainment). However, its effect on the diagnosed mass flux of plume-type clouds is shown to be small (Johnson 1977; Lord 1982).

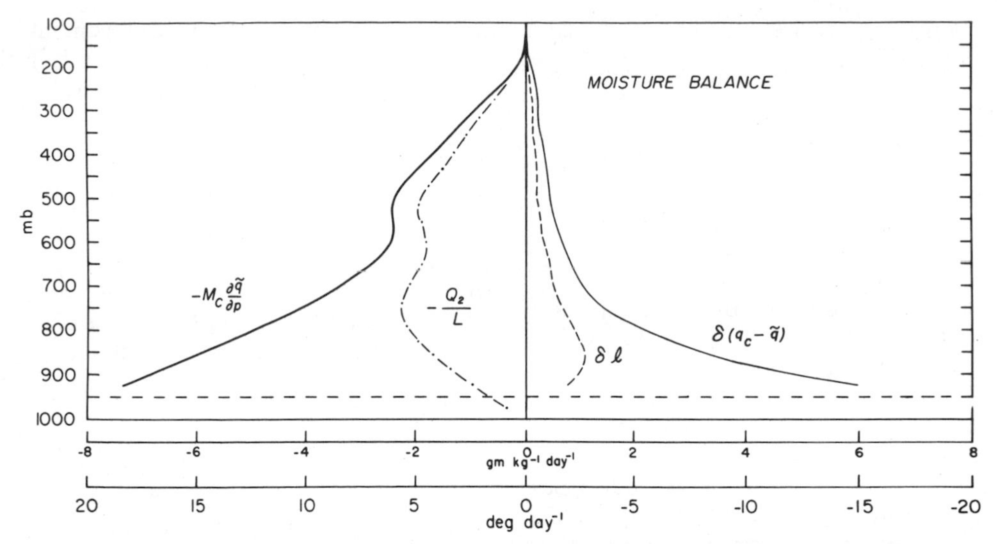

FIG. 4.7. The observed moisture source $-Q_2/L$ (dash–dot), drying due to compensating sinking motion $-M_c(\partial\bar{q}/\partial p)$ (solid), detrainment of water vapor $\delta(q_c - \bar{q})$ (thin solid), and detrainment of liquid water δl (dashed), in units of grams per kilogram per day and in equivalent heating units (K day^{-1}) (from Yanai et al. 1973).

$$M_c(p) = \int_0^{\lambda_D(p)} m_B(\lambda)\eta(p, \lambda)d\lambda, \quad (4.60)$$

where $m_B(\lambda)d\lambda$ is the subensemble mass flux at the cloud-base pressure p_B, due to the clouds whose fractional rates of entrainment are between λ and $\lambda + d\lambda$; $\eta(p, \lambda)$ is the normalized mass flux, and $\lambda_D(p)$ is the λ of clouds that detrain at pressure p. The mass detrainment $\delta(p)dp$ in the layer between p and $p + dp$ is equal to the subensemble mass flux due to clouds that have parameter λ between $\lambda_D(p)$ and $\lambda_D(p) + [d\lambda_D(p)/dp]dp$, so that we can relate δ to m_B by

$$\delta(p) = m_B[\lambda_D(p)]\eta[p, \lambda_D(p)]\frac{d\lambda_D(p)}{dp}. \quad (4.61)$$

Using (4.60) and (4.61), we write (4.59) as

$$Q_1 - Q_2 - Q_R = (h_D - \bar{h})m_B[\lambda_D(p)]\eta[p, \lambda_D(p)]\frac{d\lambda_D(p)}{dp} - \frac{\partial\bar{h}}{\partial p}\int_0^{\lambda_D(p)} m_B(\lambda)\eta(p, \lambda)d\lambda. \quad (4.62)$$

Equation (4.62) is the basic equation for obtaining $m_B(\lambda)$.

Now we briefly discuss the procedures to obtain λ_D, η, and h_D contained in (4.62). Introduction of constant fractional rate of entrainment for each cloud type simplifies the subensemble mass budget equation [corresponding to (4.28)]. The equation governing the normalized subensemble mass flux is

$$\frac{\partial\eta(p, \lambda)}{\partial p} = -\frac{\lambda}{\rho g}\eta(p, \lambda), \quad (4.63)$$

where ρ is the density of the air. The moist static energy of the subensemble $h_c(p, \lambda)$ is governed by an entrainment equation [corresponding to (4.31)]

$$\frac{\partial}{\partial p}[\eta(p, \lambda)h_c(p, \lambda)] = -\frac{\lambda}{\rho g}\eta(p, \lambda)\bar{h}(p). \quad (4.64)$$

The solution of (4.64) is

$$h_c(p, \lambda) = \frac{1}{\eta(p, \lambda)}\left[h_B + \frac{\lambda}{\rho g}\int_p^{p_B}\eta(p', \lambda)\bar{h}(p')dp'\right], \quad (4.65)$$

where h_B is the moist static energy of clouds at the cloud base.

Terms p_B and h_B in (4.65) must be chosen properly. Arakawa and Schubert (1974) placed p_B at the top of the mixed layer and used the mixed-layer value of $\bar{h}$ as h_B. Nitta (1975) also placed p_B at the top of the mixed layer but clouds were assumed to share the same virtual temperature with the environment. The sensitivity of $M_c(p_B)$ on various assumptions on h_B has been discussed by Nitta (1975) and Esbensen (1975). In Yanai et al. (1973) h_B was estimated from the heat and moisture budgets of the subcloud layer and it was close to $\bar{h}^*$ at 950 mb. Yanai et al. (1976) adopted a simple condition,

$$h_B = \bar{h}^*(p_B), \quad (4.66)$$

and p_B is taken as the pressure of the vertical finite-difference level closest to the lifting condensation level. Equation (4.66) has observational support (Simpson and Wiggert 1971).

The first step of the computations is to obtain $\lambda_D(p)$, for all p_D, as roots of (4.65) satisfying the condition of vanishing buoyancy at $p = p_D$,

$$h_c(p_D, \lambda) = \bar{h}^*(p_D). \qquad (4.67)^7$$

Knowing the values of λ for all possible values of p_D, we obtain explicit forms of $\eta(p, \lambda)$ and $h_c(p, \lambda)$ for all types of clouds from (4.63) and (4.65). Therefore, all terms in (4.62) except $m_B(\lambda)$ are known. Equation (4.62) becomes a Volterra integral equation of the second kind with respect to $m_B(\lambda)$. Inverting (4.62), we obtain $m_B(\lambda)$. Terms $M_c(p)$ and $\delta(p)$ are then obtained from (4.60) and (4.61).

In the application of the spectral diagnostic method to the data given at discrete vertical levels, we use discrete forms of (4.62), (4.63), and (4.65). We divide the whole cloud layer (from the cloud base to the tropopause) into many layers. Then we assign a cloud type that detrains at the middle level of each layer. Because we deal with a finite number of cloud types (the same as the number of divided layers), (4.62) becomes a set of algebraic equations for $M_B(p_D)$ that are the subensemble cloud-base mass fluxes of the clouds detraining in the layer centered at discrete p_D. The finite-difference version of (4.62) may be written in a form

$$\mathbf{N}(p, p_D)\mathbf{M}_B(p_D) = \mathbf{Q}(p), \qquad (4.68)$$

where $\mathbf{M}_B(p_D)$ and $\mathbf{Q}(p)$ are column vectors of M_B and $Q_1 - Q_2 - Q_R$, respectively; $\mathbf{N}$ forms a triangular matrix whose elements involve $\partial\bar{h}/\partial p$ and $(h_D - \bar{h})$ at all levels and $\eta(p, \lambda)$ for all cloud types (see Nitta 1975; Yanai et al. 1976; Cheng 1989b for details).

The spectral method has many advantages over the bulk method. It provides not only $M_c(p)$, $\delta(p)$, and other bulk properties of the total ensemble, but also the spectrum of cloud-base mass flux $M_B(p_D)$ according to the detrainment pressure. The solution of (4.68) is straightforward and needs no iterative procedure. The obtained cloud mass flux is free from assumptions on the cloud physical processes. However, the method introduces an explicit model of a cloud subensemble that is characterized by constant λ. The physical reality of the chosen cloud model must be judged by careful comparisons of the results with those from the bulk method and inferences from available direct observations.

Ogura and Cho (1973) used composite data from the Marshall Islands area (Nitta 1972) and showed bimodal distribution of the cloud spectrum as did Yanai et al. (1973). Cho and Ogura (1974) applied the same method to the dataset of Reed and Recker (1971) and demonstrated differences in spectral distribution of the mass flux in various sectors of a composite easterly wave. Nitta (1975) applied the spectral diagnostic method to the data taken during BOMEX (Barbados Oceanographic and Meteorological Experiment) phase III (section 4.2d). He showed the dominance of shallow clouds in this generally subsiding trade-wind region, and penetration of deep clouds during a day disturbed by a cloud cluster.

[7] A more accurate nonbuoyancy condition may be obtained by setting the virtual temperature excess of cloud air to be zero (see Arakawa and Schubert 1974; Nitta 1975).

2) Comparison of results of the two methods

Yanai et al. (1976) compared the time averages of the results of the total cumulus mass flux M_c and the mass detrainment $\delta\Delta p$ obtained by the spectral diagnostic method (386 cases) with the corresponding results of the bulk method (Yanai et al. 1973). Even though the two methods share similar physical principles, there are substantial differences in the formulation and the technical detail of numerical solutions. The most important differences of the spectral method from the bulk method are the explicit assumption on the subensemble mass and heat budgets using constant λ and independence of the solution from the precipitation and evaporation processes. Therefore, it is not obvious whether they will produce the identical results.

They applied the spectral method to the same dataset used earlier by Yanai et al. (1973) and obtained the solutions of $M_B(p_D)$, thus $M_c(p)$, for all 386 cases. In Fig. 4.4 the average vertical profiles of the cloud mass flux M_c obtained by the spectral method (thick solid line), the large-scale mass flux $\bar{M} = -\bar{\omega}$, and the residual mass flux in the environment $\tilde{M} = \bar{M} - M_c$ are shown. The previous estimates of M_c obtained by the bulk method are shown by dots. Except near the cloud base and near the tropopause, the agreement between the results of the two methods is remarkable. Very large cumulus mass flux at the cloud base and compensating sinking motion in the environment are again demonstrated.

Figure 4.5 compares the average vertical profiles of the mass detrainment $\delta\Delta p$ obtained by the two methods. The general agreement is again clear. We find the maxima of mass detrainment at the lowest layer and near 200 mb. Figure 4.5 also shows the average vertical distribution of the mass entrainment $\epsilon\Delta p$ obtained by the spectral method, where

$$\epsilon(p) = -\int_0^{\lambda_D(p)} m_B(\lambda)\frac{\partial\eta(p, \lambda)}{\partial p}\,d\lambda. \qquad (4.69)$$

Because of the coexistence of a spectrum of cloud subensembles, the entrainment and detrainment can take place at the same height.

The fair agreement between the two methods illustrated above may suggest that the estimates of M_c and $\delta\Delta p$ are not too dependent on the cloud model used in the spectral representation of the cumulus ensemble so long as (4.59) is satisfied. It shows also that the assumptions made on precipitation and evaporation in the bulk method had little effects upon these estimates.

3) Estimation of Q_{1c} and Q_{2c}

Because the spectral method uses only the combination $Q_1 - Q_2 - Q_R$ as input to (4.62), there is a freedom of choice as to how we use the original and separate observations of $Q_1 - Q_R$ and Q_2.

If detrainment occurs only at the cloud top, (4.43) and (4.44) become

$$Q_{1c} = -M_c \frac{\partial \bar{s}}{\partial p} + \delta(s_D - \bar{s} - Ll_D), \quad (4.70)$$

$$Q_{2c} = LM_c \frac{\partial \bar{q}}{\partial p} - L\delta(q_D - \bar{q} + l_D). \quad (4.71)$$

Nitta (1975) assumed that $Q_{1c} = Q_1 - Q_R$ (thus $Q_{2c} = Q_2$) and obtained l_D from (4.70). However, recent studies indicate that the observed $Q_1 - Q_R$ and Q_2 involve the contributions from mesoscale stratiform rains (e.g., Houze 1982, 1989; Johnson 1980; Johnson and Young 1983). Therefore, it is desirable to obtain Q_{1c} and Q_{2c} independently from the observed $Q_1 - Q_R$ and Q_2. Cheng (1989a,b) and (Wu 1993a) used a simple parameterization of rainwater generation (Arakawa and Schubert 1974) to obtain l_D. The subensemble budget equation for water substance is given by

$$\frac{\partial}{\partial p}\{\eta(p, \lambda)[q_c(p, \lambda) + l(p, \lambda)]\}$$

$$= -\frac{\lambda}{\rho g}\eta(p, \lambda)\bar{q}(p) + \frac{c_0}{\rho g}\eta(p, \lambda)l(p, \lambda), \quad (4.72)$$

where the last term expresses the rate of conversion of cloud water into rainwater, c_0 an autoconversion coefficient. The numerical values of c_0 were taken from Hack et al. (1984).

Figure 4.8 shows the spectrum of $M_B(p_D)$, M_c, Q_{1c}, and Q_{2c} together with the observed $\bar{M}$, $Q_1 - Q_R$, and Q_2 during GATE phase III (Cheng 1989b). There are systematic differences between $Q_1 - Q_R$ and Q_{1c}, and between Q_2 and Q_{2c}. In the layer below the 400-mb level the diagnosed Q_{1c} and Q_{2c} show excessive heating and drying. Cheng (1989b) showed that the inclusion of convective-scale downdrafts dramatically reduces the differences in the lower layer (section 4.2c). On the other hand, in the layer above the 400-mb level Q_{1c} and Q_{2c} give insufficient heating and drying, showing that additional heating and drying are likely provided by stratiform precipitation (section 4.3).

In Fig. 4.9 we show the observed $\bar{M}$, Q_1, Q_2, and diagnostically obtained M_c, Q_{1c}, and Q_{2c} for two mesoscale convective systems (an MCC and a squall line) observed during OK PRE-STORM (Oklahoma–Kansas Preliminary Regional Experiment for STORM-Central) (Wu 1993). The diagnosed cloud mass flux M_c is larger than the large-scale mass flux $\bar{M}$ in the lower troposphere but the reverse is true in the upper troposphere. The cumulus heating Q_{1c} and drying Q_{2c} account for most of the observed Q_1 and Q_2. However, insufficient (excessive) heating and drying appear in the upper (lower) troposphere. These suggest the presence of mesoscale stratiform heating and drying ($Q_{1m} \approx Q_{2m}$) in the upper troposphere and cooling and moistening due to downdrafts in the lower troposphere.

c. Convective-scale downdrafts

Convective clouds, both precipitating and nonprecipitating, can generate a variety of types of downdrafts (Knupp and Cotton 1985). The two most prominent of these are 1) cumulus-scale downdrafts—produced by hydrometeor loading and latent cooling due to sublimation, evaporation, and melting and 2) mesoscale downdrafts—initiated by hydrometeors detrained from convective updrafts. The first direct measurements of cumulus downdrafts, obtained during the Ohio to Florida Thunderstorm Project (Byers and Braham 1949), showed that during the thunderstorm mature stage, downdraft velocities and widths (except near cloud top) are only slightly less than those of adjacent updrafts. Recent studies have shown similar results for the tropics, despite the weaker magnitude of draft velocities by a factor of 2–3 (LeMone and Zipser 1980; Zipser and LeMone 1980; Jorgensen et al. 1985; Jorgensen and LeMone 1989). Mesoscale downdrafts were first shown to be prominent features of tropical convective systems in observations from the Line Islands Experiment (Zipser 1969) and GATE (Houze 1977; Zipser 1977).

Observational studies indicate that the properties of both downdraft types depend in a complex way on the environmental stratification, wind shear, and cloud microphysical processes. Downdrafts that descend to or penetrate into the boundary layer have an important impact on surface fluxes and the boundary-layer structure. These impacts, in turn, influence future convective development. The influence of downdrafts on the subcloud layer as well as the effects of mesoscale circulations will be treated in more detail in subsequent sections.

The heat and moisture budget equations written to include the specific effects of cumulus updrafts and downdrafts are (Johnson 1976; Nitta 1977; Cheng 1989b)

$$Q_{1c} = \delta_u(s_{Du} - \bar{s} - Ll_{Du}) - M_u \frac{\partial \bar{s}}{\partial p}$$

$$+ \delta_d(s_{Dd} - \bar{s} - Ll_{Dd}) - M_d \frac{\partial \bar{s}}{\partial p} \quad (4.73)$$

$$-Q_{2c} = L\delta_u(q_{Du} - \bar{q} + l_{Du}) - LM_u \frac{\partial \bar{q}}{\partial p}$$

$$+ L\delta_d(q_{Dd} - \bar{q} + l_{Dd}) - LM_d \frac{\partial \bar{q}}{\partial p}, \quad (4.74)$$

where the subscripts u and d denote values for updrafts and downdrafts, respectively. Combining (4.26), (4.73), and (4.74) yields

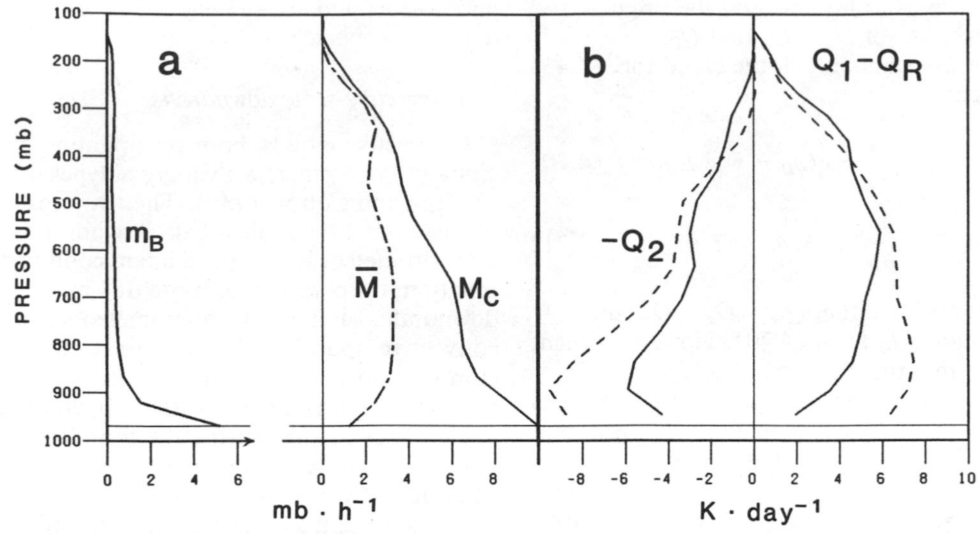

FIG. 4.8. (a) Vertical profiles of the cumulus mass flux M_c (solid) diagnosed from the GATE phase III–mean data using the model without downdrafts and the phase III–mean large-scale mass flux $\bar{M}$ (dashed–dotted). Also shown on the left is the cloud-base mass flux M_B as a function of the cloud-top pressure. Units are millibars per hour. (b) The observed $Q_1 - Q_R$ and $-Q_2$ (solid) and diagnosed Q_{1c} and $-Q_{2c}$ (dashed). Units are kelvins per day (from Cheng 1989b).

$$Q_1 - Q_2 - Q_R = \delta_u(h_{Du} - \bar{h}) - M_u \frac{\partial \bar{h}}{\partial p} + \delta_d(h_{Dd} - \bar{h}) - M_d \frac{\partial \bar{h}}{\partial p}. \quad (4.75)$$

There have been two general approaches for diagnosing *convective-scale* updraft and downdraft properties from heat and moisture budgets. In one, the moist static energy equation (4.75) is used (Johnson 1976; Cheng 1989b). This procedure has the advantage of directly using, without any ambiguity, the observed Q_1 and Q_2, which contain the sum of both cumulus and mesoscale effects. Another approach is to make use of the two independent heat and moisture budget equations (4.73) and (4.74) by solving any two of (4.73)–(4.75) (Nitta 1977). However, in this case the assumptions $Q_1 - Q_R = Q_{1c}$ and $Q_2 = Q_{2c}$ are required. If mesoscale effects are important, the observed $Q_1 - Q_R$ or Q_2 may be quite different from Q_{1c} or Q_{2c},[8] thereby introducing errors into the diagnosis of cumulus transports. In this section we will review studies involving both procedures to determine convective-scale transports.

Using the first approach, Johnson (1976) assumed that each cumulus updraft has an accompanying downdraft and both have the same fractional rate of entrainment (Fig. 4.10). A downdraft-originating level mass flux distribution function $m_0(\lambda)$ was introduced such that downdraft mass flux is given by

$$M_d = \int_0^{\lambda_D(p)} m_0(\lambda)\eta_d(p, \lambda)d\lambda, \quad (4.76)$$

where

$$\eta_d(\lambda, p) = \exp\left[\int_{p_0(\lambda)}^{p} \frac{\lambda}{\rho g} dp\right] \quad (4.77)$$

is the normalized downdraft mass flux.

To solve (4.75), Johnson (1976) made several assumptions: detrainment from downdrafts was assumed to occur only below cloud base ($\delta_d = 0$); the intensity of downdrafts was related to updrafts by introducing a parameter ϵ, where

$$\epsilon(\lambda) \equiv \frac{m_0(\lambda)}{m_B(\lambda)}; \quad (4.78)$$

and downdrafts were assumed to originate at a level $p_0(\lambda)$ given by a fraction β of the pressure depth of the corresponding updraft. With these assumptions, (4.75) becomes

$$Q_1 - Q_2 - Q_R = m_B(\lambda_D)\eta_u(p, \lambda_D) \times \frac{d\lambda_D}{dp}[h_u(p, \lambda_D) - \bar{h}] - \frac{\partial \bar{h}}{\partial p}\int_0^{\lambda_D(p)} m_B(\lambda) \times [\eta_u(p, \lambda) + \epsilon(\lambda)\eta_d(p, \lambda)]d\lambda. \quad (4.79)$$

Equation (4.79) is an integral equation [similar to (4.62)] that can be solved for the single unknown function $m_B(\lambda)$, provided that $\epsilon(\lambda)$ and $p_0(\lambda)$ are

[8] Cheng and Yanai (1989) make use of these differences to determine the mesoscale contributions to heat and moisture budgets. This subject will be treated in section 4.3.

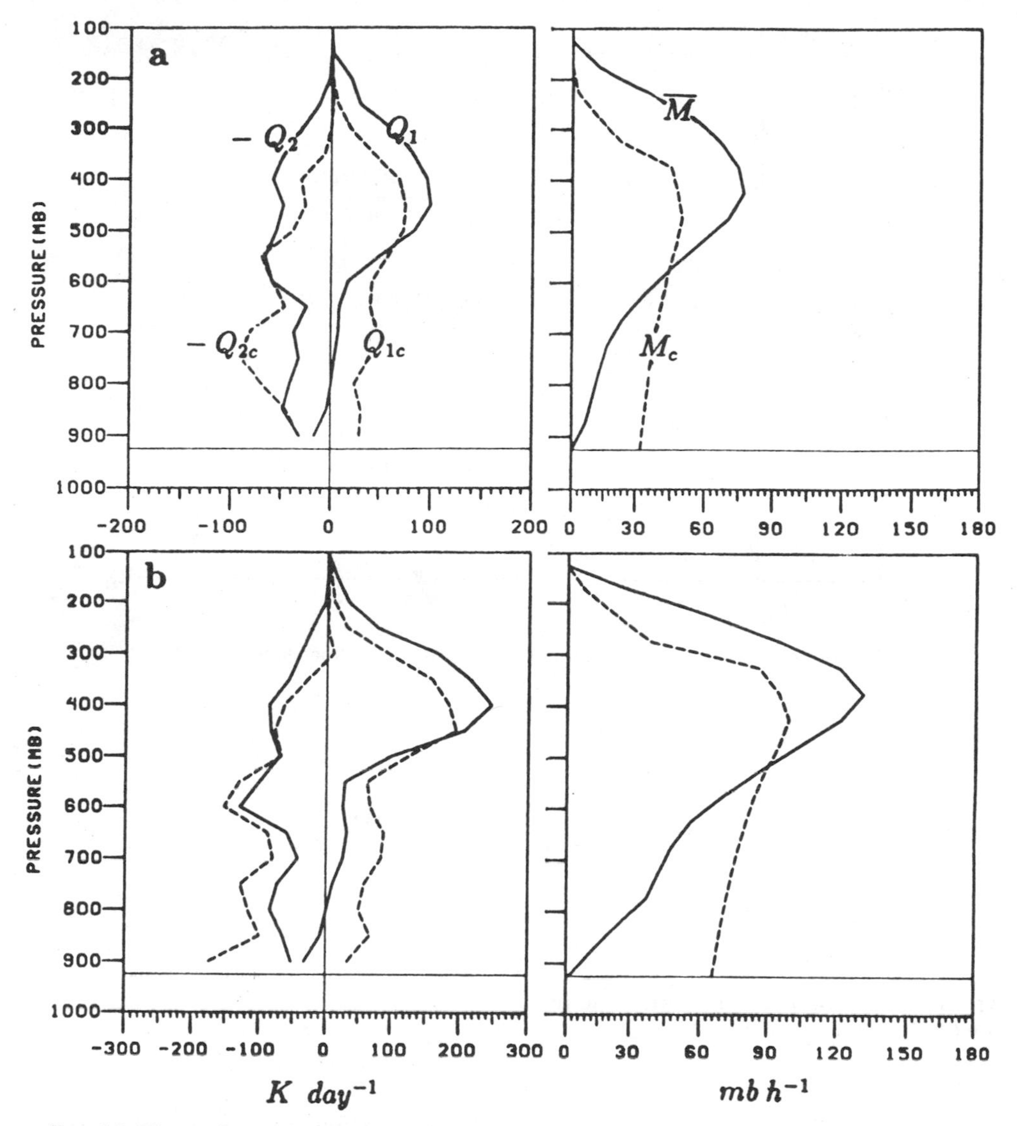

FIG. 4.9. The vertical profiles of (left) the observed Q_1 and $-Q_2$ (solid) and diagnosed Q_{1c} and $-Q_{2c}$ (dashed) (units: K day^{-1}), and (right) the large-scale mass flux $\bar{M}$ (solid) and diagnosed cloud mass flux M_c (dashed) (units: mb h^{-1}) for (a) an MCC (0000 UTC 4 June 1985) and (b) a squall line (0300 UTC 11 June 1985) (from Wu 1993a).

known. In the study of Johnson (1976) an optimum value of ϵ was determined based on two procedures: (i) a comparison of diagnosed precipitation with observed and (ii) a computation of the subcloud-layer water vapor budget. The value of β was specified based on results of observational studies. It was shown by Houze et al. (1980) that the parameter $\epsilon(\lambda)$, while physically complex in its full interpretation, is directly proportional to the ratio of condensate reevaporated in convective-scale downdrafts to the total water condensed in convective-scale updrafts.

Results from the solution of (4.79) for the Reed and Recker (1971) easterly wave trough are shown in Fig. 4.11. Here it can be seen that the inclusion of downdrafts (using an optimum value of ϵ of -0.2) significantly reduces the diagnosed environmental subsidence ($\tilde{M}$). It was concluded by Johnson (1976) that the neglect of cumulus downdrafts and their associated rainfall evaporation leads to the diagnosis of excessively large populations of shallow cumulus clouds in highly convective situations. It was further suggested that cumulus parameterization schemes employing cloud models that neglect cumulus downdrafts would likely incur excessive warming and drying in the lower troposphere [consistent with the results of Cheng (1989b), wherein Q_{1c} and Q_{2c} are computed with and without the effects of downdrafts; see section 4.2b and Fig. 4.16].

Using the second procedure described above, Nitta (1977) introduced a bulk or single-cloud mass flux model for the cumulus downdraft while retaining a

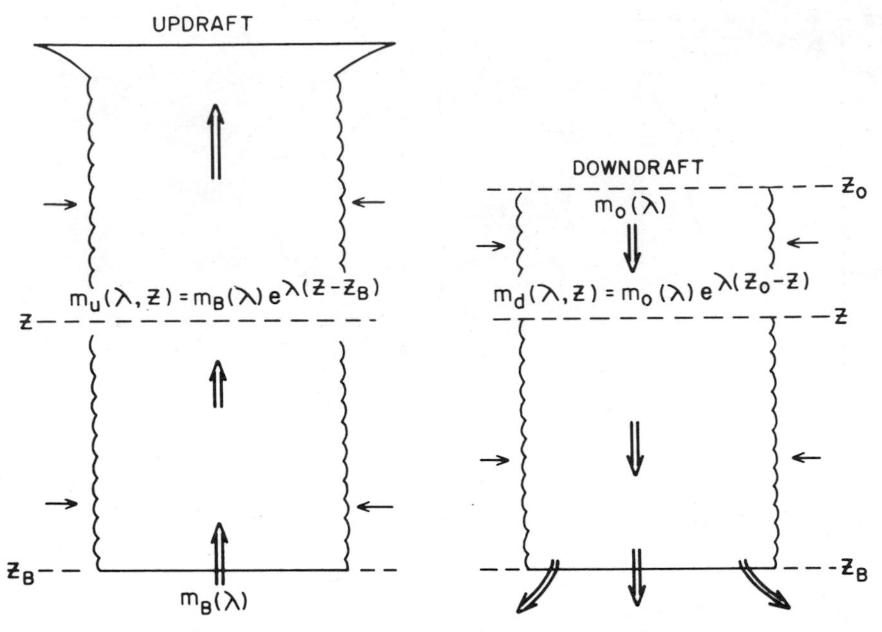

FIG. 4.10. Model for updraft and downdraft of cloud type λ (from Johnson 1976).

spectral model for the updraft. This approach does not require specification of parameters for the downdraft due to simultaneous solution of (4.73) and (4.75). However, in this instance a parameterization of the conversion of cloud droplets to raindrops [see (4.72)] is required for the determination of the detrained cloud water l_{Du} in (4.73). Also, it is assumed that Q_{1c} in (4.73) is equivalent to $Q_1 - Q_R$; that is, mesoscale effects are assumed to be small.

Conservation equations for mass and moist static energy in Nitta's (1977) downdraft model are

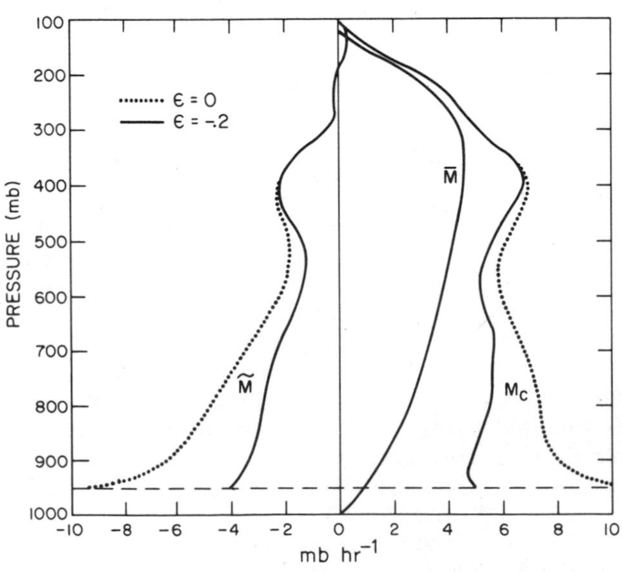

FIG. 4.11. Net cumulus, and mean and environmental mass fluxes for $\epsilon = 0, -0.2$ and $\beta = 0.75$ (from Johnson 1976).

$$\epsilon_d = 0 \quad \text{and} \quad \delta_d = \frac{\partial M_d}{\partial p}, \quad \text{when} \quad \frac{\partial M_d}{\partial p} > 0 \qquad (4.80)$$

$$\epsilon_d = -\frac{\partial M_d}{\partial p} \quad \text{and } \delta_d = 0, \text{ when } \frac{\partial M_d}{\partial p} \leq 0 \qquad (4.81)$$

$$(h' - h_d)\frac{\partial M_d}{\partial p} - M_d \frac{\partial h_d}{\partial p} = 0, \qquad (4.82)$$

where

$$h' = \begin{cases} h_d, & \partial M_d/\partial p > 0 \quad \text{(detrainment)} \\ \bar{h}, & \partial M_d/\partial p < 0 \quad \text{(entrainment)}. \end{cases}$$

It is assumed that air inside the downdraft is saturated and does not contain cloud drops ($l_d = 0$) and the thermodynamic properties of detrained air are equal to the mean values inside the downdraft. Assuming a direct conversion of cloud water to precipitation (using a constant autoconversion coefficient $c_0 = 2 \times 10^{-3}$ m^{-1}), the cloud water budget equation (4.72) was then used to determine l_{Du}.

Results of the application of the procedure of Nitta (1977) to Marshall Islands data (Yanai et al. 1976) are illustrated in Fig. 4.12. Despite considerably different assumptions from those of Johnson (1976), very similar results are obtained: a pronounced reduction in the net cumulus mass flux M_c and environmental subsidence $\tilde{M}$ near cloud base is observed when the effects of downdrafts are included. It was also found that the neglect of downdrafts leads to an overestimation of the mass flux within shallow clouds during convectively disturbed conditions.

Recently, Cheng (1989a) has developed a cloud model that takes into account the transfer of hydro-

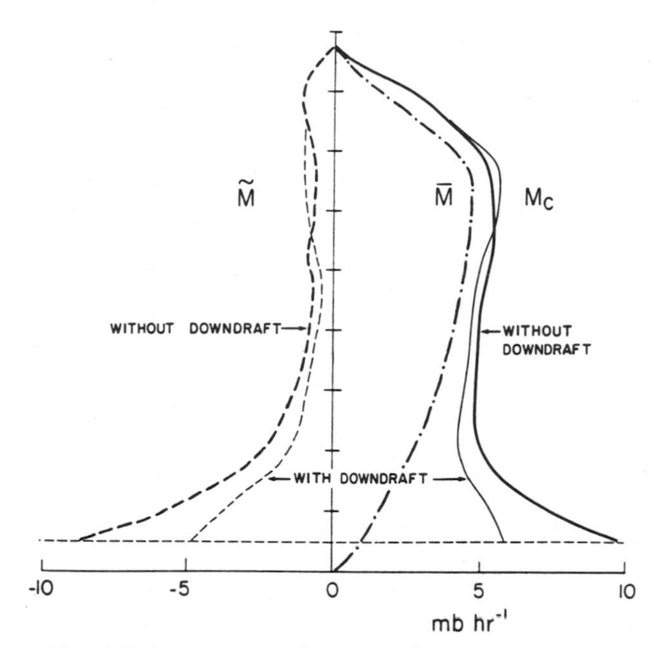

FIG. 4.12. Net cloud mass flux M_c, large-scale mass flux $\bar{M}$, and environmental mass flux $\tilde{M}$ in the Marshall Islands region obtained from a diagnostic model with and without downdrafts (from Nitta 1977).

meteors from the updraft to the downdraft, thereby allowing for the solution of (4.75) without the specification of parameters for the downdraft. The approach is to consider a tilted updraft (Fig. 4.13) that allows rainwater to freely fall into a subsaturated cloud environment. In order to determine the rainwater flux, the vertical momentum budget of the updraft must be considered. The downdraft is assumed to be initiated and maintained by loading and evaporation of precipitation from the updraft. The development of a dynamical model for the downdraft removes the need to specify properties of the downdraft (e.g., as in Johnson 1976, where the downdraft mass flux is assumed to be directly proportional to the updraft mass flux for each cloud type). In fact, it has been found by Cheng (1989b) that the contribution of the downdraft mass flux M_d to the total cumulus mass flux M_c can change substantially with time.

Information from observed thermodynamic fields has been used by Cheng (1989a) to diagnose tilting angles for GATE phase III deep cumulus updrafts. Results show that the tilting angles have maxima in convectively active regions and are much less in undisturbed regions (Fig. 4.14). Tilting angles are also greater for squall clusters than nonsquall clusters. Although the wind shear is not explicitly used in the determination of the tilting angle, Cheng and Yanai (1989) have found a close correspondence between large tilting angles and strong wind shear. From these findings it is argued that the thermodynamic and dynamic fields are closely coupled so that there is information in the thermodynamic field regarding the effects of wind shear on updraft tilt.

Cheng (1989b) has applied a form of (4.75) to diagnose updraft and downdraft mass fluxes by using the rainwater budget in the tilted updraft model to determine the available rain flux for the downdraft. Results of this method for phase III of GATE are shown in Fig. 4.15. It can be seen that while the updraft mass flux is prominent throughout most of this period, downdrafts are significant only when considerable deep upward motion exists or when cloud clusters are active within the GATE A/B array. The downdraft mass flux associated with shallow cumulus clouds is negligibly small.

Cheng (1989b) subsequently used the diagnosed updraft, downdraft, and detrainment properties of cumulus ensembles to reconstruct from (4.73) and (4.74) the cumulus contributions to the large-scale heat and moisture budget residuals, Q_{1c} and Q_{2c}. To do so, a parameterization of the rainwater generation process was used, as described in section 4.2b. The results, shown by the dashed curves in Fig. 4.16 (compare with those in Fig. 4.8), reveal that the inclusion of convective-scale downdrafts significantly improves the reconstruction of the residuals in the lower troposphere. The differences between Q_{1c} and $Q_1 - Q_R$, and Q_{2c} and Q_2 that remain were used by Cheng and Yanai (1989) to infer mesoscale effects from (4.11) and (4.12). These results will be presented in section 4.3.

d. Coupling between cloud and subcloud layers

Cumulus clouds, whether precipitating or nonprecipitating, normally have their roots in the atmospheric

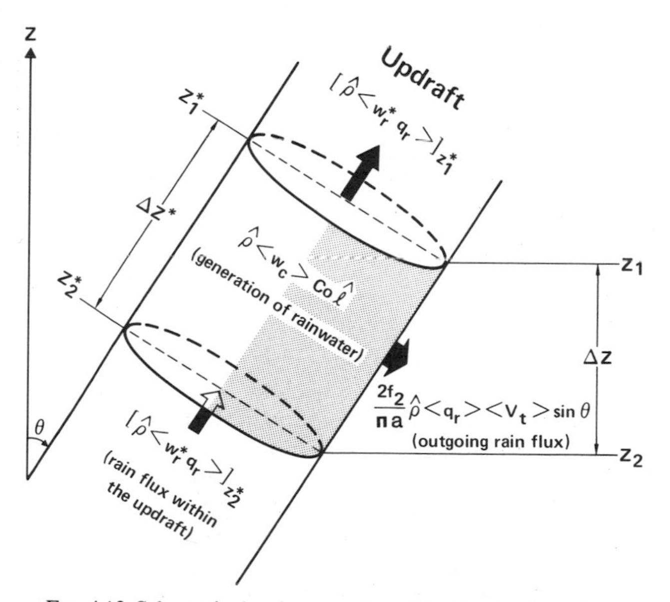

FIG. 4.13. Schematic showing a portion of the tilted updraft. Arrows indicate the rain flux within the updraft and the outgoing rain flux. Shaded area indicates the region from which rainwater detrains. (See Cheng 1989a for details.)

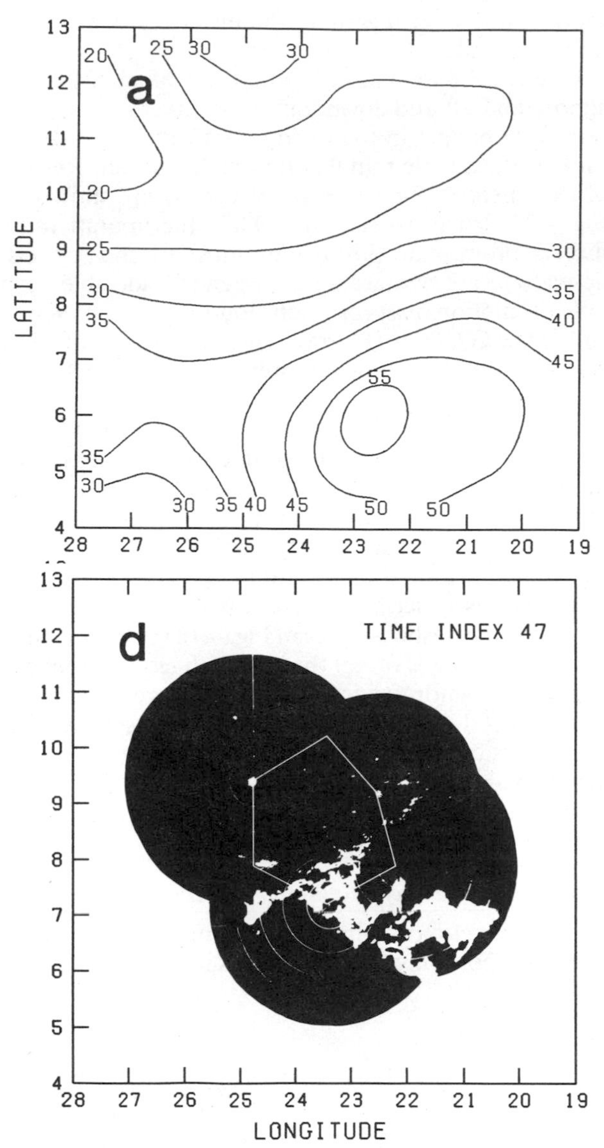

FIG. 4.14. (a) Horizontal distribution of the updraft tilting angle (deg) at time index 47 (1800 UTC 4 September 1974). (b) Radar echoes at time index 47 taken from the *GATE International Meteorological Radar Atlas* (Arkell and Hudlow 1977) (from Cheng 1989a).

boundary layer. Therefore, their properties depend importantly on boundary-layer air characteristics. Cumulus clouds, in turn, feed back to alter the boundary layer through induced subsidence, modulation of radiative fields, and direct transports. For precipitating clouds, significant boundary-layer modification can occur through the downward transport and spreading at the surface of low-h air from aloft (e.g., Betts 1976; Barnes and Garstang 1982). The proper treatment of convective downdrafts is one of the central remaining problems in cumulus parameterization.

Before considering the effects of precipitation, let us first examine shallow, nonprecipitating clouds. In a conditionally unstable atmosphere, trade-wind cumuli are the most dominant clouds of this type. Two experiments in the tropical Atlantic in 1969, ATEX (Atlantic Tradewind Experiment) and BOMEX, were designed to investigate the nonprecipitating, trade-wind cumulus boundary layer. Heat and moisture budgets for this layer during BOMEX have been determined by Nitta and Esbensen (1974). Their results (see Q_1 and Q_2 in Fig. 4.17) show cooling throughout most of the cloud layer and moistening throughout the entire cloud and subcloud layers. Notice that the strongest cooling and moistening are in the upper part of the cloud layer. By referring to (4.4) and assuming the convergence of eddy heat flux to be small, the profile of $Q_1 - Q_R$, characterized by a cooling aloft and a warming below in the cloud layer, can be explained by an excess of evaporation over condensation in the upper half of the cloud layer and the reverse below. As illustrated in Fig. 4.17 and confirmed in the diagnostic study of Nitta (1975), this behavior occurs as cloud droplets, condensed in the lower part of the cloud layer, are carried aloft, where they are detrained and evaporated (also see Betts 1975). The pronounced moistening in the upper part of the cloud layer is consistent with this explanation. The cloud and subcloud layers are closely coupled because it is the surface evaporation that provides the source for the moistening of the entire layer. For the steady-state trade-wind boundary layer, this convective moistening is offset by subsidence drying.

The importance of including effects of trade-wind cumulus in large-scale prediction models is evident upon examination of the magnitude of the cooling and moistening rates in the cloud layer (ranging 1–4°C day^{-1} and 1–5 g^{-1} kg^{-1} day^{-1}). As noted, this cooling and moistening are largely offset by subsidence warming and drying (see profile of $\bar{\omega}$ in Fig. 4.17). If a large-scale model predicts the correct $\bar{\omega}$ but does not include trade-wind cumulus effects, then excessive warming and drying would be expected in a matter of days. There is evidence to indicate that such effects have been observed in models (Albrecht et al. 1986; Tiedtke et al. 1988).

Another complication that shallow cumulus introduces for the problem of cumulus parameterization relates to their distinctly different dynamic, thermodynamic, entrainment, and microphysical properties, as compared to deep cumulus (see chapter 3). For example, cloud-top entrainment may have a pronounced effect on the properties of shallow cumulus but less of an effect on deep, nonentraining cumulonimbus. Additionally, the near-neutral buoyancy and small stratification of shallow cumulus suggest that the adjustment time for an ensemble of shallow clouds to come into quasi equilibrium with large-scale processes (Arakawa and Schubert 1974) is significantly longer than the adjustment time for deeper clouds (Esbensen 1978). These different properties may limit the appli-

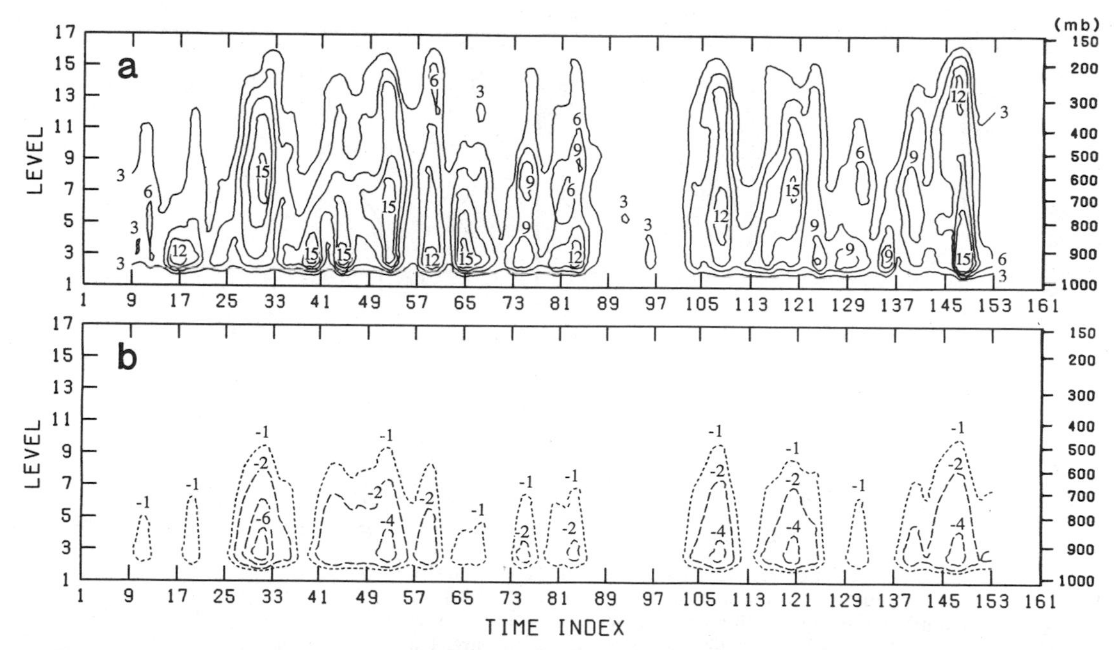

FIG. 4.15. Time–height sections of (a) the total updraft mass flux and (b) the total downdraft mass flux averaged over the 3° × 3° area at the center of the GATE network from 0000 UTC 31 August (index 9) to 0000 UTC 18 September 1974 (index 153). Units are millibars per hour (from Cheng 1989b).

cability of one-parameter cumulus models for the full spectrum of cloud types. Recent progress has been made on understanding some of the distinctive features of shallow cumulus in undisturbed convective boundary layers using mixing-line diagrams for conserved thermodynamic variables (e.g., Betts and Albrecht 1987), but more work is needed.

When precipitation develops, convective downdrafts often occur, which transport cold air to the surface that spreads out over areas much greater than that of the clouds themselves (Zipser 1977; Simpson and van Helvoirt 1980). During GATE, downdraft-modified boundary layers or "wakes" accompanying precipitating convective systems were found to cover approximately 30% of the total area (Gaynor and Ropelewski 1979). The temperature, moisture, and wind stratification within wakes is significantly modified from its structure in undisturbed conditions and recovery of the boundary may take up to approximately 10 h (Betts 1976; Echternacht and Garstang 1976; Houze 1977; Zipser 1977; Brümmer 1978; Emmitt 1978; Fitzjarrald and Garstang 1981a,b; Johnson 1981). Locally, surface sensible and latent heat fluxes can be enhanced by an order of magnitude or more in the vicinity of precipitating systems.

As an example of the impact of precipitation downdrafts, the surface sensible and latent heat fluxes associated with a GATE squall line are shown in Fig. 4.18 (from Johnson and Nicholls 1983). The analyses are based on bulk aerodynamic computations using ship boom data over a 9-h period in the squall-line life cycle. Enhancements in sensible and latent heat fluxes by factors of 10 and 5, respectively, can be seen immediately behind the leading convective line. It can also be seen that the effects of the squall line are felt over an area several hundred kilometers on a side. The collective effects of many of these wakes contribute importantly to the total surface fluxes in disturbed tropical regions.

In addition to direct modification of the boundary layer, downdrafts have an effect on boundary-layer structure in the vicinity of convective systems by influencing environmental subsidence. This effect may be seen by considering a simplified equation for the mixed-layer height z_i in the environment between clouds:

$$\frac{\partial z_i}{\partial t} = \frac{\tilde{M}_i}{\rho g} - \frac{F_i}{\Delta_i}, \qquad (4.83)$$

where F_i is the buoyancy flux $\overline{w's_v'}$ at z_i and Δ_i is the jump in virtual static energy s_v across the inversion at z_i. This situation is depicted schematically in Fig. 4.19, where $\tilde{w} = \tilde{M}/\rho g$. The buoyancy flux F_i is usually downward (negative) as a result of overshooting eddies in the mixed layer and is often parameterized in terms of the surface buoyancy flux F_0; that is, $F_i = -kF_0$, where k is a positive constant (typically between 0.1 and 0.4) and subscript 0 refers to the surface value (Betts 1973b; Tennekes 1973).

Since $\tilde{M}_i$ is less than 0, subsidence restricts the growth of the mixed layer. It was noted earlier that the neglect of downdrafts leads to the diagnosis of excessive environmental subsidence near cloud base (Johnson

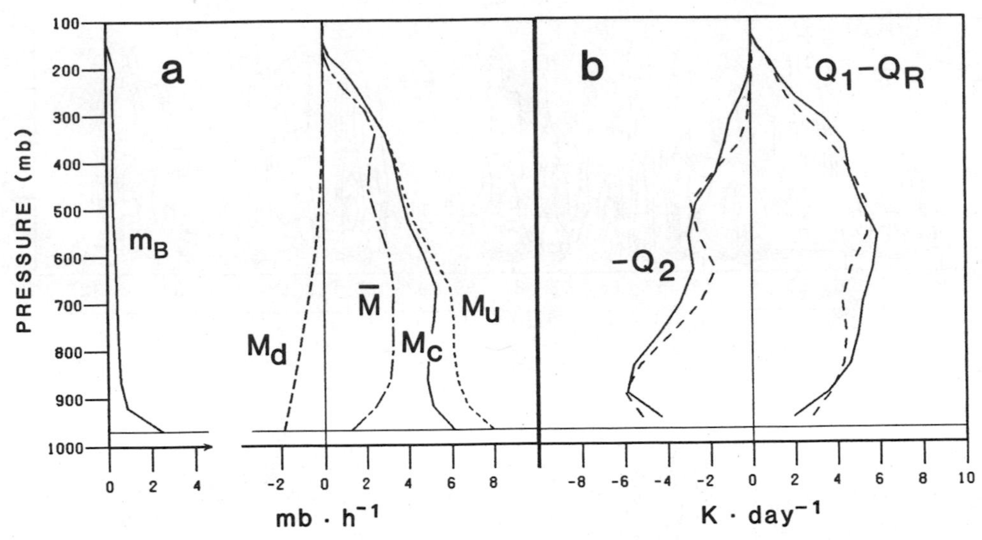

FIG. 4.16. (a) Vertical profiles of the updraft mass flux M_u (dotted), the downdraft mass flux M_d (dashed), and the total cumulus mass flux M_c (solid) diagnosed from the GATE phase III–mean data using the model with downdrafts, and the phase III–mean large-scale mass flux $\tilde{M}$ (dash–dot). Also shown on the left is the cloud-base mass flux M_B as a function of the cloud-top pressure. Units are millibars per hour. (b) The observed $Q_1 - Q_R$ and $-Q_2$ (solid) and diagnosed Q_{1c} and $-Q_{2c}$ (dashed). Units are kelvins per day (from Cheng 1989b).

1976; Nitta 1977; Cheng 1989b). Therefore, considering the effects of subsidence in (4.83), predicted mixed layers in the vicinity of convective systems during disturbed conditions may be unrealistically shallow if convective-scale downdrafts are neglected ($|\tilde{M}_i|$ will be too large).

Direct effects of the outflow of cool air on the mixed layer in regions between clouds can also be seen from (4.83). For example, enhancements of the surface fluxes (for a given Δ_i) will promote the growth of the mixed layer. On the other hand, transport of cool air from the rainy areas can retard the mixed-layer growth

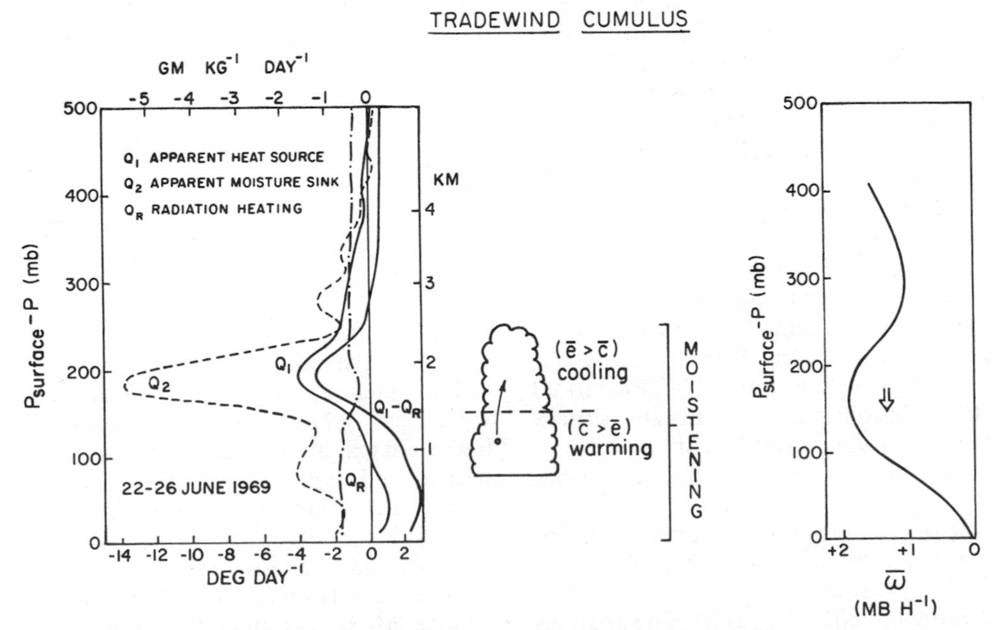

FIG. 4.17. (Left) The observed Q_1, Q_2, Q_R, and $Q_1 - Q_R$ for the undisturbed BOMEX period 22–26 June 1969 (from Nitta and Esbensen 1974). (Center) Schematic of trade-wind cumulus layer showing effects of condensation and evaporation on the heat and moisture budgets. (Right) Mean vertical p velocity $\bar{\omega}$ over budget area.

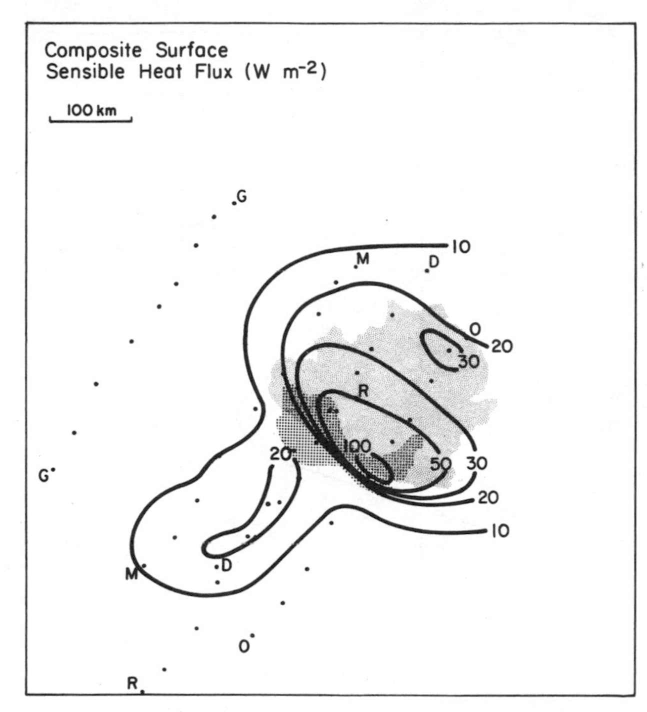

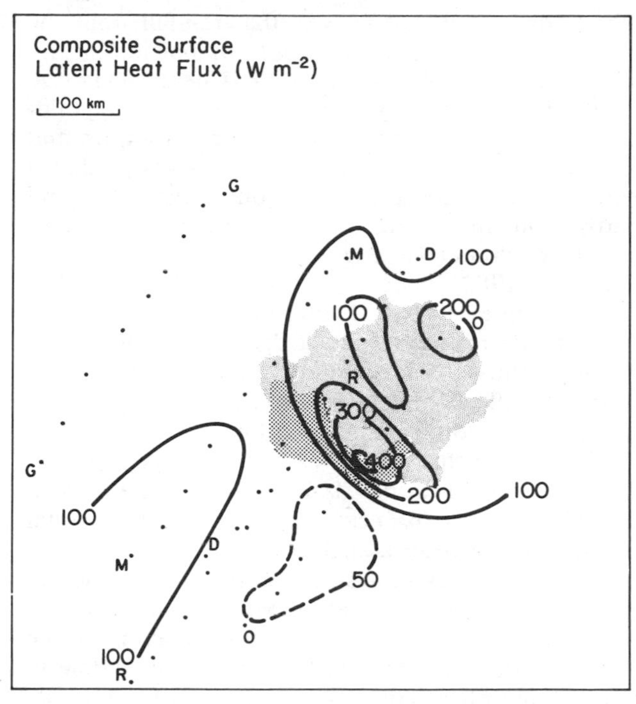

FIG. 4.18. (Left) Composite surface sensible heat flux (W m^{-2}) in region of squall-line system. Dark shading refers to leading convective line, and light shading to trailing stratiform region. Hourly positions of *Gilliss* (G), *Meteor* (M), *Dallas* (D), *Researcher* (R), and *Oceanographer* (O) are indicated by dots. (Right) As in left except for composite surface latent heat flux (from Johnson and Nicholls 1983).

through a strengthening of the inversion Δ_i (Johnson 1981). Of course, the use of (4.83) assumes a well-mixed boundary-layer structure and horizontal uniformity, conditions that may quickly become invalid with the onset of strong, cold outflows. Nevertheless, it is the net effect of all of these processes—subsidence, surface fluxes, and local advection—that determines the actual evolution of the mixed layer.

The existence of downdraft effects on the boundary layer over areas significantly larger than the convective clouds themselves presents a formidable problem for cumulus parameterization. Some treatment of downdraft effects on the boundary layer has been included in the parameterization scheme of Fritsch and Chappell (1980). Modifications of the Arakawa–Schubert scheme to include downdrafts and wake effects have been made by Payne (1981) and Cheng and Arakawa (1990), although results are preliminary and further work is under way.

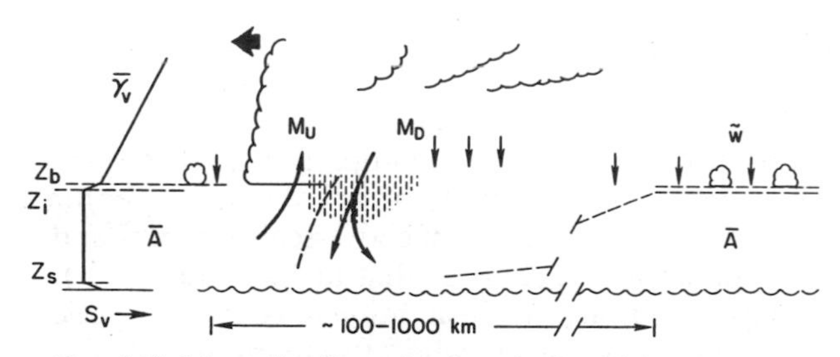

FIG. 4.19. Schematic of boundary layer in the vicinity of mesoscale convective systems. In region $\bar{A}$ away from convection the boundary layer is depicted as well mixed; M_U and M_D denote updraft and downdraft mass fluxes, and $\bar{w}$ environmental subsidence. The structure of the idealized vertical profile of virtual static energy s_v away from convection is determined by z_s (the top of the surface layer), z_i (the top of the mixed layer), z_b (the top of the transition or entrainment zone), and $\bar{\gamma}_v$ (the lapse rate in the free atmosphere) (from Johnson 1981).

4.3. Mesoscale organization

One of the most prominent features of precipitating systems both in the tropics and midlatitudes is their tendency to organize on the mesoscale (e.g., Maddox 1980; Houze and Betts 1981; Houze and Hobbs 1982). The degree and character of mesoscale organization appear to be strongly influenced by the environmental wind shear and stability; however, the specific nature of the environmental controls is not well understood. Details concerning the cloud and precipitation structures of mesoscale convective systems are found in chapter 3. As noted in section 4.1a, if mesoscale organization is important, then the simplest approaches at parameterizing cumulus effects, which assume a spectral gap [see (4.6)] between the cumulus and large scale (Frank 1983; Arakawa and Chen 1987), need modification.

In its simplest terms, a mesoscale convective system can be considered to consist of two components: a convective region and a stratiform region. An important finding from GATE was that rainfall from the stratiform region made up a significant portion (nearly 40%) of the total rainfall for GATE (Cheng and Houze 1979). Similar findings have been reported for midlatitude mesoscale convective systems (Johnson and Hamilton 1988). The stratiform rainfall is produced partly from transport from the convective region and partly from in situ deposition within the mesoscale stratiform cloud (Rutledge and Houze 1987).

Houze (1982) made use of GATE and Winter MONEX (Monsoon Experiment) findings to develop a model to determine the vertical distributions of diabatic heating within tropical cloud clusters. His model includes the effects of condensation, evaporation, melting, and radiation. Heating rates for the Houze (1982) idealized cloud cluster are shown in Figs. 4.20 and 4.21. The convective heating (dashed curve) has a peak in the midtroposphere (near 6 km), whereas the total cluster heating peak is higher, at 8–10 km, with the total heating at low levels being rather small. The upward shift of the total heating peak is a consequence of depositional and/or condensational heating in the stratiform region, which has a peak in the middle to upper troposphere (thin curve, Fig. 4.21), and a net radiative heating in the anvil for this daytime situation. At low levels the reduction in the total heating over that in the convective region is a consequence of cooling by evaporation and melting in the stratiform region (Fig. 4.21). Figure 4.21 also contains computations of the heat source for the stratiform region using rawinsonde data from the Winter MONEX ship network (Johnson and Young 1983). The comparison of the heating rates from the two studies shown in Fig. 4.21 shows good agreement and lends credence to the model advanced by Houze (1982).

The importance of mesoscale circulations within tropical eastern Atlantic convective disturbances can be inferred from the evolution of the heat and moisture budgets during GATE. For example, Nitta (1977), Houze (1982), Esbensen et al. (1988), and Fig. 4.3 of this chapter show that during the growing stages of GATE convective systems, heating is concentrated in the low to midtroposphere. During the mature to decaying stages, the heating shifts to the upper troposphere. This shift is consistent with the idea that in the later stages of the life cycle of convective systems, when stratiform precipitation becomes more prominent, heating profiles typical of those regions (Fig. 4.21) begin to have a greater influence on the total heating distribution.

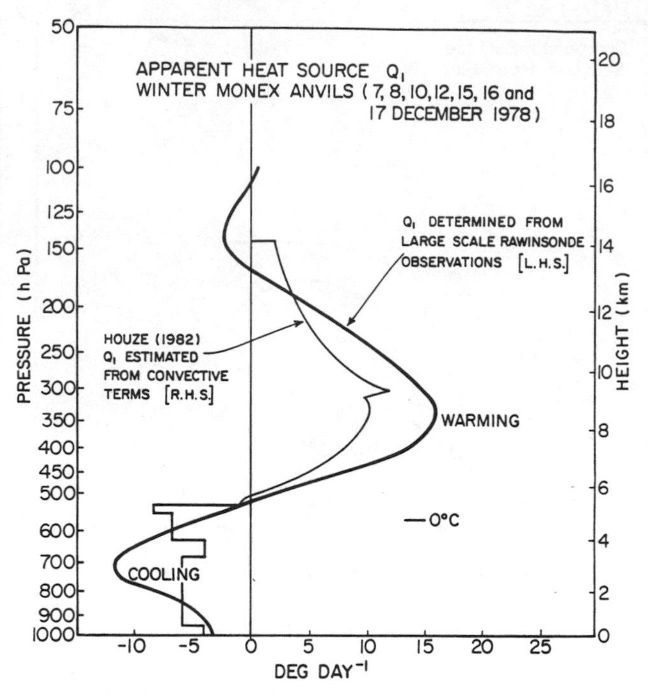

FIG. 4.21. Comparison of Q_1 obtained using Winter MONEX rawinsonde data (heavy curve) with that obtained by Houze (1982) using radar cloud structure and precipitation data (light curve). The lhs (rhs) refers to left-hand side (right-hand side) of Eq. (4.4). The curve from Houze (1982) represents the total heating by *nonradiative* processes (from Johnson and Young 1983).

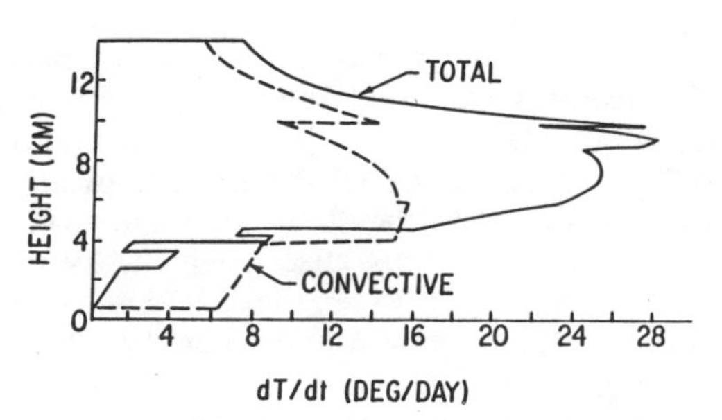

FIG. 4.20. Total heating of an idealized tropical cloud cluster (solid). Heating from convective towers alone is indicated by dashed curve (from Houze 1982).

Several diagnostic studies using GATE data suggest that mass fluxes in mesoscale updrafts and downdrafts are significant relative to convective mass fluxes (e.g., Johnson 1980; Houze and Cheng 1981; Cheng and Yanai 1989). Diagnostic results from the study of Johnson (1980) for conditions in the trough region of a composite GATE easterly wave are shown in Fig. 4.22. A best estimate of mesoscale downdraft intensity was determined based on a comparison of diagnosed environmental subsidence to observed cloud cover for the wave. The model for convective-scale updrafts and downdrafts was the same as that in Johnson (1976). The results indicate that in addition to convective-scale downdrafts, mesoscale downdrafts are important contributors to the total convective mass flux and, hence, significantly affect the diagnosed environmental mass flux. Similar conclusions were drawn by Leary and Houze (1980).

The observed apparent heat sources Q_1 and moisture sinks Q_2 of Yanai et al. (1973) and others (e.g., Fig. 4.1) commonly contain the effects of both convective-

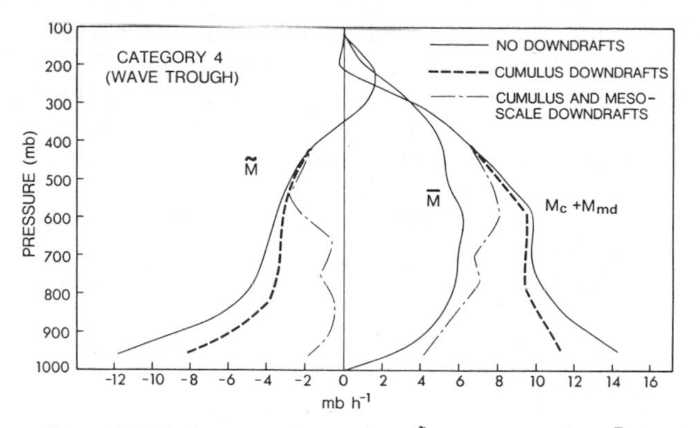

FIG. 4.22. Environmental mass flux $\tilde{M}$, mean mass flux $\bar{M}$, and net convective mass flux $M_c + M_{md}$ for GATE easterly wave trough for cases with and without downdrafts (from Johnson 1980).

scale and mesoscale motions. Johnson (1984) considered a decomposition of the total heating rate into convective-scale and mesoscale components using

$$\hat{Q}_1 = f\hat{Q}_{1m} + (1-f)\hat{Q}_{1c} + \frac{\tilde{Q}_R}{P}, \quad (4.84)$$

where the caret refers to values of normalized by rainfall rate ($\hat{Q}_1 \equiv Q_1/P$), the tilde refers to values in the cloud environment, and f is the fraction of the total rainfall produced by ascent within the mesoscale or stratiform region. It is useful to compare heating profiles from different experiments using rainfall-normalized heating rates $\hat{Q}_1$ since the areas under the heating rate curves (on plots linear in p) should be approximately equal.

The fraction f to be used in (4.84) is ordinarily less than the previously cited 40% reported by Cheng and Houze (1979) since part of the precipitation that actually falls in the stratiform region is produced in cumulus updrafts and is transferred over to the mesoscale anvil by the storm circulation. Using various values of f in the range 0.1–0.3 and assuming a mesoscale stratiform region heating profile given by Johnson and Young (1983) (Fig. 4.21), Johnson (1984) partitioned the Q_1 profile of Yanai et al. (1973) into convective-scale and mesoscale heating contributions (of course, f varies considerably over the lifetime of the system). The results for $f = 0.2$ are shown in Fig. 4.23. Within the context of (4.84), it can be seen that the total heating of Yanai et al. can be considered a consequence of two distinctly different circulation features: 1) the mesoscale anvil, which has an upper-tropospheric heating peak and a lower-tropospheric cooling peak, and 2) the cumulus, which has a peak in the lower to midtroposphere. Although the peak in the convective-scale heating curve [representing $(1-f)\hat{Q}_{1c}$ in (4.84), which is equivalent to Q_{1c} in (4.11)] is in the mid- to lower troposphere (near 600 mb), it is still separated from the diagnosed peak in $(1-f)\hat{Q}_{2c}$ (equivalent to Q_{2c}) near 750 mb (not shown). Therefore, these results indicate the presence of eddy vertical transport of moist static energy associated with cumulus convection, as discussed in connection with Fig. 4.1. The level of the cumulus heating peak is a matter of some uncertainty (Houze 1989); however, recent modeling studies of tropical squall lines (Tao et al. 1991) support the finding of a lower-tropospheric peak in the cumulus heating.

Cheng and Yanai (1989) introduced an objective procedure to diagnose mesoscale precipitation effects from observations of heat and moisture budgets. They define a quantity H by

$$H \equiv Q_2 - \kappa(Q_1 - Q_R), \quad (4.85)$$

where

$$\kappa \equiv -L\left(\frac{\partial\bar{q}}{\partial p}\right)\left(\frac{\partial\bar{s}}{\partial p}\right)^{-1}.$$

If heating and drying in the environment are solely due to subsidence compensating the cumulus mass flux, H will be zero [see (4.43) and (4.44)].

When combined with (4.11), (4.12), (4.73), and (4.74) and neglecting detrainment of liquid water by downdrafts, (4.85) becomes

$$\begin{aligned} H = {} & [Q_{2m} - \kappa Q_{1m}] \\ & - \delta_u[L(q_{Du} - \bar{q} + l_{Du}) - \kappa(s_{Du} - \bar{s} - Ll_{Du})] \\ & - \delta_d[L(q_{Dd} - \bar{q}) - \kappa(s_{Dd} - \bar{s})]. \quad (4.86) \end{aligned}$$

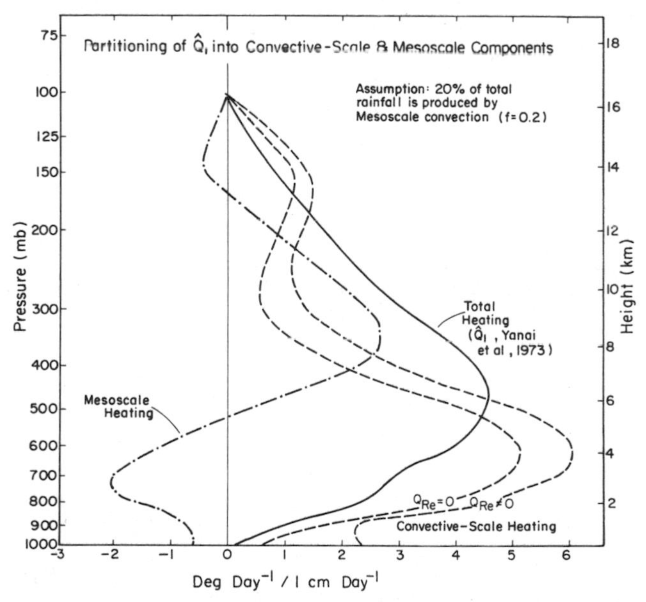

FIG. 4.23. Partitioning of the normalized apparent heat source $\hat{Q}_1$ into convective-scale (cumulus) and mesoscale components for $f = 0.2$; f is the fraction of the total rainfall produced by mesoscale anvils. Curve marked $Q_{Re} \neq 0$ is result for assumed radiative heating profile in the environment of the convective clouds (Q_{Re}) equivalent to that given by Cox and Griffith (1979) for mean phase III GATE conditions (from Johnson 1984).

This expression for H contains mesoscale effects (first term in brackets) and detrainment effects (second and third terms). In general, the detrainment terms are dominated by $-L\delta_u(q_{Du} - \bar{q} + l_{Du})$, which has a negative contribution to H. On the other hand, the mesoscale contributions to H are positive. To see this, note that if $Q_{1m} \approx Q_{2m}$ [from (4.25)], the mesoscale term in (4.86) can be written as $Q_{2m}(\partial\bar{h}/\partial p) \times (\partial\bar{s}/\partial p)^{-1}$, where $\bar{h} = \bar{s} + L\bar{q}$ has been used. In this situation both condensation associated with mesoscale updrafts ($Q_{2m} > 0$) in the upper troposphere, for which $(\partial\bar{h}/\partial p)(\partial\bar{s}/\partial p)^{-1} > 0$, and evaporation associated with mesoscale downdrafts ($Q_{2m} < 0$) in the lower troposphere, for which $(\partial\bar{h}/\partial p)(\partial\bar{s}/\partial p)^{-1} < 0$, contribute to positive values of H.

Values of H for phase III of GATE are shown in Fig. 4.24 (from Cheng and Yanai 1989). During much of the time, H is negative, showing the effects of detrainment of moisture from cumulus clouds. However, positive values are found in the lower and upper troposphere in association with the occurrence of cloud clusters [marked by solid (squall) and open (nonsquall) arrows], indicating the contributions from the mesoscale term. The maximum values of H are in the lower troposphere and are associated with squall clusters, consistent with the fact that squall lines are known to have pronounced mesoscale downdrafts (Houze 1977; Zipser 1977).

In Fig. 4.25 the horizontal distributions of H at 773 mb are shown along with radar echoes for a nonsquall cluster case (case 31), a squall cluster case (case 47), and a case with scattered convection (case 64). In both the nonsquall and squall cluster cases (Figs. 4.25a,b) positive values of H are found in the areas of organized convection. In contrast, the values of H for the case with scattered convection are negative over all of the analyzed domain (Fig. 4.25c), showing the absence of mesoscale effects.

Cheng and Yanai (1989) have further used the cumulus mass fluxes diagnosed from the observed $Q_1 - Q_2 - Q_R$ to reconstruct the cumulus residuals Q_{1c} and Q_{2c} from (4.73) and (4.74). Next, the differences between the reconstructed residuals and observed residuals (with effects of convective-scale downdrafts included) were used to determine Q_{1m}. They showed that Q_{1m} is predominantly characterized by heating in the upper troposphere and cooling in the lower troposphere. In general, their heating profiles are consistent with those determined by Houze (1982) and Johnson and Young (1983) (Fig. 4.21). Cheng and Yanai (1989) further note that the diagnosed mass flux of the cumulus environment $(M_m + \tilde{M})$, which can be regarded as a superposition of mesoscale mass flux and the mass flux in the environment of all cloudy areas, shows upward motion in the upper troposphere and downward motion in the lower troposphere, again consistent with the observed vertical motions in stratiform regions (e.g., Gamache and Houze 1982; Johnson 1982; Johnson et al. 1990).

4.4. Further considerations

As we have seen, the mathematical framework for treating heat and moisture budgets for precipitating systems can be made general enough to include effects of convective-scale updrafts and downdrafts, as well as the stratiform components of mesoscale convective systems. Several complications remain, however. One that has already been discussed is the coupling of the cloud and subcloud layers. Among others of importance are the effects of vertical wind shear and radiation.

It is generally believed that the mesoscale organization of cumulus clouds is a consequence of the interaction between convective-scale motion and vertically sheared environmental flow. Under strong wind shear, convection tends to organize in quasi-linear configurations (i.e., squall lines). In nonprecipitating convection, the tilt of convective elements, which governs the vertical transport of momentum, is determined by the orientation of convective rolls relative to the mean flow (Asai 1970). Attempts have been made to clarify the roles of vertical shear in organizing and aligning precipitating convection (e.g., Seitter and Kuo 1983; Emanuel 1986; Rotunno et al. 1988).

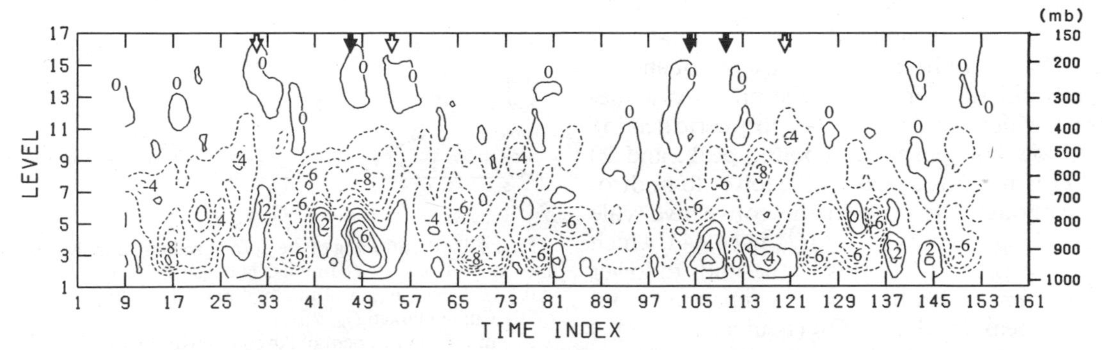

FIG. 4.24. Time–height section of H (K day^{-1}) calculated from observed $Q_1 - Q_R$ and Q_2 during GATE phase III (from Cheng and Yanai 1989).

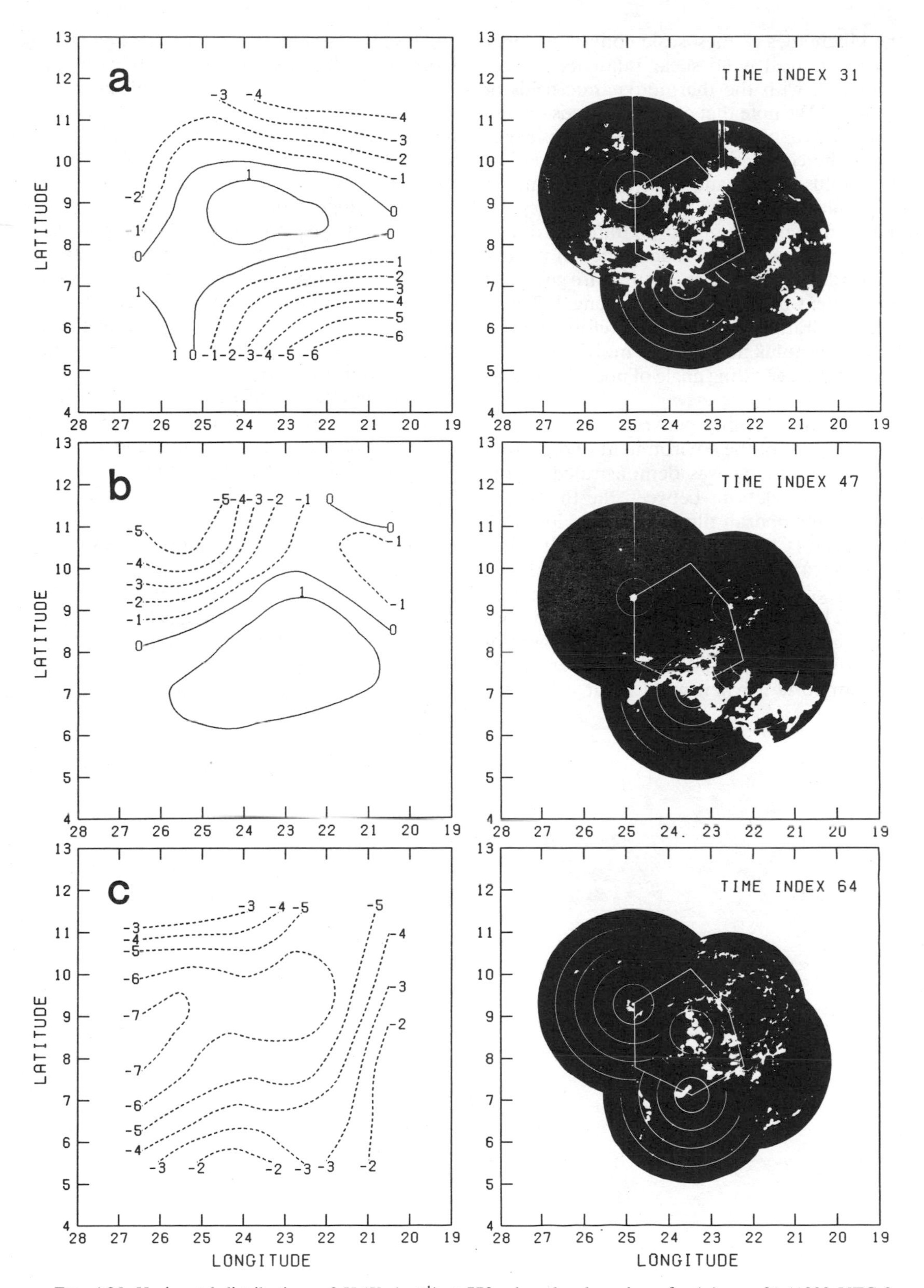

FIG. 4.25. Horizontal distributions of H (K day^{-1}) at 773 mb and radar echoes for (a) case 31 (1800 UTC 2 September 1974), (b) case 47 (1800 UTC 4 September 1974), and (c) case 64 (2100 UTC 6 September 1974) (from Cheng and Yanai 1989).

Does the kinematics of mesoscale convective organization under vertical wind shear influence the cumulus interaction with the thermodynamic fields of the environment? We note that cumulus clouds modify the large-scale environment mainly through subsidence compensating the convective mass flux. The vertical profiles of cumulus heating and moistening depend on the vertical profile of total mass flux and further depend on the relative magnitudes of the updraft and downdraft mass fluxes. The relative magnitudes of the updraft and downdraft mass fluxes, in turn, are governed (within the context of the theory of Cheng 1989a,b) mainly by the updraft tilting angle. Therefore, the vertical profiles of cumulus heating and moistening may be related to the average tilting angle of updrafts (Cheng 1989b).

There is a strong coupling between the wind and thermodynamic fields of the environment of organized convection. This coupling was demonstrated by the remarkable time correlations between the thermodynamically diagnosed updraft tilting angle and low-level vertical wind shear (Cheng and Yanai 1989). It is also noted that the shear production of convective kinetic energy is not negligible under strong shear (Wu and Yanai 1991). These considerations suggest that consistent diagnosis and parameterization of the thermodynamic and dynamic effects of a cumulus ensemble may become important when wind shear effects are considered.

Although simplified radiative heating profiles (e.g., Dopplick 1972; Cox and Griffith 1979) are often used in the diagnosis of cumulus ensemble properties, there is evidence to indicate that actual heating profiles in the vicinity of precipitating systems with extensive stratiform clouds can be very complex (Webster and Stephens 1980; Ackerman et al. 1988). For example, for a deep upper-tropospheric stratiform cloud system having the heating profiles shown in Fig. 4.21, Webster and Stephens (1980) computed a total radiative (longwave plus average daytime shortwave) heating near cloud base of approximately 10 K day^{-1} and cooling near cloud top of 5 K day^{-1}. These heating rates are comparable to those associated with the latent heating in the stratiform cloud systems (Houze 1982). Although there is considerable uncertainty in its magnitude, the shortwave heating can also induce a pronounced diurnal modulation to the total heating (e.g., Randall et al. 1991), thereby complicating the diagnosis of cumulus ensemble properties from large-scale observations.

Acknowledgments. The authors thank Drs. M.-D. Cheng, S. K. Esbensen, T. Nitta, S. A. Rutledge and Xiaoqing Wu for their careful reviews of the manuscript. Michio Yanai also thanks Mrs. Clara Wong for her expert typing of the manuscript.

This research has been supported by the National Science Foundation, Atmospheric Sciences Division, under Grants ATM-9013112 and ATM-9114229.

Chapter 5

The Nature of Adjustment in Cumulus Cloud Fields

CHRISTOPHER S. BRETHERTON

Departments of Atmospheric Science and of Applied Mathematics, University of Washington, Seattle, Washington

5.1. Introduction

By *adjustment* (in the context of cumulus parameterization) we mean the physical mechanisms whereby the internal state of a system changes so as to stay in quasi equilibrium with time-varying forcings. Built into the notion of adjustment is an assumption that the internal adjustment mechanisms act much faster than the external forcings change. Implicit assumptions about adjustment underlie all cumulus parameterizations. Since the considerations addressed in this chapter are quite general, we will not refer to specific parameterizations, except to note that several of the parameterizations discussed in this monograph address some of the issues raised in the last section of this chapter regarding horizontal inhomogeneity due to noninstantaneous adjustment within a grid area.

The nature of adjustment of a field of cumulus clouds and the unsaturated environment with which the clouds interact is subtle. Buoyant energy generation and turbulent mixing occur only in a small fraction of the volume at any one time (in and near convective clouds) but drive secondary compensative circulations everywhere. The direct and compensative effects of the cumuli produce different types of adjustment. In this chapter, we use some simple examples to illustrate the adjustment produced by the direct and compensative effects of cumulus clouds, with emphasis on the time scales for all of the air within a given area to adjust to a new cumulus cloud somewhere within that area. For cumulus parameterization, we invariably assume that the direct adjustment occurs on a horizontal scale smaller than that explicitly resolved. Depending on the horizontal resolution of the model, some broader-scale compensative motions may be explicitly resolved.

a. Adjustment in an unsaturated convecting layer

In a "dry" (or more generally, an unsaturated) convecting layer of air, the adjustment process, buoyancy-driven *turbulent mixing,* is relatively simple. If the combination of "large-scale" forcings (such as surface heat fluxes, internal radiative cooling, and differential temperature advection) would in the absence of convection increase the lapse rate to be somewhere steeper than dry adiabatic, convection starts in this layer. In the resulting circulations, potentially denser air parcels turbulently mix with and exchange positions with potentially lighter air parcels, reducing the lapse rate within the convecting layer toward adiabatic. Updrafts and downdrafts of comparable strength are organized in a cellular structure that fills the entire convecting layer. A similar type of convection can occur in an entirely saturated layer or a layer in which both updrafts and downdrafts have nearly the same saturation pressure (as is often observed in shallow stratocumulus cloud layers).

What is the time constant for adjustment by turbulent mixing? Thermally direct circulations will remove the unstable lapse rate in the *turnover time* it takes the dense air to move from the top to the bottom of the convecting layer. If the characteristic vertical velocity of the circulations is w_c and the thickness of the convection layer is H, the turnover time τ_c is H/w_c. Now suppose that at some time a horizontal sheet of dye is introduced at some height within the convecting layer. It will be rapidly distorted by the updrafts and downdrafts, and within a small number of turnover times the dye will be distributed (albeit somewhat patchily) throughout the volume of the convecting layer due to distortion and homogenization by eddies of all sizes. The dye-mixing time and the adjustment time are both comparable to the turnover time when the adjustment mechanism is turbulent mixing.

For most atmospheric convection in situations where condensation is absent (with the exception of the superadiabatic layer that forms in the lowest few meters above hot ground), the turnover time is a few minutes, and in this time in the absence of convection, the temperature changes due to larger-scale forcing processes would have been small (fractions of a kelvin). Therefore, the convecting layer maintains a nearly dry-adiabatic lapse rate. The water vapor mixing ratio is conserved in adiabatic displacements of unsaturated air, but water vapor also has forcings due to surface fluxes and differential advection with height; in many circumstances the tendency due to these forcings in the turnover time is also negligibly small, so the water vapor mixing ratio becomes "well mixed" or nearly uniform with height in an unsaturated convecting layer.

b. Adjustment in a cumulus layer

In a conditionally unstable layer undergoing cumulus convection, the adjustment process is considerably more complex, since latent heating (i.e., cloud) is required to generate buoyancy fluxes by vertical motions in the ambient stratification. Turbulent mixing is restricted to clouds and their immediate vicinity—we call such regions (which evolve rapidly in time) *cloud mixing regions.* In the stably stratified regions between clouds, turbulence is suppressed, but adjustment to the heating and motions within the clouds still occurs. The response can be viewed as a superposition of internal gravity waves propagating away from the clouds whose net effect is to induce compensating subsidence that balances the upward mass fluxes within the cloud (Bretherton and Smolarkiewicz 1989). The resulting adiabatic warming is spread over the entire atmosphere, even though the latent heating within the clouds is quite localized. Section 2 will present some examples of the paradigm of adjustment here, which is the transient response of a stably stratified atmosphere to a localized heat source. This form of adjustment, which we call *compensative adjustment,* does not produce effective mixing of an initially horizontal dye sheet except within clouds.

There is also direct exchange of air between clouds and their environment by three processes—*entrainment* of environmental air by an active updraft or downdraft of a cloud, *detrainment* of air out of the turbulent cloud-induced circulation (generally near its level of neutral buoyancy), and *decay* of clouds in which there is no longer an active updraft. None of these processes is completely well defined, because there is no unique way to define a boundary of the cloud mixing region. Furthermore, new clouds may develop in different locations than existing clouds. Even though turbulent mixing influences only a small part of the layer at any time, much or all air in the layer will ultimately be incorporated into some turbulent cloud-induced circulation if convection persists.

Evaporation of precipitation also affects the adjustment process. For example, evaporation of precipitation formed by the stratiform trailing anvil of a mesoscale convective system helps cool the subsiding rear inflow jet below (Houze and Betts 1982). The entire region affected by precipitation is best conceptually treated as part of the cloud mixing region because of the cloud source of water. Furthermore, evaporation of precipitation allows downdrafts from within the cloud layer to penetrate below cloud base and rapidly alter the subcloud layer.

In this chapter we shall concentrate on adjustment within the cumulus layer. However, the parallel adjustment of the subcloud layer to changes in large-scale forcing is equally vital to understand, and perhaps more so for understanding the initiation of new convection. The formation of cold pools within MCSs (mesoscale convective systems) is vital to sustaining the mesoscale organization of an MCS (Houze and Betts 1982), and the later modification of the boundary layer by surface fluxes, entrainment from above, horizontal advection, and convergence is a critical part of the transient response of the atmosphere to both deep and shallow convection.

c. A guide to what follows

The remainder of the paper is organized as follows. In section 5.2 we will use a simple model problem, the response of a uniformly stratified fluid layer to a time-dependent internal heat source, to illustrate how the length scales and time scales of compensative adjustment depend on cloud lifetime, the earth's rotation, and dissipative processes. In section 5.3, we will look at a two examples from the recent literature that compare and extend this idealized view of compensative adjustment to realistic convection. The second of these studies also addresses the adjustment due to the direct effect of cumulus clouds. Section 5.4 briefly discusses the ramifications of subgrid-scale adjustment processes for cumulus parameterizations.

5.2. The response to a heat source in a stably stratified fluid layer

a. Slab-symmetric case

The following analysis is taken from Bretherton and Smolarkiewicz (1989, hereafter BS). Consider a motionless horizontally unbounded Boussinesq fluid of depth D and uniform buoyancy frequency N. At time $t = 0$ a slab-symmetric "cloud" grows at $x = 0$. The cloud is idealized to be an infinitely narrow steady sinusoidal buoyancy source for $t > 0$: $Q(x, z, t) = Q_{0lm}\delta(x)\sin(mz)H(t)$. Here $\delta(y)$ is the Dirac delta function, $H(y)$ is the unit step function, and m is an integral multiple of π/H, so that the heating vanishes at the layer base and top. Note that Q_{0lm} has units of (buoyancy)(length)/(time) [$m^2\ s^{-3}$]; the l denotes a line source in the horizontal, and the m refers to the vertical wavenumber. The response to a nonsinusoidal or finite width source can be built up by superposition.

We make the hydrostatic approximation because it simplifies the analysis. The hydrostatic approximation accurately models broader horizontal wavelengths that are of most importance for the adjustment of the atmosphere far from the cloud but distorts short wavelength components of the response with horizontal wavenumber k comparable to or larger than m.

The linear hydrostatic equation for the buoyancy field $b(x, z, t)$ is (Raymond 1983)

$$b_{zztt} + N^2 b_{xx} = Q_{zzt}(x, z, t). \quad (5.1a)$$

Subscripts x, z, and t denote partial derivatives with respect to x, z, and t. Since the atmosphere is initially at rest, $b(x, z, 0) = 0$. By separating out the sinusoidal z dependence, (5.1a) can be reduced to a forced one-dimensional wave equation,

$$b_{tt} - c_m^2 b_{xx} = Q_{0lm}\delta(x)\delta(t), \quad (5.1b)$$

which has the solution

$$b(x, z, t) = b_0 \sin(mz) H(c_m t - |x|), \quad (5.2)$$

where $c_m = N/m$ and $b_0 = Q_{0lm}/c_m$. There is a discontinuous jump in buoyancy from the undisturbed atmosphere ($|x| > c_m t$) to the "adjusted" atmosphere ($|x| < c_m t$). This spreading disturbance can be interpreted as a superposition of gravity waves due to the heating. The discontinuity does not disperse, because in the hydrostatic approximation, gravity waves of all horizontal wavenumbers with vertical wavenumber m have the same horizontal phase and group speed c_m. For a source that is not sinusoidal and excites gravity waves with a variety of vertical wavelengths, the transition to an adjusted state occurs gradually in several steps as gravity waves of successively shorter vertical wavelength arrive at a given position. Nicholls et al.'s (1991a, hereafter NPC) Fig. 5 illustrates this phenomenon nicely. By computing the pressure field from hydrostatic balance and using the linearized horizontal momentum and Boussinesq continuity equations, one can deduce the velocity fields associated with the adjustment

$$u(x, z, t) = -\left(\frac{b_0}{N}\right) \cos(mz) H(c_m t - |x|)\, \mathrm{sgn}(x), \quad (5.3a)$$

$$w(x, z, t) = -\left(\frac{b_0}{mN}\right) \sin(mz) \times [\delta(c_m t - x) + \delta(c_m t + x) - 2\delta(x)]. \quad (5.3b)$$

Here sgn(y) is the sign function, 1 for $y > 0$, 0 for $y = 0$, and -1 for $y < 0$.

Figure 5.1 graphically depicts the adjustment process. The transition lines $|x| = c_m t$ are shown as thin vertical solid lines in Fig. 5.1. The arrows show the horizontal velocity field within the adjusted region and the infinitely strong vertical velocities at its edges. The dashed lines are dye lines that are initially uniformly spaced in the vertical. Inside the adjusted region, air is displaced by a height $\Delta z = -b_0 \sin(mz)/N^2$. At levels where the cloud heating is decreasing with height, there is flow away from the cloud in the "adjusted" region and the dye lines are moved apart. For a horizontally distributed source of width a, the discontinuity is smoothed over a width a.

For a transient heat source, the solution is obtained by superposition. For instance, for heating that is turned on, maintained, and turned off a time T later, the wave of adjustment to the heat source is followed a time T later by an "antiwave" that adjusts the fluid back to an unperturbed state. Thus, the transient heating leaves no permanent trace near the heat source, producing only a pair of outward-moving "adjustment ripples" of width $c_m T$.

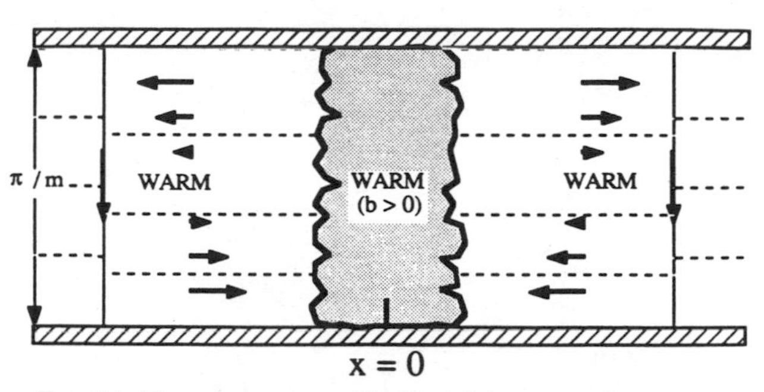

FIG. 5.1. The response to an idealized slab-symmetric narrow sinusoidal heat source in a uniformly stratified fluid layer. The arrows indicate velocities, and the dashed lines are dye lines, which are equally spaced in the undisturbed flow (adapted from Fig. 6 of BS).

b. *Cylindrically symmetric heat source*

The adjustment process in three dimensions (for a narrow, cylindrical, vertically sinusoidal heat source) is less clean. Consider the "point" buoyancy source $Q_{0pm}\delta(x)\delta(y)\sin(mz)H(t)$. Note that Q_{0pm} has dimensions of (buoyancy)(area)/(time) [m^3 s^{-3}]. As shown by BS, if $u(r, z, t)$ is the radial outward velocity at a radius r from the source, then the response, shown in Fig. 5.2, is

$$b(r, z, t) = \frac{Q_{0pm} H(c_m t - r)}{2\pi c_m^2 t[1 - (r/c_m t)^2]^{1/2}} \sin(mz), \quad (5.4a)$$

$$u(r, z, t) = -\frac{m Q_{0pm}}{2\pi r N^2 [1 - (r/c_m t)^2]^{1/2}} \cos(mz). \quad (5.4b)$$

The response remains geometrically self-similar but stretches proportional to $c_m t$. The buoyancy perturbation near the heat source actually *decreases* as t^{-1} because the cumulative heat added to the fluid is proportional to $Q_{0pm}t$ while the area affected by the response is proportional to $(c_m t)^2$. On the other hand, the outflow velocity increases near the heat source like r^{-1} to maintain a finite outward mass flux at small radii. The singular behavior at the leading edge of the wave of adjustment is smoothed out for a source of finite width.

c. *Ensemble of heat sources*

If there are other heat sources (clouds) in the area, then the subsidence due to all of the sources will add

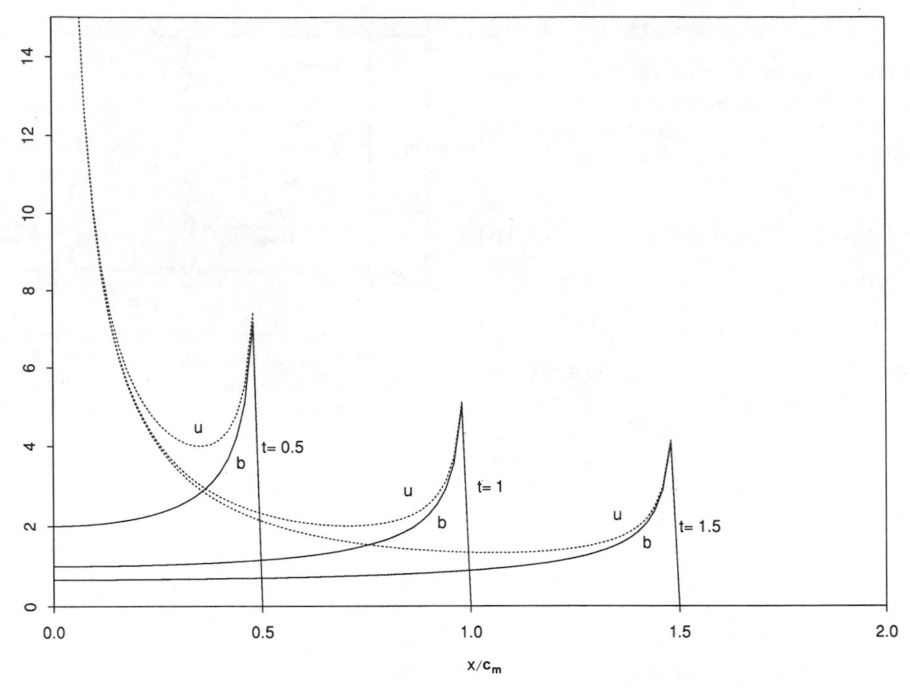

FIG. 5.2. Response to a cylindrical symmetric heat source. The buoyancy at midheight and the horizontal velocity field at the top of the domain.

to produce a fairly uniform field of subsidence once $c_m t$ is larger than the typical intercloud spacing. Figure 5.3 shows the buoyancy field produced by a doubly periodic array of sinusoidal heat sources of spacing L and finite width. It was numerically computed over one wavelength using an FFT (fast Fourier transform) to invert the Fourier transform of the extension of (5.1b) to two horizontal dimensions. At $t = 0.25\ L/c_m$ the cylindrically symmetric response of Fig. 5.2 is seen, but in time, the waves of adjustment from an increasing number of neighboring sources propagate into the domain and interfere, and the buoyancy perturbations become increasingly horizontally uniform with time.

d. Vertical propagation

So far, vertical propagation of gravity waves has been excluded by a rigid upper boundary. To examine the impact of vertical propagation in the slab-symmetric case, NPC compared the response to the localized heat source $\{Q_{0lm}a[\pi(a^2 + x^2)]^{-1}\}\ \sin(mz)H(t)$ for $z < D$ and $m = \pi/D$ with a rigid lid and with a semi-infinite domain. They obtained the solution to the semi-infinite problem in terms of an inverse Laplace transform that must be found numerically. Figure 5.4 shows the velocity and buoyancy field for the two cases, which are remarkably similar within the layer of heating. The rigid-lid fields are from NPC; the semi-infinite results are from Pandya et al. (1993), who pointed out a numerical error in the calculations for NPC's comparable Fig. 6. The upper boundary has little impact on the wave of adjustment because the adjustment is mainly a superposition of waves whose horizontal wavelength is much larger than D and that do not propagate rapidly in the vertical.

e. Dissipation

As discussed by Bretherton (1987), the effect of dissipation is to damp the wave of adjustment as it propagates away from the cloud, localizing the effect of the heating. This effect is simply illustrated by adding a Rayleigh damping with time constant τ to all perturbation fields and redoing the calculation of section 5.2a. The horizontal structure of the buoyancy field is governed by a slight modification of (5.1b):

$$(\partial_t + \tau^{-1})^2 b - c_m^2 b_{xx} = Q_{0lm}\delta(x)(\partial_t + \tau^{-1})H(t). \quad (5.5)$$

The effect of the damping is to introduce an exponential decay of the spreading adjustment wave with time constant τ and to leave in its wake a steady response that exponentially decreases with distance away from the source, with a characteristic subsidence length scale $L_{\mathrm{diss}} = c_m\tau$ equal to the wave group velocity for vertical wavenumber m multiplied by the damping time (Bretherton 1987):

$$b(x, z, t) = b_0 \sin(mz) \exp\left(\frac{-t}{\tau}\right) H(c_m t - |x|), \quad (5.6a)$$

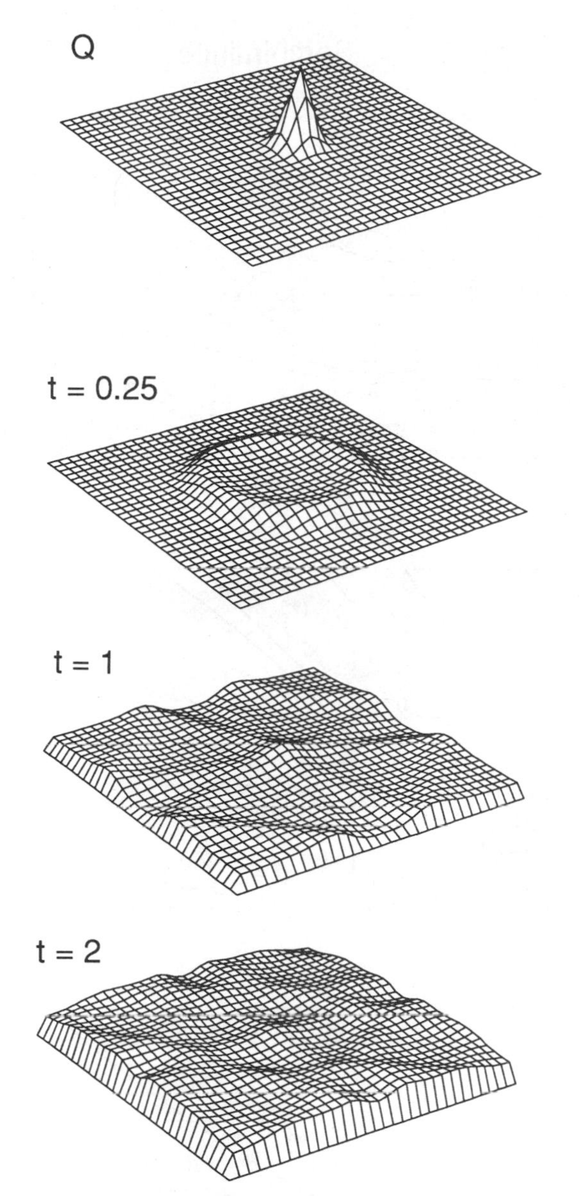

FIG. 5.3. Horizontal section at midheight of the buoyancy perturbation at three times due to a periodic array of heat sources turned on at $t = 0$. The first panel shows the Gaussian heat source; the times are in units of L/c_m.

$$u(x, z, t) = -\left(\frac{b_0}{N}\right) \cos(mz) \exp\left(\frac{-t}{\tau}\right) \times H(c_m t - |x|) \operatorname{sgn}(x). \quad (5.6b)$$

The response is compared to the undamped case in Fig. 5.5. In the damped case both the horizontal velocity and the buoyancy perturbations are everywhere weaker than for the undamped case. If only velocity perturbations were damped, however, the buoyancy anomaly due to the maintained heating would be trapped and would increase within this distance of the heat source.

f. Rotation

For heat sources that are sustained for a significant fraction of a pendulum day, Coriolis effects become important. If the calculation of section 2a for the response to a slab-symmetric heat source is repeated on an f plane, the solutions are more complicated and must be expressed as Fourier integrals, which are numerically evaluated using FFTs. The midplane buoyancy and both components of the surface horizontal velocity are shown in Fig. 5.6. The nondispersive nature of short-wavelength gravity waves, which are unaffected by rotation, maintains the initial adjustment discontinuity, but further behind, the solution starts to approach a geostrophic balance between the growing v velocity and the pressure, which is proportional in amplitude to the buoyancy b. Both fields grow linearly with time and decay away from the source with a characteristic length equal to the Rossby radius $L_{rot} = c_m/f = ND/\pi f$. The residual u also decays exponentially away from the source but asymptotes to a steady value. Hence, rotation, like dissipation, traps the response within the distance traveled by the outgoing wave of adjustment in the relevant characteristic time (f^{-1} for rotation).

Chapter 7 of Gill (1982) discusses an analogous problem involving shallow-water equations for geostrophic adjustment after a dam break. There is a deep analogy between the shallow-water equations and the horizontal structure equation for sinusoidal heating in a bounded stratified layer that has been exploited extensively in tropical meteorology—for instance, by Gill (1981).

g. Mean flow and shear

Transient heat sources in mean flow and shear can produce quite subtle and complex effects, which we will not treat here except for some brief remarks. On the whole the length scales and time scales of the gravity wave response are not greatly affected except by "strong" shear or mean flow, by which we mean flow through the heat source or shear on the order of ND/π over the depth D of heating, because this magnitude of mean flow is required to substantially modify the propagation velocity of the long gravity waves that do the adjustment (Lin 1987). However, the structure of the response within the region, and particularly near the heat source, can be quite different for different velocity profiles relative to the heating.

Uniform mean flow through a stationary heat source sets up a standing "mountain wave"–like response in and above a slab-symmetric heated region. If the heating is steady and there is no upper boundary, the wave amplitude as measured by buoyancy perturbations or

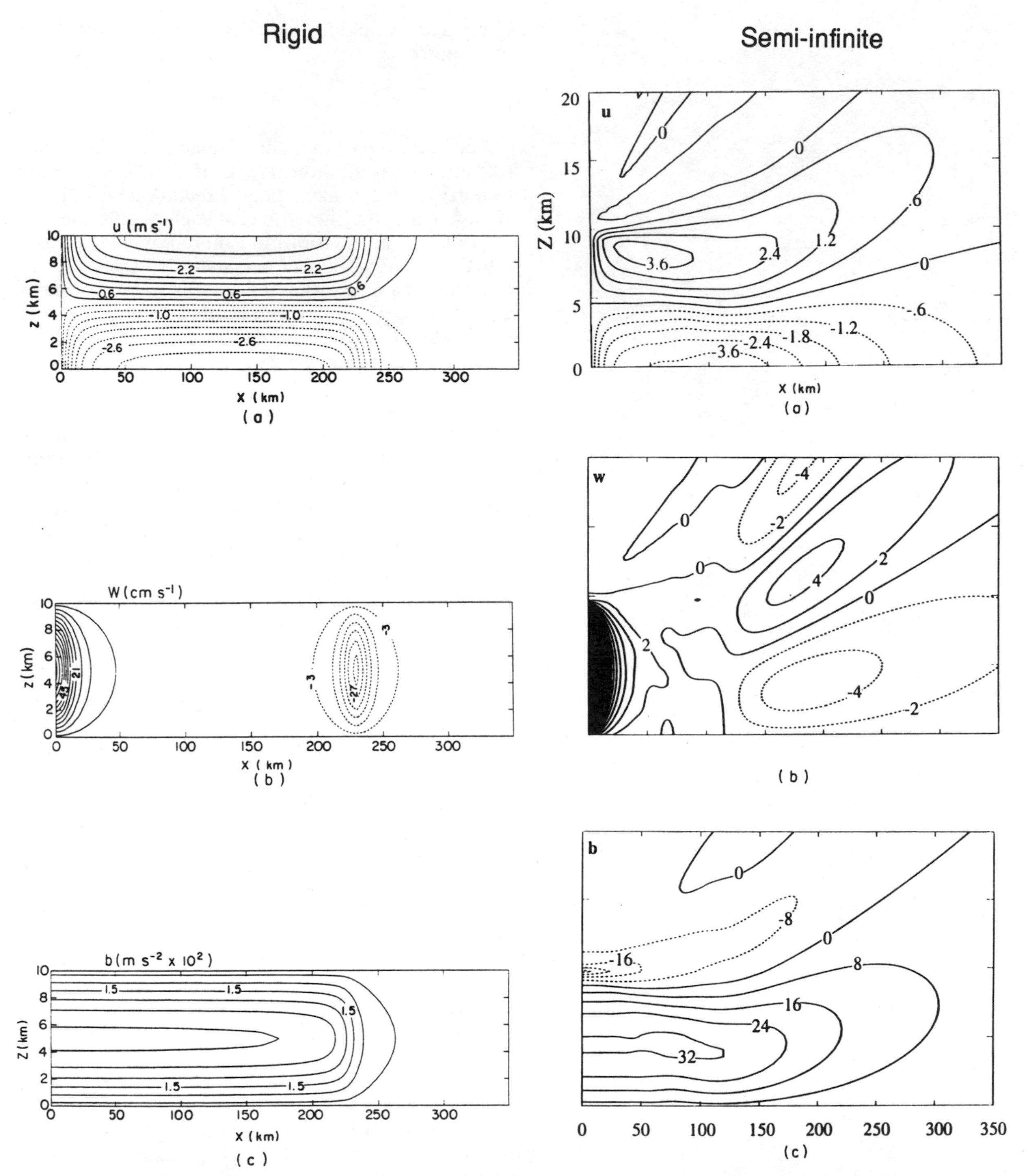

FIG. 5.4. Comparison of vertical velocity and buoyancy fields induced by a heat source for a domain with a rigid lid and a semi-infinite domain. Adapted from Fig. 3 of NPC and Fig. 1 of Pandya et al. 1993. Here the heating aspect ratio is $a/D = 1$ and the nondimensional time is $Nt = 36$.

horizontal velocities grows logarithmically with time until either dissipation, Coriolis, or non-slab-symmetric effects become important (Smith and Lin 1982; Lin and Smith 1986; Bretherton 1988). In a stratified layer of finite depth D, a strong resonant response also sets up if the mean flow is near one of the free wave propagation speeds c_m, where m is an integer multiple of π/D. Lin (1986) analyzed cylindrical heat sources in a uniform mean flow. He found that spreading V-shaped wakes in the vertical displacement field develop

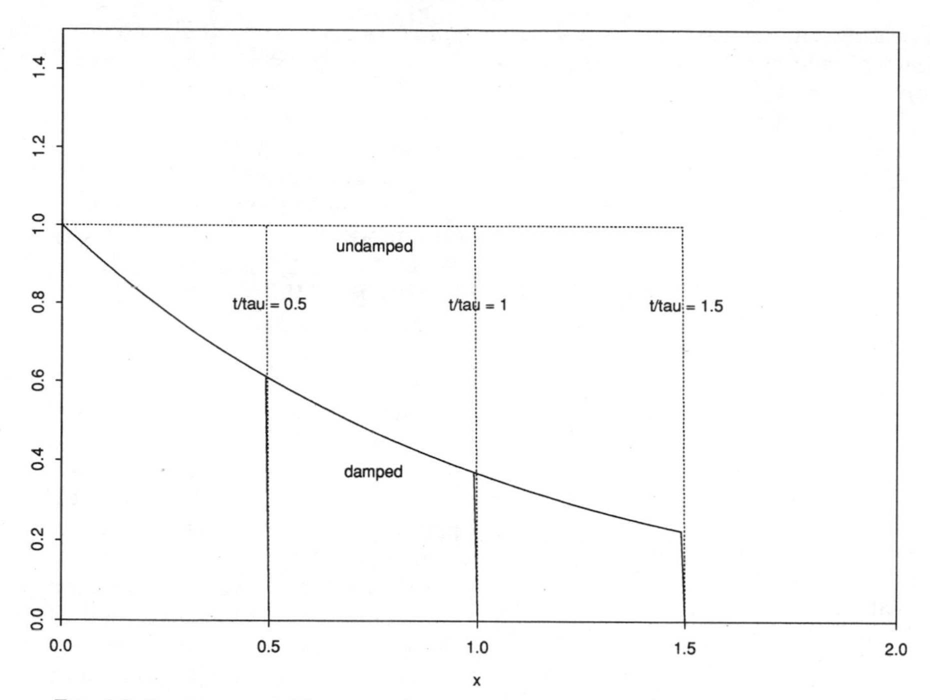

FIG. 5.5. Response to a slab-symmetric heat source in the presence of damping. The solid curves show the buoyancy at midheight (which is the same as the horizontal velocity field at $z = 0$) at times $t = \tau$ and 2τ. The dashed curves show the undamped response. The horizontal unit of length is $c_m\tau$.

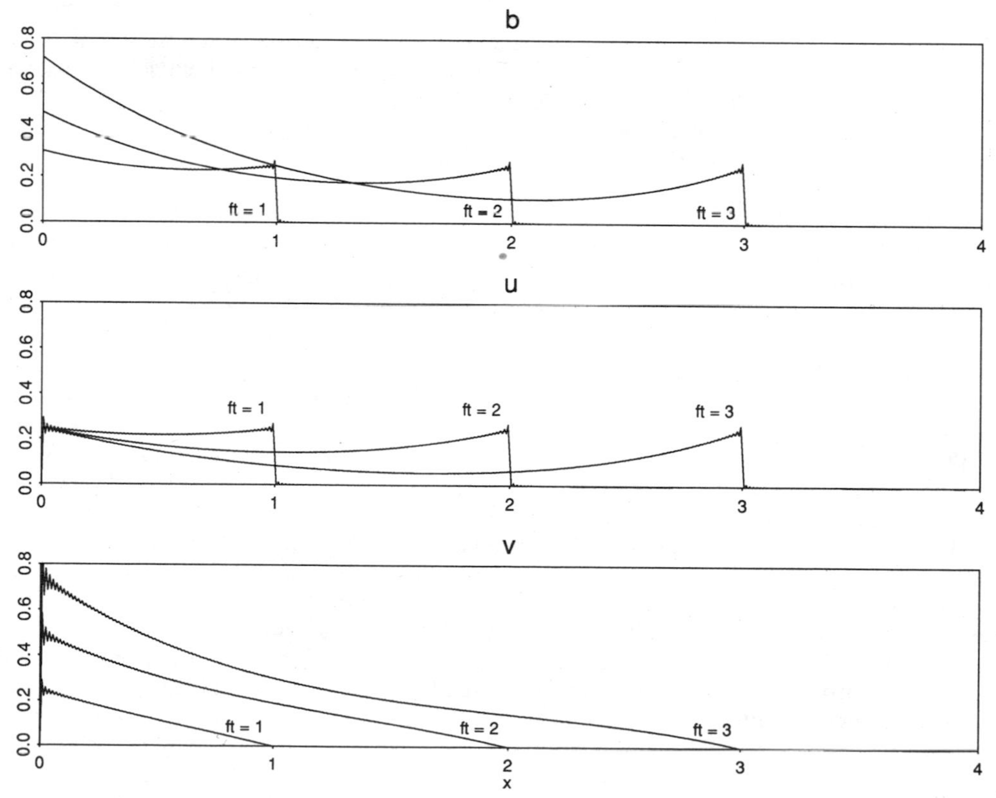

FIG. 5.6. Response to a slab-symmetric heat source on an f plane. Buoyancy b and horizontal velocity components u, v at times $t = 1, 2, 3\,f^{-1}$. The horizontal unit of length is c_m/f.

after the onset of heating, and related these to V-shaped regions of low cloud-top temperature often seen in satellite views of thunderstorms.

Lin (1987) and Lin and Li (1988) also examined two- and three-dimensional heat sources in uniform vertical shear. Their results show that the effects of a sheared mean flow at a given level are qualitatively similar to the effect at that level of the same mean flow, but vertically extended without shear, as long as the shear is not "strong" as defined at the beginning of this section.

5.3. Some applications

In this section, we will consider two numerical modeling studies that illustrate the space and time scales of cumulus-driven adjustment processes. The first study is by NPC, who considered the mesoscale effect of diurnal convection on the Florida peninsula. This study describes the response to an isolated convective heat source (as discussed in section 2) in a less idealized context. A second study, by BS, considers adjustment and mixing processes around an isolated cumulus cloud. It not only illustrates the compensative adjustment processes that force the environment toward a moist-adiabatic buoyancy profile, but also addresses the direct adjustment process, detrainment of air from the cloud itself, that is crucial to adjustment of the moisture fields by the convection.

a. Mesoscale perturbations driven by Florida sea-breeze convection

Heating over the Florida peninsula drives sea-breeze circulations on both sides of the peninsula, which move inland. In the afternoon, deep convection develops along the sea-breeze fronts, peaking in late afternoon when the two fronts converge. The resulting convective heating can be thought of as a pulse with an approximately hour-long maximum intensity, extending along the length of the interior of the peninsula. On some days the situation approaches the ideal case of a transient slab-symmetric localized heat source in no mean flow.

Using a two-dimensional numerical model, Nicholls et al. (1991b) performed simulations of the diurnal cycle using a broad enough domain to capture a large-amplitude outward propagating wave of adjustment to the convection. Unfortunately, corroborative offshore observations were not available. The results were interpreted using the ideas of section 5.2a (NPC). NPC found that the response is qualitatively quite similar to the adjustment wave produced by an hour-long pulse of vertically sinusoidal heating. In terms of section 5.2a, the idealized heating induces an adjustment wave produced by a positive step function when the heating turns on, followed by an adjustment wave of opposite sign due to the canceling negative step function of heating an hour later when the convection decays.

Figure 5.7 (from NPC) shows several fields about 2.5 h after the peak of convection. The strongest perturbations due to convection occur for x in the range 190–250 km. Within this region, the fields all have the sign and vertical structure predicted by the adjustment wave solution given in (5.2) and (5.3). Rough calculations using the known convective heating rate and a 60-km width of the heating region in the idealized model agree (within a factor of 2) with the simulated amplitudes of the perturbations. The horizontal velocity perturbations of 7 m s^{-1}, pressure perturbations of 2 mb, and temperature perturbations of over 2 K are sizable.

Both observational (Hoxit et al. 1976) and numerical (Nicholls 1987) work has shown that ahead of squall lines there is also an adjustment ahead of the line, including a mesolow that is thought to be due to compensating subsidence. However, the use of prescribed heat sources to model the overall dynamics of squall lines has not yet proved as illuminating as in the Florida convection case, partly because the heat source is not localized to the leading convective line, but also must include significant anvil heating and low-level cooling distributed through the stratified region. Without the anvil heating, the velocity, buoyancy, and pressure perturbations would be symmetric about the leading edge of the squall line if the squall line were moving at the same speed as the midlevel flow. Instead, all perturbations are much larger behind the leading edge. Future research should aim to better theoretically elucidate the connection of the mesoscale dynamics of a squall line to its internal heat source distribution.

b. Interaction of compensative adjustment with cloud structure

To this point, we have treated convection as a known heat source that is independent of the adjustment response. However, motions in and near the cloud must take up or supply the mass necessary to produce the horizontal motions demanded by the compensative adjustment process at each level. This takes place through entrainment and detrainment of air from the cloud mixing region. A numerical experiment done by BS illustrates this point. It is difficult to compare details of numerical experiments with observations of individual clouds because measurements of horizontal mass fluxes around evolving clouds are rather uncertain. Studies of mass fluxes in New Mexico cumulus and cumulonimbi using aircraft (Raymond and Wilkening 1982) and radar (Raymond et al. 1991) and in Hawaii trade-wind cumuli (Raga et al. 1990) do broadly support our conclusions.

Bretherton and Smolarkiewicz considered a slab-symmetric cumulus cloud (in which precipitation was artificially suppressed to facilitate conserved variable analysis) growing in a conditionally unstable sounding (Fig. 5.8a) with a stable layer about 2.5 km above cloud

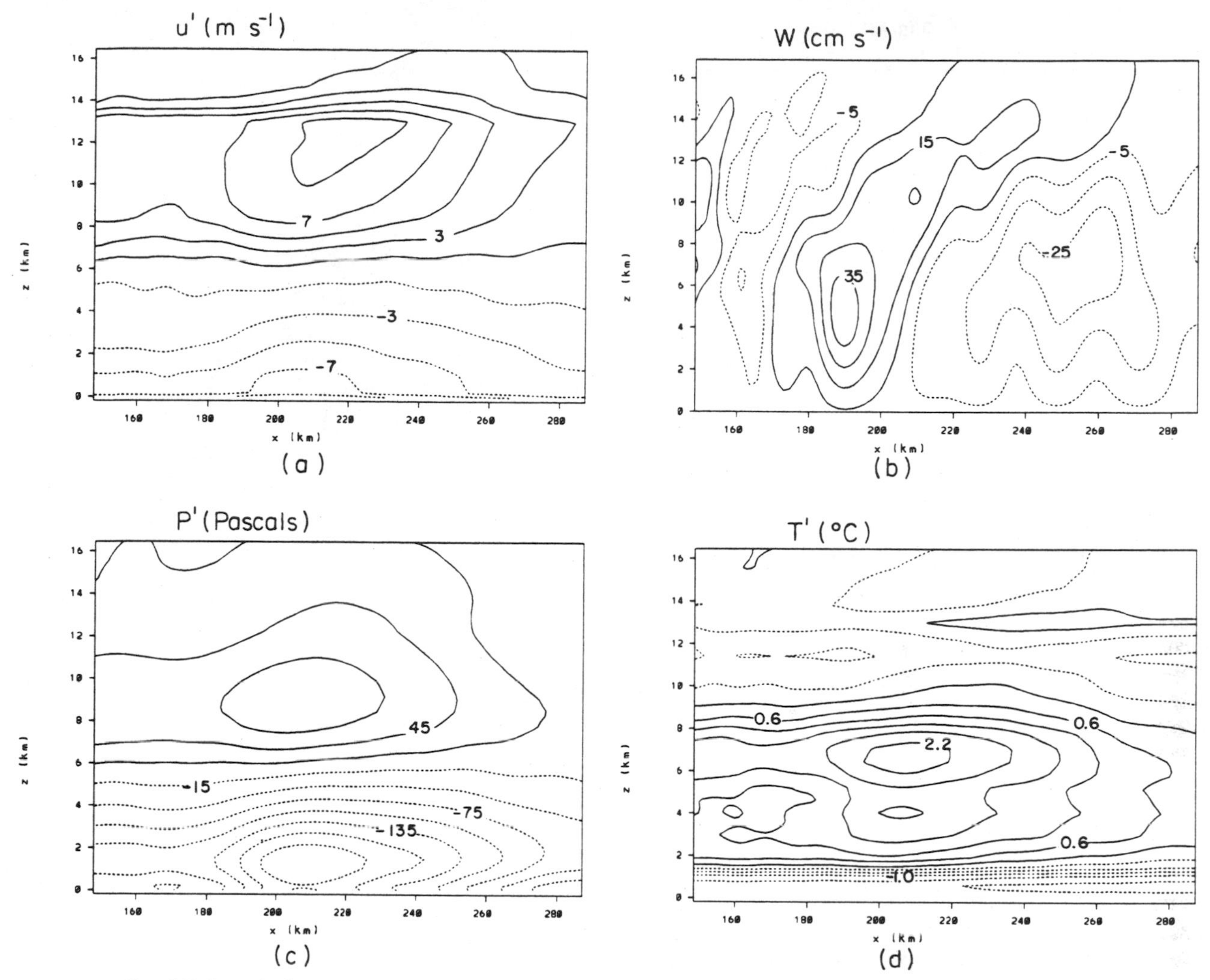

FIG. 5.7. Perturbations created by gravity wave produced in the NPC numerical simulation of Florida convection about 2.5 h after the maximum of convective heating (from NPC, Fig. 11).

base through which the cloud penetrated. The cloud was initiated by a strong localized surface heat flux perturbation that was removed after 4 min. A mean heat flux was maintained throughout the run. This had little effect on the cloud but generated a convective boundary layer in the lowest kilometer. Although the air near the ground was highly conditionally unstable, the cloud base and buoyancy of air within the cloud were better correlated over most of the cloud lifetime with a moist adiabat (dotted line in Fig. 5.8a) derived from the environmental properties at 1.8 km just below the mean cloud base. Over the period from 30 to 90 min, the maximum updraft strength did not show any significant trends, and the cloud appeared to be acting as a nearly steady heat source.

Figure 5.8b shows the cloud at 96 min into the simulation and the associated horizontal velocity and buoyancy fields. There is a nearly stagnant upper cloud mass produced primarily in the early stages of the cloud evolution. There are also hints of a cloud shelf or detrainment zone at 4 km. At this time small cumuli are just starting to grow in the heated boundary layer. In Fig. 5.8c, strong outflow of up to 3 m s^{-1} from the cloud is seen between heights of 4 and 5 km. Although not visible on this figure, the outflow becomes much weaker at distances of 20 km and more from the center of the cloud. In Fig. 5.8d, the buoyancy field from the same time shows that the region less than 20 km from the cloud (including the cloud itself) is close to horizontally uniformly buoyant below cloud top at 9 km. An extended region of uniform buoyancy and of associated horizontal motions is exactly the compensating adjustment that we would expect from section 5.2a. The net effect of the cloud circulations is to increase the temperature at heights between 2 and 4 km, decrease it slightly between 4 and 6 km, and increase it slightly between 6 km and the cloud top. The long-dashed initially vertical dye lines in Fig. 5.8d bow out

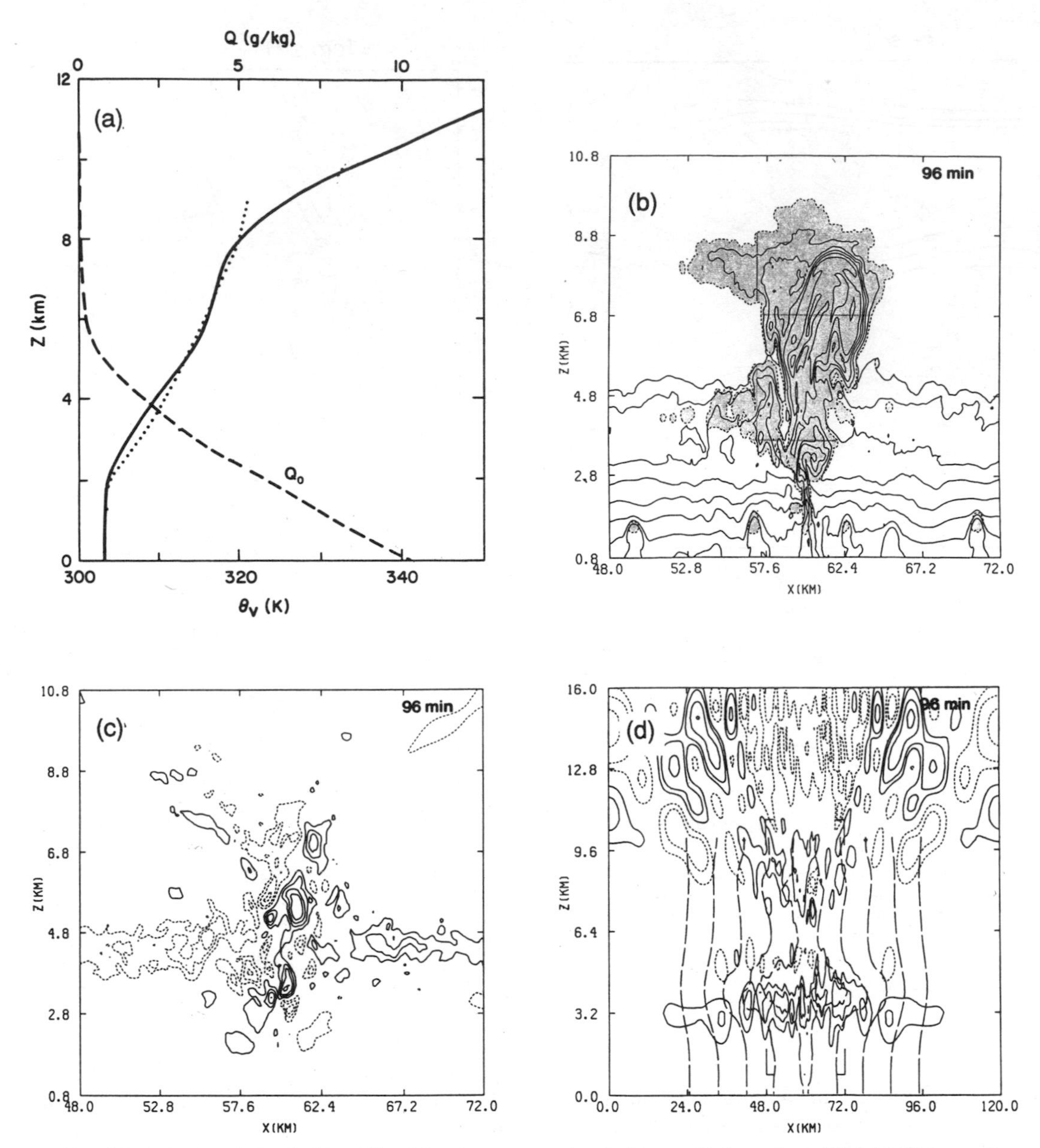

FIG. 5.8. A simulated cloud with midlevel detrainment associated with a stable layer, from BS. (a) Initial sounding. Solid line is θ_v; long dashes are Q; dots indicate a moist adiabat for air originating at 1.8 km whose θ_{v2} was very close to the mean cloud θ_v over most of the cloud lifetime. (b) Contours of total water. Cloud is shaded. (c) Horizontal velocity. Contours at ± 1, ± 3, ± 5 m s^{-1}; negative contours are dashed. (d) Virtual potential temperature perturbation from initial sounding. Contours at ± 0.5, ± 1, ± 2, ± 3 K. Negative contours are dashed. Long dashes show position of initially vertical dye lines. Note that the scale is different from in Figs. 5.9a and 5.9b.

where the horizontal outflow is strongest above the strongest temperature increase, just as in the solution of section 5.2a. Bretherton and Smolarkiewicz showed that the speed of the adjustment front and the magnitude of the associated outflow agree well with the idealized solution of section 5.2a. Above cloud top, gravity wave energy is radiating radially outward.

The average virtual temperature averaged over the cloudy grid points follows quite closely a virtual moist adiabat. The spreading wave of adjustment vertically displaces air from its original height so as to become neutrally buoyant with respect to the cloud and hence to a moist adiabat.

Bretherton and Smolarkiewicz went on to look at the mass balance of the cloud, which must be consistent with the horizontal motions in the adjusted region. In adiabatic motions, the conserved variables, the total water mixing ratio q_T and liquid water potential temperature θ_l, are nonlinearly related to each other as determined by the initial sounding. Mixing in the cloud

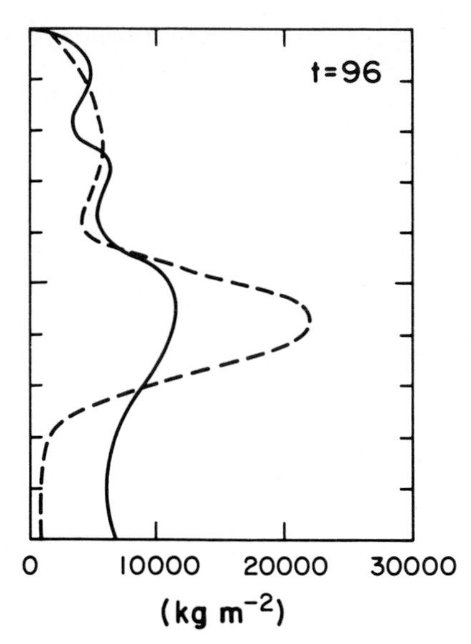

FIG. 5.9. The vertical distributions of entrained mass (solid) and detrained mass (dashed) per unit height.

mixing region will create mixtures whose q_T and θ_l do not obey this relationship, demarcating air that has been "processed" in the cloud mixing region. Combining this analysis with information on horizontal displacements of the vertical dye lines, BS deduced the total mass of air that had been entrained and the total mass detrained between any pair of pressure levels. They found (Fig. 5.9) that the total entrainment into the cloud mixing region is not strongly dependent on height but that detrainment is concentrated in levels in which the cloud buoyancy relative to the far environment was decreasing with height, that is, stable layers in the environment. In the simulation described above, the imbalance between detrainment and entrainment feeds the prominent outflow seen at the 4-km level that is required by the wave of adjustment. Intriguingly, the cumulus parameterization of Raymond and Blyth (1986) generates the same result for a different reason. They hypothesize that parcels of mixed cloud and environmental air detrain where the parcel virtual potential temperature equals that of the environment. Hence, a larger fraction of parcels will detrain where virtual temperature changes rapidly with height.

This study illustrates the intimate relation between cloud structure, entrainment, detrainment, vertical mass fluxes, and cloud heating that makes the mass flow produced by clouds consistent with the compensative subsidence response. Over short periods, levels of enhanced detrainment will likely also be levels of greater moistening and greater concentration of any pollutants or tracers that may be advected up from the subcloud layer.

5.4. Implications for cumulus parameterization

For each horizontal grid point and at each time step, a cumulus parameterization calculates a mean tendency of heating and moistening at each height level. In reality, the source of heating and moistening is concentrated in the small fraction of the volume in and near active cumulus clouds. However, the calling model generally assumes that at each height the source is uniformly averaged over the entire horizontal grid area represented by that grid point.

We have seen in the previous two sections that the dynamical processes that smear out the sources over the area between clouds do not act instantaneously. Compensative adjustment is quite rapid but only horizontally homogenizes the buoyancy field. Let $c = ND/\pi$ be an estimate of the horizontal speed of the dominant gravity wave for a convecting layer of depth D. For deep convection, c lies between 30 and 50 m s^{-1}. Suppose d_{conv} is an estimate of the distance between the dominant convective elements (whether trade cumuli or mesoscale convective systems). Then compensative adjustment over the width Δx of a grid cell will take place within a time $T_{\mathrm{comp}} = \min(\Delta x, d_{\mathrm{conv}})/c$. For a grid spacing of 100 km, and convection in the form of mesoscale convective systems several hundred kilometers apart and 15 km deep, T_{comp} is quick, on the order of 30 min.

The direct adjustment is much slower. If an estimate of the horizontally averaged cloud mass flux at some representative height in the convecting layer is ρw_{cld}, and if this is detrained in a depth D_d, then the time scale for horizontal moisture mixing is $T_{\mathrm{dir}} = D_d/w_{\mathrm{cld}}$. Typical values for w_{cld} for a 100-km grid spacing might vary from 1 m s^{-1} for a grid area almost entirely within an MCS downward to 1 cm s^{-1} for shallow trade cumulus convection. The detrainment depth can vary from as much as 8 km for an MCS down to a few hundred meters for shallow trade cumuli, and leads to direct adjustment time scales T_{dir} of hours to more than a day for moisture mixing. Thus, an instantaneous change in forcing may rapidly produce changes in the moisture fields near active cumulus convection but takes much longer to become horizontally homogenized through a grid area. This is particularly relevant to radiation and microphysical schemes for tropical convection (or any scheme that carries liquid water as a prognostic variable). MCS anvils do not spread out across entire grid areas of mesoscale or forecast models in a single time step of the model and will tend to lag behind the convective activity within each grid cell.

We stress that an understanding of the adjustment of the subcloud layer, which we have not discussed, is just as important for improvement of convective parameterization schemes as an understanding of the adjustment of the cloud layer. The transient dynamics of the subcloud layer in response to deep convection are more subtle to model than the adjustment of the cloud layer. While the ingredients of subcloud layer dynamics—surface heat, moisture and momentum fluxes, interacting density currents of evaporatively cooled air, some of which originated well above the boundary layer, and unsaturated convection—have been studied, an integrated perspective on the transient dynamics of the subcloud layer remains an important task for the future.

The tropics, and in summer, the midlatitudes, would be a dull place indeed were it not for the transient phenomena deriving from the rich nature of moist convective adjustment, blending buoyancy-driven turbulence, internal gravity wave dynamics, and microphysics. Within 20 years, when horizontal and vertical resolutions of global forecast models reach 10 km or less, the compensating adjustment process will primarily take place on resolved scales. Clearly, such a fine resolution significantly changes the goals and formulation of deep convective parameterization. Nevertheless, mesoscale model simulations of midlatitude MCSs (e.g., Zhang and Fritsch 1986) with 25-km or smaller resolution have been quite successful. The increase in computing power, which will allow use of nonhydrostatic cloud ensemble models over larger domains, along with new mesoscale observations over the tropical oceans, should clarify the exciting issues raised in this section and promises a very fruitful decade of research advances in our understanding and ability to model adjustment and the mesoscale organization of convection.

Acknowledgments. This work was supported by the National Science Foundation Meteorology Program under Grant ATM 8858846 and benefited from several useful suggestions from Brian Mapes, Dave Randall, and Victor Pan.

Chapter 6

Momentum Transport by Convective Bands: Comparisons of Highly Idealized Dynamical Models to Observations

MARGARET A. LEMONE AND MITCHELL W. MONCRIEFF

National Center for Atmospheric Research, Boulder, Colorado*

6.1. Introduction

The importance of the vertical transport of horizontal momentum by convective clouds has been documented in studies of momentum budgets (e.g., Shapiro and Stevens 1980; Sui and Yanai 1986). Furthermore, it is obvious that the boundary-layer wind itself—and hence surface heat, moisture, and momentum transports—is affected by deep convection. Most of the air-sea transfer in tropics is concentrated in regions with deep convection.

Traditionally, the transport has been thought of as a mixing process—the wind profile should be smoothed out in the presence of widespread deep cloudiness. In this regard, there is a conceptual analogy to molecular diffusion; hence, the term "eddy diffusion" has been used to describe this process. Although "random" or "ordinary" convective clouds may in fact transport momentum in this way, we shall see that convection lines or bands have a distinctively different behavior. In fact, the vertical shear of the horizontal wind is commonly observed to *increase* in the direction normal to deep linear convection.

The advent of radar and satellites showed that deep convection is mostly organized into lines. Even the tropical oceanic "cloud clusters" that appeared in satellite pictures as large cloud shields were discovered to be the spreading mass of stratiform cloud from convection that was typically organized into lines {see review of GATE [GARP (Global Atmospheric Research Program) Atlantic Tropical Experiment] convection in Houze and Betts 1981}. In these so-called mesoscale convective systems (MCSs), convective cloud is continually generated along a moving band of deep convection that can be quite narrow compared to the full extent of the system that is composed of the band plus a giant anvil (stratiform region), most commonly to the rear of the line. Such systems can last for several hours, even though individual convective elements typically have much shorter lifetimes. Although squall lines have long been recognized in the middle latitudes, Maddox (1981) reminded us that MCSs with enormous cloud shields—so-called mesoscale convective complexes—are common and significant. Studies of organized midlatitude convection by Bluestein and Jain (1985), Bluestein et al. (1987), and Blanchard (1990) show the linear form of deep convection to be common; Houze et al. (1990) estimates that two-thirds to three-quarters of the midlatitude systems they studied were linear during some part of their life cycle.

We shall review the ways in which momentum flux by deep convection has been represented mathematically and show how transport and structure are linked in section 2. With this background, section 3 presents arguments why the Moncrieff (1992) model is the most appropriate representation of the vertical transport of line-normal momentum by quasi-two-dimensional convective bands, and then shows how the model is applied to real atmospheric case studies. In section 4, we compare predictions of the Moncrieff model and the Schneider–Lindzen (1976) momentum transport representation to observed cases. Based on the behavior of the observed fluxes and their match to the model predictions, recommendations are made for potential improvements.

6.2. Mathematical representation of the vertical transport of horizontal momentum

a. Eddy diffusion or K theory

Following an analogy to molecular viscosity and turbulence mixing-length theory, the momentum flux $\bar{\rho}\overline{u'w'}$ can be represented by minus the product of an eddy exchange coefficient $K > 0$ and the vertical shear of the horizontally averaged horizontal wind $\partial U/\partial z$. That is,

$$\overline{u'w'} = -K\frac{\partial \bar{U}}{\partial z},$$

where $\bar{U}$ is the area-averaged wind at a given level, $u' = u - \bar{U}$, and $w' = w - \bar{W}$. The momentum flux in

* The National Center for Atmospheric Research is sponsored by the National Science Foundation.

this case is downgradient (of opposite sign to the vertical shear) at all levels. The acceleration of the mean horizontal wind due to mixing alone is given by

$$\frac{\partial \bar{U}}{\partial t} = -\frac{\partial \overline{u'w'}}{\partial z} = \frac{\partial}{\partial z} K \frac{\partial \bar{U}}{\partial z}. \quad (6.1)$$

For K constant, (6.1) becomes

$$\frac{\partial \bar{U}}{\partial t} = K \frac{\partial^2}{\partial z^2} \bar{U}. \quad (6.2)$$

From (6.2), mixing-length theory tends to smooth out the maxima and minima in the wind profile.

b. *Cumulus friction*

The Schneider–Lindzen (1976) scheme adopts the mixing-length concept, with the mixing length equal to the depth of the cloud. If we assume that the fractional coverage of transporting clouds is much less than unity, and the momentum of the air rising within the clouds is modified only by entrainment, the vertical transport of u momentum by the clouds and by compensating downward motion in the environment, $\bar{\rho}\,\overline{(u'w')_c}$, can be expressed as

$$\bar{\rho}\,\overline{(u'w')_c} = M_c(u_c - \bar{U}),$$

where M_c is the vertical mass flux in the transporting clouds and u_c is the u component of the wind in the clouds. The wind speed in the cloud u_c can be found from a cloud model. If $\bar{U}$ increases monotonically with height, $u_c < \bar{U}$, and momentum transport is generally negative (downgradient). However, when the U profile is more irregular, local gradients will locally produce countergradient fluxes.

The acceleration of the average wind $\bar{U}$ in the x direction ($F_{c,x}$) due to the subgrid-scale flux is represented by the expression

$$F_{c,x} = -\frac{1}{\bar{\rho}} \frac{\partial}{\partial z} M_c(u_c - \bar{U}). \quad (6.3)$$

Schneider and Lindzen point out that setting u_c to a constant, say, the wind at cloud base, makes (6.3) equivalent to the formulations of Ooyama (1971) and Arakawa and Mintz (1974).

The momentum flux divergence given by (6.3) can lead to local "kinks" in the wind profile, but it generally tends to have a smoothing effect on the wind profile through the depth of the cloud. Thus, Schneider and Lindzen refer to $F_{c,x}$ as "cumulus friction."

c. *Dynamical (archetypal) flow models for 2D convection*

Recognizing that deep convection is very often organized into lines, Moncrieff and collaborators have developed a series of nonlinear models idealizing the flow across two-dimensional convective bands of various types (Moncrieff and Green 1972; Moncrieff and Miller 1976; Moncrieff 1978, 1981; Thorpe et al. 1982). Each model idealizes the flow with a small number of distinctive branches and matches inflow to outflow through a set of equations for the conservation of mass, thermodynamic energy, total energy, and domain-integrated momentum. This type of representation considers the convective system as a physical entity, as distinct from the eddy perturbations discussed above. Since the entire flow is specified, the *total* vertical flux of horizontal momentum is predicted, and the total (and mean) vertical mass flux is nonzero.

This class of model can be illustrated by its simplest version, described in Moncrieff (1992), to which the reader is referred for elaboration. Figure 6.1, adapted from these papers, shows the model's essential features. Superimposed on this figure is the cross section of an idealized convective band. Three branches of *relative* flow are represented: 1) a current that enters the storm from the front (right) and rises through it before exiting at a higher level (Moncrieff's "jump updraft"); 2) an overturning updraft current in front of the system; and 3) an overturning current to the rear of the system. The speed U_0 of the inflowing jump current is related to its exit speed U_1 through mass conservation. The speed at the top of the rear overturning current is set equal to U_1 to fit the requirement for continuity in pressure (kinetic energy). This speed can be positive (in a downdraft, as shown in the figure) or negative (in a density current). The inflow pressure perturbation at the front of the line is assumed zero. The pressure $\delta\phi^*$ at the outflow end of the jump current is found from Bernoulli's energy equation; it is equal to the pressure to the rear of the overturning current from the pressure continuity requirement.

The two-dimensional steady-state u-momentum equation gives an exact relationship among the pressure difference across the system, the inflow and outflow momentum fluxes, and the vertical divergence of horizontal momentum:

$$-\frac{\partial \overline{u^*\omega^*}}{\partial p^*} = (U_f^{*2} - U_r^{*2}) - \delta\phi^*, \quad (6.4)$$

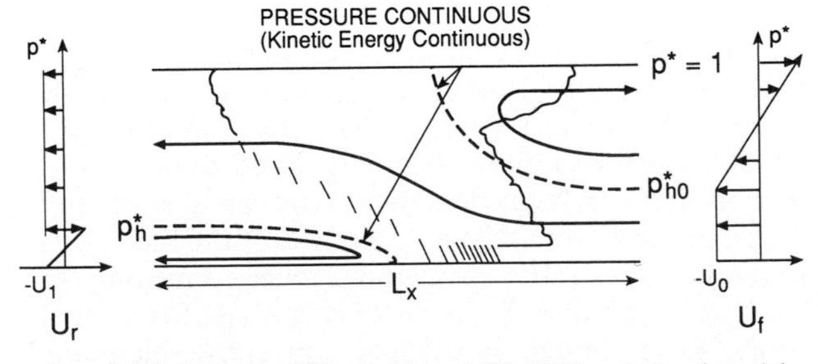

FIG. 6.1. Schematic of flow in Moncrieff's 1992 archetypal model. The nondimensional height $p^* = (p_{sfc} - p)/(p_{sfc} - p_{top})$, where p is pressure, p^*_{h0} is depth of inflow jump updraft, p^*_h is depth of rear overturning current (here shown as downdraft); L_x is the distance, across the convective line, between horizontal flows.

where the asterisk denotes a normalization, $u^* = u/U_0$ is the line-relative horizontal velocity directed positive from the rear to the front of the line, U_f^* and U_r^* are the line-relative wind speeds to the front and rear of the line, respectively, and U_0 is the line-relative inflow speed of the air feeding the rising front-to-rear updraft ("jump updraft") in Fig. 6.1. Term $\delta\phi^*$ is the normalized perturbation pressure change across the system at any pressure $p^* = (p_{sfc} - p)/(p_{sfc} - p_{top})$, and $\omega^* = dp^*/dt$. Note that p^* is the mass coordinate defined in Moncrieff (1985) that facilitates a transformation between incompressible (constant density) and compressible (variable density) model atmospheres. It is also important to note that (6.4) predicts the total flux, whereas K theory and Schneider and Lindzen deal with the flux due to the fluctuating (primed) quantities only.

The mean vertical mass flux $\bar{\omega}^*$ is derived by applying mass continuity over the length scale $x^* = x/L_x$, which spans the environment to the front and to the rear of the line. In the absence of other stresses and if there is no momentum input from the top or bottom, the volume integral of the momentum flux from (6.4) across the system must be zero. This is used to give a powerful bulk momentum constraint on the model variables.

The momentum flux divergence associated with (6.4) and Fig. 6.1 produces generally positive (rear-to-front) u accelerations in the lower levels and negative u accelerations aloft (Fig. 6.2). The associated profile of momentum flux $\overline{u^*\omega^*}$ is negative through most of the depth of the system, with a minimum at middle levels. This is consistent with the flow trajectories in Fig. 6.1, since u^* and ω^* tend to be of opposite sign, except in the portions of the overturning currents near the upper and lower boundaries.

The heights of the three branches in the archtypal model are defined by the solution of its governing "characteristic regime equation" [Eq. (3) in Moncrieff (1992), with no buoyancy differences], which includes the necessary conservative properties. There are two classes of solution. In the *symmetric* case, for which the heights of the overturning currents are equal, the normalized pressure height p_{h0}^* of the jump updraft is related to the normalized pressure height p_h^* of the rear overturning current by

$$p_{h0}^* = 1 - p_h^*. \tag{6.5a}$$

In the *asymmetric* case, the overturning currents are of unequal depth, and

$$p_{h0}^* = \frac{(1 - p_h^*)}{(3 - 4p_h^*)}. \tag{6.5b}$$

The pressure change across the asymmetric case provides the single nondimensional parameter of dynamical significance; namely, $E^* = 2\delta\phi/U_0^2$. Here E^* is a measure of the work done by pressure, normalized

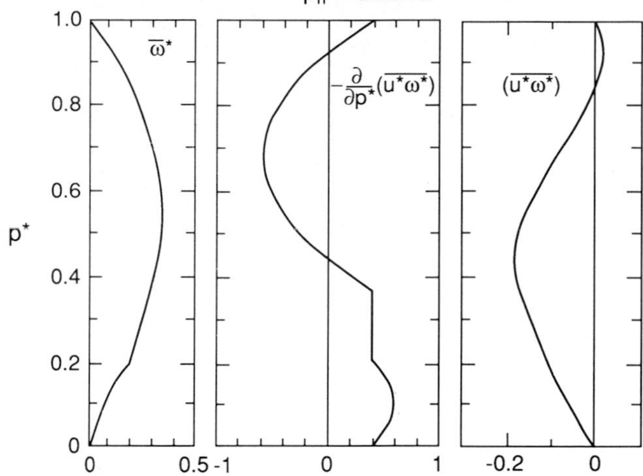

FIG. 6.2. For the schematic of Fig. 6.1: (left) vertical mass flux ω^*; (center) u^* tendency due to vertical divergence of horizontal momentum $\partial \overline{u^*\omega^*}/\partial p^*$; (right) vertical flux of horizontal momentum $\overline{u^*\omega^*}$. All quantities are dimensionless. In the heading, a denotes asymmetric, and the arrow denotes that rear overturning current is a downdraft.

by the jump updraft inflow kinetic energy per unit mass. It can be shown that

$$E^*(p_h^*) = \begin{cases} \dfrac{(1 - p_h^*)(1/2 - p_h^*)}{(3/4 - p_h^*)^2}, & \text{asymmetric} \\ 0, & \text{symmetric.} \end{cases} \tag{6.6}$$

For the asymmetric case, the asymptotic heights of the interfaces between the component branches of the system can be expressed solely in terms of E^*:

$$p_h^* = \frac{1}{4}[3 - (1 - E^*)^{-1/2}];$$

$$p_{h0}^* = \frac{1}{4}[1 + (1 - E^*)^{1/2}]. \tag{6.7}$$

The variation of p_h^* and p_{h0}^* with E^* satisfies the condition $p_h^* \leq p_{h0}^*$ for all values of E^*. From (6.5a), this restriction does not apply to the symmetric solutions.

The velocity profiles to the front (U_f^*) and rear (U_r^*) are given by

$$U_f^*(p^*) = \begin{cases} C_2\left(p^* - \dfrac{p_h^*}{2}\right), & 0 \leq p^* \leq p_h^* \\ \dfrac{-p_{h0}^*}{(1 - p_h^*)}, & p_h^* \leq p^* \leq 1 \end{cases}$$

$$U_r^*(p^*) = \begin{cases} -1, & 0 \leq p^* \leq p_{h0}^* \\ C_1[p^* - 1/2(1 + p_{h0}^*)], & p_{h0}^* \leq p^* \leq 1. \end{cases} \tag{6.8}$$

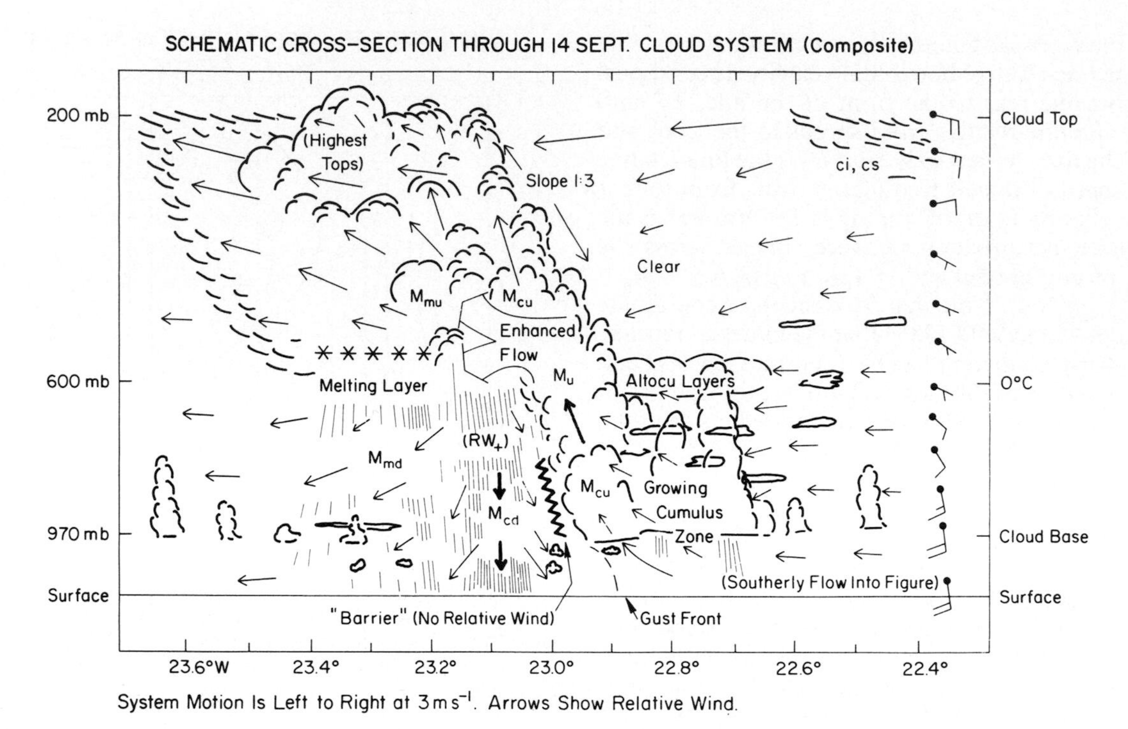

FIG. 6.3. Schematic of flow for 14 September 1974 GATE convective band (from Zipser et al. 1981).

The quantities $C_1 = 2/(1 - p^*_{h0})$ and $C_2 = 2p^*_{h0}/p^*_h(1 - p^*_h)$ are the overturning updraft and downdraft inflow shears, respectively.

More realism can be added to this model, including momentum generation associated with modification of the buoyancy field through evaporation, condensation, and advection; environmental temperature and humidity profiles; and shear in the jump updraft inflow.

6.3. Observations

a. Applicability of the flux representations

The validity of the three aforementioned approaches can be examined at least qualitatively from observational evidence. The *K* theories implicitly assume that the transport is essentially diffusive; that is, the transport is by scales small compared to the scale over which the vertical velocity gradient is measured. There are examples even in the boundary layer (e.g., Pennell and LeMone 1974) for which the flux is locally countergradient but downgradient through the depth of the boundary because the transporting eddies extend through the depth of the convecting layer. This weakness can be overcome by increasing the mixing length, as in Schneider and Lindzen. Here, however, three assumptions are critical: namely, that the transport is by convection covering a small portion of the area of interest; that the convection is statistically homogeneous; and that entrainment is the only process that influences the horizontal momentum in the cloud. In the Schneider–Lindzen scheme, it is assumed that the pressure gradient across the cloud is zero. Furthermore, the Schneider–Lindzen scheme is designed to estimate $\bar{\rho}\overline{u'w'}$ rather than $\bar{\rho}\overline{uw}$.

Although pressure perturbations are generated as clouds form (e.g., LeMone et al. 1987), observations show that the flux by randomly distributed or three-dimensional convection tends to be downgradient over much of its depth (LeMone et al. 1984). This seems to be true even for numerically simulated intense thunderstorms (Lilly and Jewett 1990).

Dramatic departures from downgradient transport, as represented by either *K* theory or the Schneider–Lindzen representation, occur when convection is organized in quasi-two-dimensional bands. The GATE convective band documented in LeMone (1983) is a good example. A schematic is shown in Fig. 6.3, while the environmental wind profile, the average wind profile, and the momentum transports appear in Fig. 6.4. In the figure, u and v are in a right-handed coordinate system, with u perpendicular to the line, and positive in the direction of line motion. Note that u is relative to line (leading edge) motion, to be consistent with section 6.2 and the schematic. It is obvious from Fig. 6.4 that neither the sign of the vertical u-momentum flux nor the ratio of $\bar{\rho}\overline{uw}$ to $\bar{\rho}\overline{vw}$ is simply related to the environmental wind shear. Furthermore, comparison of $\bar{U}$ and $\bar{\rho}\overline{uw}$ indicates a strongly countergradient flux. Figure 6.3 gives the impression that the flux is mainly by the rearward jump updraft; analyses in LeMone (1983)

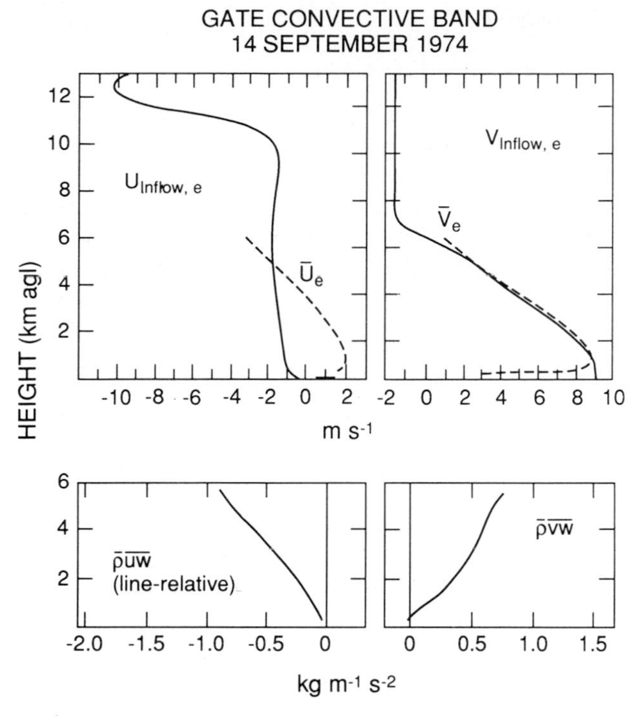

FIG. 6.4. (Top) Earth-relative inflow wind components $U_{\mathrm{inflow},e}$ and $V_{\mathrm{inflow},e}$ (solid) and earth-relative wind components $\bar{U}_e$ and $\bar{V}_e$, averaged over Δx. To get line-relative values of u, subtract 2.5 m s^{-1}. (Bottom) Line-relative vertical flux of U and V momentum.

confirm that impression. Note, however, that the flux $\bar{\rho}\,\overline{vw}$ is downgradient.

Other GATE cases and subsequent observational and numerical studies reveal similar behavior for convective bands, with broadly similar across-band structure and negative u-momentum fluxes through much of the storm's depth. We will now examine these cases in detail below, using the archetypal Moncrieff model of Fig. 6.1 as a framework.

b. Application of the Moncrieff model

1) MODEL INPUT PARAMETERS

Our objective is to apply the model to specific cases and compare the resulting model-predicted momentum flux divergence and vertical velocity to those observed. Consequently, we must first find the values of the model input parameters in Fig. 6.1. These are p^*_{h0}, the depth of the air feeding the front-to rear updraft; U_0, the line-relative inflow current speed; p^*_h, the depth of the overturning current to rear (left) of the system; H, the depth of the system; and L_x, which is called the "dynamical scale," namely, the distance required for the perturbed flow to become quasi-horizontal. The flow in the rear overturning current can either represent a downdraft or be reversed to represent a density-current circulation. For the flow in Fig. 6.1, p^*_h and p^*_{h0} are related by (6.5a) or (6.5b).

As in the foregoing discussion, u is defined relative to the convective band, normal to its axis, and positive in the direction of its movement; H fits most closely the maximum height with significant echo. Overshooting tops, if detectable, are ignored, which is consistent with the theoretical assumptions because the vertical scale is the parcel equilibrium level. The variables p_h and p_{h0} can be obtained from airflow averaged along the axis of the convective band for some cases; for others, we resort to single cross sections. These heights are converted from physical (height) units to model (p^*) coordinates to find the values that best conform to (6.5a) or (6.5b) using data from an inflow sounding. After choosing a value of p^*_{h0}, U_0 is obtained from a mass-weighted vertical average of convective line–relative U from either an inflow sounding or aircraft data taken in the inflow area. Model heights are related to geometric heights through constructing a hydrostatic standard atmosphere for each case with (a) a boundary layer with a dry-adiabatic lapse rate, (b) a layer from boundary-layer top to 11 km with a lapse rate of -0.006 K m^{-1}, and (c) isothermal atmosphere above. This is similar to the NACA (National Advisory Committee for Aeronautics) standard atmosphere except that a dry-adiabatic boundary layer is included, and the surface height, surface virtual temperature, and boundary-layer depth are specified for each case.

2) DIMENSIONALIZATION

The model is formulated in terms of nondimensional vertical fluxes of mass and storm-relative horizontal momentum in mass coordinates and in terms of $\omega^* \equiv dp^*/dt$, while momentum flux estimates based on observational data are typically given in terms of vertical velocity w and as a function of geometric height. The conversion of model output to the appropriate physical units is a two-step process: 1) conversion from nondimensional to dimensional units, and 2) conversion from mass-coordinate variables to geometrical-coordinate variables. Thus, the conversion from dimensionless mass flux ω^* to the dimensional vertical mass flux $\bar{\rho}\,\bar{W}$ is as follows:

$$\bar{\rho}\,\bar{W} \approx \frac{-\bar{\omega}}{g} = \frac{U_0 \Delta P}{L_x g}\,\overline{\omega^*} \qquad (6.9)$$

where $\Delta P = p_{\mathrm{sfc}} - p(H)$ and $\bar{\rho}$ is the air density. We then divide (6.9) by $\bar{\rho}$ based on the above-described standard atmosphere.

To a good approximation, the momentum flux divergence $\partial(\overline{u\omega})/\partial p$ in p coordinates is equal to $\bar{\rho}^{-1}\partial(\overline{\rho u w})/\partial z$ in geometric coordinates. Since $u = U_0 u^*$ and $\bar{\omega} = (U_0 \Delta P / L_x)\omega^*$,

$$\frac{1}{\bar{\rho}}\frac{\partial}{\partial z}(\bar{\rho}\,\overline{uw}) = \frac{U_0^2}{L_x}\frac{\partial}{\partial p^*}(\overline{u^*\omega^*}). \qquad (6.10)$$

The flux divergence in earth-relative coordinates $\bar{\rho}^{-1}\partial(\bar{\rho}\,\overline{u_e w})/\partial z$ is given by

$$\frac{1}{\bar{\rho}}\frac{\partial}{\partial z}\bar{\rho}\overline{u_e w} = \frac{1}{\bar{\rho}}\frac{\partial}{\partial z}\bar{\rho}\overline{uw} + \frac{c}{\bar{\rho}}\frac{\partial}{\partial z}\bar{\rho}\bar{W}, \quad (6.11)$$

where c is the system travel speed relative to the earth.

For the observational studies, Δx is the horizontal distance, measured normal to the convective band, over which the fluxes were calculated. Ideally Δx should be equal to or greater than the dynamical scale L_x of the circulation as previously defined; that is, observations should extend into where flow becomes quasi-horizontal. Equivalently, there should be neither vertical mass flux nor vertical flux of horizontal momentum for values of x outside the observational domain. If these criteria are satisfied, the vertical fluxes of mass and horizontal momentum across Δx and L_x are simply related:

$$(\bar{\rho}\overline{uw})_{L_x} = \frac{\Delta x}{L_x}(\bar{\rho}\overline{uw})_{\Delta x}; \quad (6.12a)$$

similarly,

$$\bar{W}_{L_x} = \frac{\Delta x}{L_x}\bar{W}_{\Delta x}. \quad (6.12b)$$

As long as $\Delta x \geqslant L_x$, (6.12a) and (6.12b) imply that Δx can be used in place of L_x to obtain model predictions of the measured fluxes.

This is a very different physical situation from that encountered in the horizontally homogeneous and stationary planetary boundary layer, for which the average vertical fluxes of mass and horizontal momentum are independent of sampling length (provided the transporting eddies have been adequately sampled). Since $\bar{W}$ is nonzero and $\bar{U}$, like $\bar{W}$, varies with L_x, means and departures from means have little significance. Furthermore, the model predicts total fluxes across L_x. Thus, primarily total fluxes ($\bar{\rho}\overline{uw}$) are dealt with here, and observed fluxes evaluated as the mean product of fluctuating quantities [e.g., $\bar{\rho}\overline{u'w'}$, where the prime denotes departure from a mean] are converted to total fluxes by adding fluxes due to means ($\bar{\rho}\bar{U}\bar{W}$).

The comparison of the observed vertical fluxes of mass and horizontal momentum to results from the model is complicated by the fact that the observed convective bands are rarely isolated because they often have massive stratiform cloud shields, and often form in an environment where there is mean ascent. Thus, although a zero or small vertical flux of horizontal momentum is plausible outside a convective area (unless the stratiform shield included in the sample is extensive and active), nonzero vertical velocities probably occur. To get an idea of what sort of impact might be expected, if a convective system had an average vertical velocity of 1 m s^{-1}, compensating subsidence of 5 cm s^{-1} would have to occur over 20 times the area of the system for a net zero vertical mass flux. Fortunately, momentum fluxes are typically sampled over a domain not much larger than the major convective transporters.

6.4. Results

a. Dataset

A brief description of convective bands for which momentum transport has been documented appears along with references in Table 6.1. There are other known cases—for example, the majority of cases in LeMone et al. (1984) and a case described in Rao and Hor (1991)—but needed detail is lacking. The cases are broadly grouped according to Δx, the horizontal scale over which the measurements were made. Four types of data are represented: from aircraft in situ instruments; airborne Doppler radars; ground-based Doppler radars; and radiosonde budget studies. Note that five cases have been simulated with numerical models. Table 6.2 summarizes the characteristics of the environment and convective systems for the cases in Table 6.1 for which there are sufficient data for comparison with each other and with Moncrieff's model. For the table, the convective available potential energy (CAPE) is defined as

$$\text{CAPE} = \int_{z_{\text{lcl}}}^{z_{\text{equil}}} \frac{g}{\bar{T}_v} T'_v dz,$$

where T'_v is the difference between the virtual temperature of an undilute parcel and its environment, $\bar{T}_v$ is the environmental virtual temperature, z_{lcl} is the height of the condensation level, z_{equil} is the height at which the parcel virtual temperature matches that of the environment, and g is the acceleration of gravity. Although some of the authors may have used T or θ instead of T_v, the differences resulting are of the order of 200 J kg^{-1}, fairly small considering the uncertainty typical for CAPE estimates and the range of CAPEs in the table.

The cases selected are from a variety of locations. The range of sizes and CAPEs will enable some comparisons to be made below.

b. Data accuracy

It is impossible to assess with any rigor the accuracy of the data for the ten case studies described in Tables 6.1 and 6.2. The case studies represent four categories of data: namely, multiple-Doppler radar; airborne Doppler radar; aircraft in situ data; and radiosonde budgets. As to measurements of momentum fluxes, the correlations involved make rigorous error estimates difficult, and to our knowledge none have been done in reference to momentum transport by deep convection. It is reassuring that five of the ten cases have been simulated in numerical cloud models and the consistency between model data and observed data is good.

However, some general statements can be made. (a) Multiple-Doppler radar will generally give more accuracy than dual-Doppler radar (although this must be interpreted with caution because multiple-Doppler

TABLE 6.1. Documented cases of momentum transport for quasi-two-dimensional convective bands.

Case	Description	References used (M—some numerical modeling involved)
COPT 22 June 1981	Tropical MCS, squall-line convection 2D (single- and dual-Doppler radar)	Roux et al. 1984 Roux 1985 Chong et al. 1987 Lafore and Moncrieff 1989 M Chong and Hauser 1990 Chong 1992, (personal communication)
COPT 23 June 1981	Tropical MCS, squall line cellular (dual-Doppler radar)	Redelsperger and Lafore 1988 M Lafore et al. 1988 M Lafore and Moncrieff 1989 M Roux 1988
PRE-STORM 10–11 June 1985	Midlatitude squall line/MCS (radiosonde budget, single- and dual-Doppler)	Zhang et al. 1989 M Gao et al. 1990 M Trier et al. 1991 Biggerstaff and Houze 1991 Gallus and Johnson 1992 Gallus 1992, personal communication
GATE 14 September 1974	Tropical oceanic convective band (aircraft in situ data)	Zipser et al. 1981 LeMone 1983 LeMone et al. 1984
22 May 1976	Midlatitude MCS (dual-Doppler radar)	Ogura and Liou 1980 Smull and Houze 1985 Smull and Houze 1987a Smull 1992, personal communication Lipps and Hemler 1991 M
CCOPE 20 June 1981	Midlatitude squall line (multiple-Doppler radar)	Fankhauser et al. 1992
CCOPE 17 July 1981	Midlatitude MCS (multiple-Doppler radar)	LeMone et al. 1990 Matejka and LeMone 1990
TAMEX 17 May 1987	Subtropical marine squall line (dual-Doppler radar)	Lin et al. 1990 Wang et al. 1990 Tao et al. 1991 M
TAMEX 16 June 1987	Subtropical convective band (airborne Doppler radar)	Jorgensen et al. 1991 LeMone and Jorgensen 1991
12 June 1985	Hawaiian rainband (aircraft in situ data)	LeMone and Jensen 1990

TABLE 6.2. Cases analyzed.

Mesoscale convective system		CAPE (J kg^{-1})	U_0 (m s^{-1})	Ri $\frac{(\mathrm{CAPE})^{1/2}}{0.5U_0^2}$	H (km)	Δx (km)	p_h^*	c (m s^{-1})
COPT	22 June 1981	1328	13.73	14	12.0	40, 250	↓0.66s[†]	19.0
COPT	23 June 1981	2810	11.45	43	14.0	30	↓0.12a	14.5
PRE-STORM	0300 UTC 10–11 June 1985	~2000[††]	20.54	9	14.0	170, 50	↓0.75s	16
PRE-STORM	0600 UTC 10–11 June 1985	~2000[††]	18.54	12	14.0	190, 60	↓0.75s	14
GATE	14 September 1974	1000	3.90	132	12.0	140	↓0.43a	2.5
Oklahoma	22 May 1976	1140	14.30	11	12.0	40	↓0.12a	15
CCOPE	20 June 1981	600	7.54	21	9.6	30	0.00a	12
CCOPE	17 July 1981	1000(E)	13.75	11	10.0	50	0.00a	11.4
TAMEX	17 May 1987	1369	6.85	58	9.0	25	↓0.51a	15.5
TAMEX	16 June 1987	1000–1500	11.59	15–22	12.0	30	0.00a	0
Hawaii	12 July 1985		3.2		2.6	10.7	↑0.79s	4.6

[†] a—asymmetric; s—symmetric; ↓ indicates rear overturning current is a downdraft; ↑ indicates it is an updraft.

[††] There was considerable variation in CAPE ahead of the line (Trier et al. 1991; Biggerstaff and Houze 1991; Biggerstaff 1992, personal communication).

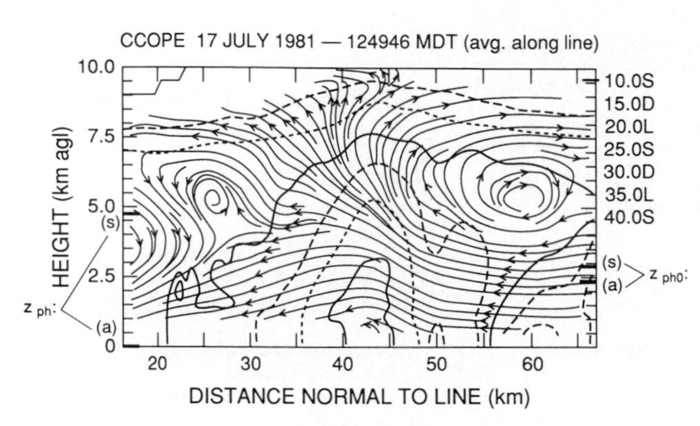

FIG. 6.5. Cross section of streamlines of the flow normal to the 17 July CCOPE convective band, averaged along the band. Heavy horizontal lines show z_{p_h} and $z_{p_{h0}}$; symmetric case denoted by s; asymmetric case by a (figure courtesy of T. Matejka).

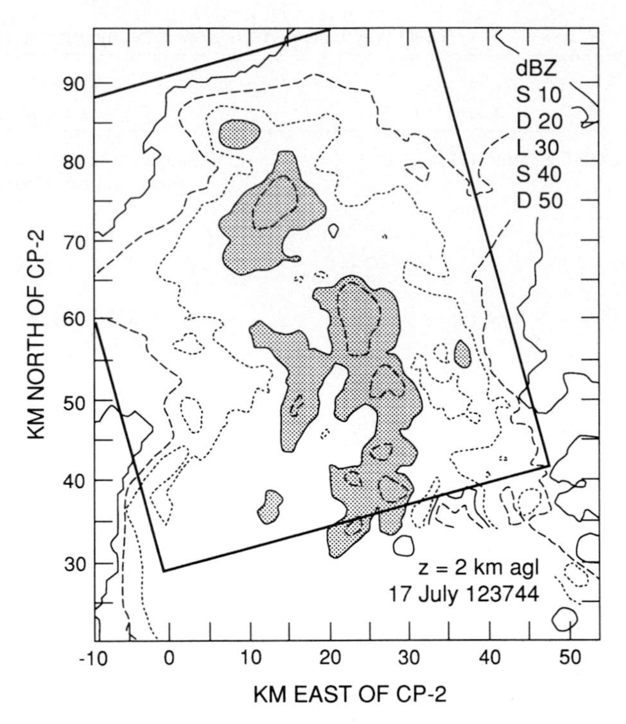

FIG. 6.6. Cloud-base reflectivity field associated with 17 July CCOPE system. Rectangle outlines area over which averages were taken (figure courtesy of T. Matejka).

studies also cover a larger area, parts of which have dual-Doppler coverage). (b) The airborne Doppler case probably does not have the accuracy of the ground-based Doppler studies because of a broad beamwidth and the time required to obtain samples. (Note the sampling time will be vastly improved in the newer airborne Doppler systems, which will scan more rapidly in frontward and rearward cones, affording crossing beams and thus dual-Doppler coverage without a change in aircraft heading.) (c) The accuracy of the radiosonde budget is increased with scale; that is, the mass fluxes and budgets for the entire system are probably more accurate than those for the convective line. (d) Doppler radar estimates of mass and momentum fluxes near the top and bottom of the domain are compromised by a paucity of data. (e) Aircraft gust-probe data probably do not provide accurate measurements of weak vertical motions on scales of 10 km and greater.

c. An example

We choose the 17 July case study of a convective system in the 1981 Cooperative Convective Precipitation Experiment (CCOPE) as an example of model–observation comparison because of the quality of the dataset and the thorough documentation of the structure. Figure 6.5 shows a cross section averaged along the line, and the depths z_{p_h} and $z_{p_{h0}}$. Since the downdraft is only partly sampled we selected two values for p_h^* (around 0.6 and 0.0) and determined for each whether the system was symmetric or asymmetric by the fit of p_{h0}^* to Fig. 6.5. Each set of values was then adjusted to minimize departure from the figure. The vertical pressure change through the system ΔP corresponds to the pressure change from the surface to the system top H = 10 km. Figure 6.6 is a depiction of the reflectivity pattern associated with the line, showing the rectangle over which the fluxes were averaged. Since the system is assumed two-dimensional, only Δx = 50 km is needed for conversion of model output to dimensional quantities in (6.10)–(6.12). Figure 6.7 (left) shows the profile of momentum flux divergence, $-\bar{\rho}^{-1}\partial(\bar{\rho}\overline{uw})/\partial z$, for two observation times superimposed on the flux divergence predicted by the Moncrieff model. Figure 6.7 (right) shows the profile of vertical velocity, averaged over Δx. In both cases, the comparisons are surprisingly good. The shape of both curves is reproduced best for the symmetric p_h^* = 0.6 case, while the p_h^* = 0 asymmetric case reproduces the amplitude.

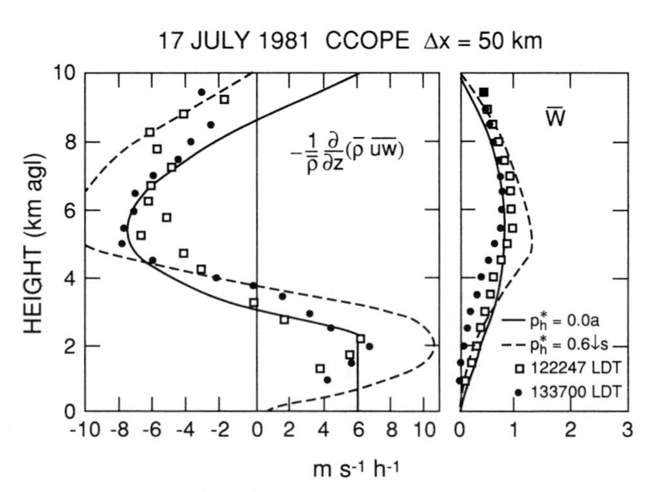

FIG. 6.7. For 17 July CCOPE system, observed and model-predicted values of (left) U-momentum flux divergence; (right) mean vertical velocity (data courtesy of T. Matejka).

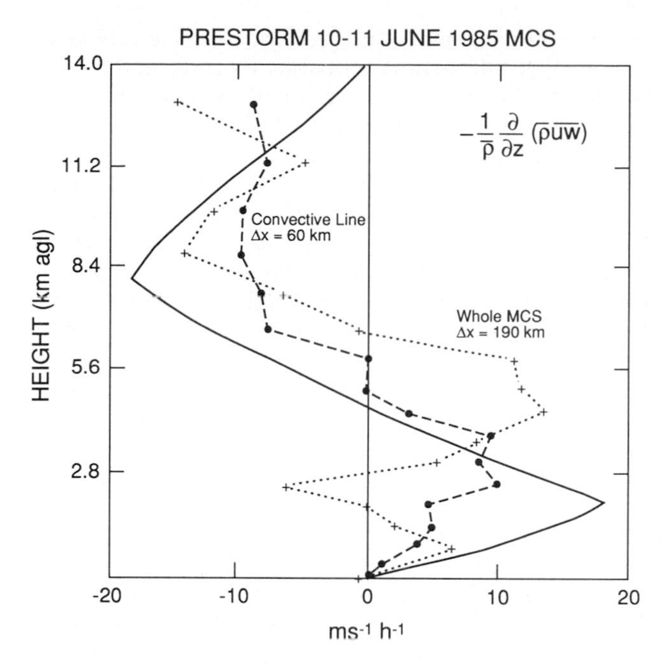

FIG. 6.8. For 10–11 June PRE-STORM squall-line system, observed momentum-flux divergence at 0600 UTC for the entire system (+, multiplied by 2) and for the convective portion (●), and predicted flux divergence (solid line) for the convective band.

The amplitudes of the curves differ mainly because the normalized flux profiles differ: the amplitude of the flux divergence associated with the symmetric case is about twice that of the asymmetric case. This is compensated for by a larger U_0 for the asymmetric case (13.75 m s^{-1}) compared to 12.13 m s^{-1} for the symmetric case.

It follows that determining the parameters p_h^* and p_{h0}^* is a matter of judgement. An attempt has been made to find those that best fit both updraft and downdraft. However, it is reassuring that the form of the momentum flux is essentially the same throughout the entire range of archetypal solutions. Furthermore, differences in the detailed shape of the profiles that occur seem related in a systematic way to differences in system structure and environment, as we shall see in the discussion that follows.

d. *Variation of the momentum flux divergence profile with $\Delta x/H$*

Figure 6.8 shows the momentum flux divergence curve for the complete 10–11 June squall line in PRE-STORM (Preliminary Regional Experiment for STORM-Central), based on radiosonde budgets of Gallus and Johnson (1992). Note that the height of the crossover from positive to negative U acceleration is much higher for the entire system (Δx = 170 km) than for either the Moncrieff prediction or the crossover for the budget restricted to the convective part of the system (Δx = 60 km). (Equivalently, the height of the largest negative u-momentum flux is higher when the average is over the entire line.)

The change of the height of the crossover from rearward to frontward acceleration with the fraction of the system considered is not surprising. Figure 6.9 is a composite cross section of the 10–11 June convective system from Biggerstaff and Houze (1991). Note that the boundary between the rearward-flowing air and the

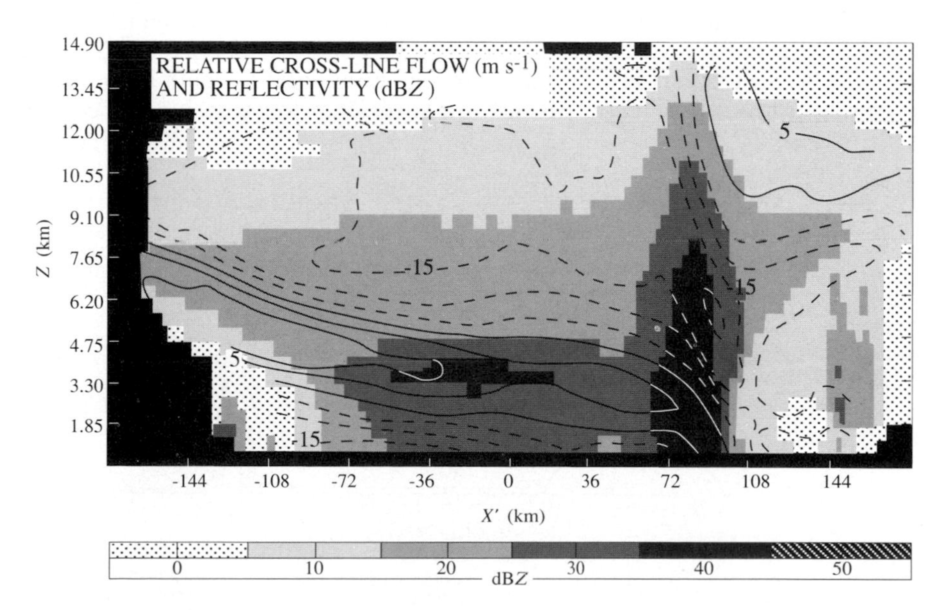

FIG. 6.9. For 10–11 June PRE-STORM squall line, composite storm-relative cross-line flow u and reflectivity, from Biggerstaff and Houze (1993).

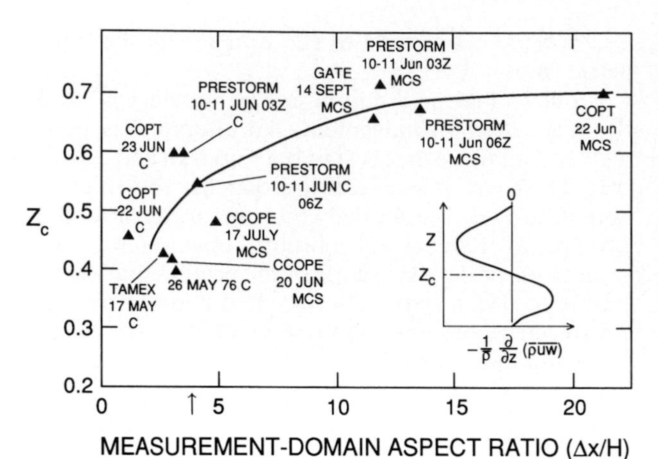

FIG. 6.10. Height at which momentum flux divergence changes sign, or height of extreme value of $\overline{\bar{\rho} u w}$ as a function of measurement domain aspect ratio $\Delta x/H$. MCS denotes x average across most or all of MCS; C denotes x average that includes convective line but excludes much of the MCS. Vertical arrow denotes the aspect ratio ($\sim$4) of the two-dimensional solution for the Moncrieff (1992, Fig. 4) archetypal model.

frontward-flowing air becomes higher toward the rear of the line. We can get a crude estimate of the time rate of change of $\bar{U}$ by comparing the average flow before the line was there to the average flow in the figure. Clearly, the average forward acceleration would occupy a smaller depth in a volume covering only the convective part of the system than if we include the entire system.

The depth of the crossover from positive to negative U acceleration is plotted as a function of domain aspect ratio in Fig. 6.10. Although there is considerable scatter, the effect of the deepening of the overturning downdraft is clear: the crossover is higher for larger values of $\Delta x/H$. Figure 6.11, an MCS schematic cross section from Houze et al. (1989), reflects the fact that the deepening of the rear-to-front flow layer toward the rear of MCSs is a rather common feature (see also Smull and Houze 1987b). This deepening probably depends on the ice–water phase changes, among other processes. For example, Lafore and Moncrieff (1988) showed a comparison between simulations that crudely modeled and omitted the ice phase in their Fig. 11. The horizontal scale of the system, and especially the slope of the front-to-rear/rear-to-front boundary, was quite different in these two cases. This can be explained in terms of the differing scales of the baroclinic production of vorticity, which is omitted in the Moncrieff (1992) dynamical model.

The vertical arrow on the abscissa of Fig. 6.10 marks roughly where Moncrieff's two-dimensional solutions (see Moncrieff 1992, Fig. 4) for the flow in Fig. 1 exhibit horizontal flow. That is, the cases to the left have aspect ratios that "fit" the archetypal model. These cases are of two different types. The first is represented by the convective portions of larger systems, that is, those from Convection Profonde Tropicale (COPT), the 10–11 June system, and the 22 May 1976 system. These tend to have higher crossover points than the second type, which includes the systems with small total aspect ratio.

e. Performance of the Moncrieff representation of mass flux and momentum flux divergence

Another measure of the validity of the Moncrieff scheme is the faithfulness with which it reproduces the mass flux and momentum flux divergence amplitudes. In the case shown in Fig. 6.7, the comparison was good—the amplitudes of the observed mean vertical velocity and momentum flux divergence were comparable to those predicted. This was a particularly good case also because data were collected from up to five Doppler radars with a corresponding increase in the confidence in the measurements. Figure 6.12 compares the ratios of the Moncrieff model predictions to observed amplitudes for both quantities. The square at the lower left of the figure outlines cases we might define

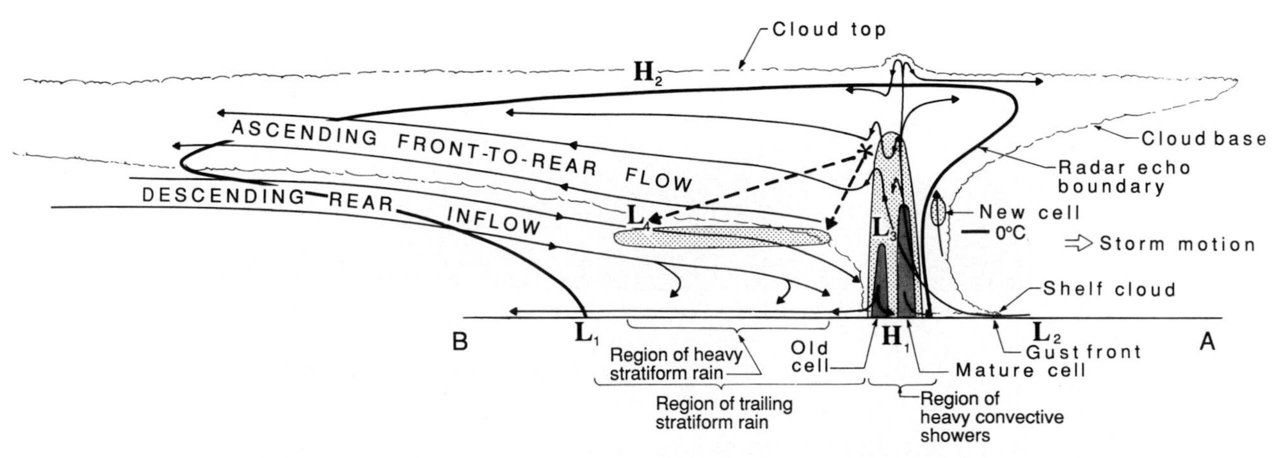

FIG. 6.11. Schematic cross section of a midlatitude mesoscale convective system, from Houze et al. (1989).

as "good agreement." Within the square, the predicted amplitude is between two-thirds and three-halves that measured. These bounds were selected so that the same answer would be achieved if the ratio of measured to predicted amplitudes were used.

The cases within the square include the three small-aspect-ratio systems of Fig. 6.10—20 June and 17 July from CCOPE, 17 May from TAMEX (Taiwan Area Mesoscale Experiment), and the Hawaiian rainband (aspect ratio 4.1). The latter does not appear in Fig. 6.10, because turbulence effects at midlevels prevent identification of the crossover point. In addition, the total MCS data for the 10–11 June 1985 system fall within the square for 0300 UTC. The worst predictions are for the 10–11 June convective band for both 0300 and 0600 UTC, the GATE system, and the COPT 23 June system.

The source of the discrepancy can be partly found from the relationship between the $\bar{W}$ and the momentum flux divergence ratios for the worst comparisons. If $\bar{W}$ is predicted correctly, the graph suggests that the momentum flux divergence would be underestimated by the Moncrieff flux scheme (points on the lower left side of the graph). Conversely, $\bar{W}$ has to be overpredicted for these cases for the momentum flux divergence to be correctly predicted. These relationships suggest that for the *outlying cases* the horizontal component of velocity in the model are not strong enough for a given mass flux.

This discrepancy is probably due to the neglect of the buoyancy distribution and its effect on the pressure and wind fields. Examination of the GATE system, for which $\bar{W}$ is predicted accurately, illustrates this point. Figure 6.13 compares the observed and predicted (solid lines) momentum flux divergence and $\bar{W}$. Note that the momentum flux divergence is underpredicted. According to LeMone (1983), the momentum transport was mainly by the jump updraft current, which was being strongly accelerated rearward at the same time it was being accelerated upward by buoyancy.[1] The rearward acceleration was due to the (mostly hydrostatic) rearward pressure gradient generated in the wedge-shaped volume of buoyant air at the leading edge of the convective system; the minimum pressure was just above the cloud base and cold pool, and probably slightly forward of where the convection reached its maximum depth (Fig. 6.14). LeMone et al. (1984) represented the amplitude of the earth-relative momentum transport by the variable N, given by

$$N = (U_{0,e} - c)\left[U_{0,e} + \frac{\Delta p}{\bar{\rho}(U_{0,e} - c)}\right],$$

where the storm-relative inflow speed ($U_{0,e} - c$) represents the mean vertical velocity, assuming all the inflow ascends, and the expression in parentheses represents the horizontal velocity, altered by Δp [equal to the pressure perturbation for the inflow (assumed zero) minus the largest negative hydrostatic pressure perturbation induced by the buoyant air overhead, normally observed just above cloud base toward the rear of the convective band]. For momentum flux divergence in storm-relative coordinates,

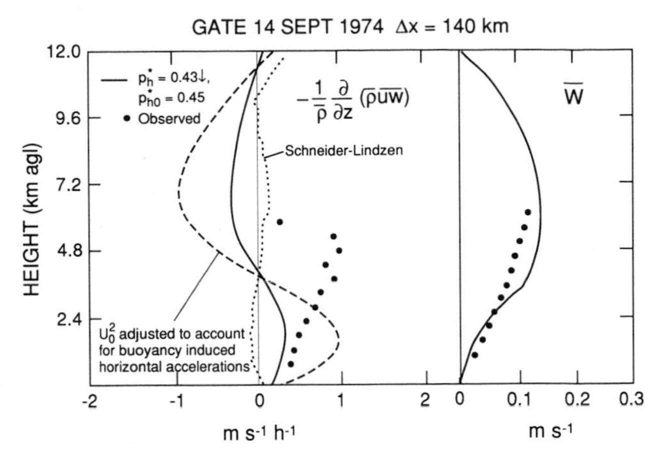

FIG. 6.13. For GATE 14 September 1974 convective band, observed and predicted momentum flux divergence and mass flux. Model curve for asymmetric case with a downdraft of normalized depth $p_h^* = 0.43$: solid line, U_0 estimated from inflow wind profile; dot–dash line using N from Eq. (6.13) instead of U_0^2 in Eq. (6.10). Dotted line is prediction from Schneider–Lindzen scheme Eq. (6.3) with u_c set equal to inflow U at 500 m.

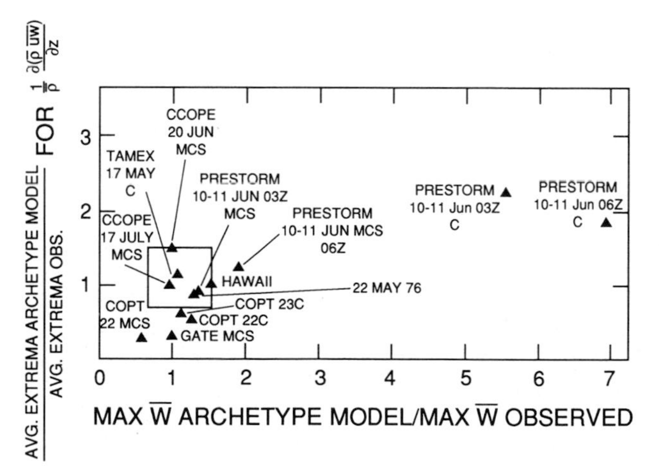

FIG. 6.12. Ratio of predicted to observed momentum flux divergence, $\bar{\rho}^{-1}\partial(\bar{\rho}\overline{uw})/\partial z$, as a function of the ratio of predicted to observed vertical mass flux. Curve magnitudes defined by averages of the two extrema for the momentum flux divergence and by maxima for the mass fluxes; C and MCS symbols as in Fig. 6.10.

[1] There is evidence in Zipser et al. (1981) of a downdraft as well. In section 6.4b, it was pointed out that aircraft gust-probe data might "miss" weak motions of scales of the order of 10 km and greater; if the rear inflow downdraft was not detected by the aircraft, we expect the measured u-momentum flux to be weaker than actually occurred. This would make the discrepancy even worse.

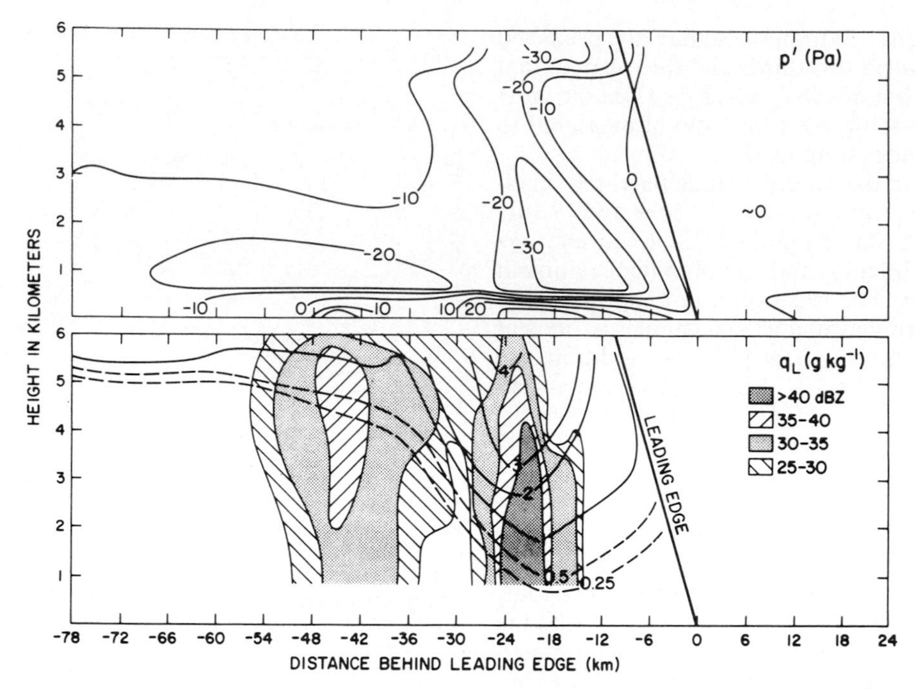

FIG. 6.14. For 14 September 1974 GATE convective band: (top) pressure perturbation field; (bottom) reflectivity field and condensed water content from NOAA C-130 total water sensor. "Leading edge" indicates front of the main cloud; vertical exaggeration 6:1. Figure is from LeMone (1983).

$$N = U_0\left(U_0 + \frac{\Delta p}{\bar{\rho} U_0}\right). \qquad (6.13)$$

Using N instead of U_0^2 to convert the amplitude for the normalized momentum flux divergence curve to dimensional units, we obtain the dashed curve in Fig. 6.12—a better match for the amplitude than the prediction not accounting for buoyancy-generated pressure gradients. This suggests that allowing the jump updraft to be buoyant (baroclinic vorticity generation) will improve the model prediction.

We can get a crude idea of when horizontal accelerations due to buoyancy are important by assuming that the low pressure is the hydrostatic effect of the wedge of buoyant air at the leading edge, as in LeMone (1983). The maximum magnitude of the perturbation pressure is thus represented by LeMone's equation (7):

$$-\frac{p'(z_{cb})}{\bar{\rho}} = \int_{z_{cb}}^{z_{ct}} \frac{g}{\bar{T}_v} T_v' dz \sim \text{CAPE}, \qquad (6.14)$$

where z_{cb} and z_{ct} are the heights of cloud base and cloud top.

The effect of this pressure gradient on u is greater for small inflow wind speeds (as in the GATE case) than for the case of strong inflow (as in, for example, the CCOPE 17 July case). Referring to Fig. 6.1 and applying the Bernoulli equation to the air flowing into the convective line from the front along the lower boundary and up and over the front of the rear overturning current (cold pool) to the location of the lowest pressure, we have

$$\frac{1}{2}U^2 - \frac{1}{2}U_0^2 \approx -\frac{p'}{\rho},$$

provided the release of buoyant energy is small for the air parcel. Clearly, the pressure gradient force (and hence CAPE) has the most significant effect for the smallest values of U_0. Using Eq. (6.14) we have

$$\frac{1}{2}U^2 - \frac{1}{2}U_0^2 \sim \text{CAPE}. \qquad (6.15)$$

Equations (6.14) and (6.15) suggest the relevance of a Richardson number similar to the one that appears in Moncrieff and Green (1972),

$$\text{Ri} = \frac{\text{CAPE}}{1/2\, U_0^2},$$

could give an indication of the fit of the data to the Moncrieff (1992) archetypal model prediction. Better agreement is expected for smaller values of Ri, assuming that the convection remains two-dimensional. This is one type of Richardson number that characterizes the convective band; others are appropriate. For example, the work of Chisholm and Rennick (1972) and Weisman and Klemp (1982) suggests that in the presence of large CAPE, small Richardson numbers defined using the shear in both horizontal wind components

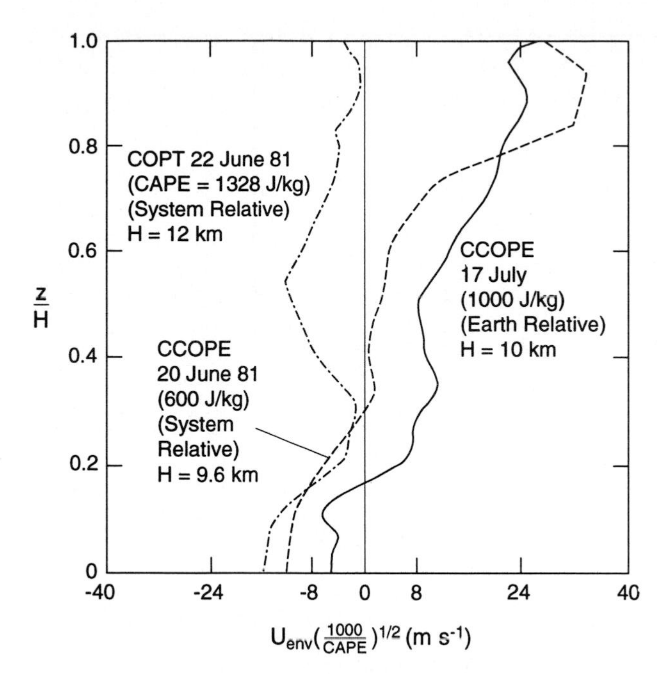

FIG. 6.15. For three of the cases of Table 6.1, the environmental wind profile, normalized by $(\mathrm{CAPE})^{1/2}$.

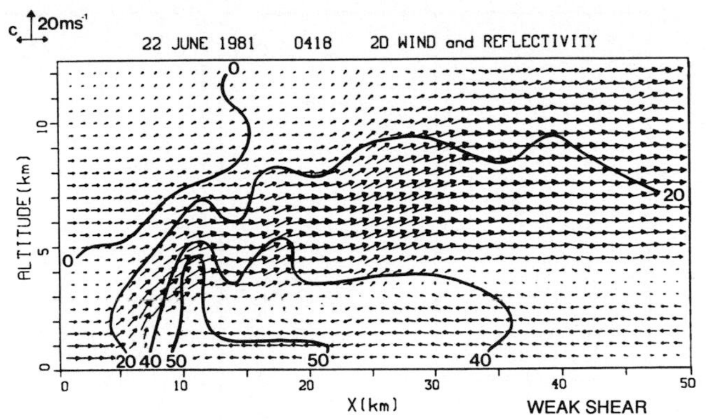

FIG. 6.17. As in Fig. 6.15 but for a portion of the convective line associated with the 22 June COPT convective band. (Figure is from Roux 1985.)

indicate supercells; and lines of supercells probably act as independent transporters.

Table 6.2 shows Richardson numbers for five of the six cases in the "good agreement" box of Fig. 6.12. Four of the six have Richardson numbers of 20 or less; one has a Richardson number of 58. This is contrast to the GATE case, for which Ri = 132. However, this is not the whole story: Ri = 14 for the COPT 22 June case, which falls in the "poor agreement" category, and

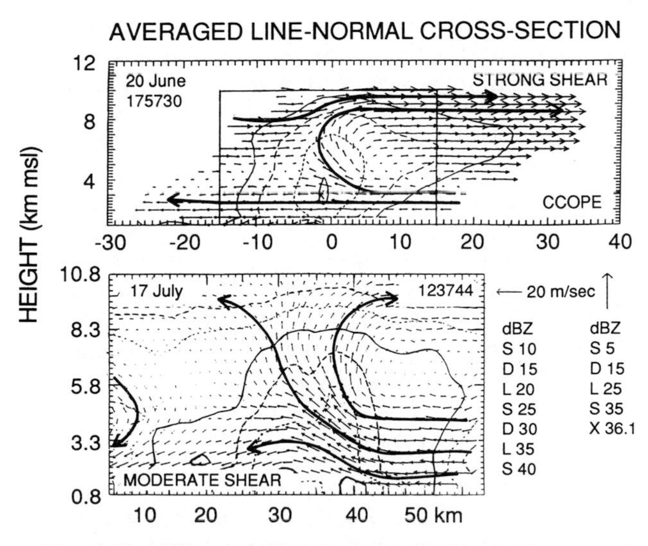

FIG. 6.16. Airflow normal to the convective band axis, averaged along the band, for (a) 20 June CCOPE case, and (b) 17 July CCOPE case, from LeMone et al. (1990).

the data for the 10–11 June system are quite scattered. The scatter in the latter case could be a function of data quality: particularly for the outliers representing the convective band only, the errors from the radiosonde budget can be significant (Gallus 1992, personal communication). For both cases, it is likely that the downdraft contributes significantly to the momentum transport. Lafore and Moncrieff (1989) and Weisman (1992) all emphasize the important role of the buoyancy distribution of the downdraft as well as the updraft air in determining the rear-to-front descending current (commonly called the "rear inflow jet").

f. Effect of the wind profile on convective structure and momentum transport

There are several ways to estimate Richardson numbers for deep convection. Typically, they involve taking the vertical shear over different layers. A general definition of the band-normal convective Richardson number is $\mathrm{CAPE}/(\Delta U)^2$. We can circumvent the problem of selecting a given layer for evaluation of ΔU and normalize U by $(\mathrm{CAPE})^{1/2}$ to isolate effects of the tropospheric wind profile. Figure 6.15 shows the inflow U wind profiles for three cases with different shear. Clearly the COPT case has the smallest normalized shear, while the CCOPE case of 20 June has the largest, particularly in the upper troposphere. The differences in the inflow wind profiles are clearly reflected in the cross sections for these three cases. Figure 6.16 shows average cross sections for the two CCOPE cases, while Fig. 6.17 shows an along-line average through part of the COPT 22 June convective band. Note the effect of the shear on the depth of the front overturning current: that of 20 June is the deepest, that for 17 July is of intermediate depth, and there is no identifiable front overturning current for the COPT case, because the wind tends to flow entirely from front to rear, as in the GATE case (Fig. 6.3).

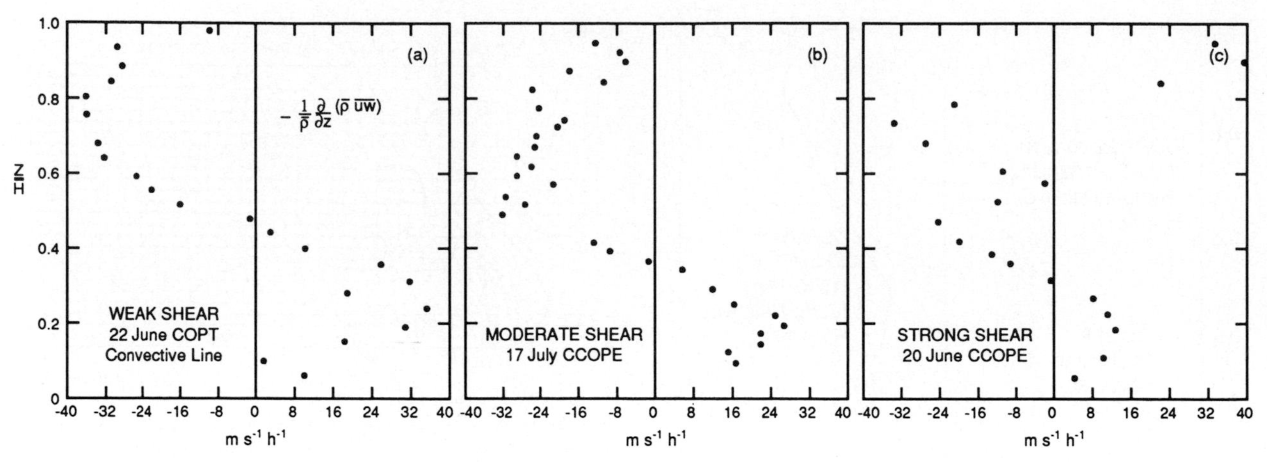

FIG. 6.18. Comparison of momentum flux divergences for the three cases as a function of wind normalized by $(\mathrm{CAPE})^{1/2}$. (a) Convective portion of the 22 June COPT system (weak shear); (b) 17 July CCOPE system (moderate shear); (c) 20 June CCOPE line (strong shear). (22 June data courtesy Michel Chong; 17 July data courtesy T. Matejka, 20 June data courtesy of J. C. Fankhauser.)

This shear-related change in structure has a measurable effect on the momentum flux divergence profiles compared in Fig. 6.18. Note that the strongest shear case has significant positive flux divergence near H. The flux divergence for the moderate shear case is near zero at H, while that for the COPT case remains somewhat negative. For the asymmetric case (Fig. 6.19), the Moncrieff model reflects this change if we interpret a deepening overturning updraft (or decreasing p_{h0}) as more shear aloft: a decrease in p^*_{h0} increases the amount of positive momentum flux divergence near H.

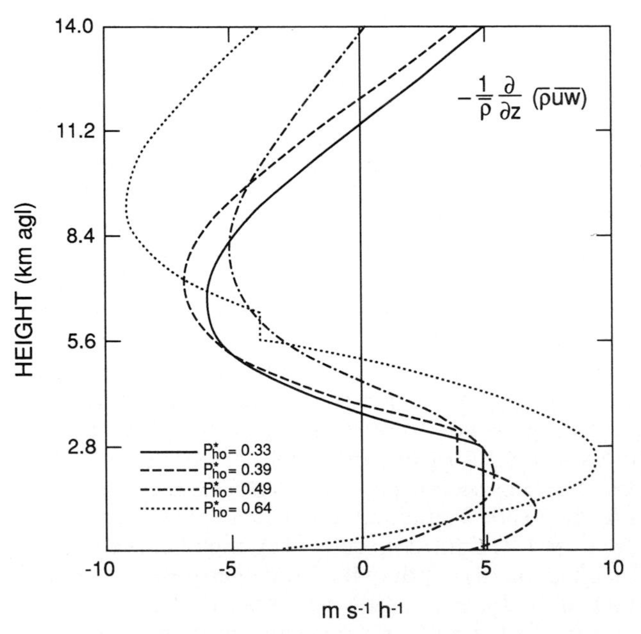

FIG. 6.19. The change of model predictions as a function of change in p^*_{h0}. Since the depth of the front overturning updraft is $1 - p^*_{h0}$, smaller values of p^*_{h0} correspond to deeper overturning updrafts and deeper layers of shear.

g. *Application of the Schneider–Lindzen method*

We have focused on the vertical transport of u momentum, but v-momentum transport is equally important. From the foregoing, both measurements and numerical models suggest that this component of momentum is transported downgradient, suggesting that a Schneider–Lindzen-type parameterization (6.3) would be appropriate, provided allowance is made for the fact that it is designed for estimation of fluxes by fluctuating (primed) quantities. Furthermore, we should ask how useful the Schneider–Lindzen scheme would be in representing line-normal momentum transport in view of its neglect of the pressure field. For simplicity, we will assume that air parcels within the cloud conserve their momentum and use the inflow wind to represent the mean wind $\bar{U}$. Although in many cases this is clearly not so, using the environmental wind to represent the mean wind is consistent with the assumption that actively transporting clouds cover a small fraction of the area of interest. We have only the mean mass flux over the averaging domain for most of the cases, so we will assume that M_c is proportional to the area-averaged mass flux. For simplicity, we set the proportionality constant to unity and recognize that we can predict the shape, but not the magnitude, of the momentum flux divergence curve. In the discussion below, we focus on the GATE system, for which $\bar{\rho}\overline{u'w'} \approx \bar{\rho}\overline{uw}$ (LeMone 1983), and the 10–11 June Oklahoma squall line, for which $\bar{\rho}\overline{u'w'}$ was calculated by Gallus and Johnson (1992).

1) LINE-NORMAL TRANSPORT

The success of this flux representation is variable. For the GATE system (Fig. 6.13), the scheme cannot produce the correct sign of the momentum flux divergence curve at most altitudes, reflecting the fact that the observed flux is countergradient. Nevertheless, there are some successes. For the 10–11 June MCS at 0600 UTC, the Schneider–Lindzen representation does better than the Moncrieff scheme in reproducing the level of zero flux divergence (Fig. 6.20). (Recall that the Moncrieff scheme applies to the total fluxes, while the Schneider–Lindzen scheme applies to the primed fluxes.) The irregularities in the Schneider–Lindzen flux divergence profile result from amplification of fluctuations in the wind profile by taking the vertical derivative. Lipps and Hemler (1991) used in-cloud data from their numerical simulation of the 22 May 1976 case to represent u_c and obtained good agreement of modeled flux divergences with the Schneider–Lindzen scheme.

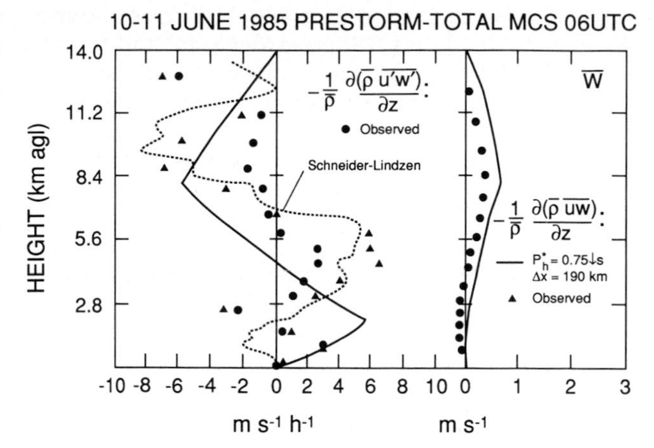

FIG. 6.20. For 10 June 1985 PRE-STORM MCS at 0600 UTC, comparison of $-\partial(\bar\rho\overline{u'w'})/\partial z$ from Schneider–Lindzen scheme (dotted lines) to observed data (●). Moncrieff archetypal model representation and observed estimates (▲) of $-\partial(\overline{\rho uw})/\partial z$ shown for comparison. Data courtesy of W. Gallus and R. Johnson.

2) LINE-PARALLEL TRANSPORT

One would expect greater successes for Schneider–Lindzen applied to line-parallel transport, particularly if realistic physics (entrainment) is used in the determination of v_c. This would mask but not eliminate the sensitivity to small-scale deviations in the environmental V profile. Methods of using this method to represent the line-parallel fluxes are being explored.

6.5. Discussion and conclusions

There is a remarkable universality to the vertical flux of horizontal momentum by quasi-two-dimensional lines of deep convection. If we define U perpendicular to a convective line and positive in the direction of line motion, the flux $\overline{\rho uw}$ is mostly negative, with a minimum at midlevels. The resulting U acceleration is positive at low levels and negative at high levels. This is consistent with the rising front-to-rear updraft and sinking rear-to-front downdraft long associated with squall lines (Newton 1950). The flux profiles retain their shape when we consider entire mesoscale convective systems—the convective line plus associated stratiform precipitation—as well. The flux profiles are also nearly independent of environmental shear, since convective bands can be parallel, perpendicular, or at an angle to it. These cannot be consistently represented by either K theory or mixing-length approaches such as Schneider and Lindzen (1976).

Thus, an approach that idealizes two-dimensional convection is necessary. Moncrieff and his colleagues have developed a series of models that idealize the flow across a convective band in terms of a minimal number of flow branches, whose properties are determined by a set of conservation equations. The most recent and archetypal (most idealized) model appears in Fig. 6.1. There are three branches: a rising front-to-rear updraft called the "jump updraft," a rear overturning current that can correspond to the sinking downdraft or to a density current depending on choice of flow direction, and a front overturning updraft over the jump updraft. The relationship between the heights of the rear overturning current and the jump updraft is obtained from a characteristic regime equation that ignores vorticity generation and low-level environmental shear but includes the necessary conservative properties [Eq. (3) in Moncrieff (1992) with no density differences]. The inflow and outflow speeds of the jump updraft are related by mass continuity; the peak velocities of the overturning currents by kinetic energy (pressure) continuity. The inflow pressure perturbation in front of the line is assumed zero. The pressure difference across the line is found from the inflow and outflow of the jump current by means of Bernoulli's equation; it is equal to the pressure to the rear of the overturning current from the pressure continuity requirement. Accelerations due to buoyancy distribution are not accounted for. In spite of these simplifications, the model reproduces the momentum flux divergence curve remarkably well. However, the comparison with observations suggests that inclusion of the baroclinic vorticity generation through convective buoyancy and diabatic heating in the stratiform region would be advantageous.

Our understanding of momentum flux by organized convection and any future application of this understanding to parameterization schemes relies on our ability to predict how the magnitude and shape of the curve changes as a function of the structure of the mesoscale convective system and its environment. We have compared ten separate cases to examine some of these changes and used them to evaluate the Moncrieff model depicted in Fig. 6.1. Now we summarize the major findings, the success of the model, and how shortcomings may be reduced. We will close by discussing some issues too important to neglect, even though they are beyond the scope of this paper.

a. Summary

The widths of the domains over which the convective systems were sampled, as measured normal to their convective bands, varied considerably, from around 11 km for the 12 July 1985 Hawaii rainband to 250 km for the COPT 22 June mesoscale convective system. This size variation was reflected in the height of largest negative u-momentum transport—the larger the system, the greater the altitude of the extremum. The rise in this height, when associated with the switch from positive acceleration at low levels to negative accelerations higher up fits well with the rise in the crossover point between observed mesoscale updrafts and downdrafts toward the rear of larger systems (see, e.g., Smull and Houze 1987b; Houze et al. 1989).

Buoyancy distribution and associated acceleration of air parcels as they traverse the convection region also have an important influence on momentum flux divergence. If we define a Richardson Ri = CAPE/($\frac{1}{2}U_0^2$), most of the cases for which the Moncrieff model reproduces the amplitude of the divergence curve fairly well have quite low Ri values. The GATE case, where agreement is poor, has the highest value of the ten cases. If we account for the effects of buoyancy in the updraft as in LeMone et al. (1984), the amplitude match improves markedly. The other poor-agreement cases were associated with large systems and deep downdrafts, whose negative buoyancy could have been important.

The vertical U shear also seems to have an important influence on the structure and vertical transport of u momentum by convective systems studied. When normalized by $(\text{CAPE})^{1/2}$ to isolate shear effects, strong shear at higher levels is associated with deeper rear-to-front flow aloft in the front overturning current, associated positive u momentum flux ($u > 0$, $w > 0$) and positive U accelerations near H. These effects are at least qualitatively reproduced by the model since deeper overturning updrafts provide a better fit to strong higher-level shear than do shallow overturning updrafts.

These findings suggest that the Moncrieff model is fairly successful at replicating the momentum flux divergence for convective bands with low values of Ri and without extensive stratiform precipitation ($L_x/H \leqslant 4$–5). The quality of the model prediction could be improved through inclusion of effects of the buoyancy of the updraft and downdraft circulations.

b. Important neglected issues

Besides a method to activate the scheme and determine the grid-scale amplitude (see Moncrieff 1992), application of the Moncrieff model as a parameterization scheme will require 1) a method for determining p_h, p_{h0}, and H a priori, 2) a method for determining line orientation and speed, and 3) a method for calculating line-parallel momentum transport.

1) *Determination of* p_h, p_{h0}, H. In this study, these values were determined from convective band cross sections. However, some patterns have emerged: p^*_{h0} seems to become smaller with greater shear at higher levels. The height corresponding to p_h could be associated with a minimum in the environmental equivalent potential temperature θ_e (e.g., Normand 1946; Fankhauser 1971). Convective systems with high humidities aloft tend to have smaller rear downdrafts or none at all (Jorgensen et al. 1991; Fankhauser et al. 1992). The neutral buoyancy level of an undilute air parcel might be a good estimate of the system depth H.

2) *Orientation and speed.* Though results are mixed, some patterns emerge. Some authors report that squall lines are normal to jetlike profiles (e.g., Betts et al. 1976; Barnes and Sieckman 1984). Extensive work has been done on squall lines perpendicular or nearly perpendicular to low-level shear (e.g., Thorpe et al. 1982; Rotunno et al. 1988). However, lines parallel to the shear are also found (Barnes and Sieckman; LeMone et al. 1984). Bluestein and collaborators (Bluestein and Jain 1985; Bluestein et al. 1987) note that Oklahoma convective lines tend to parallel to low-level (<1 km) shear and 40°–50° to the left of the tropospheric shear, the smaller number being for severe events.

The lines do not travel with the mean wind throughout their depth. Tropical systems have been documented with inflow from the front at all levels (e.g., Zipser et al. 1981). Since the downdraft leads to generation of new cells, its speed is the speed of the system. Under certain conditions, the speed of the downdraft relative to the inflow air can be found from density-current dynamics (e.g., Rotunno et al. 1988; Weisman et al. 1988), while Moncrieff and Green (1972) and Moncrieff and Miller (1976) derived formulas for the propagation speed in terms of a Richardson number and CAPE.

3) *Determination of along-line momentum transport.* A representation of vertical fluxes of v momentum by systems having a nonzero line-parallel flow component ($v \neq 0$) but zero line-parallel gradients ($\partial/\partial y = 0$) can be defined as an extension of the strictly two-dimensional theory as suggested in Moncrieff (1990). This type of weakly three-dimensional flow satisfies the momentum equation $Dv/Dt = 0$ because the along-line pressure gradient ($\partial p/\partial y$) is zero, showing that the v component of momentum is conserved as air parcels flow through the convective system. Since the Schneider–Lindzen method requires this property, it could be applied to the v-momentum transport if allowance is made for the fact that only $\bar{\rho}\overline{v'w'}$ is estimated. Inclusion of a simple cloud model to determine in-cloud y velocity be a useful improvement.

Finally, there have been no comparisons of observed pressure changes across the convective systems to those predicted by the Moncrieff model (6.6). This is because of the great difficulty in estimating pressure changes across the line from data—expected changes are of the order of a few pascals, and observed convective lines typically occur in environments with horizontal pressure gradients of comparable or even greater magnitudes.

Acknowledgments. This paper makes use of the experimental results of many investigators. These results were not only gleaned from the literature but were sent to us by individuals. We are grateful to William Gallus, Michel Chong, James C. Fankhauser, and Thomas Matejka for generously supplying us with data not yet in the literature. Michael Biggerstaff, Stan Trier, and Brad Smull pointed out needed details in the literature.

APPENDIX

Application of the Moncrieff (1992) Archetypal Model

In order to enable the reader to study the behavior of the Moncrieff (1992) model in more detail, the formulas for the u-momentum flux divergence, $\partial\overline{u^*\omega^*}/\partial p^*$, and the nondimensional mass flux ω^* are provided. The momentum flux $\overline{u^*\omega^*}$ is obtained by integrating $\partial\overline{u^*\omega^*}/\partial p^*$ subject to the boundary condition that $\partial\overline{u^*\omega^*}/\partial p^* = 0$ at $p^* = 0$. The formulas here have been written in mass (p^*) coordinates. The fluxes can be put into dimensional form as in section 6.3. The reader is referred to Moncrieff (1992) for further discussion.

a. Momentum flux divergence

In the asymmetric model $p_h^* \leqslant p_{h0}^* = (1 - p_h^*)/(3 - 4p_h^*)$ for all values of E^*. The system-relative momentum flux divergence is given by

$$\frac{\partial}{\partial p^*}(\overline{u^*\omega^*}) = \begin{cases} \dfrac{(2p^* - p_h^*)^2}{p_h^{*2}(3 - 4p_h^*)^2 + 1/2E^* - 1}, & 0 \leqslant p^* \leqslant p_h^* \\ \dfrac{1}{(3 - 4p_h^*)^2} + 1/2E^* - 1, & p_h^* \leqslant p^* \leqslant p_{h0}^* \\ \dfrac{1}{(3 - 4p_h^*)^2} - \dfrac{(2p^* - 1 - p_{h0}^{*2})}{(1 - p_{h0}^*)^2} + 1/2E^*, & p_{h0}^* \leqslant p^* \leqslant 1, \end{cases} \tag{A.1}$$

where E^* is given by (6.6). The symmetric model is independent of E^* (in fact $E^* = 0$) and the solutions span $p_{h0}^* \in [0, 1]$ and $p_h^* \in [0, 1]$. The solution for $p_h^* \leqslant p_{h0}^* = 1 - p_h^*$,

$$\frac{\partial}{\partial p^*}(\overline{u^*\omega^*}) = \begin{cases} \dfrac{(2p^* - p_h^*)^2}{p_h^{*2}} - 1, & 0 \leqslant p^* \leqslant p_h^* \\ 0, & p_h^* \leqslant p^* \leqslant p_{h0}^* \\ 1 - \dfrac{(2p^* - 1 - p_{h0}^*)^2}{(1 - p_{h0}^*)^2}, & p_{h0}^* \leqslant p^* \leqslant 1, \end{cases} \tag{A.2}$$

while for $p_h^* \geqslant p_{h0}^*$,

$$\frac{\partial}{\partial p^*}(\overline{u^*\omega^*}) = \begin{cases} \dfrac{(2p^* - p_h^*)^2}{p_h^{*2}} - 1, & 0 \leqslant p^* \leqslant p_{h0}^* \\ \dfrac{(2p^* - p_h^*)^2}{p_h^{*2}} - \dfrac{(2p^* - 1 - p_{h0}^*)^2}{(1 - p_{h0}^*)^2}, & p_{h0}^* \leqslant p^* \leqslant p_h^* \\ 1 - \dfrac{(2p^* - 1 - p_{h0}^*)^2}{(1 - p_{h0}^*)^2}, & p_h^* \leqslant p^* \leqslant 1. \end{cases} \tag{A.3}$$

b. Mass flux

The vertical mesoscale mass flux is obtained from integration horizontal mass convergence using the "environmental" wind profiles of Fig. 6.1 [Eq. (6.8)]. For the *asymmetric model* that has $p_h^* \leqslant p_{h0}^* = (1 - p_h^*)/(3 - 4p_h^*)$,

$$\omega^* = \begin{cases} p^* + \dfrac{p^*(p^* - p_h^*)}{p_h^*(3 - 4p_h^*)}, & 0 \leqslant p^* \leqslant p_h^* \\ \mathcal{M}_1 + \dfrac{(1/2 - p_h^*)(p^* - p_h^*)}{(3/4 - p_h^*)}, & p_h^* \leqslant p^* \leqslant p_{h0}^* \\ \mathcal{M}_2 - \dfrac{(p^* - p_{h0}^*)}{(3 - 4p_h^*)} - \dfrac{[(p^* - p_U^*)^2 - (p_{h0}^* - p_U^*)^2]}{(1 - p_{h0}^*)}, & p_{h0}^* \leqslant p^* \leqslant 1. \end{cases} \quad \text{(A.4)}$$

For the *symmetric model* and $p_h^* \leqslant p_{h0}^* = 1 - p_h^*$,

$$\omega^* = \begin{cases} p^* + \dfrac{p^*(p^* - p_h^*)}{p_h^*}, & 0 \leqslant p^* \leqslant p_h^* \\ \mathcal{M}_1, & p_h^* \leqslant p^* \leqslant p_{h0}^* \\ \mathcal{M}_1 - (p^* - p_{h0}^*) - \dfrac{[(p^* - p_U^*)^2 - (p_{h0}^* - p_U^*)^2]}{(1 - p_{h0}^*)}, & p_{h0}^* \leqslant p^* \leqslant 1, \end{cases} \quad \text{(A.5)}$$

while for $p_h^* \leqslant p_{h0}^*$,

$$\omega^* = \begin{cases} p^* + \dfrac{p^*(p^* - p_h^*)}{p_h^*}, & 0 \leqslant p^* \leqslant p_{h0}^* \\ \mathcal{M}_3 - \dfrac{[(p^* - p_U^*)^2 - (p_{h0}^* - p_U^*)^2]}{(1 - p_{h0}^*)} + \dfrac{[(p^* - p_D^*)^2 - (p_{h0}^* - p_D^*)^2]}{p_h^*}, & p_{h0}^* \leqslant p^* \leqslant p_h^* \\ \mathcal{M}_4 - (p^* - p_h^*) - \dfrac{[(p^* - p_U^*)^2 - (p_h^* - p_U^*)^2]}{(1 - p_{h0}^*)}, & p_h^* \leqslant p^* \leqslant 1. \end{cases} \quad \text{(A.6)}$$

In these formulas, $p_U^* = 1/2(1 + p_{h0}^*)$ and $p_D^* = 1/2 p_h^*$ are the stagnation points in the overturning updraft and downdraft relative flow, respectively, while $\mathcal{M}_1 = p_h^*$; $\mathcal{M}_2 = \mathcal{M}_1 + (1/2 - p_h^*)^3(3/4 - p_h^*)^{-2}$; $\mathcal{M}_3 = (1 - p_h^*)^2/p_h^*$; and

$$\mathcal{M}_4 = \mathcal{M}_3 - \frac{(p_h^* - p_U^*)^2 - (p_{h0}^* - p_U^*)^2}{1 - p_{h0}^*} + \frac{(p_h^* - p_D^*)^2 - (p_{h0}^* - p_D^*)^2}{p_h^*}.$$

Chapter 7

Cumulus Effects on Vorticity

STEVEN K. ESBENSEN

College of Oceanic and Atmospheric Sciences, Oregon State University, Corvallis, Oregon

7.1. Introduction

Parameterization of the effects of cumulus convection on the vorticity fields of a large-scale dynamical model is a practical problem requiring knowledge of the physical mechanisms of interaction and their cumulative effects. Although the ultimate test of any cumulus parameterization scheme is the validity of the large-scale fields produced by the model, simple empiricism is not a satisfying or efficient approach. In practice, the modeler not only is unsure of the parameterization scheme but has difficulty quantifying the extent of the agreement between the model and the real world. Physically based and internally consistent schemes provide a rational basis for evaluation and model improvement.

Attempts to parameterize cumulus vorticity effects have generally followed one of two approaches. In the first, the cumulus effects on large-scale vorticity are obtained by taking the curl of parameterized cumulus effects on the large-scale momentum field, along the lines first suggested by Pearce and Riehl (1969). In the second, cumulus effects are parameterized from a consideration of vorticity dynamics alone (e.g., Cho and Cheng 1980; Shapiro and Stevens 1980; Yanai et al. 1982). The second approach has now been abandoned by most investigators because of the difficulty of successfully reducing the complex three-dimensional processes acting on vortex tubes in clouds to a small number of parameters.

The equation for the local rate of change of the large-scale momentum of air, for a frame rotating with angular velocity Ω, can be written in the form

$$\frac{\partial \bar{\mathbf{u}}}{\partial t} = -\bar{\mathbf{u}} \cdot \nabla \bar{\mathbf{u}} - 2\Omega \times \bar{\mathbf{u}} - \frac{1}{\bar{\rho}} \nabla \bar{p} - \overline{\nabla \Phi} + \mathbf{X}_m + \mathbf{X}_c, \quad (7.1)$$

where $\mathbf{u}$ is the three-dimensional velocity vector, ρ and p the air density and pressure, respectively, Φ the apparent gravitational potential, $\mathbf{X}_m$ the force due to molecular friction, and $\mathbf{X}_c$ the apparent force exerted on the large-scale flow by the inviscid convective-scale processes associated with cumulus clouds. In general $\mathbf{X}_c$ also includes the effects of convection, turbulence, and gravity waves not directly related to cumulus clouds. The overbar represents a mathematical operator that defines what is meant by large scale.

Since it is the vertical component of vorticity that plays a central role in dynamics of nearly all large-scale atmospheric disturbances, the quantity of interest for our discussion is Z_c, the vertical component of the curl of $\mathbf{X}_c$. Here Z_c is calculated directly from

$$Z_c = \mathbf{k} \cdot \nabla \times \mathbf{X}_c, \quad (7.2)$$

where $\mathbf{k}$ is the unit normal in the vertical direction. The parameterization problem thus reduces to determining $\mathbf{X}_c$, or, more specifically, its horizontal component. Note that the consistency between Z_c and $\mathbf{X}_c$ is assured by a simple mathematical operator.

We will discuss the parameterization problem in more detail in sections 7.2 and 7.3, but some important characteristics of Z_c may be anticipated solely on the basis of Eq. (7.2). For example, cumulus clouds can affect the vertical component of the vorticity only when there is a horizontal variation in $\mathbf{X}_c$—the cumulus momentum effects. Apparent force $\mathbf{X}_c$ will vary only if there is spatial inhomogeneity in the population of cumulus clouds, the large-scale wind field, or some combination of the two. In regions of weak horizontal wind shear on the large scale, Eq. (7.2) focuses our attention on properly representing the horizontal gradient of cumulus activity, rather than its presence or absence.

The purpose of this chapter is to introduce the reader to the observational basis and framework for parameterization of Z_c. Section 7.2 discusses the methods for obtaining observational estimates of Z_c and reviews the observational evidence for physically significant values of Z_c in large-scale atmospheric vorticity budgets. Sections 7.3 and 7.4 present and discuss a consistent framework for the parameterization of Z_c.

7.2. Observed vorticity budget residuals

Since no existing observational system can provide a direct determination of Z_c over a large-scale area, we must estimate Z_c as the residual of the large-scale vorticity budget. The accuracy of the diagnosed Z_c depends primarily on the quality of the upper-air wind data and the spatial analysis of the wind field and its spatial derivatives.

Following Chu et al. (1981), we may write the vertical component of the curl of Eq. (7.1) in the pressure coordinate form

$$Z_c = \frac{\partial \bar{\zeta}}{\partial t} + \bar{\mathbf{V}} \cdot \nabla \bar{\zeta}_a + \bar{\zeta}_a \nabla \cdot \bar{\mathbf{V}} + \bar{\omega} \frac{\partial \bar{\zeta}}{\partial p} + \mathbf{k} \cdot \nabla \bar{\omega} \times \frac{\partial \bar{\mathbf{V}}}{\partial p}, \quad (7.3)$$

where $\mathbf{V}$ is the horizontal velocity vector, ζ and ζ_a the vertical components of the relative and absolute vorticity, and ω the time rate of change of pressure following an air parcel (the proxy for vertical motion in pressure coordinates). The local temporal and horizontal derivatives in (7.3) are evaluated along isobaric surfaces. Effects of molecular friction can be neglected to a high degree of accuracy over the interior of the fluid, and the accuracies of the pressure and density fields are of secondary importance for the vorticity budget calculation.

The usual approach to evaluating the terms on the rhs of (7.3) is to perform a spatial analysis of $\bar{\mathbf{V}}$ from available atmospheric soundings and other sources of wind information. The analysis can vary in sophistication from a linear interpolation between upper-air wind soundings to the data assimilation scheme of a numerical weather prediction model. Values of $\bar{\omega}$ can be obtained by vertically integrating the mass continuity equation in the form

$$\nabla \cdot \bar{\mathbf{V}} + \frac{\partial \omega}{\partial p} = 0, \quad (7.4)$$

with the divergent part of $\bar{\mathbf{V}}$ adjusted to allow $\bar{\omega}$ to satisfy physically realistic boundary conditions at the top and bottom of the cloud domain.

Great care must be taken in constructing the observed vorticity budget. Upper-air wind observations are typically accurate to no better than 1 m s^{-1}. An error in the divergent wind component of 1 m s^{-1} over a distance of 250 km gives a divergence error of 4 $\times$ 10^{-6} s^{-1}, which implies a divergence term error in the tropics of nearly the same order of magnitude as the budget residual itself. The advection and twisting terms involve second-order derivatives of the wind field and are subject to errors of similar magnitude.

Much of the difficulty is associated with the fact that upper-air wind data is spatially and temporally sparse on the synoptic scale over much of the earth. The analysis of $\bar{\mathbf{V}}$ involves converting these sparse upper-air wind observations into continuous, three-dimensional, time-varying wind fields. Sophisticated data assimilation schemes (e.g., Bengtsson et al. 1982) can extrapolate and interpolate wind data into data-sparse domains by dynamically constraining the analyzed wind field, but care must be taken to assure that the dynamical constraints do not essentially determine the values of Z_c.

Sparse data create aliasing problems. In the vicinity of strong cumulus convection, the spatial and temporal variability of the wind field can be as strong as the large-scale signal. Without additional information, the undersampled variability due to the cumulus clouds cannot be separated from the large-scale fields. In other words, the cloud-scale variability goes under the alias of large-scale variability in the analyzed field unless the analyst can use experience in the form of dynamical or statistical information to filter out contributions from the undersampled variations.

Even large-scale fields that are well sampled in either the spatial or the temporal domain can present serious representational problems for the analyst. Cumulus activity that is well sampled and persistent in time may be undersampled in space. Typical examples where representational problems may become severe are cumulus convection in the vicinity of a midlatitude front and cumulus convection along the intertropical convergence zone. In these examples, the cloud patterns projected onto a horizontal plane are effectively one-dimensional from a large-scale point of view. Nonzero values of Z_c in the vicinity of these line structures may be simply due to representational limitations of the analysis method. A closely related problem is the validity of Reynolds averaging procedures in the presence of disturbances with no clear separation of space or time scales (e.g., Cotton 1986).

Finally, the estimation of horizontal advection and twisting terms requires accurate and consistent second-order spatial derivatives of $\mathbf{V}$. If $\bar{\mathbf{V}}$ is represented by a twice-differentiable analytical function, the value of Z_c can be obtained directly and without ambiguity. If finite differences are used, a consistent scheme must be developed. As noted by Haynes and McIntyre (1987), the divergence and twisting terms on the rhs of Eq. (7.3) should not be viewed as independent physical processes that act as sources or sinks of vorticity in the same way that evaporation and condensation act as sources or sinks of atmospheric water vapor. Indeed, the rhs of Eq. (7.3) may be written in "flux" form

$$Z_c = \frac{\partial \bar{\zeta}}{\partial t} + \frac{\partial}{\partial x}\left(\bar{u}\bar{\zeta}_a + \bar{\omega}\frac{\partial \bar{v}}{\partial p}\right) + \frac{\partial}{\partial y}\left(\bar{v}\,\bar{\zeta}_a - \bar{\omega}\frac{\partial \bar{u}}{\partial p}\right). \quad (7.5)$$

Finite-difference forms of Eqs. (7.3) and (7.5) must be consistent, to avoid large errors in Z_c due to the incomplete cancellation of terms.

The difficulties in analyzing the large-scale vorticity budget can nevertheless be overcome using research quality data and analysis techniques. Figure 7.1, from Sui and Yanai (1986), demonstrates the statistical significance of vorticity budget residuals based on a careful analysis using the wind fields of Ooyama (1987) and a finite-differencing scheme that maintains the consis-

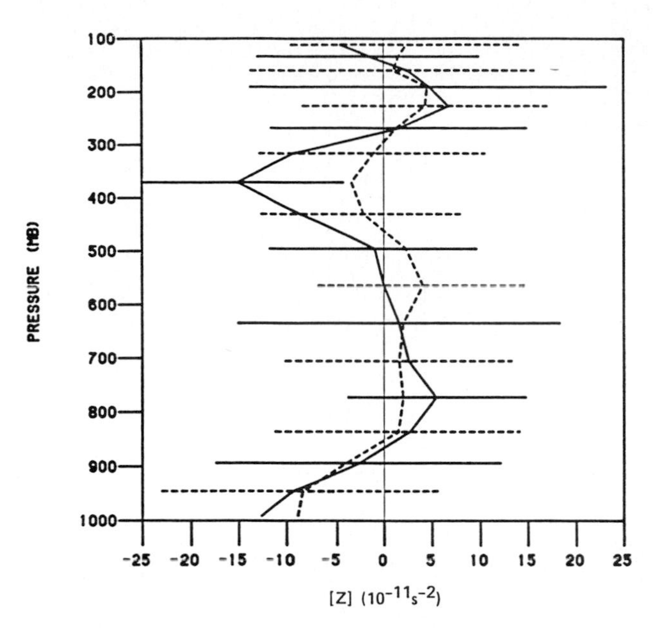

FIG. 7.1. Composited vertical profiles of Z_c with standard deviations for disturbed (solid) and undisturbed (dashed) conditions over eastern tropical Atlantic Ocean during GATE (from Sui and Yanai 1986).

tency among Eqs. (7.3)–(7.5). The composited values of Z_c in the vicinity of active convection depart from zero at some elevations by amounts on the order of the standard deviation or greater, while the values in undisturbed conditions above the boundary layer typically do not.

On the synoptic scale, observed Z_c values are found to be as large as any of the explicitly calculated large-scale terms on the right-hand side of Eq. (7.3) (see Sui and Yanai 1986 and references therein). Figure 7.2, from Chu et al. (1981), shows an example of the vertical profiles of terms in the time-mean, large-scale vorticity budget for a tropical region with deep cumulus cloud activity. As often occurs, the divergence term in the vicinity of an atmospheric disturbance with deep convection tends to be large ($\sim 10^{-10}$ s^{-2}) and acts to generate cyclonic vorticity in the lower troposphere and anticyclonic vorticity in the upper troposphere. To the extent that low-level convergence and upper-level divergence are characteristic of deep cumulus cloud systems, we can take the profile in Fig. 7.1 to be characteristic of the divergence term, since $\bar{\zeta}_a$ tends to be cyclonic for large-scale flows. The local change and advection terms in Fig. 7.2 are the same order of magnitude as the divergence term, while the twisting term is somewhat smaller. Unlike the divergence term, however, the vertical structure of the local change, horizontal and vertical advection, and twisting terms cannot be generalized across large-scale systems. In synoptic-scale waves that are strongly constrained by the rotation of the earth, for example, the local change term tends to compensate the horizontal advection of absolute vorticity (e.g., Shapiro 1978).

The Z_c field may have significant horizontal variations in the vicinity of deep cumulus convection. Figure 7.3, from Tollerud and Esbensen (1983), shows the apparent production by cumulus convection of a large-scale vorticity couplet in the upper troposphere over the eastern tropical Atlantic Ocean during GATE [GARP (Global Atmospheric Research Program) Atlantic Tropical Experiment]. This pattern was interpreted by Tollerud and Esbensen (1983) and Sui et al. (1989) as observational evidence for the importance of vertical momentum mixing by cumulus convection on upper-tropospheric vorticity fields.

At the planetary scale, recent theoretical and observational studies indicate that cumulus effects are relatively unimportant (e.g., Sardeshmukh and Hoskins 1985). Haynes and McIntyre (1987) suggest the apparent inconsistency between synoptic- and planetary-scale vorticity budget studies may be simply a matter of scale, since the second and third terms on the rhs of (7.5) have the form of a divergence. Averaging Eq. (7.5) over a planetary-scale area and applying the divergence theorem, we anticipate that all three terms will be much smaller than the observationally estimated, synoptic-scale values of Z_c in the vicinity of strong cumulus convection. In particular, the local change term is small because the planetary-scale features are quasi-stationary; the second and third terms on the rhs are small because their sum tends to decrease with averaging regardless of the magnitude of the divergence and twisting terms [see Eq. (7.3)] on the interior of the averaging domain; and the average of Z_c is small because it is the residual of a budget with small terms.

7.3. Interpretation of vorticity budget residuals

Based on the observational evidence, we conclude that cumulus clouds can have important effects on the large-scale vorticity budget in the vicinity of strong cumulus activity. It is of interest therefore to interpret the residual Z_c in terms of a simplified model of cumulus convection that might provide the basis for a cumulus parameterization scheme. As a demonstration, we present one possible framework for parameterization based on the momentum budget approach (Tollerud and Esbensen 1983; Esbensen et al. 1987).

We visualize the effects of the convection on the large-scale vorticity by considering the changes in large-scale circulation defined over a horizontal area A. In the context of a gridpoint model, A would be the gridbox area; in a spectral model, the area A would be related to the smallest explicitly resolved horizontal scales of the model. The area should be chosen large enough to contain an ensemble of cumulus cloud elements but small enough so it can be considered a small fraction of the area covered by the large-scale disturbance of interest.

We assume that it is possible to define a value of $\bar{\mathbf{V}}$, and the statistical properties of a cumulus cloud en-

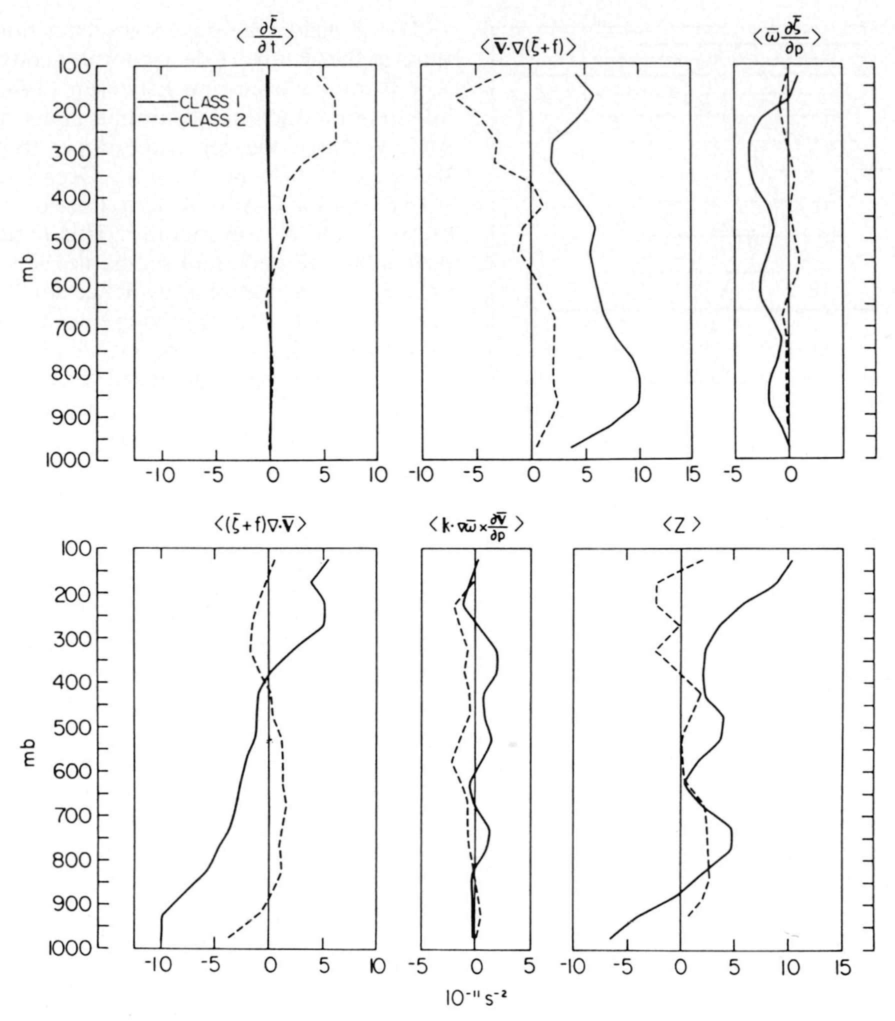

FIG. 7.2. Composited large-scale vorticity budgets for disturbed (solid) and undisturbed (dashed) cases over the western tropical Pacific Ocean (from Chu et al. 1981).

semble, at every point along the perimeter of A. The large-scale horizontal velocity $\bar{\mathbf{V}}$ may be defined by a running spatial average over an area that is approximately the same size as A, or by some other spatial filter. We define the large-scale circulation C around A as

$$C = \oint \bar{\mathbf{V}} \cdot d\mathbf{l}, \tag{7.6}$$

where $d\mathbf{l}$ is an incremental unit distance in the direction of the counterclockwise path around the perimeter of area A.

The average circulation C/A is a good approximation in most cases to the value of the large-scale vorticity $\bar{\zeta}$ $(= \mathbf{k} \cdot \nabla \times \bar{\mathbf{V}})$ at the center of A. Cumulus-induced changes in the circulation C, and hence $\bar{\zeta}$, may therefore be viewed as the cumulative effects of clouds acting *locally and independently* on $\bar{\mathbf{V}}$ in vertical columns at each point along the path. In the case of a gridpoint model, the change in circulation occurs from the accumulation of changes in $\bar{\mathbf{V}}$, computed in each grid-box column without benefit of information about wind shear or cumulus activity in neighboring grid boxes.

To explicitly introduce information about the structure of the elements in the cumulus cloud ensemble, we adopt the framework proposed by Ooyama (1971). Following this scheme, we model cumulus clouds as independent buoyant elements sharing a common dynamical environment described by three parameters: the large-scale horizontal velocity vector $\bar{\mathbf{V}}$, the large-scale vertical p velocity of the cloud environment ω_e, and the large-scale geopotential field $\bar{\phi}$. For notational convenience, we will use a "single-cloud-type" formulation to describe the bulk effects of all cloud types.

The mass conservation law for the large-scale environment of the clouds is given by the expression

$$\nabla \cdot \bar{\mathbf{V}} + \frac{\partial \omega_e}{\partial p} = D - E, \tag{7.7}$$

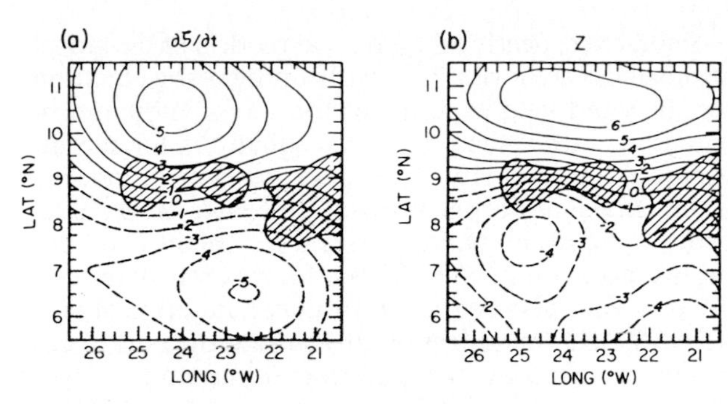

FIG. 7.3. Composited local vorticity change and Z_c fields at 227 mb in advance of easterly wave troughs over the eastern tropical Atlantic Ocean during GATE. Units are 10^{-10} s^{-2}. In hatched regions, composited cloud cover at 300 mb is 30% or greater (from Tollerud and Esbensen 1983).

where D and E are the mass entrainment and detrainment rates, respectively, summed over all cumulus cloud elements. Note that the total mass flux is the sum of the convective and environmental values; that is, $\bar{\omega} = -M_c + \omega_e \approx -g\overline{\rho w}$, where w is the vertical component of the velocity vector.

Within this framework, the momentum equation for the large-scale environment of the cumulus clouds is written

$$\frac{\partial \bar{\mathbf{V}}}{\partial t} = -\bar{\mathbf{V}} \cdot \nabla \bar{\mathbf{V}} - \omega_e \frac{{}_e\partial \bar{\mathbf{V}}}{\partial p} - f\mathbf{k} \times \bar{\mathbf{V}} - \nabla \bar{\phi} + D(\mathbf{V}_D - \mathbf{V}) + \mathbf{F}_c, \quad (7.8)$$

where $\mathbf{V}_D$ is the value of the velocity detrained to the large-scale environment of the clouds, f the Coriolis parameter, and $\mathbf{F}_c$ all convectively generated forces (such as the dynamic pressure exerted by the cloud on the large-scale flow) that cannot be explicitly represented as a cloud transport.

If we use the definition of $\bar{\omega}$ to rewrite the vertical advection term as $\bar{\omega}\partial\bar{\mathbf{V}}/\partial p$, an additional virtual source of momentum (Ooyama 1971, the compensating subsidence term) must be added to the rhs of (7.8). The equation governing $\bar{\mathbf{V}}$ then becomes

$$\frac{\partial \bar{\mathbf{V}}}{\partial t} = -\bar{\mathbf{V}} \cdot \nabla \bar{\mathbf{V}} - \bar{\omega} \frac{\partial \bar{\mathbf{V}}}{\partial p} - f\mathbf{k} \times \bar{\mathbf{V}} - \nabla \bar{\phi} - M_c \frac{\partial \bar{\mathbf{V}}}{\partial p} + D(\mathbf{V}_D - \bar{\mathbf{V}}) + \mathbf{F}_c. \quad (7.9)$$

Determination of $\mathbf{V}_D$ and $\mathbf{F}_c$ is the crucial step in closing the parameterization and requires a model of the momentum budget of the cloud. It is assumed that M_c can be determined from closure hypotheses that maintain certain equilibrium states or by relating the cumulus mass flux to the large-scale horizontal convergence of an explicit property of the large-scale flow (see discussion of closure hypotheses by Arakawa in this volume). The mass detrainment is related to M_c through the cloud mass budget. For observational validation of the parameterization scheme represented by Eq. (7.9), both M_c and detrainment may be obtained diagnostically from the mass, heat, and moisture budgets (see chapter 4).

The large-scale vorticity budget is readily obtained from (7.9) by taking the curl

$$\frac{\partial \bar{\zeta}}{\partial t} = -\bar{\mathbf{V}} \cdot \nabla \bar{\zeta}_a - \bar{\zeta}_a \nabla \cdot \bar{\mathbf{V}} - \bar{\omega} \frac{\partial \bar{\zeta}}{\partial p} - \mathbf{k} \cdot \nabla \bar{\omega} \times \frac{\partial \bar{\mathbf{V}}}{\partial p} - M_c \frac{\partial \bar{\zeta}}{\partial p} - \mathbf{k} \cdot \nabla M_c \times \frac{\partial \bar{\mathbf{V}}}{\partial p} + \mathbf{k} \cdot \nabla \times [D(\mathbf{V}_D - \bar{\mathbf{V}})] + \mathbf{k} \cdot \nabla \times \mathbf{F}_c. \quad (7.10)$$

Consistency with Eq. (7.3) requires that

$$Z_c = -M_c \frac{\partial \bar{\zeta}}{\partial p} - \mathbf{k} \cdot \nabla M_c \times \frac{\partial \bar{\mathbf{V}}}{\partial p} + \mathbf{k} \cdot \nabla \times [D(\mathbf{V}_D - \bar{\mathbf{V}})] + \mathbf{k} \cdot \nabla \times \mathbf{F}_c. \quad (7.11)$$

The physical statement made by Eqs. (7.9)–(7.11) is that the momentum transport effects of clouds can change the large-scale vorticity only by changing the momentum field that defines the large-scale environment of the clouds. All cloud parameters can be computed without knowledge of the vorticity in the cloud or its environment. The Z_c term contains terms having the form of vertical vorticity advection and twisting, but the effects of the three-dimensional advection and twisting of vortex tubes in the clouds and their surroundings are not explicitly represented. Some formulations of Z_c obtained by detailed consideration of the vorticity budgets of the cloud and its environment (e.g., König and Ruprecht 1989 and references therein) reduce to essentially the same form as (7.11) but require numerous untested assumptions to achieve consistency with (7.11).

The reader should note that the appearance of pressure effects in the large-scale vorticity equations comes from viewing changes in $\mathbf{V}$ from the point of view of the large-scale environment of convection. Combining (7.8) with a proper budget of in-cloud momentum (e.g., Shapiro and Stevens 1980) results in the cancellation of the excess pressure gradient forces in the clouds. In other words, the excess pressure force exerted by a cloud on its environment should be exactly balanced by an opposing pressure force of the environment on the cloud. From the point of view of the momentum budget averaged over both the cloud and its environment, the excess pressure forces play no direct role. Thus, the appearance of pressure effects in the last term on the right-hand side of Eq. (7.11) does not violate physical principles when the in-cloud momentum budget is properly formulated.

7.4. Concluding remarks

A consensus appears to be emerging on the observational evidence and the framework for parameterization of cumulus effects on vorticity. The consistent formulation of Z_c obtained by taking the curl of cumulus momentum effects has been tested diagnostically in tropical systems by Sui et al. (1989) and appears to provide a simple alternative to parameterization schemes based on direct consideration of vorticity dynamics. This is an encouraging result since the explicit representation of subgrid-scale vorticity dynamics can significantly increase the number of parameters that must be specified by calibration with either observed data or models (e.g., Cho and Clark 1981). If the details of the cumulus vorticity budget are important for a specific modeling application, increasing the spatial resolution of the model may be the only practical alternative for the near future.

In practice, nearly all numerical models of the global atmosphere are formulated with a prognostic equation for the wind field in terms of the vector momentum equation, rather than a pair of equations for the divergence and vorticity. The questions facing modelers of the global atmosphere therefore center on the adequacy of current parameterization schemes for cumulus momentum effects. Even for theoretical models of large-scale disturbances formulated in terms of vorticity, the parameterization of Z_c using the curl of cumulus momentum effects appears to have advantages over a parameterization based on a direct consideration of the vorticity dynamics of clouds.

Finally, parameterizing cumulus effects on vorticity by means of Eq. (7.2) focuses attention on the momentum budget of cumulus clouds. Clouds do not simply mix horizontal momentum between adjacent layers in the atmosphere. Momentum transfers by convective clouds may be either upgradient or downgradient depending on the atmospheric stratification and wind shear (e.g., LeMone 1983; LeMone et al. 1984). Progress depends on being able to parameterize the interaction between the momentum and pressure fields in the vicinity of convection to determine quantities such as $\mathbf{V}_D$ or $\mathbf{F}_c$ in Eq. (7.11). Additional discussion of the effects of cumulus convection on the large-scale momentum fields can be found in chapter 6.

Acknowledgments. The author thanks Drs. Michio Yanai, Xiaoqing Wu, and Edward Tollerud for numerous helpful suggestions and comments. The preparation of this chapter was supported by the National Science Foundation under Grant ATM-8718475.

Part II

Schemes for Large-Scale Models

Chapter 8

Convective Adjustment

WILLIAM M. FRANK

Department of Meteorology, The Pennsylvania State University, University Park, Pennsylvania

JOHN MOLINARI

Department of Atmospheric Science, State University of New York at Albany, Albany, New York

8.1. Basic principles

The term *convective adjustment* describes any of several procedures used in numerical models to simulate the effects of dry and/or moist convection by adjusting the lapse rates of temperature and moisture to specified profiles within the local grid column. The origins of convective adjustment lie in simple concepts of local mixing within unstable layers. For example, most numerical models assume that dry superadiabatic lapse rates overturn spontaneously, mixing to a neutral lapse rate. The original form of the moist convective adjustment cumulus parameterization (Manabe et al. 1965) applied the same logic to conditionally unstable layers that become saturated in models. The physical properties of the clouds are entirely implicit. The characteristic of convective adjustment schemes that distinguishes them from other cumulus parameterizations is their direct specification of the adjusted vertical structure of the column without attempting to simulate the explicit convective processes.

There are several approaches to convective adjustment, but all of those currently in use function by adjusting the lapse rate of a density-related variable toward a reference profile that is related to the stability of saturated parcels. Examples of such variables are the equivalent potential temperature θ_e, virtual potential temperature θ_v, and virtual temperature T_v.

The primary arguments in favor of using a convective adjustment scheme are as follows.

1) Convection tends to reduce the large-scale CAPE (convective available potential energy). It is plausible to assume that it will do so by adjusting the vertical stratification toward a state that is approximately neutral for moist convection. The time scale of convection is much smaller than that of circulations resolved in large-scale models, so an assumption of instantaneous adjustment to a neutral state can be supported.

2) Convective cloud ensembles in nature are complex subgrid-scale phenomena. Attempts to simulate their integrated effects by defining the actual properties of subgrid-scale clouds require the use of many arbitrary parameters whose values are poorly known in nature, and require enormous computing power. Convective adjustment is a conceptually simple and straightforward approach that uses relatively little computer time.

By the above arguments, convective adjustment would seem to be best suited for large-scale models and for physical problems whose study requires considerable computational resources. Three-dimensional general circulation or climate models are an obvious example. It will be argued, however, that instantaneous moist convective adjustment is not suitable for any model except under certain narrowly defined circumstances.

8.2. The observational basis

Convective adjustment makes sense as a cumulus parameterization only if it is possible to identify some sort of reference profile toward which the vertical stratification adjusts as a result of convection. It is reasonable to examine observations from regions of heavy convection to see if they have common stratification characteristics. It is best to examine data from regions with weak baroclinicity, where horizontal advection of temperature does not obscure the effects of the convection. Vertical profiles of θ_v computed with data from two tropical maritime datasets are shown in Fig. 8.1. Two of the four profiles represent mean soundings averaged over the entire third phase of the GATE [GARP (Global Atmospheric Research Program) Atlantic Tropical Experiment] in the eastern Atlantic and the entire AMEX (Australian Monsoon Experiment) in the Gulf of Carpentaria. The mean soundings are representative of long-term average conditions in these

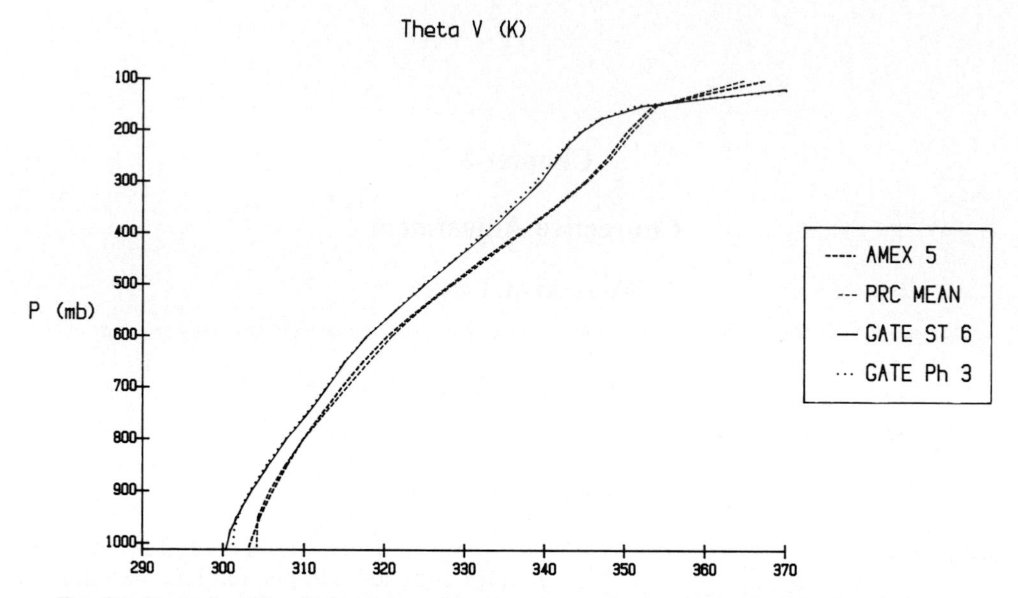

FIG. 8.1. Vertical profiles of virtual potential temperature θ_v for four tropical soundings. The rightmost two curves are from AMEX; the left two are from GATE. AMEX 5 is a composite sounding from the late decay stages of four cloud clusters, while PRC mean is a six-week mean sounding from a ship during AMEX. GATE ST 6 is a composite for the decay stages of eight GATE cloud clusters, and GATE Ph 3 is the phase III mean for the GATE array.

heavily disturbed regions. (Both experiments were centered on the mean locations of the intertropical convergence zone). The other two profiles are composite profiles from the same experiments computed from soundings taken only during the decay stages of cloud clusters. The composite cluster soundings represent the stratification over areas a few hundred kilometers in diameter immediately after major deep convective events.

It is immediately obvious that there is a qualitative similarity between the sets of profiles in Fig. 8.1. The shapes of θ_v profiles in other tropical convective regimes are generally similar. Further, the profiles resemble those of moist adiabats to some degree. Betts (1986) argues that lapse rates below the freezing level tend to resemble reversible adiabats of lifted low-level air, while Cohen and Frank (1989) show that the mean stratification tends to lie parallel to ice pseudoadiabats for levels higher than the freezing level. Xu and Emanuel (1988) related soundings in convective tropical regions to reversible adiabats of lifted low-level air.

A reference moisture profile is more difficult to define from observations. Although water vapor is of great importance in determining conditional instability of air parcels or layers, it has only a small effect upon the virtual temperature. Thus, a specified lapse rate of θ_v could exhibit a wide range of relative humidities. Moisture tends to be much more spatially variable than does temperature, particularly in convective regions. As a result, it is hard to define or observe a characteristic adjusted moisture profile for postconvective environments.

When changes in the atmospheric stratifications within mesoscale convective systems are examined on time scales as short as a few hours, the values of θ_v above the boundary layer usually change by less than 1°C regardless of the amount of latent heat that is released by the convection. (The one common exception is that cooling near cloud tops is sometimes large.) This is illustrated for an average of four AMEX cloud clusters in Fig. 8.2. As in many other studies, virtually all of the change in stability occurs due to changes in the θ_e of the boundary layer, as colder, drier air from downdrafts replaces the warmer, moister air of the undisturbed environment. Changes similar to those of Fig. 8.2 are typically observed in midlatitude continental convective systems as well. The explanation for the small temperature changes is just that the latent heat release is occurring over horizontal scales that are small relative to the Rossby radius of deformation. Therefore, the response to the mass perturbation of the heating is almost completely divergence, and the realization of the heating as temperature change is extremely small except at the top and bottom of the column, where the divergence is constrained by the stability of the stratosphere and the material boundary, respectively.

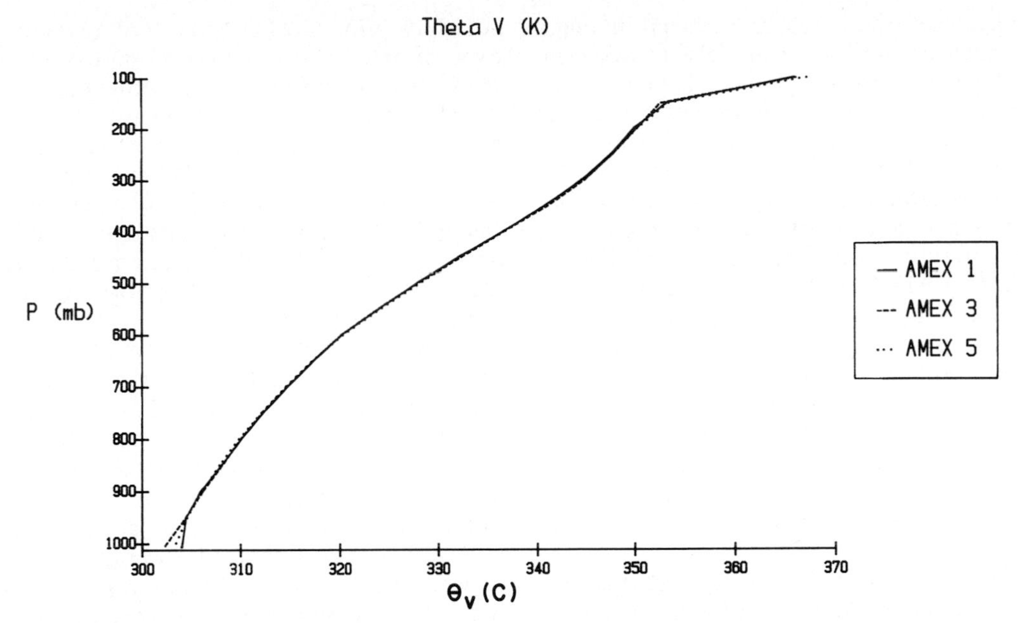

FIG. 8.2. Vertical profiles of θ_v for a composite of four maritime cloud clusters during AMEX. AMEX 1 is the initial stage, AMEX 3 is the time of maximum rainfall, and AMEX 5 is the late decay stage.

8.3. Classical convective adjustment schemes

a. Hard adjustment

Hard adjustment is the name given to the original Manabe et al. (1965) procedure. Convective adjustment is invoked only within layers that are saturated and convectively unstable. The basic procedure is to mix contiguous saturated layers iteratively, beginning with the lowest, until they are neutrally stratified (i.e., θ_e is constant with height). The final state is saturated in the adjusted layer. The iterative mixing process may extend the depth of the adjustment layer upward somewhat by increasing θ_e at the top of the initially unstable layer. The process of adjusting a layer of saturated air from a conditionally unstable lapse rate to a neutral one lowers the mean saturation mixing ratio in the layer, producing supersaturation. This supersaturation is removed by condensation, which in turn produces latent heat release and also an estimate of rainfall. The adjustment is assumed to occur in a single time step.

When an unstable layer becomes saturated and hard adjustment is invoked, very large rainfall rates may occur. This can happen even during a numerical forecast because, as noted by Molinari and Dudek (1986) and Emanuel (1991), unrealistically large convective instability can build prior to the occurrence of saturation on the grid scale. In nature, convective rain begins before relative humidity reaches 100%. The saturated final state in hard adjustment is also rarely observed in nature after convection has occurred.

b. Soft adjustment

Soft adjustment attempts to remedy the weaknesses of hard adjustment by assuming that saturation occurs only over a specified fraction of the grid area, with the air between the clouds remaining unchanged. Alternatively, one may specify the final mean relative humidity in the grid column. This assumption introduces an additional empirical parameter, but it allows the parameterization to activate before entire layers become saturated. The earliest example of soft adjustment was proposed by Miyakoda et al. (1969) in which saturation was simply defined as 80% relative humidity. This assumption allowed convective adjustment to activate prior to grid-scale saturation, and the final state was unsaturated as well.

8.4. Shortcomings of convective adjustment

The use of instantaneous convective adjustment to simulate the effects of convection, particularly deep, precipitating convection, has significant drawbacks.

1) Since convective adjustment makes no attempt to simulate the physical properties of the clouds themselves, some convective effects will be estimated with little accuracy. The most obvious problem is with the moisture field. Convective effects upon moisture depend upon compensating subsidence between clouds and detrainment of both cloud air and hydrometeors. These influences cannot be estimated realistically by simply specifying the relative humidity of the adjusted state. The evolution of the moisture field is important

in most applications, and it becomes crucial in longer simulations, such as in climate models. In addition, convective effects upon the momentum field and upon the production and transport of hydrometeors cannot be described from simple lapse-rate adjustments.

2) The timing and amount of convection estimated by convective adjustment are very sensitive to the conditions under which convection is assumed to begin.

3) Moist convective adjustment tends to influence only the layers in which θ_e decreases with height, or those immediately above them. Deep convection in nature depends upon parcel instability and can influence upper-tropospheric levels that are convectively stable. Moist convective adjustment invariably underestimates the depth of the adjusted layer.

4) The physical basis of the reference profile of θ_v to which the atmosphere is adjusted is not known. Current schemes use empirical profiles or moist pseudoadiabats.

5) By choosing a saturated final state, hard adjustment chooses the coldest possible adiabat. When the adjustment layer includes the surface, the resultant cold surface temperatures produce unrealistically large surface fluxes locally. Global energetics can be significantly distorted by these large energy inputs, as was noted by Ooyama (1982b).

6) Hard adjustment has an additional problem due to the fact that the final adjusted state is saturated in the adjustment layer. Subsequent condensation will occur in explicit clouds. Such explicit clouds have considerably different vertical profiles of heating than do convective clouds.

7) J. McBride (BMRC—Australia; 1993, personal communication) argues that moist convective adjustment procedures in general should not be used as a cumulus parameterization in models with grids much finer than the radius of deformation. Briefly, convective adjustment computes both the amount and vertical distribution of convective heating based on assumed adjustment to a reference profile that is determined from local lower-tropospheric θ_e values. However, the observed reference profiles (e.g., Figs. 8.1 and 8.2) reflect a large-scale equilibrium resulting from interactions between surface fluxes, radiation, large-scale circulations, and moist processes occurring over very large areas. On the other hand, θ_e and CAPE vary significantly on small horizontal scales due to large gradients of temperature and moisture, particularly in the boundary layer. As a result, any reference profile defined using local θ_e values in a small grid column is unlikely to reflect the large-scale equilibrium profile.

8.5. Possible improvements

Krishnamurti et al. (1980) compared several forms of cumulus parameterization in semiprognostic calculations using GATE data. Hard adjustment greatly overestimated precipitation, as noted above. Krishnamurti et al. optimized their soft adjustment by specifying various adjusted state relative humidities, then choosing a fractional cloud coverage that minimized the errors in calculated rainfall rates for each specified relative humidity. The optimal values for the GATE case were 82% relative humidity and a fractional cloud area of 3.7%. Nevertheless, this optimized soft adjustment was one of the worst-performing cumulus parameterizations in their tests.

Several attempts have been made to improve on convective adjustment while retaining the concept of a neutral final state. Kuo (1974), Kurihara (1973), and Betts and Miller (1986) compute the depth of the adjusted layer using parcel concepts rather than localized grid-scale instability. Each of these can be viewed as a time-dependent convective adjustment, in which it is not assumed that adjustment occurs instantaneously. In the Kuo and Kurihara schemes the time scale at which neutrality is approached depends upon moisture convergence in the column, while Betts and Miller (1986) specify the adjustment time explicitly. The adjusted vertical profiles of temperature and water vapor are specified. The reference relative humidity is saturated in the Kuo and Kurihara approaches and unsaturated in the Betts–Miller scheme. The Kuo and Betts–Miller schemes are described in more detail in other chapters.

Several cumulus parameterizations combine an explicit cloud model with the assumption that convection acts to restore the local grid column to a stratification based on moist parcel stability. The cloud model is used to estimate the properties of the convection in terms of the resolved conditions, and a closure assumption specifies the amount of convection that occurs in order to achieve the desired rate of stabilization. Early examples of this approach are the Arakawa and Schubert (1974) and Fritsch and Chappell (1980) schemes, which are discussed in other chapters. Cloud model schemes with stability-based closures represent an attempt to estimate convective properties in a more physically consistent manner than is possible using the empirical techniques of convective adjustment. There are tradeoffs, since the increased complexity of the cloud model schemes introduces new problems and requires additional assumptions. Increased complexity does not always lead to increased accuracy. In climate models, drift of the moisture field away from observed values can be a major problem. Modified moist convective adjustment, like that of Betts and Miller (1986), allows the relative humidity of the final state to be controlled. On balance, however, the authors feel that the cloud model approach is more promising than convective adjustment for parameterizations intended for research applications.

8.6 Conclusions

Instantaneous moist convective adjustment provides an effective approach for removing local instability within saturated layers, along the lines proposed by Suarez et al. (1983). When used in the same manner to parameterize deep, boundary-layer-based convection, it has many shortcomings. For mesoscale research models, in which computer demands are not an overriding consideration, the actual properties of the clouds should be included in the parameterization in order to model realistically the convective effects on the moisture field. Large-scale models are more consistent with the equilibrium assumptions that form the basis of convective adjustment schemes but such models are particularly sensitive to the forecasts of moisture, which are not physically based in these schemes. Traditional hard or soft moist convective adjustment does not appear to be a prudent choice for a complete cumulus parameterization scheme in numerical models of any scale.

Acknowledgments. This work was supported by National Science Foundation Grants ATM-8817915 and ATM-9115950 (W. Frank, PI), and ATM-8902487 (J. Molinari, PI).

Chapter 9

The Betts–Miller Scheme

ALAN K. BETTS

Atmospheric Research, Pittsford, Vermont

MARTIN J. MILLER

European Centre for Medium-Range Weather Forecasts, Reading, Berkshire, United Kingdom

9.1. Introduction

The impetus for the development of this simple lagged convective adjustment scheme came from the series of tropical field experiments in the decade 1969–79 [VIMHEX, the Venezuela International Meteorological the Hydrological Experiment in 1969 and 1972; GATE, the GARP (Global Atmospheric Research Program) Atlantic Tropical Experiment in 1974, and MONEX, the Monsoon Experiment in 1979]. Deep convection is the dominant vertical transport process in the tropics. In conjunction with the radiation field and the subsiding branches of the tropical circulations, convective processes maintain a vertical thermal structure, which is quite close to the moist adiabat through the equivalent potential temperature θ_e of the subcloud layer in the regions of deep convection. This was the basis of early cumulus parameterization schemes. Manabe (1965) proposed adjustment toward a moist-adiabatic structure to remove conditional instability in large-scale models. Kuo (1965, 1974) proposed a simple cloud model for deep convection that adjusted the atmosphere toward the saturated moist pseudoadiabat in the presence of grid-scale moist convergence. However, the mean tropical atmosphere is always cooler by several degrees in the middle troposphere than this reference moist adiabat, even in regions of vigorous convection (see Figs. 9.5–9.7). At the same time, the deep convective transports also maintain the water vapor and cloud distributions in the tropics, which in turn play a crucial role in the radiative fluxes.

One of the key objectives of GATE in 1974 was to study organized deep convection in order to test and develop convective parameterizations for numerical models (Betts 1974a). Meanwhile, the work of Ooyama (1971), Arakawa and Schubert (1974), Ogura and Cho (1973), and Yanai et al. (1973) had sparked much research to parameterize and model tropical deep convection in terms of spectral cloud ensembles with a simple entraining cloud model. Betts (1973a) had formulated deep convective transports in terms of an updraft and downdraft mass circulation using conserved variables. Diagnostic studies from GATE showed the importance of mesoscale updrafts and downdrafts in addition to convective-scale processes and the microphysical effects of freezing, melting, and water loading (Houze and Betts 1981). Similar studies from MONEX confirmed the importance of processes on many scales (Houze et al. 1981). Frank (1983) concluded from these phenomenological studies that cloud models of greater complexity might be needed to parameterize cumulus convection. Work has continued on these lines (Frank and Cohen 1987; Krueger 1988), but progress has been slow. The introduction of every new degree of freedom in a parametric cloud model requires a new closure assumption, and it remains impossible to integrate a numerical cloud model of much realism at every grid point in a global model. Typically, simplified mass flux schemes have been developed as parameterizations for operational global models—for example, Tiedtke (1989) at the European Centre for Medium-Range Forecasts (ECMWF) and Gregory and Rountree (1990) at the U.K. Meteorological Office. The reader is referred to these original papers, since this work is not summarized in this volume.

The Betts–Miller scheme was designed to represent directly the quasi-equilibrium state established by deep convection, so as to avoid the uncertainties involved in attempting to determine this state indirectly using increasingly complex cloud models, whose closure parameters can themselves ultimately be determined only by comparison with atmospheric observables. The concept of a quasi equilibrium between the convective field and the large-scale forcing was introduced for shallow convection by Betts (1973b) and for deep convection by Arakawa and Schubert (1974). The idea, in fact, has a long history (e.g., Ludlam 1966). On larger space and longer time scales, quasi equilibrium has been well established (Lord and Arakawa 1980; Lord 1982; Arakawa and Chen 1987). Quasi equilibrium means that the convective cloud field tightly constrains the temperature and moisture structure of the atmosphere. Arakawa and Schubert (1974) closed the problem by constraining the cloud work function. Betts and Miller (1986) chose to independently constrain the temperature and moisture structure. In the presence of extensive deep convection, the atmospheric thermal structure rather than approaching the pseudoadiabat seems to remain slightly unstable to the wet virtual adiabat at least up to the freezing level (Betts 1982, 1986; Xu and Emanuel 1989; Binder 1990). The theoretical basis for this is not completely clear. The wet virtual adiabat (Betts 1983; Betts and Bartlo 1991) includes the effect of liquid water loading, which clearly plays an important role in reducing buoyancy and updraft velocities in the weaker updrafts found in tropical cumulonimbi (Zipser and LeMone 1980). Cloud modeling studies (Cohen 1989; Cohen and Frank 1989) have shown that the quasi-equilibrium thermal state does indeed approach this wet virtual adiabat. In addition, atmospheric soundings in convective regions typically show a stable kink in the thermal structure near the freezing level, suggesting that the freezing–melting process, which is closely tied to the 0°C isotherm, also has an important control on the midtropospheric equilibrium thermal structure. Thus, it seems likely that below the freezing level both water loading and, perhaps to a lesser extent, melting of falling frozen precipitation may contribute to the observed quasi-equilibrium thermal structure, unstable to the pseudoadiabat but close to the wet virtual adiabat. Similarly, above the freezing level, the observed more stable structure is probably associated with both the freezing process and the fallout of precipitation, which reduces the liquid and ice loading on a parcel. In the Betts–Miller scheme, we chose to use the wet virtual adiabat as a reference process up to the freezing level, to link the complex physics of cloud fields to a well-defined thermodynamic reference process that appears to have an observational and some (albeit incomplete) theoretical basis. Although it can be argued with some justification that this is a semiempirical approach, our objective is a simple parameterization of complex processes on a wide range of scales. Since the observations and cloud-scale models suggest that both water loading (and therefore the microphysics) and the freezing–melting process are important in determining the thermal structure, if we do not parameterize the structure directly we must include all these processes with some realism in parametric cloud models. So far, few cumulus parameterization schemes have faced this daunting task.

Since its introduction, the scheme has been tested at ECMWF (Heckley et al. 1987) and in data assimilation studies (Puri and Miller 1990; Puri and Lonnberg 1991), and used in limited-area models (Janjic 1990), theoretical studies of hurricane development (Baik et al. 1990), and climate (Lawsen and Eliasen 1989). Here we discuss some of the improvements that have been made to the scheme since these works were published.

9.2. Observational basis

The convection scheme involves a lagged adjustment toward calculated reference profiles. The procedure selected for calculating these reference profiles was influenced by observational studies of convective equilibrium. Betts and Miller (1986) showed examples of equilibrium structure for deep and shallow convection, and subsequent papers have generally provided supporting evidence. In this section, we shall illustrate the basis of the parameterization scheme using examples from observations. The formal details of the parameterization scheme are left to sections 9.3 and 9.4.

a. Shallow scheme

Figure 9.1 shows a parametric idealization of the coupling of a temperature and dewpoint T_D structure of a convective layer to a mixing line from Betts (1985, 1986). The four points, A, B, C, and D, on the heavy dashed mixing line (a line of constant $\partial\theta/\partial q$) at a saturation pressure p^* are connected with a corresponding pair of (T, T_D) points at a pressure level p (the dashed lines) by lines of constant θ and q. Although Fig. 9.1 shows a single mixing line, the convective boundary layer (CBL) has three distinct layers, which have different characteristic values of the gradient of saturation pressure p^* with p. We define a parameter

$$\beta = \frac{\partial p^*}{\partial p}. \tag{9.1}$$

Note that the gradient with p of θ and q are linked by β to the slope of the mixing line

$$\frac{\partial \theta}{\partial p} = \beta\left(\frac{\partial \theta}{\partial p^*}\right)_M \tag{9.2a}$$

$$\frac{\partial q}{\partial p} = \beta\left(\frac{\partial q}{\partial p^*}\right)_M, \tag{9.2b}$$

where the suffix M denotes the mixing line. The parameter β represents in some sense a measure of the mixing within and the coupling between convective layers. The insert on the right of Fig. 9.1 shows β for the three layers of the schematic CBL. Here $\beta = 0$ would represent a well-mixed layer: the subcloud layer often approaches this structure. In Fig. 9.1, the subcloud layer has $0 < \beta < 1$, which represents a layer not quite mixed, in which the profiles of θ, q converge with height toward saturation on the mixing line. The lower part of the cloud layer is drawn with $\beta = 1$; this is a partially mixed structure in which the θ, q (or T, T_D) profiles are approximately parallel to the mixing line. Near cloud top $\beta > 1$; this represents the divergence

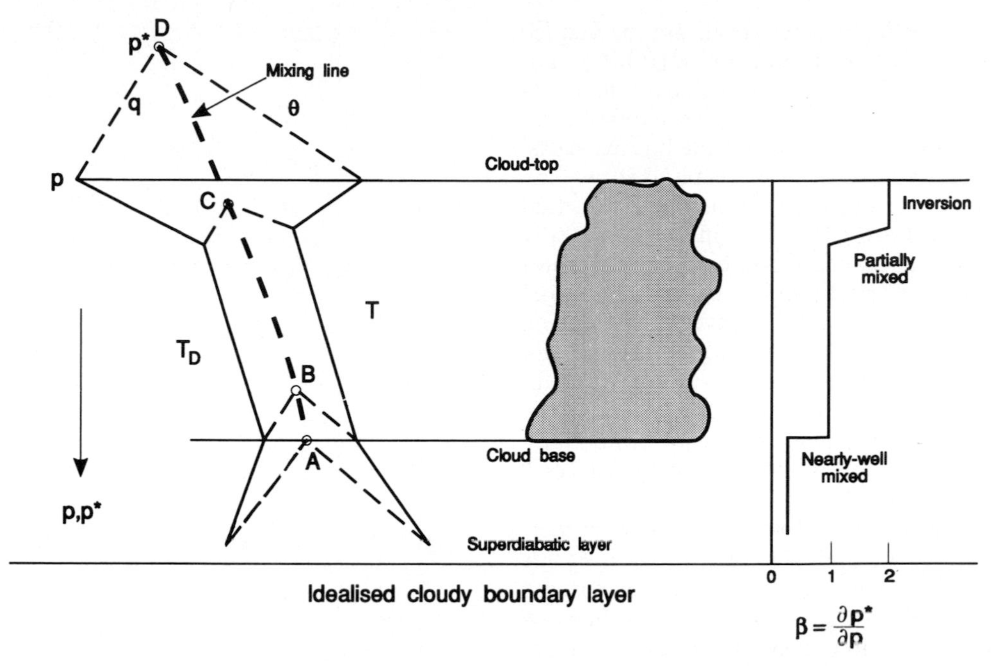

FIG. 9.1. Relationship between mixing line, temperature, and dewpoint, and a mixing parameter β for an idealized convective boundary layer. The light dashed lines are lines of constant potential temperature θ and mixing ratio q (from Betts 1986).

of θ and q from the mixing line that is characteristic of the transition through an inversion at the top of a convectively mixed layer to the free atmosphere. Figure 9.2 shows an actual mean CBL sounding for the equatorial Pacific from Betts and Albrecht (1987) as profiles of saturation equivalent potential temperature θ_{es} and equivalent potential temperature θ_e with pressure.

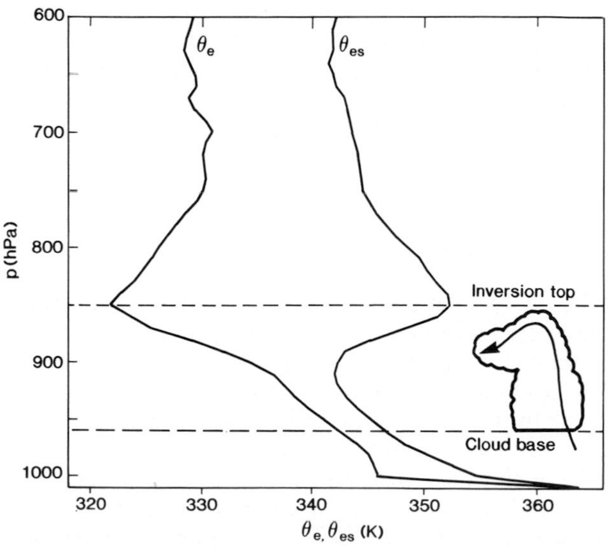

FIG. 9.2. Mean profiles of equivalent potential temperature θ_e and saturation equivalent potential temperature θ_{es} through an oceanic convective boundary layer (data from Betts and Albrecht 1987).

Cloud base is near 960 hPa, and the main trade-wind inversion is between 890 and 850 hPa in this average: the minimum in θ_{es} is a little lower at 910 hPa. The top of the inversion is marked by a maximum in θ_{es} and a minimum in θ_e, which means that the air just above the CBL has a minimum in relative humidity. This is a typical structure in the trade winds of the central equatorial Pacific. Figure 9.3 shows the same mean sounding on a conserved variable diagram: a θ^*–q^* plot, which for this unsaturated sounding is identical to a θ–q plot. The open circles are (θ, q) plotted every

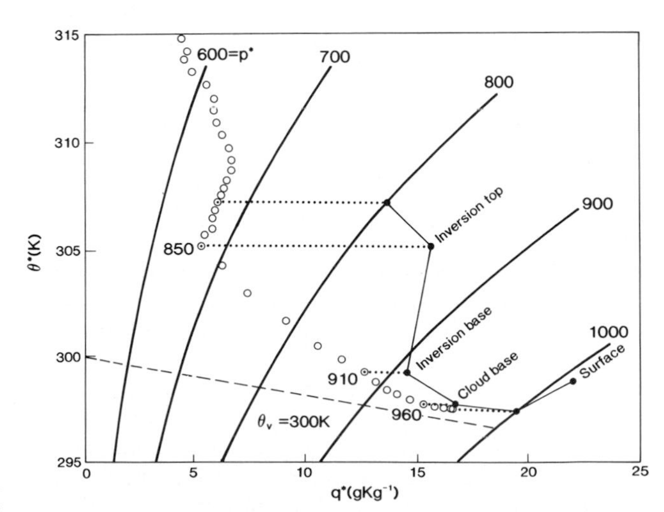

FIG. 9.3. Conserved variable plot (θ^*, q^*) of data from Fig. 9.2.

10 hPa from 1000 hPa. Selected levels are marked for cloud base, (960 hPa), θ_{es} minimum (910 hPa), and inversion top (850 hPa). There is a sharp kink at inversion top that marks the top of the convectively mixed layer. Within the CBL the profile has two parts. Below cloud base the θ–q structure is parallel to a line of neutral density for dry convection, the dry virtual adiabat (Betts and Bartlo 1991); a line of $\theta_v = 300$ K is shown (dashed). Between cloud base and inversion top the profile can be approximated by a linear mixing line, signifying a layer in which convection is mixing air sinking through the trade inversion with air in the subcloud layer, itself coupled to the ocean. In this cloud layer the profile is actually slightly curved, because the time scale of the convective mixing is of order many hours, during which time radiative cooling reduces θ at constant q (Betts 1982).

The open circles in Fig. 9.3 are data plotted at their saturation level p^*, θ^*, and q^*. The profile of $\theta(p)$ is also shown on this figure by solid circles at selected levels, joined by light lines. Each $\theta(p)$ point shown is connected by a dotted line at constant θ to the corresponding saturation point at p^* (the corresponding lifting condensation level). The solid circle marked surface represents saturation at the ocean surface temperature and pressure (1010 hPa). Note that the trade inversion, which is very marked on a $\theta(p)$ or $\theta_{es}(p)$ plot (Fig. 9.2), appears only on a θ^*–q^* plot as a wider spacing of points on the same mixing line. The air above the CBL has its own characteristic structure, which here has almost constant p^*. Betts and Albrecht (1987) suggested that this air had sunk over many days, originating from deep convective outflows near the freezing level.

Figure 9.4 shows the actual plot of $\beta = \partial p^*/\partial p$ for this observed profile. The inversion layer has a large value of β peaking near 9.5, while the cloud layer has nearly constant β near 1.15. The subcloud layer has smaller values of β, but the subcloud values are probably a little large, because of averaging of the poor vertical resolution dropsonde data (Betts and Albrecht 1987).

The shallow convection scheme in the present ECMWF code does not treat convection in the subcloud layer; this is done by a separate diffusion scheme. We then specify $\beta = 1.2$ from cloud base to cloud top, as a reasonable approximation to the observational studies in Betts and Albrecht (1987). Cloud top is defined as the level below buoyancy equilibrium (see section 9.3c). This means that the subsaturation parameter $\mathcal{P} = p^* - p$ increases slowly (in magnitude) in the cloud layer since $\partial\mathcal{P}/\partial p = \beta - 1 = 0.2$. The value of $\mathcal{P}$ is not specified but implicitly determined by the two separate integral energy constraints on water vapor and enthalpy (see section 9.3f), since we assume shallow convection does not precipitate. A linear approximation to the slope $\partial\theta/\partial p$ of the mixing line is computed between low-level air and air from two levels

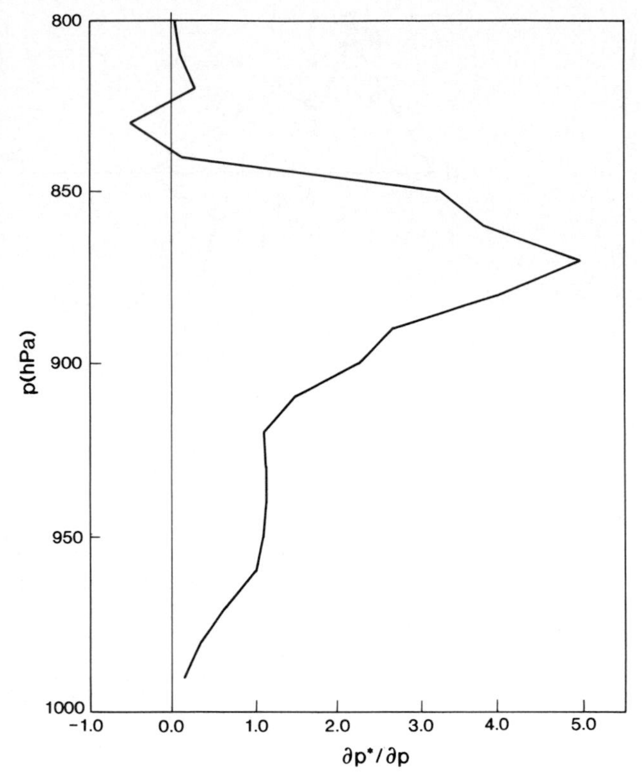

FIG. 9.4. Profile of parameter $\beta = \partial p^*/\partial p$ (the gradient of saturation pressure with pressure) through the convective bondary layer shown in Fig. 9.2.

above cloud top. One can see in Fig. 9.3 that a low vertical resolution model using data perhaps 30–50 hPa above the inversion top will tend to overestimate the mixing-line slope. Therefore, we have introduced a weighting parameter (presently 0.85) to reduce the mixing-line slope to compensate for this (see section 9.3f). At the model level just above cloud top, an intermediate value of β is estimated to smooth the transition at cloud top. A generalization of the parametric model may well be useful in which β is a function of mixing-line slope (Betts 1985), but this has not been implemented.

b. Deep scheme

The deep scheme adjusts toward reference profiles characteristic of deep convective equilibrium. Figure 9.5 shows average profiles of the temperature difference ΔT_w from the moist pseudoadiabat through cloud base for the hurricane eyewall composite (solid) and 2° radius composite (dashed) from Frank (1977). The mean sounding for this extreme convective situation is cooler than the pseudoadiabat throughout the troposphere, reaching a maximum in ΔT_w, corresponding to a minimum in θ_{es}, at 500 hPa, a little above the freezing level. The wet virtual adiabat is shown dotted. This is the neutral density curve for the reversible adiabat in

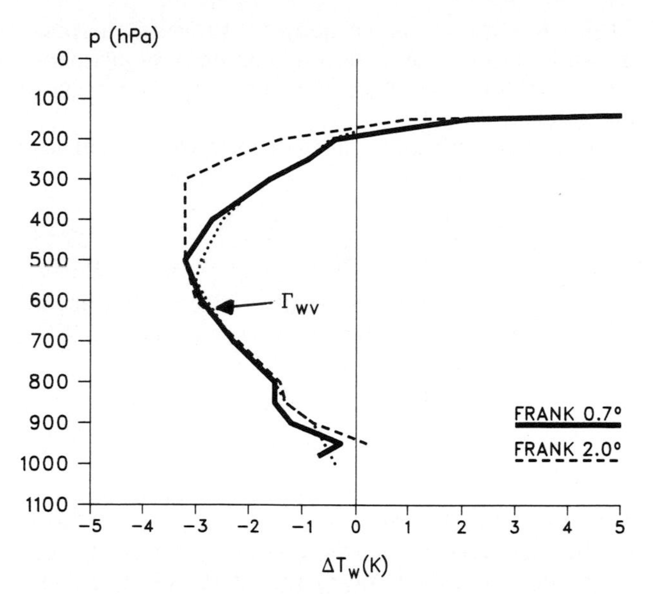

FIG. 9.5. Profile of ΔT_w (temperature difference between sounding and moist pseudoadiabat through cloud base) against pressure for two hurricane averages (data from Frank 1977).

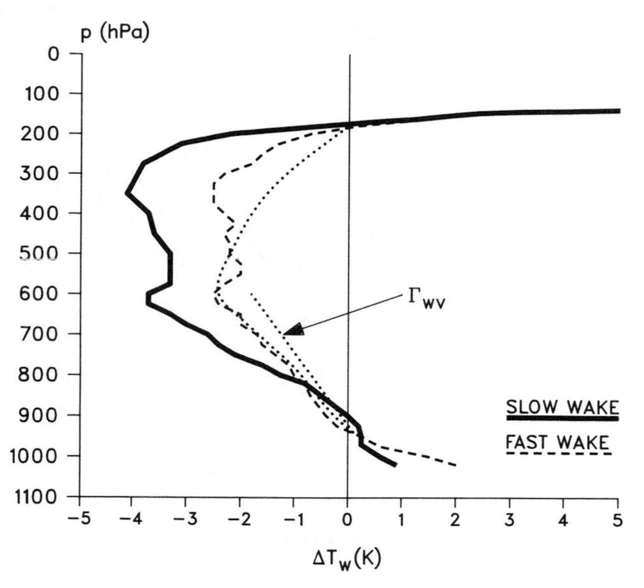

FIG. 9.6. As Fig. 9.5 for two GATE averages (data from Barnes and Sieckman 1984).

which the liquid water loading is included. The temperature profile up to the freezing level is quite close to this reference adiabat. The dotted profile above the freezing level is a quadratic fit between the freezing level and the outflow equilibrium level. The dashed profile, at 2° radius from the storm center, is a little more unstable below the freezing level, and it is cooler above than the eyewall composite. Figure 9.6 shows a second pair of curves for the wake of slow-moving (solid) and fast-moving (dashed) GATE lines from Barnes and Sieckman (1984). They are similar to the hurricane composites. The slow-moving composite has a slope above cloud base (near 950 hPa) roughly 1.3 times that of the wet virtual adiabat (dotted), while the fast-moving composite has a more unstable structure with a slope twice that of the wet virtual adiabat in the lower troposphere. Figure 9.7 shows a composite of 24 soundings within and in the vicinity of eight large storm systems (area greater than 2000 km^2) in Venezuela (see Betts 1976). Again the profiles are similar to Figs. 9.5 and 9.6, although the cloud base over land is higher, near 850 mb. The slope of ΔT_w up to the freezing level is 2.2 times that of the wet virtual adiabat. The data in Binder (1990) over Switzerland are similar. The profile above the freezing level is not far from the quadratic curve shown dotted.

Figure 9.8 shows the mean profiles of $\mathcal{P}$ for the five averages shown. Unlike Figs. 9.5–9.7 for the thermal structure, which have some qualitative similarities, there is a large variation in moisture between the averages for different types of storms and regimes. The fast-moving storms tend to have stronger low-level downdrafts, which dry out the lower troposphere, but inject more moisture into the upper troposphere, while slow-moving systems over the ocean (and over land, not shown) tend to have moister outflows in the lower troposphere. In the present version of the scheme, we use a simple mean profile (heavy line), which has a minimum at the freezing level and a linear variation above and below.

9.3. Original Betts–Miller scheme

The scheme was designed to adjust the atmospheric temperature and moisture structure back toward a ref-

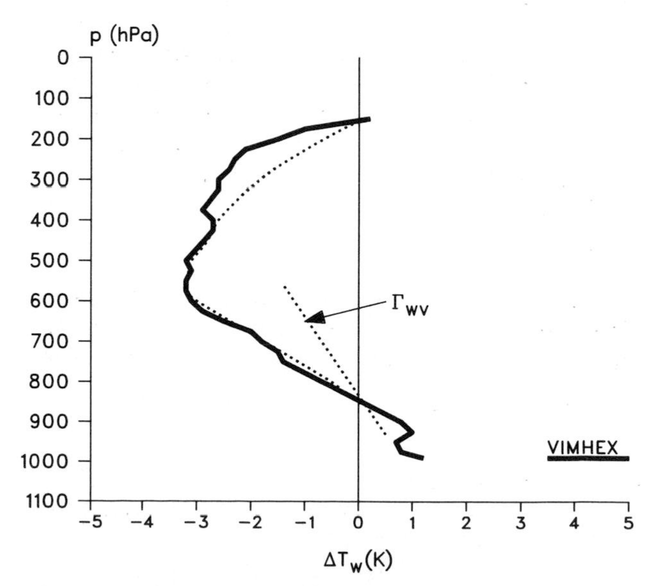

FIG. 9.7. As Fig. 9.6 for average of 24 soundings in the outflow of eight large storms over Venezuela (data from Betts 1976).

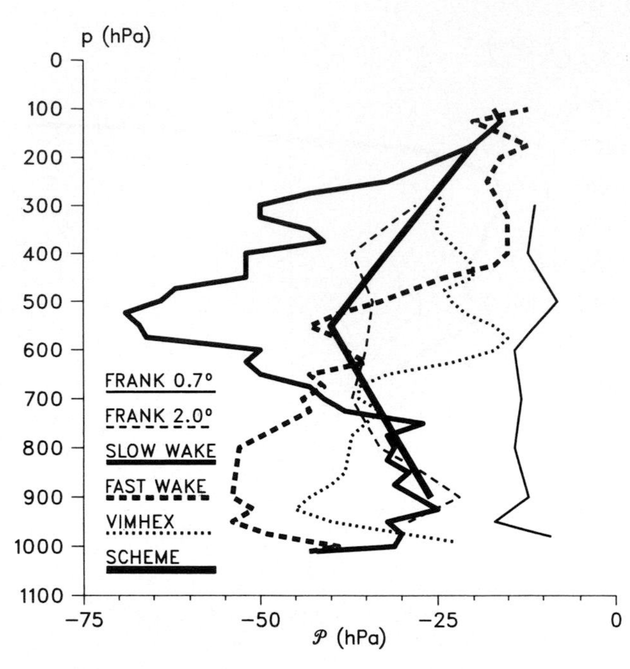

FIG. 9.8. Profile of saturation pressure departure ($\mathcal{P} = p^* - p$) with pressure for the five convective averages, and the profile used by the scheme.

erence quasi-equilibrium thermodynamic structure in the presence of large-scale radiative and advective processes. Two distinct reference thermodynamic structures (which are partly specified and partly internally determined) are used for shallow and deep convection. We have subsequently introduced a revised low-level adjustment to simulate a downdraft mass flux into the boundary layer (BL): this is discussed in section 9.4. Formally the convection scheme involves four parts: the specification of τ the adjustment time scale, finding cloud base and cloud top, determining the reference profiles for deep and shallow convection, and the method of distinguishing between deep and shallow convection. Figure 9.9 shows a flow diagram for the overall scheme. We shall discuss the formal structure of a lagged adjustment scheme first and then outline the components.

a. Formal structure

If we denote a saturation point (θ^*, q^*) by a two-dimensional vector $\mathbf{S}$ (Betts 1983), the large-scale thermodynamic tendency equation can be written as

$$\frac{\partial \bar{\mathbf{S}}}{\partial t} = -\mathbf{V} \cdot \nabla \mathbf{S} - \bar{\omega}\frac{\partial \bar{\mathbf{S}}}{\partial p} - g\frac{\partial \mathbf{N}}{\partial p} - g\frac{\partial \mathbf{F}}{\partial p}, \quad (9.3)$$

where $\mathbf{N}$, $\mathbf{F}$ are the net radiative and convective fluxes (including the precipitation flux). The convective flux divergence is parameterized as

$$-g\frac{\partial \mathbf{F}}{\partial p} = \frac{\mathbf{R} - \bar{\mathbf{S}}}{\tau}, \quad (9.4)$$

where $\mathbf{R}$ is the reference quasi-equilibrium thermodynamic structure and τ is a relaxation or adjustment time representative of the convective and unresolved mesoscale processes.

Simplifying the large-scale forcing to the vertical advection and combining (9.3) and (9.4) gives

$$\frac{\partial \bar{\mathbf{S}}}{\partial t} = -\bar{\omega}\frac{\partial \bar{\mathbf{S}}}{\partial p} + \frac{\mathbf{R} - \bar{\mathbf{S}}}{\tau}. \quad (9.5)$$

If the large-scale forcing is steady on time scales longer than τ, then the atmosphere will reach a quasi equilibrium with $\partial \bar{\mathbf{S}}/\partial t \approx 0$. Then

$$\mathbf{R} - \bar{\mathbf{S}} \approx \bar{\omega}\left(\frac{\partial \bar{\mathbf{S}}}{\partial p}\right)\tau. \quad (9.6)$$

In a T-106 global model we use $\tau = 1$ h (see below). This means that $\mathbf{R} - \bar{\mathbf{S}}$ corresponds to one hour's forcing by the large-scale fields, including radiation. For deep convection the atmosphere will therefore remain slightly cooler and moister than $\mathbf{R}$. Furthermore, for small τ, the atmosphere will approach $\mathbf{R}$, so that we may substitute $\mathbf{S} \approx \mathbf{R}$ in the vertical advection term, giving

$$\mathbf{R} - \bar{\mathbf{S}} \approx \omega\tau\frac{\partial \mathbf{R}}{\partial p}, \quad (9.7)$$

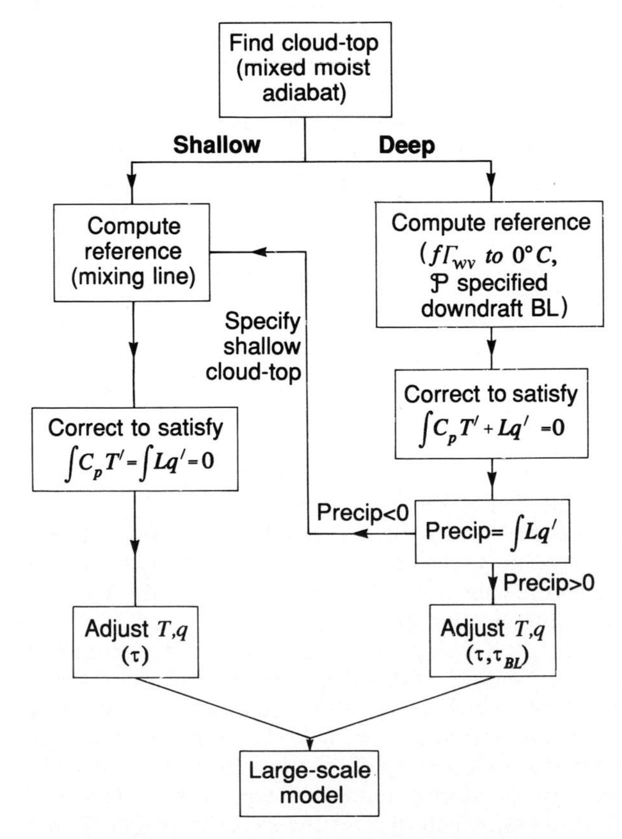

FIG. 9.9. Flowchart for Betts–Miller convection scheme.

from which the convective fluxes can be approximately expressed using (9.2) as

$$\mathbf{F} = \int \frac{\mathbf{R} - \bar{\mathbf{S}}}{\tau} \frac{dp}{g} \approx \int \bar{\omega} \frac{\partial \mathbf{R}}{\partial p} \frac{dp}{g}. \qquad (9.8)$$

Equation (9.8) shows that the structure of the convective fluxes is closely linked to the structure of the specified reference profile **R**. By adjusting toward an observationally realistic thermodynamic structure **R**, we simultaneously constrain the convective fluxes including precipitation to have a structure similar to those derived diagnostically from (9.3), or its simplified form (9.8), by the budget method (Yanai et al. 1973).

Substituting p and p^* in (9.7) gives (suffix R for the reference profile)

$$\mathcal{P}_R - \bar{\mathcal{P}} = p_R^* - \bar{p}^* \approx \omega\tau \frac{dp_R^*}{dp} \approx \omega\tau, \qquad (9.9)$$

since $1 < dp_R^*/dp < 1.1$ for the deep reference profiles that are used. Rearranging gives an approximate value for

$$\bar{\mathcal{P}} \approx \mathcal{P}_R - \omega\tau. \qquad (9.10)$$

This means that while the deep convection scheme is operating the mean vertical advection (if steady for time periods longer than τ) will shift the grid-scale value of $\bar{\mathcal{P}}$ away from the specified reference state $\mathcal{P}_R$ toward saturation by approximately $\omega\tau$ mb. Thus, although we specify in the present simple scheme a constant global value of the reference structure $\mathcal{P}_R$, $\bar{\mathcal{P}}$ does have a spatial and temporal variability in the presence of deep convection related to that of $\bar{\omega}$.

b. Choice of τ

The role of convective parameterization in a global model is to produce precipitation before grid-scale saturation is reached, both to simulate the real behavior of atmospheric convection and also to prevent grid-scale instability associated with a saturated conditionally unstable atmosphere. We can see from (9.10) that if the convection scheme is to prevent grid-scale saturation ($\bar{\mathcal{P}} = 0$), there is a constraint on τ—$\tau < \mathcal{P}_R/\omega_{\max}$, where $\omega_{\max}$ is a typical maximum ω in, say, a major tropical disturbance. With $\mathcal{P}_R \sim -40$ mb in the middle troposphere, we have found that this suggests an upper limit on τ, which is approximately 2 h for the ECMWF T-63 spectral model and 1 h for the T-106 model, with smaller values at higher resolutions. We recommend that τ should be set so that the model atmosphere nearly saturates on the grid scale in major convective disturbances. This is an empirical approach to the adjustment time scale by convection. In a numerical model a lagged adjustment, rather than a sudden adjustment at a single time step, has the advantage of smoothness, with less of a tendency to "blink" on and off at grid points in a physically unrealistic way. We have subsequently introduced a distinct boundary layer, $\tau_{\rm BL}$, connected physically to downdraft mass flux and evaporation. Typically $\tau_{\rm BL} > \tau$ (see section 9.4). However, we have as yet no satisfactory physical model for τ, since in current global or regional models the convection scheme is parameterizing processes on both the cloud scale and the unresolved mesoscale, processes that themselves have a range of time scales. We use $\tau = 2$ h for shallow convection.

c. Cloud base and cloud top

Cloud base is found by lifting air from the lowest model level to saturation and then testing for buoyancy at the next model level. We look for the lowest cloud base. Air is lifted from higher model levels if air from the lowest level is not buoyant. Convection from middle-tropospheric levels is also permitted if air lifted from the boundary layer is not buoyant. The moist adiabat corresponding to buoyant ascent is computed. The temperature of a partially mixed parcel (see section 9.3d) is then compared with the sounding at each level until a level of negative buoyancy is found. One level below is the initial choice of cloud top, p_T.

d. Cloud-top mixing algorithm

Subsequent to Betts and Miller (1986), we have introduced an algorithm to find cloud top that combines the moist adiabat with cloud-top mixing in an effort to pick up weak trapping inversions, which are not well resolved in the vertical.

Figure 9.10 shows the construction following Fig. 5 of Betts (1982). We linearize the mixing line (heavy dashed line) between the saturation point of cloud-base air at presure p_B and the environment at a pressure level p_K. We then find a cooler cloud parcel ascent

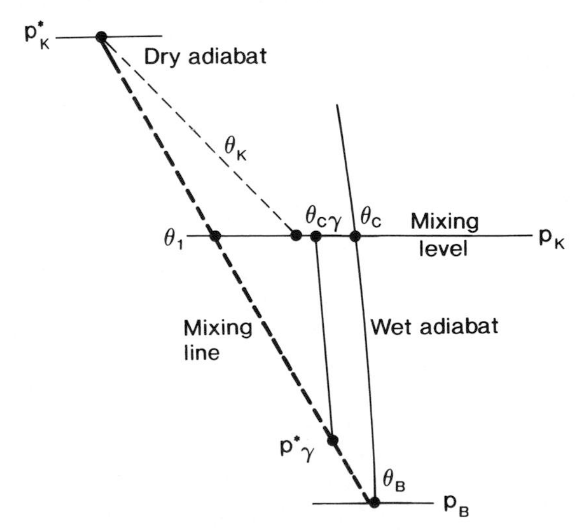

FIG. 9.10. Schematic showing thermodynamics of cloud-top mixing algorithm.

temperature $\theta_{c\gamma}$ at level K by mixing a fraction γ of the environmental air at that level with air that has risen adiabatically from cloud base (θ_c). This mixture has a saturation point at p_γ^* on the mixing line

$$p_\gamma^* = p_B - \gamma \Delta p^*, \tag{9.11}$$

where Δp^* is the difference in saturation level $p_B - p_K^*$ between cloud base and air at p_K. The potential temperature $\theta_{c\gamma}$ of this cloudy mixture at p_k can be written

$$\theta_{c\gamma} = \theta_c + (\theta_1 - \theta_c)\left(\frac{\gamma}{\gamma_c}\right), \tag{9.12}$$

where $\gamma_c = \Delta p / \Delta p^*$, $\Delta p = p_B - p_K$, and θ_1 is the cool potential temperature of the mixture that is just saturated at p_K. For the mixture to be cloudy, we require $\gamma < \gamma_c$. Finally θ_1, which lies on the same mixing line, is given by

$$\theta_1 = \theta_B + (\theta_K - \theta_B)\gamma_c. \tag{9.13}$$

Combining (9.11) and (9.12) gives the temperature of the mixture

$$\theta_{c\gamma} = \theta_c\left(1 - \frac{\gamma}{\gamma_c}\right) + \gamma\theta_K + \theta_B\left(\frac{\gamma}{\gamma_c} - \gamma\right). \tag{9.14}$$

We perform this mixing calculation *separately* at each level, and look for a level of negative buoyancy, by comparing the mixed cloud temperature with the sounding temperature: that is, we look for a level where $\theta_{c\gamma} - \theta_K < 0$. We have used a value of $\gamma = 0.2$.

e. Selection of deep or shallow convection

Shallow nonprecipitating convection is first separated from deep precipitating convection by using a cloud-top threshold: $p_{\text{SHAL}} \sim 700$ mb (strictly the hybrid coordinate of approximately 0.7). If $p_T > p_{\text{SHAL}}$, then it is a shallow convection point. If $p_T < p_{\text{SHAL}}$, then we call this initially a deep convection point. At this stage the separation of deep and shallow convection is based purely on a static thermodynamic criterion (equilibrium cloud top: see Fig. 9.9), not on any dynamic constraint (such as moisture convergence as in Kuo's scheme). We proceed to set up deep or shallow reference profiles and compute the convective adjustments and precipitation. A dynamic constraint enters at this point, because often the convective adjustment for a deep convective point gives "negative precipitation," which is clearly physically unrealistic. These points arise because in regions of weak subsistence the model may not resolve the low-level inversions that in the tropics cap the convective boundary layer, so that the scheme first attempts a deep convective adjustment. We then swap these points from the deep to the nonprecipitating shallow scheme as indicated in Fig. 9.9, with a specified shallow cloud top, and compute new reference profiles.

f. Reference thermodynamic profiles for shallow convection

Shallow convection is parameterized in terms of an approach toward a mixing-line structure. The scheme computes the slope of the mixing line from cloud base to two levels above cloud top (p_{T+2}). An exact method would use saturation level θ^*, q^* to compute $(\partial\theta^*/\partial q^*)_M$, but for computational simplicity, we compute a linearized slope with respect to pressure $(\partial\theta^*/\partial p^*)_M = (\theta_{T+2} - \theta_B)/(p_B^* - p_{T+2}^*)$. Because of the relatively poor vertical resolution in the global model above the CBL, we multiply $(\partial\theta^*/\partial p^*)_M$ by a coefficient (currently 0.85) to reduce the tendency to overestimate $(\partial\theta^*/\partial p^*)_M$ using model level data (see Fig. 9.3, for example). We define

$$M_\theta = 0.85\left(\frac{\partial\theta^*}{\partial p^*}\right)_M, \tag{9.15}$$

and then we construct first-guess profiles (superscript 1) of θ_R, q_R from

$$\theta_R^1(p) = \bar{\theta}(p_B) + \beta M_\theta(p - p_B), \tag{9.16}$$

where $\beta = \partial p^*/\partial p$ is specified. The current value is $\beta = 1.2$. In the current (experimental) ECMWF code, we continue the reference profiles to one higher level p_{T+1} using a value of β calculated for the inversion, in order to smooth the transition through the cloud-top inversion. The first-guess moisture reference profile is then computed from T_R and $p_R^* = p_B^* + \beta(p - p_B)$. These first-guess profiles are then corrected to satisfy the two separate energy constraints

$$\int_{p_B}^{p_{T+1}} c_p(T_R - \bar{T})dp = \int_{p_B}^{p_{T+1}} L(q_R - \bar{q})dp = 0, \tag{9.17}$$

so that the condensation (and precipitation) rates are zero when integrated from cloud base p_B to one level above cloud top p_{T+1}. This implies that the shallow convection scheme does not precipitate but simply redistributes heat and moisture in the vertical. This is done by correcting the first-guess values of T_R, q_R at each level by

$$\Delta T = \frac{1}{p_B - p_{T+1}} \int_{p_{T+1}}^{p_B} (T - T_R^1)dp \tag{9.18a}$$

$$\Delta q = \frac{1}{p_B - p_{T+1}} \int_{p_{T+1}}^{p_B} (q - q_R^1)dp. \tag{9.18b}$$

By making this correction independent of pressure, we preserve (to sufficient accuracy) the slope of the reference profiles, and a value of $\mathcal{P}$ independent of pressure. Since we have two constraints, it is not necessary to constrain $\mathcal{P}$ (unlike for deep convection where only one energy constraint is available; see next section). Instead, after we apply the corrections ΔT, Δq, the adjustment closely conserves the vertically averaged value of $\mathcal{P}$ through the shallow convective layer.

g. Reference profiles for deep convection

We first construct a first-guess thermal profile, followed by a first-guess moisture profile, and then these are corrected to satisfy moist static energy balance.

The reference profile for θ is computed up to the freezing level as a fraction of the slope of the moist pseudoadiabat. Defining $\Gamma_m = \partial\theta/\partial p$ for the moist adiabat, we set the first-guess profile at

$$\theta_R^1(p) = \bar{\theta}_B + 0.85\Gamma_m(p_B - p) \quad (9.19)$$

for $p_B < p < p_F$.

A coefficient of 0.9 corresponds to the slope of the wet virtual adiabat: the coefficient of 0.85 is a more unstable profile equivalent to ΔT_w having a slope 1.5 times that of the wet virtual adiabat in Figs. 9.6 and 9.7. In the original scheme we extended this profile down to one level above the surface and chose a value of $\bar{\theta}_B$ near cloud base. In the current revised scheme (see section 9.4), the deep reference profile near the surface is computed differently from an unsaturated downdraft profile, and $\bar{\theta}_B$ is at a level just above this new boundary layer.

Above the freezing level, the profile returns to the moist pseudoadiabat at cloud top. The interpolation is done quadratically in terms of the temperature difference from the wet adiabat, so that

$$T_R^1(p) = T_c(p) + [T_R(p_F) - T_c(p_F)](1 - y^2), \quad (9.20)$$

where $y = (p_F - p)/(p_F - p_T)$. This involves several small changes from Betts and Miller (1986), in which the reference profile returned linearly to the environmental temperature at cloud top, and θ rather than T was used for the interpolation.

The moisture profile q_R is then computed from the temperature profile by specifying $\mathcal{P} = (p^* - p)$ at three levels—cloud base $\mathcal{P}_B$, the freezing level $\mathcal{P}_F$, and cloud top $\mathcal{P}_T$—with linear gradients between.

For $p_B > p > p_F$, this gives

$$\mathcal{P}_R(p) = \frac{(p_B - p)\mathcal{P}_F + (p - p_F)\mathcal{P}_B}{p_B - p_F}, \quad (9.21a)$$

and for $p_F > p > p_T$,

$$\mathcal{P}_R(p) = \frac{(p_F - p)\mathcal{P}_T + (p - p_T)\mathcal{P}_F}{p_F - p_T}. \quad (9.21b)$$

In the present version of the model, the values chosen are $(\mathcal{P}_B, \mathcal{P}_M, \mathcal{P}_T) = (-25\text{ mb}, -40\text{ mb}, -20\text{ mb})$. The first-guess profiles of (T_R^1, q_R^1) are then modified until they satisfy the total enthalpy constraint

$$\int_{p_0}^{p_T} (k_R - \bar{k})dp = 0, \quad (9.22)$$

where $k = c_pT + Lq$ and the integral is through the depth of the convective layer.

The procedure is to calculate

$$\Delta k = \frac{1}{\Delta p_c}\int_{p_0}^{p_T} (k_R - \bar{k})dp, \quad (9.23)$$

where Δp_c is the depth of the deep convective layer included in the integral; T_R is then corrected at each level, at constant $\mathcal{P}$, so as to change k_R by Δk, independent of pressure. This energy correction is iterated once. In Betts and Miller (1986), this correction was applied at all levels except cloud top and a shallow surface layer. In the present scheme the correction is applied at all levels above a model boundary layer, where the adjustment is linked to the downdraft thermodynamics. This involves a modification to (9.23) to include the two adjustment time scales [Eq. (9.30) in section 9.4].

h. Convective tendencies and precipitation

The convective adjustment, $(\mathbf{R} - \mathbf{S})/\tau$, is then applied to the separate temperature and moisture fields as two tendencies (suffix cu for cumulus convection):

$$\left(\frac{\partial T}{\partial t}\right)_{cu} = \frac{T_R - \bar{T}}{\tau}, \quad (9.24a)$$

$$\left(\frac{\partial q}{\partial t}\right)_{cu} = \frac{q_R - \bar{q}}{\tau}. \quad (9.24b)$$

The precipitation rate is given by

$$\text{PR} = \int_{p_0}^{p_T}\left(\frac{q_R - \bar{q}}{\tau}\right)\frac{dp}{g} = -\frac{c_p}{L}\int_{p_0}^{p_T}\left(\frac{T_R - \bar{T}}{\tau}\right)\frac{dp}{g}. \quad (9.25)$$

No liquid water is stored in the present scheme, and the deep convective adjustment is suppressed if it ever gives PR < 0. These terms are slightly modified in the present scheme, which has a distinct adjustment in the BL (see section 9.4). If PR < 0, the shallow cloud scheme is called. Since a shallow convective cloud top has not previously been found from a buoyancy criterion, at present we specified a shallow cloud top. We intend to determine the depth of the CBL from a second thermodynamic criterion, by setting shallow cloud top at the base of the layer that has a maximum of $\partial p_K^*/\partial p_K$. This identifies the top of any moist layer that is capped by a sharp increase of θ or decrease of q. This has not yet been implemented.

Note that this scheme handles the partition between moistening of the atmosphere and precipitation in a quite different way from, say, Kuo's scheme, which specifies a partitioning of the moisture convergence. Given moisture convergence and, say, mean grid-scale upward motion, the model atmosphere moistens with no precipitation until a threshold is reached, qualitatively related to a mean value of $\mathcal{P}_R$, when precipitation starts. However, the model atmosphere continues to moisten (given steady-state forcing) until (9.10) is satisfied, when the moistening ceases, and the "converged moisture" is then all precipitated. If the forcing ceases ($\omega \rightarrow 0$), then precipitation continues for a while, until the atmosphere has dried out to the reference profile again. We are essentially controlling (through $\mathcal{P}_R$) the relative humidity of air leaving convective disturbances. In medium range, and particularly climate integrations of a global model, this seems a better way of maintaining the long-term moisture structure of the atmosphere than through constraints on the partition of moisture convergence. The water vapor distribution in tropics is crucial to the long-term radiative balance and at present we have insufficient data, particularly in the upper troposphere, to develop improved models. Clearly the parameterization of the moisture transport or the equilibrium moisture structure is a difficult problem. It depends on the moisture transport from primarily the subcloud layer, and the efficiency of the precipitation process, which in turn depends on the cloud and mesoscale dynamics and microphysics. We know qualitatively that more energetic convective systems typically in sheared flows and over land tend to have lower precipitation efficiencies (Fritsch and Chappell 1980), both because more condensed water is evaporated in downdrafts and because more ice is ejected at anvil levels. Figure 9.8 gives some indication of this complexity. However, the Betts–Miller scheme at present simplifies this complexity for deep convection by specifying a single reference moisture structure in terms of the saturation pressure deficit $\mathcal{P}_R$.

9.4. Modification to include downdraft mass flux boundary layer

One of the weaknesses of the original Betts–Miller (1986) scheme was that the adjustment near the surface was not well constrained. The reference profile was defined to produce a stabilization near the surface to mimic downdrafts. In practice, in the original scheme, the first-guess reference profile was started with the mean temperature at the next to lowest level, so that the adjustment at this level then depended solely on the energy correction. This was typically negative, but it was not well constrained by any physical model. In fact the deep convective interaction with the BL is crucial, especially over the oceans, where the surface temperature is constrained. As a result the scheme was modified to explicitly introduce the low-level cooling and drying by a downdraft mass flux. Observational studies have long shown the crucial role of these downdrafts in the subcloud layer interaction (e.g., Zipser 1969, 1976; Betts 1973a, 1976). To keep the parameterization simple, we define a simple unsaturated downdraft thermodynamic path and inject downdraft air with constant divergence into the three lowest model levels. Thus, for the three lowest model levels we introduce a different reference profile, related to a simple unsaturated downdraft. We also introduce a different adjustment time scale for this BL, related to the divergence of the downdraft mass flux, which is taken independent of height in the BL. The BL time scale (related to downdraft mass flux) is determined by one new closure that couples the evaporation in the downdraft to the precipitation: this represents a measure for precipitation efficiency.

a. Downdraft thermodynamics

Evaporation into downdrafts depends on complex dynamical and microphysical processes (e.g., Kamburova and Ludlam 1966; Betts and Silva Dias 1979). These we shall simplify to an unsaturated downdraft reference profile, which starts at a downdraft inflow level with the mean properties at that level and descends at constant θ_e and constant subsaturation: that is, the temperature and moisture paths are parallel to a moist adiabat. In the present code, the downdraft originates at a single level near 850 mb. Figure 9.11 shows the thermodynamics schematically. The reference profile, T_R, q_R, for the three lowest model levels (K to $K-2$ in Fig. 9.11) are set equal to the downdraft outflow properties:

$$T_R = T_{\rm DN} = \bar{T}_{\rm IN} + \Delta T_c, \qquad (9.26a)$$

$$q_R = q_{\rm DN} = \bar{q}_{\rm IN} + \Delta q_c, \qquad (9.26b)$$

where $\bar{T}_{\rm IN}$, $\bar{q}_{\rm IN}$ are the grid-mean values at the downdraft inflow level and ΔT_c, Δq_c are the changes of T, q along the downdraft descent path (defined positive). For computational efficiency we used the profiles of T_c, q_c on the ascent moist adiabat through cloud base, which are available at all levels. The approximation of

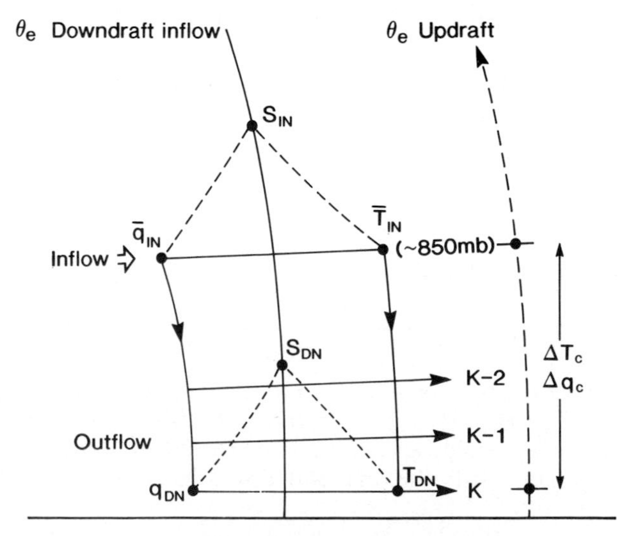

FIG. 9.11. Schematic showing thermodynamics of unsaturated downdraft model.

using a different (warmer) moist adiabat is small compared with the uncertainties in the precipitation efficiency (see below). This simple downdraft parameterization has other advantages besides being computationally efficient. It couples the relative humidity in the BL to the constrained subsaturation $\mathcal{P}$ at higher levels and gives tendencies toward a subsaturated moist-adiabatic structure [see Eq. (9.28)]. In the BL, the tendencies due to cumulus convection are

$$\left(\frac{\partial T}{\partial t}\right)_{\rm cu} = \frac{T_R - \bar{T}}{\tau_{\rm BL}} = \frac{\Delta T_c - \Delta \bar{T}}{\tau_{\rm BL}} \qquad (9.27a)$$

$$\left(\frac{\partial q}{\partial t}\right)_{\rm cu} = \frac{q_R - \bar{q}}{\tau_{\rm BL}} = \frac{\Delta q_c - \Delta \bar{q}}{\tau_{\rm BL}}, \qquad (9.27b)$$

where $\tau_{\rm BL}$ is the adjustment time of BL, discussed in the next section, and $\Delta\bar{T}$, $\Delta\bar{q}$ are the vertical differences in the mean structure between downdraft inflow and BL outflow levels. The BL tendencies become zero if

$$\Delta\bar{T} = \Delta T_c,$$

$$\Delta\bar{q} = \Delta q_c, \qquad (9.28)$$

that is, if the mean profiles become parallel to the moist adiabat. (The downdraft will not be saturated, however, unless the downdraft inflow is saturated.) Typically, unless (9.28) is achieved, $\Delta\bar{T} > \Delta T_c$ and $\Delta\bar{q} > \Delta q_c$, and the downdraft cools and dries the BL.

b. Boundary-layer adjustment time $\tau_{\rm BL}$

This is computed by coupling the evaporation into the downdraft to the precipitation rate (PR). We define

$$\text{EVAP} = \int_{p_0}^{p_{\rm BL}} (\nabla \cdot \mathbf{V})_D \Delta q_c \frac{dp}{g} = \alpha \text{PR}, \qquad (9.29)$$

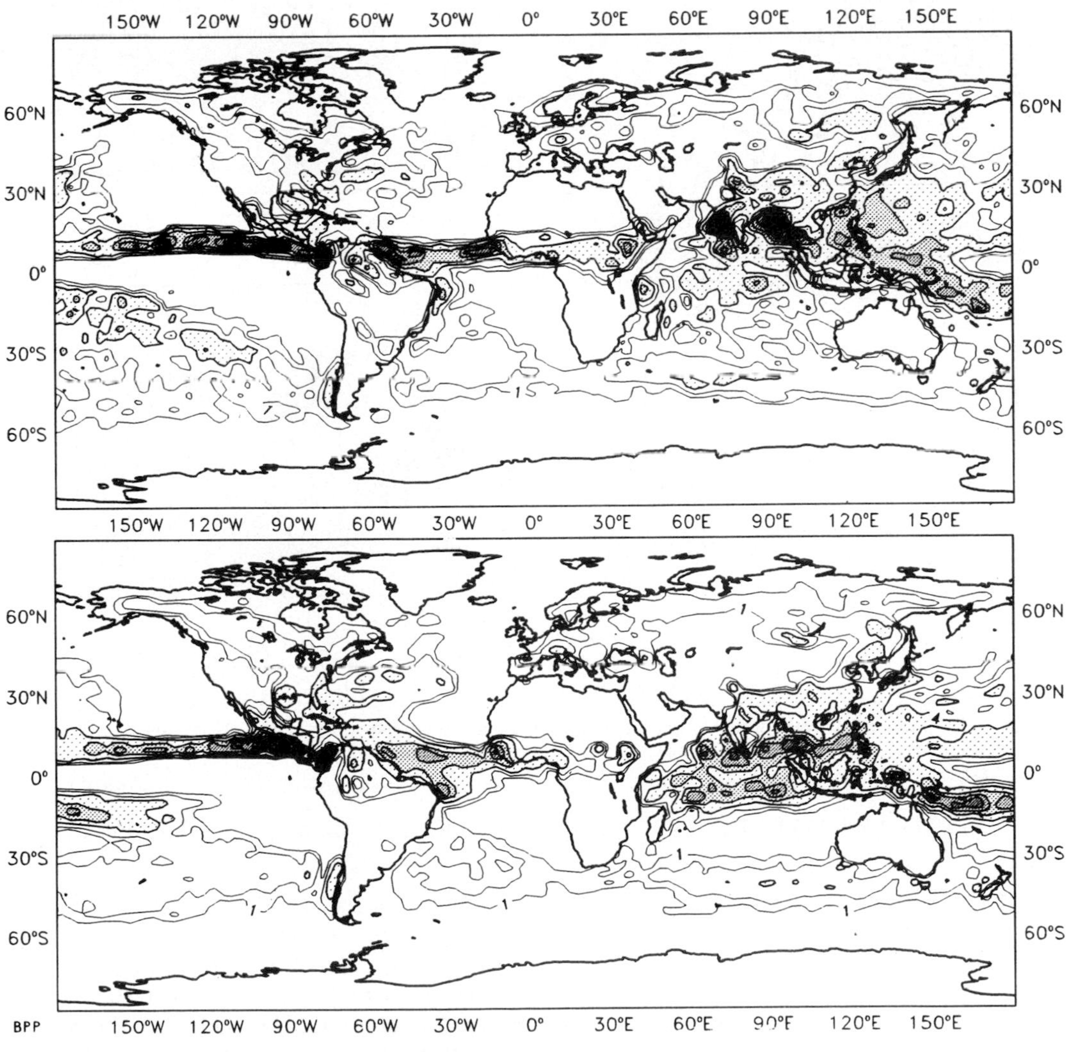

FIG. 9.12. Global map of 90-day convective precipitation for adjustment scheme (upper panel) and operational model (lower panel).

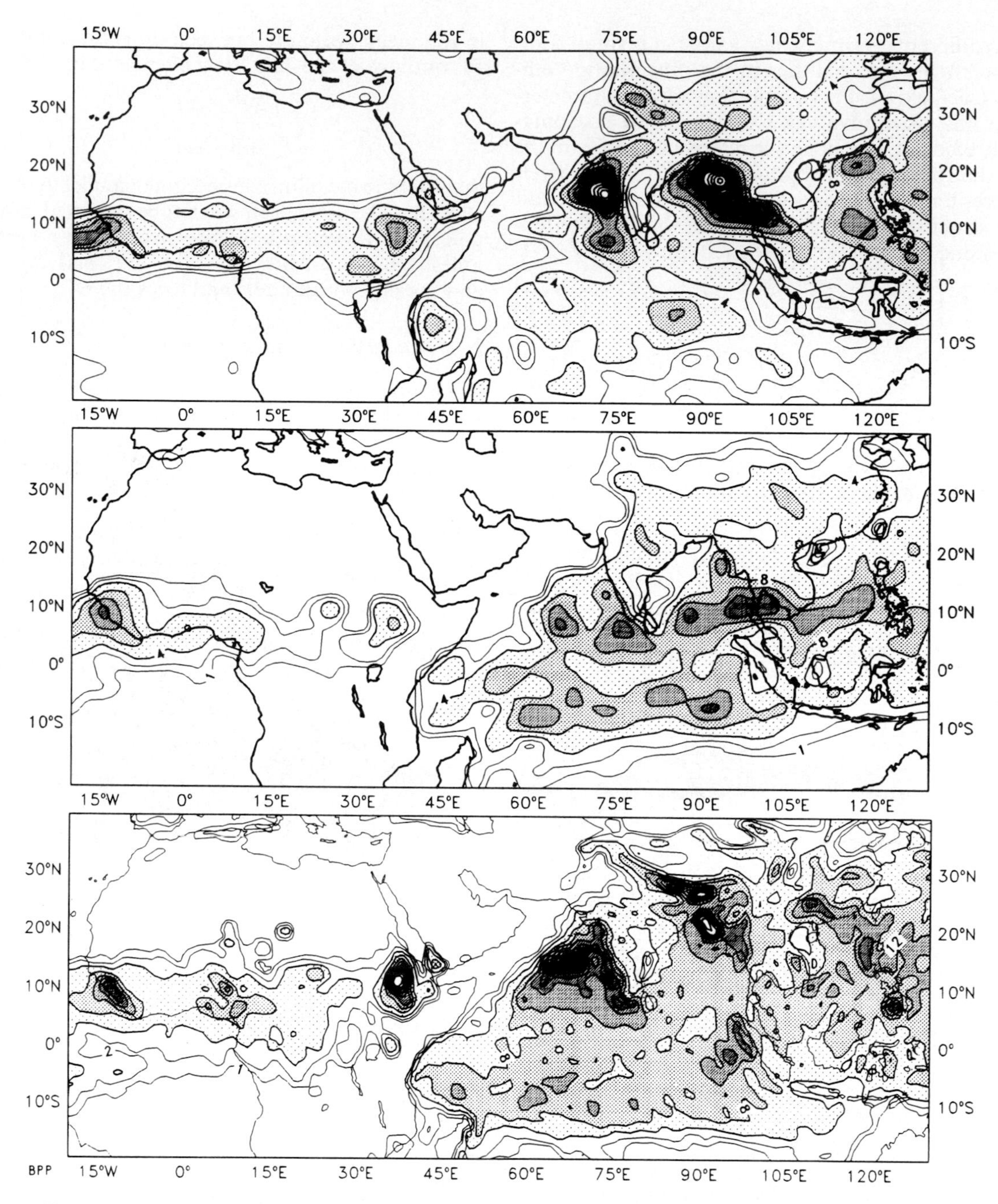

FIG. 9.13. 90-day convective precipitation for monsoon region for adjustment scheme (upper panel), operational model (center), and average of the short-range (12–36 h) forecasts (lower panel) for the same period from the operational model (T-106).

where we assume constant divergence $(\nabla \cdot \mathbf{V})_D$ of the downdraft in the BL and a constant of proportionality α. The BL time scale τ_{BL} is given by

$$\frac{1}{\tau_{BL}} = (\nabla \cdot \mathbf{V})_D = \frac{\alpha PR}{\int_{p_0}^{p_{BL}} \Delta q_c dp/g}. \quad (9.30)$$

This couples the BL time scale to the precipitation-driving downdraft processes. We set $\alpha = -0.25$ globally to represent a precipitation efficiency of order 0.80, consistent with tropical budget studies (Betts 1973a). The parameter could be made a function of wind shear (e.g., Fritsch and Chappell 1980). Typically τ_{BL} is longer than τ, so that the boundary-layer adjustment

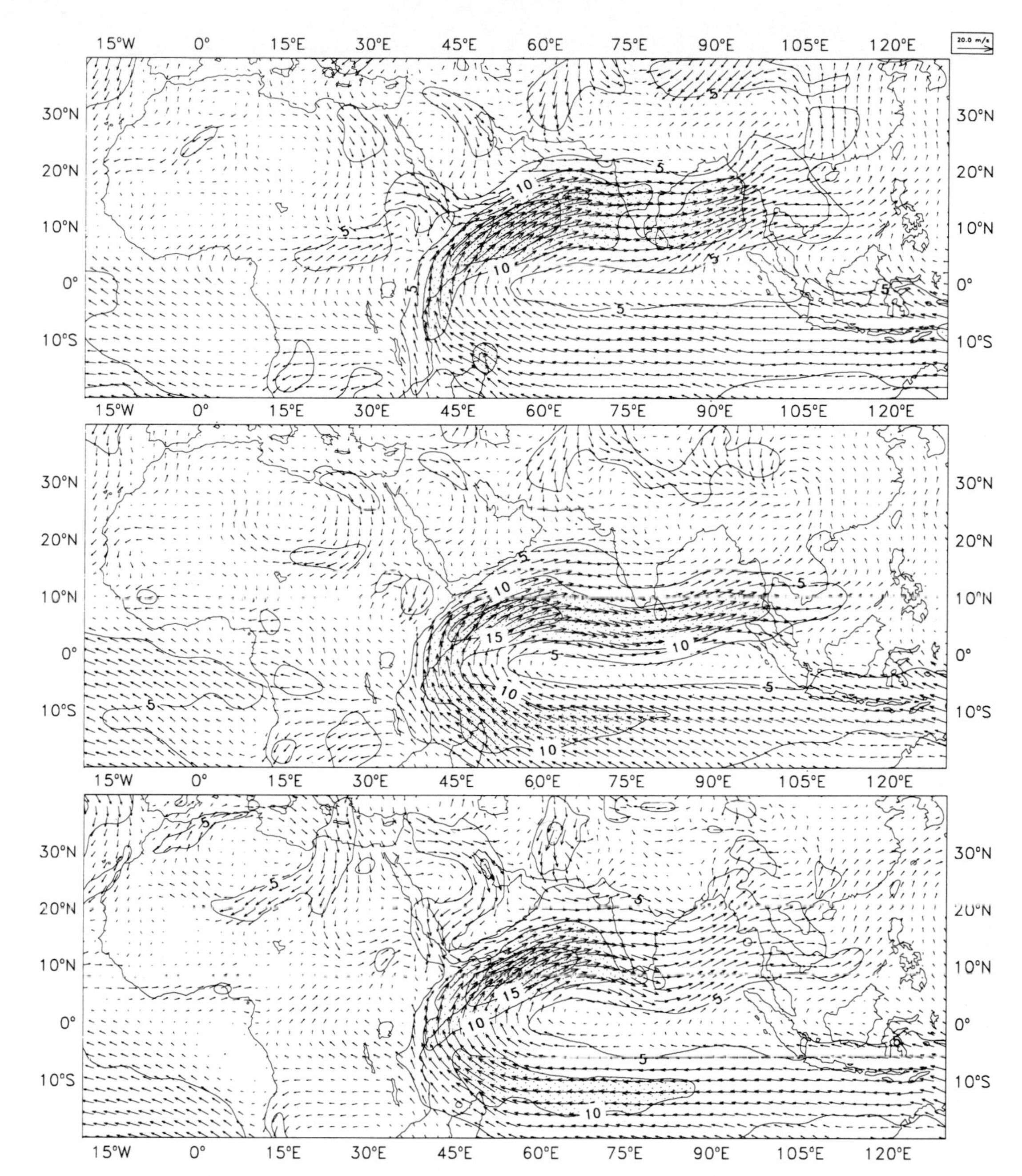

FIG. 9.14. 90-day mean flow near 850 mb for monsoon region for adjustment scheme (upper panel), operational model (center), and analysis (lower panel).

is slower and smoother than in the original version of the scheme, as well as being well defined in terms of a physical process. We have found much smoother precipitation patterns in this revised version of the scheme, presumably because the convection scheme is less apt to be shut off by rapid changes of θ_e in the BL.

c. *Modification to energy correction*

Equation (9.23) is modified to include the two adjustment time scales,

$$\Delta k = \frac{\tau}{\Delta p_c}\left[\int_{p_0}^{p_{\rm BL}}\left(\frac{k_R-\bar{k}}{\tau_{\rm BL}}\right)dp + \int_{p_{\rm BL}}^{p_T}\left(\frac{k_R-\bar{k}}{\tau}\right)dp\right], \tag{9.31}$$

where $p_{\rm BL}$ separates the model BL from the rest of the deep convective layer. In fact, since $\tau_{\rm BL}$ depends on the precipitation rate PR in (9.28), it is possible to formally eliminate $\tau_{\rm BL}$ from (9.31) using (9.30) and find Δk from the reference profiles *above* the BL, as follows. The downdraft reference profiles are left un-

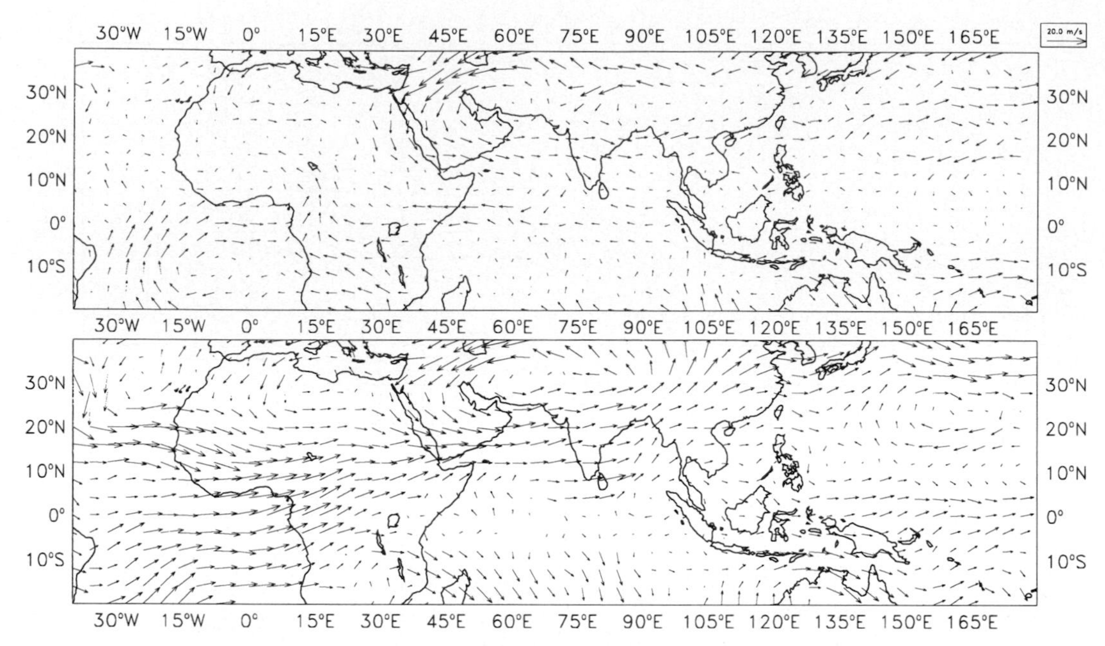

FIG. 9.15. Difference from analysis for the 200-mb flow for the adjustment scheme (upper panel) and the operational model (lower panel).

changed. Expanding (9.31) and substituting from (9.30) and (9.33) gives, after rearrangement,

$$\Delta k = \frac{1}{\Delta p_c}\left[\int_{p_{\rm BL}}^{p_T} C_p(T_R - \bar{T})dp + \frac{1+e}{1-f}\int_{p_{\rm BL}}^{p_T} L(q_R - \bar{q})dp\right], \quad (9.32)$$

where

$$e = \alpha C_p \int_{p_0}^{p_{\rm BL}} \frac{(T_R - \bar{T})dp}{LE}$$

$$f = \alpha \int_{p_0}^{p_{\rm BL}} \frac{(q_R - \bar{q})dp}{E}$$

and $E = \int_{p_0}^{p_{\rm BL}} \Delta q_c dp$ is part of (9.29). Note that the terms f and e (which are both defined positive) can be zero if either $\alpha = 0$ (the prescribed ratio of evaporation to precipitation) or if (9.28) are satisfied in some integral sense (an internal adjustment).

Once the energy correction is made, the precipitation can be calculated:

$$\mathrm{PR} = \int_{p_0}^{p_{\rm BL}} \frac{(q_R - \bar{q})}{\tau_{\rm BL}}\frac{dp}{g} + \int_{p_{\rm BL}}^{p_T} \frac{(q_R - \bar{q})}{\tau}\frac{dp}{g} = \left(\int_{p_{\rm BL}}^{p_T} \frac{q_R - \bar{q}}{\tau}\frac{dp}{g}\right)\frac{1}{1-f}. \quad (9.33)$$

Typically, for $\alpha = -0.25$, f and e are small with $f \leq 0.2$ and $e < 0.1$.

9.5. Impact of scheme on the tropical summer climate in the ECMWF model

We shall illustrate the impact of the scheme using a 90-day summer run made at T-42 triangular truncation with cycle 38 of the ECMWF model. We show comparisons between the operational model, which uses Tiedtke's (1989) mass flux scheme with a boundary-layer moisture convergence closure, and a parallel run with the current version of the Betts–Miller scheme discussed in sections 9.3 and 9.4. This comparison will illustrate both the impact of a different convection scheme on the tropical climate of a model and the complexity of understanding the interactions between convection and the large-scale fields in a global model.

This 90-day T-42 forecast was initialized with atmospheric data from 1 June 1988 and with observed sea surface temperatures updated every five days. We will present averages from days 1 to 90 for selected fields and compare them with a parallel run from the same initial conditions using the operational model (cycle 38). Figure 9.12 shows the global pattern of convective precipitation from the two simulations: for the adjustment scheme (upper panel) and for the operational scheme (lower panel). We see significant differences in several areas of the globe. In the northeastern Pacific the maximum in the precipitation in the tropical convergence zone is shifted off the coast with the Betts–Miller scheme. Over the tropical continents, the precipitation is somewhat enhanced. The biggest difference, however, is in the Indian monsoon circulation, where the climate with the adjustment scheme has a strong monsoon flow giving the typical monsoon precipitation on the northwest coast of India (e.g., Grossman and Durran 1984; Grossman and Garcia 1990). Figure 9.13 shows the Indian monsoon region on a larger scale with three panels. The lower panel is an average of 90 short-range precipitation forecasts (12–36 h) from the operational forecast model at T-106 for the relevant period. The upper panel is from the 90-day T-42 forecast with the adjustment scheme. The agreement with the average of the short-range forecasts for the same period is striking. The middle panel is the 90-day T-42 forecast with the operational model: we see that it clearly drifts to a tropical climate that has a weak monsoon. Figure 9.14 compares the flow near 850 mb for the 90-day average of the analysis (lower panel) with the two 90-day forecasts. The adjustment scheme 90-day climate (upper panel) has a much stronger southwesterly monsoon, somewhat stronger than the analysis, which impinges on the Western Ghat Mountains, producing the precipitation pattern shown in Fig. 9.13. In contrast the 850-mb flow in the Indian Ocean for the climate of the operational model (middle panel) has appreciable differences from the analysis, with the main branch of the southwesterly monsoon flow passing to the south of India and an increase in the flow across 20°S. This change in the low-level monsoon circulation has a big impact on the upper-level flow. Figure 9.15 shows the difference from the analysis for these 90-day averages for the 200-mb flow for this region and farther east. The operational model, which has too weak a monsoon circulation (in the 90-day climate), has large wind errors at 200 mb. With the adjustment scheme the upper-level wind errors are greatly reduced because the monsoon circulation and precipitation are improved. From a climate modeling viewpoint these changes are significant, but what features of the different convection schemes are responsible for the differences in climate? This is a difficult question to answer. The interaction between the climatic scale, the synoptic-scale disturbances, and the convection scheme occurs within a few days in the tropics, and the model circulation shifts to a different state.

Despite almost two decades since the planning of the GATE experiment (Betts 1974a), more research is still needed to understand the interaction between convection and the larger scales in the tropics.

Acknowledgments. This work was supported by the National Science Foundation under Grant ATM-9001960, the NASA Goddard Space Flight Center under Contracts NAS5-30524 and NAS5-31738, and by ECMWF, while the first author was a visiting scientist.

Chapter 10

The Arakawa–Schubert Cumulus Parameterization

AKIO ARAKAWA

Department of Atmospheric Sciences, University of California, Los Angeles, Los Angeles, California

MING-DEAN CHENG

Central Weather Bureau, Taipei, Taiwan, Republic of China

10.1. Introduction

The motivation of the paper by Arakawa and Schubert (1974) (hereafter referred to as A–S) was to present a theoretical framework that can be used for understanding the physical and logical basis for cumulus parameterization. More specifically, the paper attempted to answer the following questions:

1) How can cumulus clouds modify their environment while condensation takes place only inside clouds. Does the modification occur only through mixing of cloud air with the environment as clouds decay? If not, how can clouds in their mature phase modify the environment?

2) Since the vertical structure of that modification depends on cloud type (primarily cloud-top height) and typically more than one type of clouds coexist, a parameterization scheme should determine the spectral distribution of clouds instead of assuming a particular cloud type (or particular cloud types) a priori. What is an appropriate framework for doing this?

3) How does the subcloud layer control cumulus activity? What is the nature of the feedback in this link?

4) Finally, how can the intensity of overall cumulus activity and that of each cloud type be determined for a given large-scale condition? It is obvious that we can only parameterize cumulus activity in balance with large-scale processes. Then, how can we quantitatively define such balance?

The parameterization by Arakawa (1969) presents preliminary answers to these questions. It was designed for a three-layer (or three-level) model, in which the lowest layer represents the planetary bounday layer (PBL). Corresponding to this vertical structure, three types of clouds are considered: deep and shallow clouds originating from the PBL and "middle-level" clouds originating above the PBL. Despite its simplified vertical structure, this parameterization explicitly formulates the effects of cloud-induced subsidence and cloud air detrainment on the large-scale environment. For type I closure (see chapter 1), the parameterization introduces a measure of the bulk buoyancy of cloud air, given by the difference between lower-level moist static energy and upper-level saturation moist static energy. When large-scale processes tend to increase the bulk buoyancy for a certain cloud type or cloud types, cloud-base mass flux is determined for each cloud type to restore the bulk buoyancy toward an equilibrium.

The A–S parameterization generalizes and elaborates the basic ideas of Arakawa (1969) described above. Sections 10.2, 10.3, 10.4, and 10.5 describe how the parameterization addresses questions 1, 2, 3, and 4, respectively. Section 10.6 outlines standard procedures in practical application of the parameterization. Section 10.7 gives further comments. The Appendix describes an important recent revision of the parameterization through the inclusion of convective-scale downdraft effects. Readers are encouraged to read chapter 1 by Arakawa on closure assumptions as an introduction to this chapter.

10.2. Modification of the environment by cumulus clouds

This part of the A–S parameterization formulates the vertical distributions of cumulus heating and moistening in terms of the vertical distributions of mass flux through clouds, mass detrainment from clouds, and thermodynamical properties of detraining cloud air.

Let us consider an ensemble of clouds in a "large-scale" area, as shown in Fig. 10.1. The area is assumed to be large enough so that the cloud ensemble can be considered as a statistical entity but small enough so that the cloud environment is approximately uniform horizontally. The existence of such an area is an idealization. In practice, cloud effects averaged over the area may not be statistically significant, especially when the

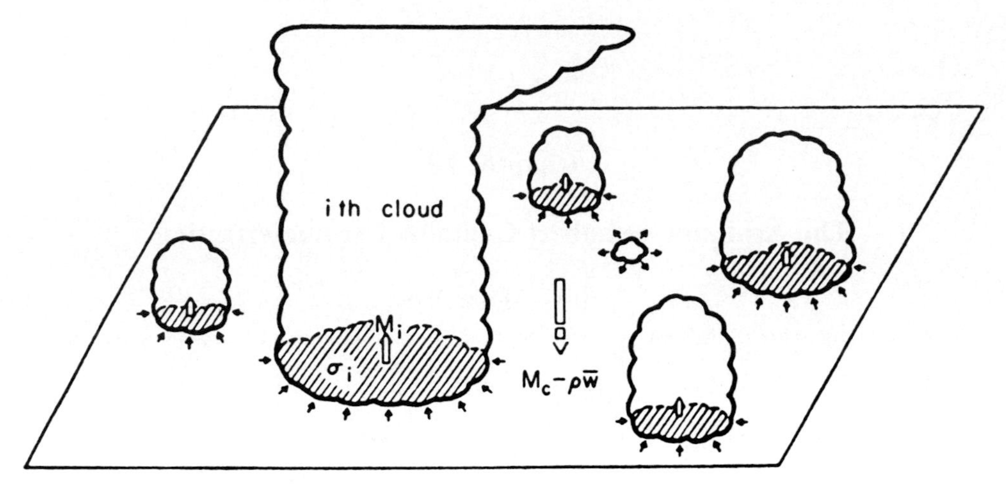

FIG. 10.1. Clouds within a large-scale horizontal area, some of which are penetrating this level while others are detraining, from Arakawa and Schubert (1974).

area is small. Also, intermediate scales of cloud organization should be considered especially when the area is large. Even for that idealized situation, however, the physical and logical basis for parameterizability is by no means obvious.

In Fig. 10.1, M_i is the vertical mass flux through the ith cloud, σ_i is the fractional area covered by the ith cloud, and $M_c \equiv \sum_i M_i$ is the total vertical mass flux through clouds. Let an overbar denote the average over the large-scale area. The net vertical mass flux through the unit large-scale horizontal area is given by

$$\rho\bar{w} = M_c + \tilde{M}, \quad (10.1)$$

where $\tilde{M}$ is the vertical mass flux through the environment. When $\tilde{M}$ is negative, there is subsidence between clouds.

We now consider the budgets of dry static energy $s \equiv c_pT + gz$ for the environment and for clouds. [Static energy is approximately conserved under adiabatic processes and is used here in place of the potential enthalpy $c_p\theta \equiv c_pT(p_0/p)^{R/c_p}$ in formulating cloud effects.] Continuing to use the tilde to denote a value in the environment and an overbar to denote the average over the large-scale area,

$$\frac{\partial}{\partial t}\rho(1-\sigma_c)\tilde{s} = -\overline{\nabla\cdot(\rho\mathbf{v}s)} - \frac{\partial}{\partial z}(\tilde{M}\tilde{s}) - \sum_i\left(\frac{\partial M_i}{\partial z} + \rho\frac{\partial\sigma_i}{\partial t}\right)s_{ib} - LE + \tilde{Q}_R, \quad (10.2)$$

$$\frac{\partial}{\partial t}\rho\sum_i\sigma_i s_i = -\frac{\partial}{\partial z}\left(\sum_i M_i s_i\right) + \sum_i\left(\frac{\partial M_i}{\partial z} + \rho\frac{\partial\sigma_i}{\partial t}\right)s_{ib} + \sum_i(LC_i + Q_{Ri}). \quad (10.3)$$

Here $\sigma_c \equiv \sum_i \sigma_i$, s_i is the mean dry static energy of the ith cloud, s_{ib} is the dry static energy of the air entraining into or detraining from the ith cloud, E is the evaporation of liquid water in the environment, C_i is the condensation in the ith cloud, Q_R is the radiation heating, all per unit large-scale horizontal area and per unit height, and ρ is a function of z only as in the standard anelastic approximation. The first term in the rhs of (10.2) represents the net flux of s through the lateral boundary of the large-scale area. The $\sum_i(\partial M_i/\partial z + \rho\partial\sigma_i/\partial t)s_{ib}$ terms in (10.2) and (10.3) represent the exchange of s between the environment and the clouds. (Note that from the continuity of mass the quantity inside the parentheses must be equal to the mass flux into the ith cloud through its lateral boundary.) We may define and choose

$$\begin{aligned}&\text{entrainment:} \quad \frac{\partial M_i}{\partial z} + \rho\frac{\partial\sigma_i}{\partial t} > 0, \quad s_{ib} = \tilde{s};\\ &\text{detrainment:} \quad \frac{\partial M_i}{\partial z} + \rho\frac{\partial\sigma_i}{\partial t} < 0, \quad s_{ib} = s_i.\end{aligned} \quad (10.4)$$

Then,

$$\sum_i\left(\frac{\partial M_i}{\partial z} + \rho\frac{\partial\sigma_i}{\partial t}\right)s_{ib} = \sum_{ec}\left(\frac{\partial M_i}{\partial z} + \rho\frac{\partial\sigma_i}{\partial t}\right)\tilde{s} + \sum_{dc}\left(\frac{\partial M_i}{\partial z} + \rho\frac{\partial\sigma_i}{\partial t}\right)s_i, \quad (10.5)$$

where $\sum_{ec}$ is the sum over entraining clouds and $\sum_{dc}$ is the sum over detraining clouds.

Using $\partial\sigma_c/\partial t \equiv \sum_i \partial\sigma_i/\partial t$, $\tilde{M} = \rho\bar{w} - M_c$, $M_c \equiv \sum_i M_i$ and (10.5), we may rewrite (10.2) as

$$\rho(1-\sigma_c)\frac{\partial\tilde{s}}{\partial t} = -\overline{\nabla\cdot(\rho\mathbf{v}s)} - \frac{\partial(\rho\bar{w}\tilde{s})}{\partial z} + M_c\frac{\partial\tilde{s}}{\partial z} - \sum_{dc}\left(\frac{\partial M_i}{\partial z} + \rho\frac{\partial\sigma_i}{\partial t}\right)(s_i - \tilde{s}) - LE + \tilde{Q}_R. \quad (10.6)$$

The third term in the rhs of (10.6) represents the adiabatic warming due to the *hypothetical* subsidence between the clouds given by $-M_c$. We call this subsidence "cloud-induced subsidence." It should be noted that this is generally different from the actual subsidence between clouds given by $\tilde{M} = \rho\bar{w} - M_c$, which does not necessarily take place in the region of cumulus activity.

When only convectively active clouds are considered, we may assume, as a first approximation,

$$\sigma_c \ll 1. \qquad (10.7)$$

Then, since $\bar{s} \equiv (1 - \sigma_c)\tilde{s} + \sum_i \sigma_i s_i = \tilde{s} + \sum_i \sigma_i (s_i - \tilde{s})$ and $\sum_i \sigma_i (s_i - \tilde{s}) \leqslant \sigma_c (s_i - \tilde{s})_{\max} < \sigma_c \tilde{s}$,

$$\bar{s} \cong \tilde{s}. \qquad (10.8)$$

Using (10.7) and (10.8) in (10.6), adding and subtracting a term, and assuming $\nabla \cdot (\rho\bar{\mathbf{v}}\tilde{s}) \cong \nabla \cdot (\rho\bar{\mathbf{v}}\bar{s})$,

$$\frac{\partial}{\partial t}\rho\bar{s} = -\nabla \cdot (\rho\bar{\mathbf{v}}\bar{s}) - \frac{\partial}{\partial z}(\rho\bar{w}\bar{s}) - \overline{\nabla \cdot (\rho\mathbf{v}s - \rho\bar{\mathbf{v}}\bar{s})} + M_c\frac{\partial\bar{s}}{\partial z} - \sum_{dc}\left(\frac{\partial M_i}{\partial z} + \rho\frac{\partial\sigma_i}{\partial t}\right)(s_i - \bar{s}) - LE + \tilde{Q}_R. \qquad (10.9)$$

The third term on the rhs can be interpreted as the net horizontal transport of s by cloud-scale circulations across the lateral boundary of the large-scale area. The lhs of (10.9), which is the sum of the lhs of (10.2) and that of (10.3), is now used as an approximation to the former. This means that the use of (10.9) involves neglecting the lhs of (10.3), which represents the storage of s in clouds.

A similar equation can be derived for the specific humidity q. The result is

$$\frac{\partial}{\partial t}\rho\bar{q} = -\nabla \cdot (\rho\bar{\mathbf{v}}\bar{q}) - \frac{\partial}{\partial z}(\rho\bar{\mathbf{w}}\bar{q}) - \overline{\nabla \cdot (\rho\mathbf{v}q - \rho\bar{\mathbf{v}}\bar{q})} + M_c\frac{\partial\bar{q}}{\partial z} - \sum_{dc}\left(\frac{\partial M_i}{\partial z} + \rho\frac{\partial\sigma_i}{\partial t}\right)(q_i - \bar{q}) + E. \qquad (10.10)$$

In the following, we neglect the net horizontal transports by cloud-scale circulations. We also assume that detrainment takes place only near the cloud top and clouds detraining at the same level share common thermodynamical properties. Then $s_i - \bar{s}$ and $q_i - \bar{q}$ in (10.9) and (10.10) are independent of i and, therefore, the terms involving $\rho\partial\sigma_i/\partial t$ can be neglected if $\sum_{dc}\rho\partial\sigma_i/\partial t$ is small. Further, *if* detrained cloud liquid water evaporates at the same level without falling through the environment, (10.9) and (10.10) may finally be written as

$$\frac{\partial}{\partial t}\rho\bar{s} = -\nabla \cdot (\rho\bar{\mathbf{v}}\bar{s}) - \frac{\partial}{\partial z}(\rho\bar{w}\bar{s}) + M_c\frac{\partial\bar{s}}{\partial z} + D(\hat{s} - L\hat{l} - \bar{s}) + \bar{Q}_R \qquad (10.11)$$

$$\frac{\partial}{\partial t}\rho\bar{q} = -\nabla \cdot (\rho\bar{\mathbf{v}}\bar{q}) - \frac{\partial}{\partial z}(\rho\bar{w}\bar{q}) + M_c\frac{\partial\bar{q}}{\partial z} + D(\hat{q} + \hat{l} - \bar{q}), \qquad (10.12)$$

where $D \equiv \sum_{dc}\partial M_i/\partial z$ is the detrainment of mass per unit height, the caret denotes the cloud-top value *before* the influence of radiational cooling near the cloud top so that $\tilde{Q}_R$ is replaced by $\bar{Q}_R = \tilde{Q}_R + \sum_i \tilde{Q}_{Ri}$ in (10.11), and l is the mixing ratio of cloud liquid water.

Equations similar to (10.11) and (10.12) were derived and used by Arakawa (1969, 1972), Ooyama (1971), Yanai et al. (1973), and Arakawa and Schubert (1974). These equations show that the effects of clouds on the environment (except the effect through $\bar{Q}_R$) depend on the vertical mass flux through clouds, M_c, the mass detrainment from clouds, D, the thermodynamical properties of detraining cloud air, $\hat{s}$, $\hat{q}$, and $\hat{l}$, all of which are functions of height.

10.3. A spectral cloud ensemble model

This part of the A–S parameterization introduces a spectral cumulus ensemble model to relate the *vertical distributions* of M_c and D to the *spectral distribution* of cloud-base mass flux and determine the vertical distributions of $\hat{s}$, $\hat{q}$, and $\hat{l}$ from large-scale thermodynamical variables.

The spectral cumulus ensemble model assumes that a single positive parameter denoted by λ fully characterizes a cloud type. The parameter can be the initial size of buoyant elements, the maximum height of cloud top, or any other physical parameter that effectively determines the properties of clouds. Depending on the value of λ, the model divides a cloud ensemble into subensembles, expressing $M_c(z)$ and $D(z)$ as

$$M_c(z) = \int_0^{\lambda_{\max}} m(z, \lambda)d\lambda = \int_0^{\lambda_{\max}} m_B(\lambda)\eta(z, \lambda)d\lambda, \qquad (10.13)$$

and

$$D(z) = -[m_B(\lambda)\eta(z, \lambda)]_{\lambda=\lambda_D(z)}\frac{d\lambda_D(z)}{dz}, \qquad (10.14)$$

where $m(z, \lambda)d\lambda$ is the mass flux of clouds that have the value of λ between λ and $\lambda + d\lambda$, $m(z, \lambda)$ is the subensemble mass flux, $m_B(\lambda) \equiv m(z_B, \lambda)$ is the cloud-base subensemble mass flux, where z_B is the height of cloud base, $\eta(z, \lambda) \equiv m(z, \lambda)/m_B(\lambda)$ is the normalized subensemble mass flux, and $\lambda_D(z)$ is λ of the clouds

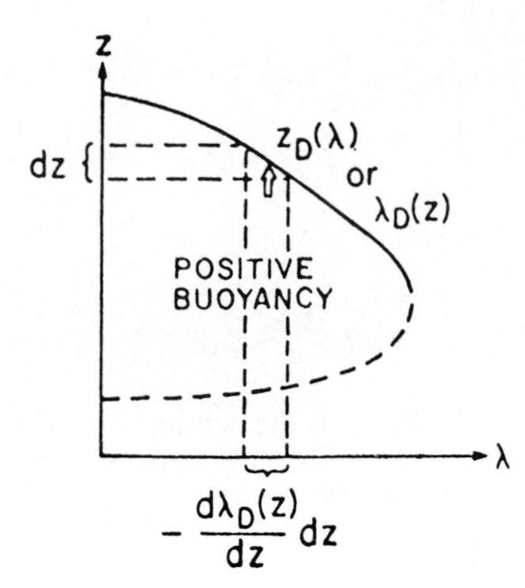

FIG. 10.2. The detrainment level z_D as a function of λ, or λ of the detraining clouds, λ_D, as a function of z. Clouds detraining in the layer between z and $z + dz$ have λ between $\lambda_D(z)$ and $\lambda_D(z) - d\lambda$ where $d\lambda = -[d\lambda_D(z)/dz]dz$, from Arakawa and Schubert (1974).

that lose buoyancy (and detrain) at level z (see Fig. 10.2).

We assume that $\eta(z, \lambda)$ is a known function of z for each subensemble. To specify this function, A–S defines the fractional rate of entrainment $\mu(z, \lambda)$ for subensemble λ by

$$\frac{\partial\eta(z, \lambda)}{\partial z} = \mu(z, \lambda)\eta(z, \lambda). \qquad (10.15)$$

Note that (10.15) is applied to the *subensemble* mass flux. [For an individual cloud identified by the index i, entrainment is defined by (10.4). Correspondingly, the fractional rate of entrainment μ_i is defined by $(\partial M_i/\partial z + \rho\partial\sigma_i/\partial t)M_i^{-1}$.] *If* the vertical variation of $\mu(z, \lambda)$ is negligible, $\mu(z, \lambda)$ becomes a function of λ alone so that μ itself can be used as the parameter λ. The A–S scheme makes this choice. Then

$$\eta(z, \lambda) = e^{\lambda(z-z_B)}. \qquad (10.16)$$

In actual applications, however, (10.16) does not have to be strictly followed (see Lord et al. 1982; Moorthi and Suarez 1992).

To compute $D(z)$ from (10.14), we must find $\lambda_D(z)$. To compute the effects of detrainment in (10.11) and (10.12), we must also find $\hat{s}(z)$, $\hat{q}(z)$, and $\hat{l}(z)$, which are the cloud-top values of s, q, and l for clouds detraining at level z. To find these unknowns, we consider the thermodynamical and moisture budget equations applied to each subensemble.

First consider the budget of the moist static energy, $h \equiv s + Lq$. We obtain

$$\frac{\partial}{\partial z}[\eta(z, \lambda)h_c(z, \lambda)] = \lambda\eta(z, \lambda)\bar{h}(z), \qquad (10.17)$$

where $h_c(z, \lambda)$ is the moist static energy of type λ clouds. The rhs of (10.17) represents the entrainment effect. If $h_c(z_B, \lambda)$ is known (see section 10.4), $h_c(z, \lambda)$ can be obtained by integrating (10.17) with respect to height. From $h_c(z, \lambda)$ thus obtained, $s_c(z, \lambda)$ and $q_c(z, \lambda)$ can be found (assuming saturation and neglecting the pressure difference between clouds and the environment) from the following approximate relations [see Eqs. (55) and (56) of A–S]:

$$s_c(z, \lambda) = \bar{s}(z) + \frac{1}{1+\gamma}[h_c(z, \lambda) - \overline{h^*}(z)], \qquad (10.18)$$

$$q_c(z, \lambda) = \overline{q^*}(z) + \frac{1}{L}\frac{\gamma}{1+\gamma}[h_c(z, \lambda) - \overline{h^*}(z)], \qquad (10.19)$$

where $q^*(T, p)$ is the saturation specific humidity, $h^* \equiv s + Lq^*$ is the saturation moist static energy, and $\gamma(z) \equiv (L/c_p)(\overline{\partial q^*}/\partial\bar{T})_p$. Assume that cloud top is the level of no buoyancy. When the virtual temperature effect is neglected, the vanishing buoyancy condition is given by $s_c(z, \lambda_D) = \bar{s}(z)$. Then, from (10.18),

$$h_c(z, \lambda_D) = \overline{h^*}(z). \qquad (10.20)$$

Since $h_c(z, \lambda)$ is already known, (10.20) can be used to find $\lambda_D(z)$. Figure 10.3 shows an example of vertical profiles of $\bar{h}$, $\overline{h^*}$, and $h_c(z, \lambda)$ for various values of λ.

Similarly, from the moisture budget equation,

$$\frac{\partial}{\partial z}\{\eta(z, \lambda)[q_c(z, \lambda) + l(z, \lambda)]\}$$
$$= \frac{\partial\eta(z, \lambda)}{\partial z}\bar{q}(z) - \eta(z, \lambda)r(z, \lambda), \qquad (10.21)$$

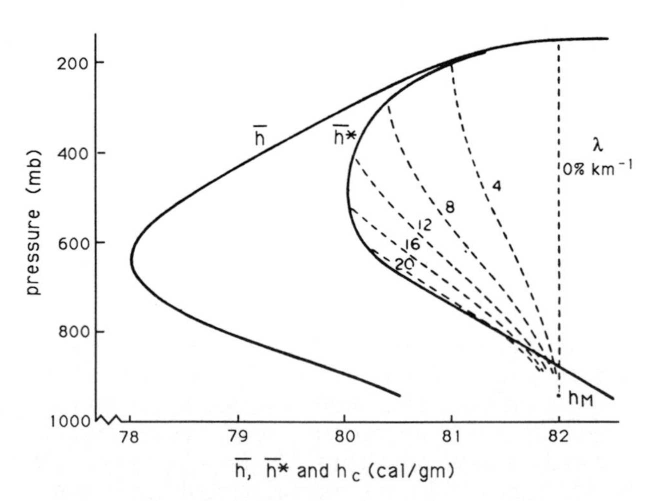

FIG. 10.3. Vertical profiles of $\bar{h}(p)$ and $\bar{h}^*(p)$ (solid lines) and $h_c(p, \lambda)$ (dashed lines). The lines for $h_c(p, \lambda)$ are labeled with the value of λ in percent per kilometer. Profiles of $\bar{h}$ and $\bar{h}^*$ are from Jordan's (1958) sounding. The top of the mixed layer is assumed to be at 950 mb; h_M is assumed to be 82 cal g^{-1}, from Arakawa and Schubert (1974).

where $r(z, \lambda)$ is the rate of production of raindrops per unit cloud mass flux and per unit height. In standard applications of this parameterization,

$$r(z, \lambda) = C_0 l(z, \lambda), \qquad (10.22)$$

where C_0 is a constant. Since $l = 0$ at cloud base and $q_c(z, \lambda)$ is already known, integration of (10.21) with respect to height gives $l(z, \lambda)$.

Finally, $\hat{s}(z)$, $\hat{q}(z)$, and $\hat{l}(z)$ are given by

$$\hat{s}(z) = s_c[z, \lambda_D(z)], \qquad (10.23)$$

$$\hat{q}(z) = q_c[z, \lambda_D(z)], \qquad (10.24)$$

$$\hat{l}(z) = l[z, \lambda_D(z)]. \qquad (10.25)$$

Thus, the thermodynamical properties of detraining cloud air can be calculated as functions of height if the thermodynamical properties at cloud base are known.

The above procedure of finding cloud properties is similar to the use of a stationary, one-dimensional, entraining plume model. As we noted earlier, this does not mean that such a model is used for individual clouds. The procedure considers subensemble averages, which can be viewed as averages over different phases of cloud development. Then the subensemble average vertical structure may simply reflect the temporal change of rising bubbles, as in the spectral model of Ooyama (1971).

10.4. Cloud-base conditions

As shown in the last section, the thermodynamical properties of clouds can be calculated *for each subensemble* if $h_c(z_B, \lambda)$ is known for each subensemble. There we assumed that level B is the condensation level. If we choose a level below the condensation level as level B to start vertical integrations, $s_c(z_B, \lambda)$ and $q_c(z_B, \lambda)$ must be specified separately.

In the original A–S, a mixed-layer model of variable depth is considered as a model for the subcloud layer and used

$$s_c(z_B, \lambda) = s_M, \qquad (10.26)$$

$$q_c(z_B, \lambda) = q_M, \qquad (10.27)$$

$$h_c(z_B, \lambda) = h_M, \qquad (10.28)$$

where the subscript M denotes the mixed-layer values. The height of mixed-layer top is taken to be z_B, which can be predicted by the mass budget equation applied to the mixed layer. In this equation, the entrainment of mass from the cloud environment appears as a source term and the mass flux into clouds appears as a sink term. The height z_B predicted in this way is generally different from the height of the condensation level, z_c. As discussed in A–S, however, z_B cannot be much lower than z_c for cumulus clouds to be maintained, since it is mainly the drier air above the mixed layer that is supplied to clouds under such situations. Thus, the choice $z_B = z_c$ should be an acceptable choice in practical applications.

In the above formulation, there is no feedback from clouds to the subcloud layer except through the change of its depth. This is partly because the (subcloud branch of the) updraft has no effect on the time change of subcloud thermodynamical properties since they have no vertical gradient. In addition, perhaps more importantly, there is no convective-scale downdraft that can penetrate into the subcloud layer (see the Appendix for inclusion of downdraft effects in this parameterization).

From the point of view of closure assumptions, the formulations presented up to this point represent type II closure (see chapter 1). In A–S, these formulations are combined with type I closure to obtain a fully closed parameterization, as described in the next section.

10.5. The cloud work function and its quasi equilibrium

We now face the problem of finding the spectral distribution of cloud-base mass flux under a given large-scale condition. The problem includes determining the existence or nonexistence of cumulus clouds and, when they exist, the overall intensity of cumulus activity and dominant cloud types. Obviously this cannot be done without introducing an additional closure assumption. One of the unique aspects of the A–S parameterization is its introduction of the concept of cloud work function (CWF) equilibrium. CWF is a generalized measure of moist convective instability, which can be used in formulating type I closure along the lines discussed in section 1.6 of chapter 1.

The A–S scheme defines CWF as the work done by the buoyancy force per unit mass flux at cloud base. For type λ clouds, it is given by

$$A(\lambda) = \int_{z_B}^{z_D(\lambda)} \eta(z, \lambda) g \frac{T_c(z, \lambda) - \bar{T}(z)}{\bar{T}} dz. \qquad (10.29)$$

Here the virtual temperature effect due to the existence of vapor and liquid phases of water has been omitted for simplicity. Arakawa and Schubert (1974) showed that, when $z_c = z_{B+}$, (10.29) can be expressed as

$$A(\lambda) = \int_{z_B}^{z_D(\lambda)} \beta(z) \left\langle h_M - \bar{h}^*_{B+} + \int_{z_{B+}}^{z} \eta(z', \lambda) \times \left\{ \lambda[\bar{h}(z') - \bar{h}^*(z')] - \frac{\partial \bar{h}^*(z')}{\partial z'} \right\} dz' \right\rangle dz, \qquad (10.30)$$

where the subscripts M and $B+$ denote the subcloud mixed layer and the level slightly above the mixed layer, respectively, and $\beta \equiv g/c_p\bar{T}(1 + \gamma)$. In deriving (10.29), (10.17) and (10.18) have been used. Note that $\bar{h} - \bar{h}^* = L\bar{q}^*(\text{RH} - 1)$ and $\partial \bar{h}^*/\partial z = -(1 + \gamma)(\Gamma - \Gamma_m)$, where $\text{RH} \equiv \bar{q}/\bar{q}^*$ is the relative humidity, Γ is the lapse rate, and Γ_m is the moist-adiabatic lapse rate.

It is important to note that (10.30) expresses CWF in terms of the *large-scale* thermodynamical variables. Consider saturated air (i.e., RH = 1), for example, and neglect the difference between h_M and $\bar{h}^*_{B+}$. Then we have $A(\lambda) = 0$ if $\Gamma - \Gamma_m = 0$ at all heights and $A(\lambda) > 0$ if an integrated measure of $\Gamma - \Gamma_m$ is positive. For a given $\Gamma - \Gamma_m$, the value of $A(\lambda)$ decreases as RH decreases. Thus, we can interpret CWF as a generalized integral measure of moist convective instability defined for each cloud type, which can be calculated from given vertical profiles of large-scale temperature and humidity.

CWF for the deepest clouds, for which $\lambda = 0$ and $\eta(z, \lambda) = 1$ for all z, becomes

$$A(0) = \int_{z_B}^{z_T} \beta(z)\left[h_M - \bar{h}^*_{B+} - \int_{z_{B+}}^{z} \frac{\partial \bar{h}^*(z')}{\partial z'} dz'\right] dz. \tag{10.31}$$

Here z_T is the height of the deepest cloud top. Let the subscript S denote the level within the mixed layer at which $h^*_S = \bar{h}^*_{B+}$ (see Fig. 10.4). Using $h_S = h_M$, we may rewrite (10.31) as

$$A(0) = (\mathrm{RH}_S - 1)L\bar{q}^*_S \int_{z_B}^{z_T} \beta(z)dz + \int_{z_B}^{z_T} \beta(z) \times \int_{z_{B+}}^{z} c_p[1 + \gamma(z')](\Gamma - \Gamma_m)dz'dz, \tag{10.32}$$

where RH_S is the relative humidity at level S. Thus, $A(0)$ increases as either RH_S or $\Gamma - \Gamma_m$ increases.

Using $A(\lambda) > 0$ as the instability criterion, A–S presented an argument similar to that in section 1.6 of chapter 1. Let the subscripts LS and C denote the effects of large-scale and cloud-scale processes, respectively. The net time change of $A(\lambda)$ is given by

$$\frac{dA(\lambda)}{dt} = \left[\frac{dA(\lambda)}{dt}\right]_{\mathrm{LS}} + \left[\frac{dA(\lambda)}{dt}\right]_C. \tag{10.33}$$

The first term on the rhs is *large-scale forcing* if

$$\left[\frac{dA(\lambda)}{dt}\right]_{\mathrm{LS}} > 0 \tag{10.34}$$

(destabilization). The second term is *adjustment,* which is assumed to satisfy

$$\left[\frac{dA(\lambda)}{dt}\right]_C < 0 \tag{10.35}$$

(stabilization). If the adjustment takes place very efficiently, we may assume that the air follows a sequence of quasi-neutral states characterized by $A(\lambda) \sim 0$ for all t, as long as (10.34) holds, so that

$$\left[\frac{dA(\lambda)}{dt}\right]_{\mathrm{LS}} + \left[\frac{dA(\lambda)}{dt}\right]_C \sim 0. \tag{10.36}$$

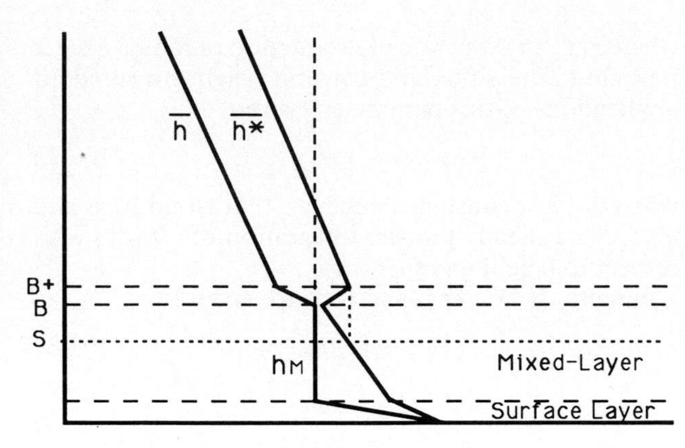

FIG. 10.4. A schematic figure illustrating vertical profiles of h^* and h and definitions of levels B and S.

The use of this approximation represents the CWF quasi-equilibrium assumption. As shown in A–S (see also chapter 1), it is valid if

$$\tau_{\mathrm{ADJ}} \ll \tau_{\mathrm{LS}}, \tag{10.37}$$

where τ_{LS} and τ_{ADJ} are time scales of the large-scale forcing and adjustment, respectively.

We may further interpret and generalize the CWF quasi equilibrium following Lord and Arakawa (1980). The cloud-scale kinetic energy equation for subensemble λ may be written as

$$\frac{dK(\lambda)}{dt} = [A(\lambda) - \delta(\lambda)]m_B(\lambda), \tag{10.38}$$

where $K(\lambda)$ is the cloud-scale kinetic energy and $\delta(\lambda)$ is the dissipation of cloud-scale kinetic energy *per unit* $m_B(\lambda)$. For the equilibrium of $K(\lambda)$, we have

$$A(\lambda) = \delta(\lambda), \tag{10.39}$$

provided that type λ clouds exist.

Equation (10.39) has an important implication. As pointed out earlier, $A(\lambda)$ can be determined from the large-scale thermodynamical vertical structure. According to (10.39), however, it is equal to $\delta(\lambda)$, which is the dissipation *normalized* by $m_B(\lambda)$. Since it is normalized, $\delta(\lambda)$ should be approximately independent of the intensity of the cloud subensemble and, therefore, it should not be very sensitive to individual synoptic situations. From (10.39), the same must be true for $A(\lambda)$. In other words, the large-scale vertical structure of temperature and that of moisture must be coupled with each other in such a way that $A(\lambda)$ is approximately constant (CWF quasi equilibrium) if type λ clouds exist. Observational verifications of the above statement are given by Lord and Arakawa (1980).

Let us consider, for example, the deepest clouds for which $\lambda = 0$. From $A(0)$ given by (10.32), we see that the CWF quasi equilibrium of the deepest clouds means that changes in the two terms in (10.32) tend to compensate each other so that RH_S and $\Gamma - \Gamma_m$ are negatively correlated. For more discussions of the cloud work functional equilibrium, see the chapter by Arakawa in this monograph.

From (10.11), (10.12), (10.13), and (10.14), we can see that the cumulus effect on the large-scale thermodynamical fields is linear with respect to $m_B(\lambda)$. Then we can write

$$\left[\frac{dA(\lambda)}{dt}\right]_C = \int_0^{\lambda_{max}} K(\lambda, \lambda')m_B(\lambda')d\lambda'. \quad (10.40)$$

Here $K(\lambda, \lambda')m_B(\lambda')d\lambda'$ is the contribution to $dA(\lambda)/dt$ by clouds with the fractional rate of entrainment in the range λ' and $\lambda' + d\lambda'$. The diagonal component of the kernel, $K(\lambda, \lambda)$, is usually negative, representing the self-stabilizing effect of type λ clouds. The nondiagonal component of the kernel, $K(\lambda, \lambda')$ with $\lambda \neq \lambda'$, represents the stabilizing (destabilizing) effect of type λ' clouds on type λ clouds when it is negative (positive) through the modification of their common environment. For further interpretations of K, which is called the *mass flux kernel,* see the original paper by A–S.

Substituting (10.40) into (10.36), we obtain

$$\int_0^{\lambda_{max}} K(\lambda, \lambda')m_B(\lambda')d\lambda' + F(\lambda) = 0, \quad \text{for} \quad m_B(\lambda) > 0, \quad (10.41)$$

where $F(\lambda) \equiv [dA(\lambda)/dt]_{LS}$. This is the *mass flux distribution function equation* proposed by A–S. If we ignore the inequality condition, this is a Fredholm integral equation of the first kind for the unknown $m_B(\lambda)$. For $m_B(\lambda) = 0$, the following condition should be satisfied:

$$\int_0^{\lambda_{max}} K(\lambda, \lambda')m_B(\lambda')d\lambda' + F(\lambda) < 0 \quad \text{for} \quad m_B(\lambda) = 0. \quad (10.42)$$

See A–S for the physical basis of (10.42).

If $m_B(\lambda)$ that satisfies (10.41) and (10.42) is found (see section 10.6), $M_c(z)$ and $D(z)$ can be obtained from (10.13) and (10.14), and the large-scale temperature and moisture fields can be predicted using (10.11) and (10.12).

10.6. Standard procedures in practical application of the parameterization

In the description presented above, which closely follows the original paper by A–S, the fractional rate of entrainment λ is used to identify cloud types. In most practical applications of the parameterization, however, cloud-top values of the vertical coordinate (such as pressure or sigma) are used instead to identify cloud types. Accordingly, the fractional rate of entrainment is computed for each cloud type by iteration from the nonbuoyancy condition applied to the cloud top (Lord 1982). For this iteration to converge, any supersaturation and superadiabatic lapse rate should be eliminated in advance.

The standard procedure of incorporating this parameterization into large-scale models is through an adjustment of CWF for each subensemble to an empirically determined critical value. Let A_i be the value of CWF for the ith subensemble at a given time step after large-scale effects (but not cumulus effects) are implemented. The large-scale forcing is then defined by

$$F_i \equiv A_i - (A_0)_i \quad (10.43)$$

where $(A_0)_i$ is the critical CWF corresponding to the dissipation δ in (10.39). In the University of California, Los Angeles (UCLA) general circulation model (GCM), $(A_0)_i = (A_N)_i(p_{N-1/2} - p_i)$, where $(A_N)_i$ is empirically determined from observations (Lord et al. 1982). We require that, for a cloud type to exist, the corresponding large-scale forcing F_i is to be positive.

Let the elements of the kernel in the integral equation (10.41) be $K_{i,j}$, which represents the change of the CWF for the ith subensemble due to a unit $m_B\Delta t$ of the jth subensemble. Here Δt is the time interval used in the model for implementing cumulus effects. To calculate $K_{i,j}$, we modify the temperature and moisture profiles by $(m_B\Delta t)_{j,\text{test},}$, which is a properly chosen amount of $m_B\Delta t$ of the jth subensemble. The amount should be neither very small, because runoff errors may contaminate the results, nor very large, because the time change of CWF is a nonlinear function of $m_B\Delta t$. We then recalculate the CWF for the ith subensemble using the modified profiles. Let the result be (A_i'). The kernel elements can then be calculated by

$$K_{i,j} = \frac{(A_i') - A_i}{(m_B\Delta t)_{j,\text{test}}}. \quad (10.44)$$

The discrete version of the equations (10.41) and (10.42) may be written as

$$\sum_j^{\text{imax}} K_{i,j}(m_B\Delta t)_j + F_i = 0, \quad \text{for} \quad (m_B\Delta t)_i > 0; \quad (10.45)$$

$$\sum_j^{\text{imax}} K_{i,j}(m_B\Delta t)_j + F_i < 0, \quad \text{for} \quad (m_B\Delta t)_i = 0. \quad (10.46)$$

If we ignore the inequality condition in (10.45), solution of the discrete integral equation simply involves inversion of the matrix $\mathbf{K}$. Experience shows that $(m_B\Delta t)_i$ obtained in this way does satisfy the inequality condition in most cases but not always. This means that there are exceptional cases for which the CWF equilibrium cannot be strictly enforced while it is generally a good approximation. The standard way of handling these cases with the A–S parameterization is to find optimum solutions by minimizing the errors in satisfying (10.45) and (10.46).

More detailed computational procedures for implementing the parameterization are given in Lord (1982), Lord et al. (1982), and Cheng and Arakawa (1990). See also Suarez et al. (1983) for the way in which this parameterization is coupled with the PBL in the UCLA GCM.

10.7. Further comments

In this chapter, we have outlined the essence of the A–S cumulus parameterization with an emphasis on its logical structure, which is basically a combination of a spectral cloud ensemble model and the cloud work function quasi-equilibrium assumption.

Lord (1982), Kao and Ogura (1988), Cheng and Arakawa (1990), and Moorthi and Suarez (1992) presented verification of this parameterization following the semiprognostic approach, in which the results of parameterization applied to each observation time are compared to those estimated from residuals in the budget equations applied to a time sequence of observations (see the Appendix for more recent results of Cheng and Arakawa). Also following the semiprognostic approach, Xu and Arakawa (1992) verified this parameterization against the data simulated by a numerical cloud ensemble model developed by Krueger (Krueger 1988; see also Xu and Krueger 1991). One of their major findings is that the CWF quasi-equilibrium assumption is better when it is applied to averages over smaller intervals so that mesoscale processes are included in the "large-scale" processes. This finding, which is consistent with that of Grell et al. (1991), means that the parameterizability increases as the mesoscale organization of cumulus activity is resolved. For more details, see chapter 22 by Xu in this monograph.

Cheng and Arakawa (1990, 1991a,b) revised the A–S parameterization by including convective-scale downdrafts. They compared the result without and with downdraft effects both semiprognostically and fully prognostically. Some of their results and a brief description of the revised parameterization are given in the Appendix.

There are attempts to make an implementation of the A–S parameterization computationally more convenient and efficient by relaxing the CWF quasi equilibrium (Moorthi and Suarez 1992; Pan and Randall 1991; see also the chapter by Randall and Pan in this monograph). These attempts very likely represent the future direction of implementing the A–S parameterization in numerical models although the conceptual importance of the CWF quasi equilibrium will remain the same as the basis for parameterizability.

Acknowledgments. The author would like to thank Mrs. Clara Wong and Ms. Kristin Mah for typing the manuscript. The authors also would like to thank Professor Kerry Emanuel for his thorough review of the original manuscript and a number of valuable suggestions for its improvement. This material is based on the research supported jointly by NSF under Grant ATM-8910564 and NASA under Grant NAG 5-789. Part of the research reported in the Appendix was performed while the second author (MDC) was supported by the Advanced Study Program, NCAR. Computations were performed at the NCAR SCD and at the Office of Academic Computing, UCLA.

APPENDIX

Inclusion of Convective Downdraft Effects

a. Introduction

Among various types of convective downdrafts, those associated with precipitation seem to be the most important for the large-scale heat and moisture budgets. These downdrafts are initiated and maintained by the loading and cooling due to evaporation of rainwater, and how rainwater detrains from the updraft is essential in determining the properties of downdraft. Cheng (1989a) presented a combined updraft–downdraft spectral cumulus ensemble model based on the rainwater budgets for updrafts and associated convective downdrafts. In the model the rainwater generated in the updraft is assumed to fall partly inside and partly outside the updraft. The mean tilting angle of the updraft, which determines this partition, is estimated assuming statistically steady states. This model has been applied to diagnostic studies of the effects of convective downdrafts and mesoscale processes (Cheng 1989b; Cheng and Yanai 1989). The model has also been incorporated into the A–S parameterization. Results from semiprognostic tests and GCM experiments using the modified A–S parameterization are presented by Cheng and Arakawa (1990, 1991a).

Cheng's model in its original form does not allow the detrainment of rainwater through cloud top due to the use of an inappropriate upper boundary condition. The rainwater detraining from cloud top, however, can be an important source for anvil precipitation in a mesoscale convective system. Following Cheng and Arakawa (1991b), this appendix summarizes a revised version of the coupled updraft–downdraft model, which has also been incorporated into the A–S parameterization. Results from semiprognostic tests of the parameterization using GATE phase III data are presented. Finally, results from the UCLA GCM with (the earlier version of) the coupled updraft–downdraft model are shown.

b. Rainwater budget for the updraft

As in the standard application of the A–S parameterization, updrafts are classified into subensembles according to their cloud-top levels. Applying the vanishing-buoyancy condition to cloud top, we determine the vertical profiles of the normalized mass flux and moist static energy for each subensemble. Then, the temperature and water vapor mixing ratio of the updraft can be obtained. The rate of rainwater generation is parameterized using an autoconversion coefficient.

The combined updraft–downdraft model introduces the rainwater budget equation, which expresses a balance between the generation of rainwater, the vertical convergence of in-cloud rain flux, and the amount of outgoing rain flux. The budget equation for an updraft subensemble can be written as

$$\frac{1}{\cos\theta}\frac{\partial F}{\partial t} = G - \cos\theta\frac{\partial H}{\partial Z}. \qquad (A.1)$$

Here

$$F \equiv \rho\sigma q_r \cos\theta, \qquad (A.2)$$

$$G \equiv -\frac{2}{\pi a}\rho\sigma q_r V_t \sin\theta + mC_0 l, \qquad (A.3)$$

$$H \equiv \rho\sigma q_r\left(\frac{w_c}{\cos\theta} - V_t\cos\theta\right), \qquad (A.4)$$

$$\rho\sigma \equiv \frac{m}{w_c}, \qquad (A.5)$$

ρ is the air density, σ is the fractional area covered by the updraft, q_r is the rainwater mixing ratio, θ is the updraft tilting angle, a is the radius of the updraft, V_t is the mean terminal fall velocity of raindrops, C_0 is the autoconversion coefficient for conversion from cloud water to rainwater, l is the cloud water mixing ratio, w_c is the updraft vertical velocity, and m is the updraft mass flux.

The mean terminal fall velocity of raindrops is calculated from

$$V_t \equiv 36.34(\rho q_r)^{0.1364}\left(\frac{\rho}{\rho_0}\right)^{-0.5} \quad (\mathrm{m\ s^{-1}}), \qquad (A.6)$$

where ρ_0 is the air density of a reference state at the ground level (Soong and Ogura 1973). To calculate the updraft velocity, we use the vertical momentum budget equation, which may be written as

$$(1+\gamma)\frac{\partial}{\partial z}(mw_c) = \rho\sigma(B - gq_r), \qquad (A.7a)$$

where B is the thermal buoyancy including the loading of cloud water and γ is the virtual mass coefficient (Simpson and Wiggert 1969). Using (A.5), (A.7a) can be rewritten as

$$\frac{1}{2}\frac{\partial}{\partial z}w_c^2 + \lambda w_c^2 = \frac{B - gq_r}{1+\gamma}, \qquad (A.7b)$$

where $\lambda \equiv (1/m)\partial m/\partial z$ is the fractional rate of entrainment. For a given vertical profile of the net buoyancy, $B - gq_r$, (A.7b) can be solved for the corresponding vertical velocity profile with a proper boundary condition of w_c at cloud base. [Cheng (1989a) used a simpler form of the vertical momentum budget equation, which has been replaced by the above form in the current model.]

c. Determination of the tilting angle

We are interested in stationary solutions of the rainwater budget equation, (A.1), coupled with the vertical momentum equation, (A.7b). Because rainwater may be carried away from cloud top through detrainment, we should use "open" boundary conditions at cloud top as well as at cloud base for the rainwater budget equation. In producing the results shown below, we used $w_c = 0.1$ m s^{-1} at cloud base. Due to the use of different boundary conditions for the rainwater and vertical momentum budget equations, it requires iterations for obtaining a solution.

Figure A1 shows stationary solutions of the rainwater budget equation for the range of θ from 0° to 60°. This particular example is taken from cloud type 4, whose top reaches 160 mb, using GATE phase III mean data (see Cheng 1989b) as inputs. In this figure we see that the vertical profiles of the rainwater mixing ratio, vertical velocity, and in-cloud and outgoing rain fluxes change smoothly as θ increases from 0° to 28° and from 30° to 60° with a discontinuity between 28° and 30°.

For convenience, we call the two groups of solutions separated by a discontinuity by *solutions for small tilting angles* (0°–28° in this case) and *solutions for large tilting angles* (30°–60° in this case), respectively. In general, the solutions for small tilting angles are characterized by relatively small vertical velocity and very small detrainment of rainwater from cloud top. On the other hand, the solutions for large tilting angles are characterized by relatively large vertical velocity and large detrainment of rainwater from cloud top. At 30° of this case, the detrainment of rainwater from cloud top accounts for about 40% of the total rainwater generated in the updraft. This ratio increases as the tilting angle increases. In contrast, the ratio between the cloud-top detrainment and the total generation of rainwater for the solutions for small tilting angle is less than 3%.

To examine the stability of the stationary solutions, we linearize (A.1) with respect to a small perturbation of q_r. The vertically discrete, linearized rainwater budget equation, consistent with the scheme used to solve for the stationary solution, has the form

$$\frac{\partial \mathbf{q}_r'}{\partial t} = \mathbf{C}\mathbf{q}_r', \qquad (A.8)$$

where $\mathbf{q}_r'$ is a column vector and $\mathbf{C}$ is a coefficient matrix. We have found that the solutions for small tilting

STATIONARY SOLUTION FOR THE UPDRAFT
Cloud type 4 (cloud top at 190 mb)

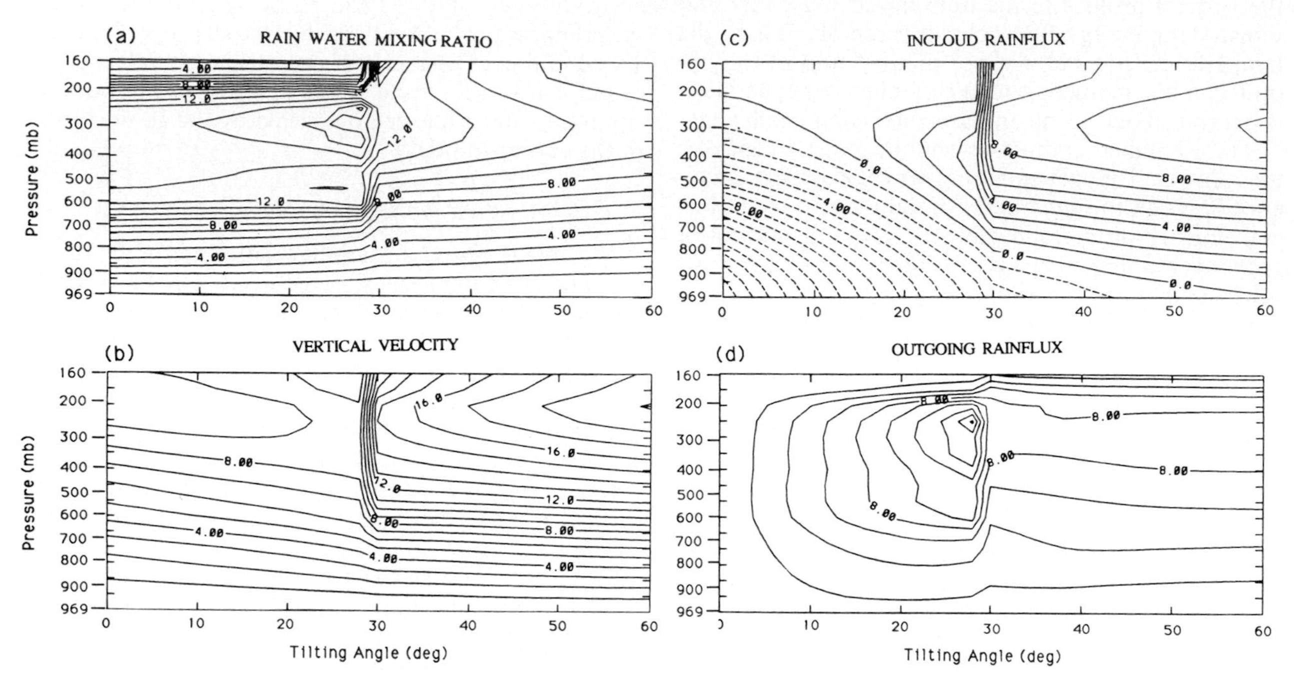

FIG. A1. Stationary solutions of the rainwater budget equation for the range of θ from 0° to 60°: (a) q_r (g kg^{-1}), (b) w_c (m s^{-1}), (c) in-cloud, and (d) outgoing rain fluxes (arbitrary unit).

angles are unstable, while the solutions for large tilting angles are stable. The existence of the instability for small tilting angles is confirmed also through time integrations (Cheng and Arakawa 1991b). We interpret that the instability is due to the coupling of the rainwater and vertical momentum equations through the rainwater loading effect.

Since the cloud-scale horizontal momentum budget equation is not used in the model, we cannot uniquely determine the updraft tilting angle θ. For a given value of θ, however, we can solve the stationary rainwater budget equation simultaneously with the vertical momentum budget equation. Then we can examine stability of the solutions for all possible values of θ. Since we do not expect a very large tilting angle to occur unless the environmental vertical wind shear is extremely strong, we assume that the subensemble mean tilting angle can be optimally represented by the smallest value in the range in which stable stationary solutions can be found.

d. Downdraft model

We assume that convective downdrafts are driven by the rainwater falling through cloud environment. Once the updraft tilting angle is determined, as described in the last section, the amount of the rainwater available for driving the downdraft can be calculated. The properties of the downdraft are then obtained by solving the budget equations for mass, moist static energy, water vapor, vertical momentum, and rainwater.

The downdraft mass flux varies in height through changes in the air density, the downdraft area, and the downdraft vertical velocity. In the model, the downdraft area is defined as the rain-covered area outside the updraft. The rain-covered area may change with height because the updraft from which the rainwater detrains is tilted. This area may also change due to airflow toward the downdraft and evaporation of the rainwater along the edge of the downdraft. The downdraft area is calculated considering these three processes, while the change of the vertical velocity is obtained from the vertical momentum budget equation. To calculate the vertical momentum budget, however, we need to know the mixing ratio of the rainwater and the thermal buoyancy of the downdraft. The rainwater mixing ratio is diagnosed from the rainwater budget, in which evaporation of the rainwater is formulated following Soong and Ogura (1973). With the water vapor mixing ratio and the moist static energy determined from their budget equations, the thermal buoyancy of the downdraft can be calculated.

e. Cloud work function quasi equilibrium

Arakawa and Schubert defined CWF as the work done by the thermal buoyancy of the updraft per unit mass flux at cloud base [see (A.9)]. There the effect

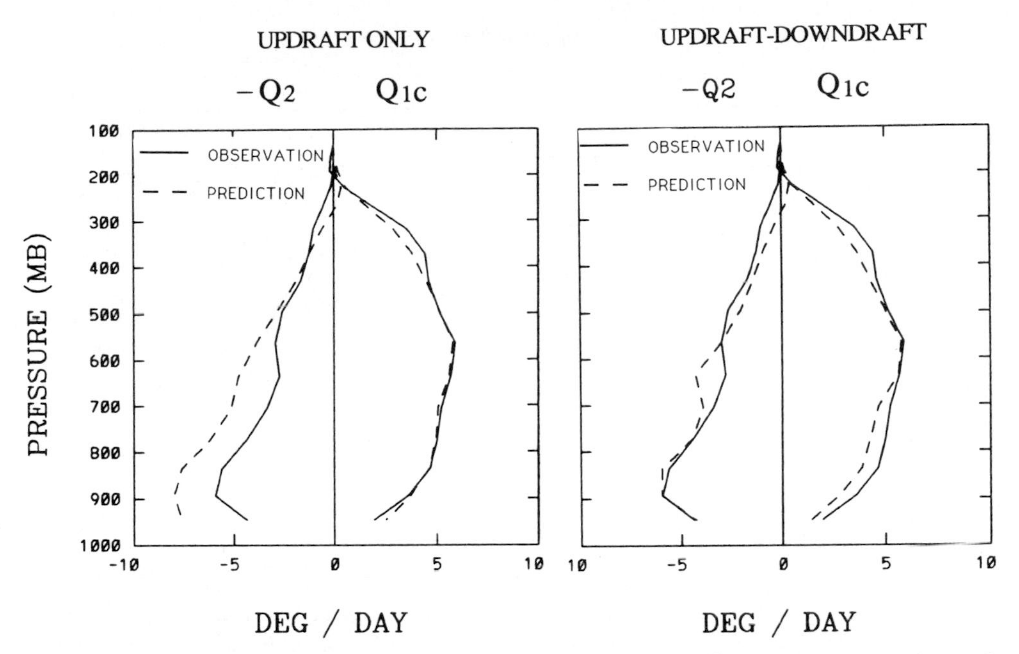

FIG. A2. Time-averaged profiles of the observed (solid lines) and predicted (dashed lines) $Q_{1C} \equiv Q_1 - Q_R$ and $-Q_2$ for GATE phase III. Left panel: updraft-only model. Right panel: updraft–downdraft model. Units are kelvins per day.

of rainwater loading is considered as a part of the dissipation mechanism. Since that effect is now explicitly calculated, we may redefine CWF as the work done by the *net* buoyancy of the updraft per unit mass flux at cloud base. We may also include the generation of cloud kinetic energy associated with convective downdrafts in the definition of CWF.

Let B_u, D_u, B_d, and D_d be the thermal buoyancy of the updraft, the rainwater drag on the updraft, the thermal buoyancy of the downdraft, and the rainwater drag on the downdraft. CWF as originally defined is

$$A_1 = \int_{z_B}^{z_T} \eta_u B_u dz, \tag{A.9}$$

where η_u is the updraft mass flux normalized by its cloud-base value. As mentioned above, we can redefine CWF by

$$A_2 \equiv \int_{z_B}^{z_T} \eta_u (B_u + D_u) dz, \tag{A.10}$$

or by

$$A_3 \equiv \int_{z_B}^{z_T} [\eta_u (B_u + D_u) dz + \eta_d (B_d + D_d)] dz, \tag{A.11}$$

where η_d is the downdraft mass flux normalized by the *updraft* mass flux at cloud base.

Cheng and Arakawa (1990) calculated A_1, A_2, and A_3 for clouds with top at 190 mb during GATE phase III. In general, A_2 is only slightly larger than one-quarter of A_1, indicating that the rainwater drag offsets about three-quarters of the thermal buoyancy for updraft. The difference between A_3 and A_2 represents the generation of cloud kinetic energy by the net buoyancy of the downdraft, which is about one-third of A_2. Nevertheless, A_1, A_2, and A_3 are approximately proportional to each other, and $dA_1/dt \cong 0$ can be justified if $dA_2/dt \cong 0$ or $dA_3/dt \cong 0$. We therefore remain to use A_1 as CWF even in the updraft–downdraft model.

f. Semiprognostic test

We have incorporated the revised version of the updraft–downdraft model into the A–S parameterization. As we did with the earlier version (Cheng and Arakawa 1990, 1991a), we have tested the parameterization semiprognostically using the GATE phase III data. To separate the problem of cumulus parameterization from that of a PBL parameterization, we performed "fixed-PBL" experiments, in which the total effect of PBL processes is assumed to balance the effect of cumulus clouds, and therefore, no net time change of PBL properties is produced.

For comparison we show in Fig. A2 the phase III averaged vertical profiles of the cumulus heating $Q_{1C} \equiv Q_1 - Q_R$ and the cumulus moistening $-Q_2$ diagnosed from observed data (solid) and those predicted by the updraft-only and (revised) updraft–downdraft models. We see that the prediction of $-Q_2$ is significantly improved by the inclusion of downdrafts. Because the downdraft air is usually drier and cooler than the environmental air in the subcloud layer, the updraft–

PRECIPITATION

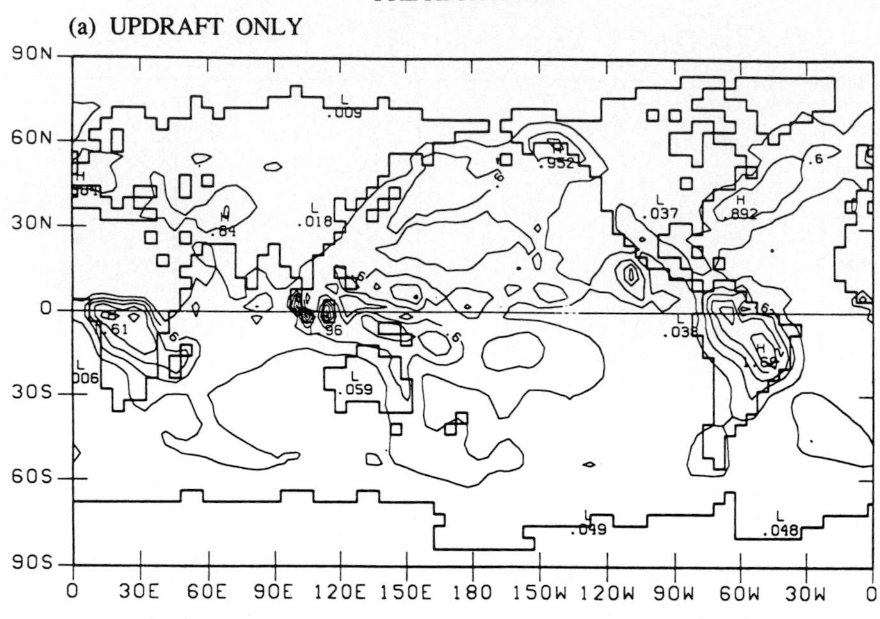

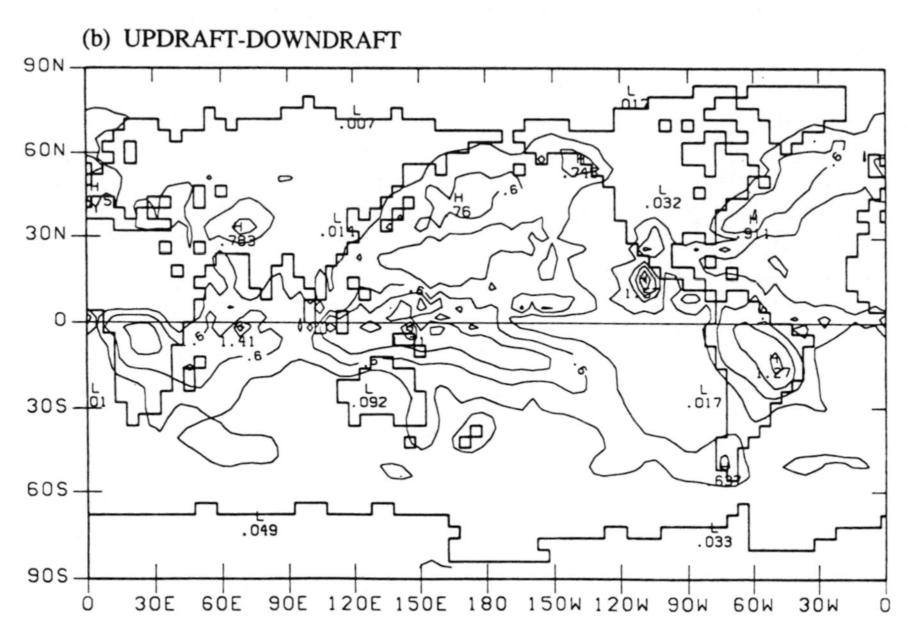

FIG. A3. January mean precipitation, (a) from the updraft-only model and (b) from the updraft–downdraft model. Units are centimeters per day.

downdraft model predicts large drying and moderate cooling below cloud base.

With the updraft–downdraft model, we can also find the cumulus contribution to anvil precipitation. In general, the contribution is a small fraction of the total cumulus precipitation but not necessarily so for individual cases.

g. Experiments with the UCLA GCM

In a fully prognostic model, we must explicitly formulate the effect of downdrafts on the PBL. We may assume that the downdraft air completely mixes with the PBL air immediately after it detrains below cloud base. Then, since the downdraft air is usually cooler and drier than the PBL air, cumulus clouds will be significantly stabilized. It is more likely, however, that the air entering the updraft from cloud base is basically the undisturbed PBL air, even after the air detrained from the downdraft has spread out over an area of significant size. We anticipate that when the horizontal resolution of the large-scale model is coarse, the complete mixing between the downdraft and PBL air is

EVAPORATION

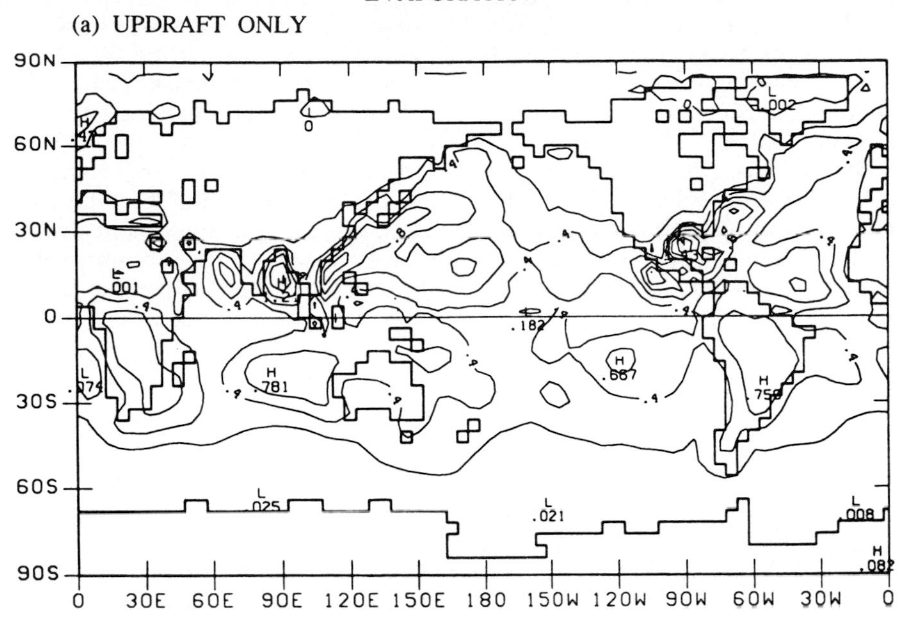

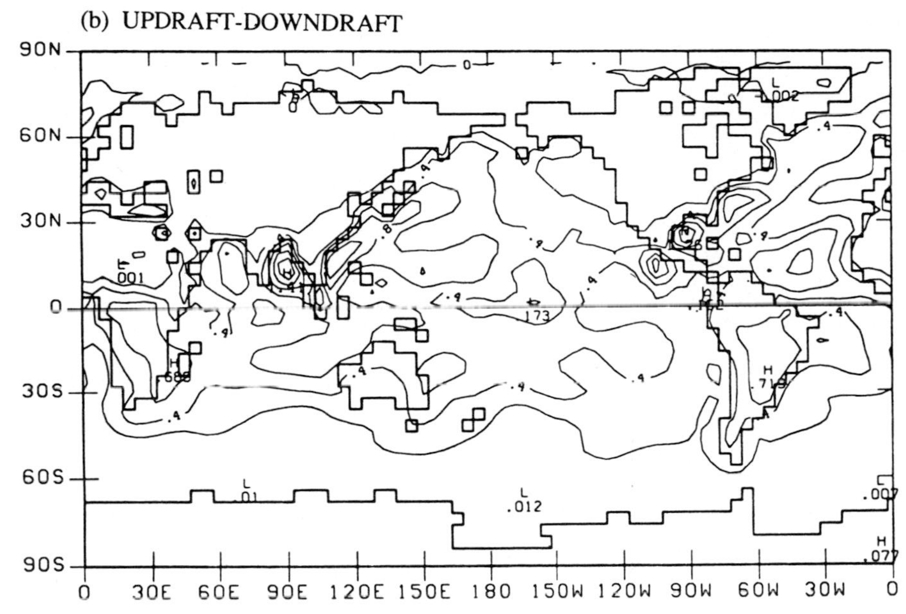

FIG. A4. January mean surface evaporation, (a) from the updraft-only model and (b) from the updraft–downdraft model. Units are centimeters per day.

unrealistic. In nature these effects are confined in relatively narrow regions near convective cells, which are typically much smaller than the grid size of GCMs. In the experiment performed with the UCLA GCM, we assumed that the effects of the downdrafts are locally and immediately compensated by enhanced latent (and sensible) heat fluxes from the underlying surface.

The initial conditions for the GCM experiments are taken from the observed fields for 1 October 1982. The integrations are performed over the four-month period ending 31 January 1983. In the following we show the January mean of the global distributions of precipitation and surface evaporation. Figures A3a and A3b show the distributions of precipitation from the updraft-only and updraft–downdraft experiments, respectively. We see that the updraft–downdraft scheme generates much less precipitation over the tropical continents and islands than the updraft-only scheme does. On the other hand, precipitation associated with the southern Pacific convergence zone (SPCZ) and the

Atlantic intertropical convergence zone (ITCZ) is enhanced in the updraft–downdraft experiment. In general, the inclusion of downdrafts generates a relatively even distribution of precipitation over continents and over oceans compared to that generated by the updraft-only model. Figures A4a and A4b show the distributions of surface evaporation from the updraft-only and updraft–downdraft experiments, respectively. For the latter, evaporation compensating the downdraft effects on the PBL is included. Although these two maps are similar in the extratropics, there are significant differences in the tropics. For example, the values are larger with downdrafts over the equatorial Indian and western Pacific oceans and along the SPCZ. On the other hand, the values are much smaller with downdraft over the Gulf of Mexico and eastern Pacific close to Central America.

We obviously need more extensive experiments to investigate the impact of this difference on general circulation of the atmosphere and air–sea interaction. At present we are working on improving computational efficiency of the revised parameterization.

Chapter 11

Implementation of the Arakawa–Schubert Cumulus Parameterization with a Prognostic Closure

DAVID A. RANDALL AND DZONG-MING PAN

Department of Atmospheric Science, Colorado State University, Fort Collins, Colorado

11.1. Introduction

The cumulus parameterization of Arakawa and Schubert (1974, hereafter referred to as A–S), as implemented by Lord et al. (1982), is currently being used in a number of general circulation models (GCMs) including those at the University of California, Los Angeles, the National Aeronautics and Space Administration's Goddard Space Flight Center, the U.S. Navy's Fleet Numerical Weather Prediction Center, the Japanese Meteorological Research Institute, and Colorado State University (CSU). Its strengths and weaknesses are well documented in the literature on the basis of many observational and numerical studies, some of which are mentioned below.

The physical basis of the A–S parameterization is summarized by Arakawa (see chapter 1) and Arakawa and Cheng (see chapter 10). Briefly, a model's grid column is assumed to contain an ensemble of cumulus clouds, which have their bases at or near the top of the planetary boundary layer. The cumulus ensemble is divided into subensembles, roughly speaking, according to cloud-top height. Each subensemble represents a collection of cumulus clouds. The soundings characteristic of the interiors of each subensemble are determined using an entraining plume model. A convective mass flux is associated with each subensemble. The vertical distribution of the heating and drying can be determined in terms of the convective mass flux. The parameterization can thus be closed by determining the convective mass flux associated with each subensemble.

Closure is achieved through a quasi-equilibrium hypothesis, which is based on consideration of the near balance between the rate of generation of moist convective instability by large-scale processes and the compensating release of moist convective instability by cumulus convection. The cumulus clouds are assumed to be very efficient consumers of the convective available potential energy (CAPE; see Randall and Wang 1992 for a recent discussion of this concept) associated with moist convective instability. Even though "large-scale processes" may tend to destabilize the column rapidly, convection prevents the instability from growing by removing the CAPE as fast as it is generated. This convective stabilization is achieved by drying through precipitation, and by warming aloft, which reduces the lapse rate of temperature.

Quasi equilibrium must be maintained for all subensembles simultaneously. All of the cumulus clouds share a common "large-scale environment," which they modify through their mass fluxes. The requirement of simultaneous quasi equilibrium for all subensembles thus leads to an integral equation, in which the integration is taken over all subensembles.

The quasi-equilibrium hypothesis has received considerable support from observations (e.g., Lord and Arakawa 1980; Lord 1982; Arakawa and Chen 1987; Xu and Emanuel 1989) and numerical simulations (e.g., Ogura and Kao 1987; Kao and Ogura 1987; Grell et al. 1991; Xu 1991). Nevertheless, its practical implementation is somewhat troublesome. The origins of these difficulties are outlined below.

Arakawa and Schubert measured the potential energy available for conversion into cumulus kinetic energy (CKE) in terms of the cloud work function A, which is an integral of temperature and moisture over the convectively active layer. The cloud work function is defined separately for each subensemble. It can be determined, for any subensemble, from a given sounding of temperature and moisture. Schematically, we can write $A = A(T, q)$, where T is temperature and q is the specific humidity. The chain rule gives

$$\frac{\partial A}{\partial t} = \frac{\partial A}{\partial T}\frac{\partial T}{\partial t} + \frac{\partial A}{\partial q}\frac{\partial q}{\partial t}. \quad (11.1)$$

Expressions for $\partial T/\partial t$ and $\partial q/\partial t$ are given by the thermodynamic energy equation and the moisture conservation equation, respectively. Each of these two equations contains both "convective" terms (e.g., convective transports) and "nonconvective" terms (e.g., advection by the mean flow, surface fluxes, and radiative cooling). By substituting into (11.1) and collecting the convective and nonconvective terms into groups, we can derive

$$\frac{\partial A}{\partial t} = JM_B + F. \qquad (11.2)$$

The first term on the right-hand side (rhs) of (11.2) represents the combination of all convective effects, and the second represents all other effects. Here M_B is the convective cloud-base mass flux; J is a negative "kernel" that represents the rate at which convection reduces A, per unit M_B; and F is the "large-scale forcing" that may tend to increase A. [The full-fledged version of (11.2), with the left-hand side (lhs) replaced by zero, is the integral equation mentioned above; see A–S for details.] Since A depends only on the sounding, and J is the tendency of A due to a unit mass flux, J also depends only on the sounding. In practice, J is usually computed by prescribing a small "trial mass flux" and evaluating the implied tendency of A. In a model, we do not actually integrate (11.2) to predict A; instead, we predict T and q, from which A can be determined.

The quasi-equilibrium hypothesis states that the actual time rate of change of A is small compared with either term on the rhs of (11.2). This allows us to determine the convective mass flux from

$$M_B \approx -\frac{F}{J}. \qquad (11.3)$$

(The physical basis of the quasi-equilibrium hypothesis is discussed later.) Once M_B is known, the convective heating and drying rates can easily be determined. Of course, negative values of M_B are physically unacceptable, so we require

$$M_B \geqslant 0. \qquad (11.4)$$

Keep in mind that, in (11.3) and (11.4), M_B schematically represents the cloud-base mass fluxes for the various subensembles of cumulus clouds rather than the cloud-base mass flux for a single subensemble. The condition (11.4) must be satisfied for all subensembles simultaneously. In practice, a direct solution of (11.3) often fails to satisfy this requirement. Various methods (e.g., linear programming) have been proposed to find a "best" solution that comes close to satisfying (11.3) without violating (11.4), but the uniqueness of such solutions is questionable (e.g., Moorthi and Suarez 1992).

We have undertaken a multistep effort to improve the A–S parameterization, both by simplifying its implementation and by increasing its physical realism. Our primary goal is to improve the physical content of the parameterization by providing a more consistent coupling between cumulus convection and the associated stratiform anvil and cirrus cloudiness (Randall 1989). Secondary goals are to allow a spectrum of cloud-base heights, so that the parameterization can apply in a unified way to convective clouds originating at various levels within the same large-scale atmospheric column, and to include the effects of convective-scale downdrafts, following the approach of Cheng (1989a) as refined by Cheng and Arakawa (1991b). A tertiary goal is to simplify and speed up the parameterization.

11.2. Approach

Our approach is to relax the quasi-equilibrium assumption without abandoning the basic concepts on which it is based. As a starting point, we take the ideas of Lord and Arakawa (1980, hereafter LA), who extended the concept of quasi equilibrium by considering the conservation equation for the CKE of a cumulus subensemble:

$$\frac{\partial K}{\partial t} = M_B A - D. \qquad (11.5)$$

Here K (also called the CKE) is the vertically integrated kinetic energy of the subensemble per unit area, and D is the rate at which K is dissipated. Strictly speaking, (11.5) should also include a "shear production" term. There is evidence that this is often negligible (Xu 1991), although it may be important in some cases (Lilly and Jewett 1990). Vertical transport terms (pressure-velocity covariance and triple-moment terms) do not appear in (11.5), because (11.5) governs the vertically integrated kinetic energy of the cumulus subensemble. One may question whether the kinetic energy budgets of the various subensembles can be formulated independently of one another; it is possible that subensembles interact directly. We have assumed, in writing (11.5), that such direct interactions are negligible, so that the subensembles interact with each other only indirectly, by modifying their common environment.

We have modified the A–S parameterization by making explicit use of (11.5) to prognostically determine the CKE. The dissipation term is determined by introducing a dissipation time scale, denoted by τ_D, so that the prognostic equation for K becomes

$$\frac{\partial K}{\partial t} = M_B A - \frac{K}{\tau_D}. \qquad (11.6)$$

To relate M_B to the CKE, we use

$$K = \alpha M_B^2, \qquad (11.7)$$

which was proposed by Arakawa and Xu (1990) and Xu (1991). The physical basis of (11.7) is discussed later. Substitution of (11.7) into (11.6), with the assumption that α is independent of time, yields a linear prognostic equation for M_B:

$$\frac{\partial M_B}{\partial t} = \frac{A}{2\alpha} - \frac{M_B}{2\tau_D}. \qquad (11.8)$$

To explore the implications of this result, combine (11.8) with (11.2) and make the temporary assumption that J is independent of time. In this way, we can derive inhomogeneous second-order ordinary differential equations for A and M_B:

$$\ddot{A} + \frac{\dot{A}}{2\tau_D} - \frac{J}{2\alpha} A = \frac{F}{2\tau_D} + \dot{F}, \quad (11.9)$$

$$\ddot{M}_B + \frac{\dot{M}_B}{2\tau_D} - \frac{J}{2\alpha} M_B = \frac{F}{2\alpha}. \quad (11.10)$$

Here we use the "dot" notation to denote a time derivative. Xu (1991) derived (11.10) and discussed its solution in some detail. Here we present only a qualitative discussion.

First, suppose that F is independent of time. Then we obtain the steady, forced solution

$$A = -\frac{\alpha}{J}\frac{F}{\tau_D} \equiv A_F, \quad (11.11)$$

$$M_B = -\frac{F}{J} \equiv (M_B)_F, \quad (11.12)$$

$$K = \alpha\left(\frac{F}{J}\right)^2 \equiv K_F. \quad (11.13)$$

This shows that positive equilibria of A, M_B, and K can occur as a result of a positive large-scale forcing, F. In (11.11)–(11.13), we have denoted these forced equilibria by the subscript F.

If A and M_B are assigned initial values that differ from the equilibria given by (11.11) and (11.12), then (11.9) and (11.10) dictate a relaxation to equilibrium, subject of course to the constraints

$$A \geqslant 0, \quad M_B \geqslant 0. \quad (11.14)$$

The time scale for this relaxation is apparently of order τ_D. Note, however, that (11.9) and (11.10) also contain a second natural time scale, namely, $(\alpha/|J|)^{1/2}$. To interpret this time scale, consider the limit $\tau_D \rightarrow \infty$, with $F = 0$ and $\dot{F} = 0$. Then (11.9) and (11.10) reduce to

$$\ddot{A} - \frac{J}{2\alpha} A = 0, \quad (11.15)$$

$$\ddot{M}_B - \frac{J}{2\alpha} M_B = 0. \quad (11.16)$$

Suppose that we initialize A and M_B at positive values. Although (11.15) and (11.16) appear to predict ensuing free oscillations about $A = 0$ and $M_B = 0$, (11.14) implies that these oscillations will halt as soon as A and M_B have decreased to zero. The time required for this to occur is proportional to $(\alpha/|J|)^{1/2}$. We can thus interpret $(\alpha/|J|)^{1/2}$ as τ_{ADJ}, the "adjustment time," defined by A–S as the time required for convective processes to reduce A to zero in the absence of large-scale forcing. This shows that the adjustment time is closely related to α.

According to (11.11), A_F approaches zero as α approaches zero, that is, as the adjustment time becomes short. Arakawa and Schubert argued that the observed values of A are in fact "small" compared to those that would occur if the large-scale forcing acted unopposed over τ_{LS}. The fact that the observed tropical atmosphere is close to neutral stability with respect to moist convection has been emphasized by Xu and Emanuel (1989).

When F varies in time, with time scale τ_{LS}, under what conditions are (11.11) and (11.12) good approximations to (11.9) and (11.10), respectively? To investigate this question, we write

$$\dot{F} \sim \frac{F}{\tau_{LS}}, \qquad \ddot{A} \simeq \ddot{A}_F \sim \frac{A_F}{\tau_{LS}^2}, \qquad \dot{A} \cong \dot{A}_F \sim \frac{A_F}{\tau_{LS}}. \quad (11.17)$$

Here we *define* τ_{LS} as the time scale on which the large-scale forcing F varies, and we *assume* that A varies on this same time scale. With this scaling, we find that the first term on the lhs of (11.9) is negligible compared with the third term on the lhs if

$$\frac{\tau_{ADJ}}{\tau_{LS}} \ll 1, \quad (11.18)$$

the second term on the lhs is negligible compared with the third term on the lhs if

$$\left(\frac{\tau_{ADJ}}{\tau_{LS}}\right)\left(\frac{\tau_{ADJ}}{\tau_D}\right) \ll 1, \quad (11.19)$$

and the second term on the rhs is negligible compared to the first term on the rhs if

$$\frac{\tau_D}{\tau_{LS}} \ll 1. \quad (11.20)$$

According to (11.18) and (11.20), both τ_{ADJ} and τ_D should be much smaller than τ_{LS}, which is easy to understand. In addition, however, (11.19) implies that τ_D should not be significantly shorter than τ_{ADJ}. An interpretation of the latter condition is that if τ_D is very small compared to τ_{ADJ}, then cumulus kinetic energy is dissipated so efficiently that the convection cannot become vigorous enough to reduce A to its equilibrium value over time scale τ_{LS}.

We conclude that when (11.18)–(11.20) are satisfied, (11.11) is a good approximation to (11.9), and the solution obtained by time integration of (11.2) and (11.6) with (11.7) should closely approximate the quasi-equilibrium solution given by (11.11)–(11.13).

In particular, the cumulus cloud-base mass flux M_B will satisfy (11.12), which does not involve α or τ_D.

TABLE 11.1. Typical numerical values for the parameters of the model.

Parameter	Typical value
A	10^3 J kg^{-1}
M_B	3×10^{-2} kg m^2 s^{-1}
K	10^4 J m^{-2}
τ_D	10^3 s
τ_{ADJ}	10^3–10^4 s
J	-10 m^4 kg^{-1} s^{-2}
α	10^8 m^4 kg^{-1}

This means that the heating and drying produced by the convection, which are determined by M_B, will be independent of the numerical values of α and τ_D that we use, provided that (11.18)–(11.20) are satisfied.

We can interpret the use of (11.11)–(11.13), that is, the quasi-equilibrium approximation, as essentially a filtering approximation, analogous to that used to derive the quasigeostrophic equations from the primitive equations. This analogy is discussed further later in this paper.

At this point, it is useful to introduce "typical" numerical values for the parameters under discussion. These are given in Table 11.1. Here our intention is only to indicate rough magnitudes typical of fairly deep tropical clouds under disturbed conditions. Shallower clouds, or different weather regimes, would be characterized by different numerical values.

To illustrate the ideas presented in this section in a concrete way, we present the following simple example. Let the large-scale forcing be given by

$$F = 1 + 0.5 \sin\left(\frac{2\pi t}{T}\right) \quad \text{J kg}^{-1}\ \text{s}^{-1},$$

where T is 3 days, and let $J = -10$ m^4 kg^{-1} s^{-2}, and $\tau_D = 1000$ s. A forcing of 1 J kg^{-1} s^{-1} is fairly strong; it can produce a cloud work function of 1000 J kg^{-1} in 1000 s. With this assumed forcing, we numerically integrate (11.2) and (11.6), using (11.7) and imposing (11.14) as necessary. The dissipation term of (11.6) is included using implicit time differencing; otherwise, the simple forward time-differencing scheme is used. The time step must be shorter than the adjustment time $(\alpha/|J|)^{1/2}$ in order to maintain computational stability. The initial conditions are arbitrarily specified as $K = 10^4$ J m^{-2}, and $A = 10^3$ J kg^{-1}. We consider three cases: $\alpha = 10^9$, 10^8, and 10^7 m^4 kg^{-1}, which correspond to τ_{ADJ} in the range 10 000–1000 s.

Figure 11.1 shows the results. In Fig. 11.1a, the convective mass flux is plotted for the three values of α, and $(M_B)_F$ is also shown for comparison. After the first few simulated hours, the results for $\alpha = 10^7$ m^4 kg^{-1} are so close to $(M_B)_F$ that it is difficult to depict the differences graphically. For $\alpha = 10^9$ m^4 kg^{-1}, the differences are much larger, and for $\alpha = 10^8$ m^4 kg^{-1} the differences are noticeable although considerably reduced. This demonstrates that, as expected, the cumulus cloud-base mass flux is asymptotically independent of α for sufficiently "small" α, that is, for sufficiently small τ_{ADJ}. The results also suggest that $\alpha = 10^7$ m^4 kg^{-1} is small enough to give a very close approximation to quasi equilibrium, at least for the parameters used in this example. Further experiments

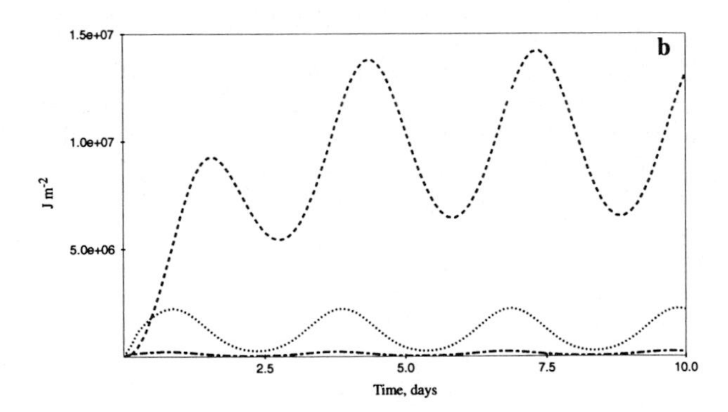

FIG. 11.1. Results of numerical integration of Eqs. (11.2) and (11.6), with (11.7), for three different values of α. (a) Cumulus mass flux; (b) cumulus kinetic energy; (c) cloud work function. In each panel, the dashed curve shows the results obtained for $\alpha = 10^9$ m^4 kg^{-1}, the dotted line for $\alpha = 10^8$ m^4 kg^{-1}, and the dash–dot line for $\alpha = 10^7$ m^4 kg^{-1}. The solid curve in panel (a) shows the quasi-equilibrium result, for comparison.

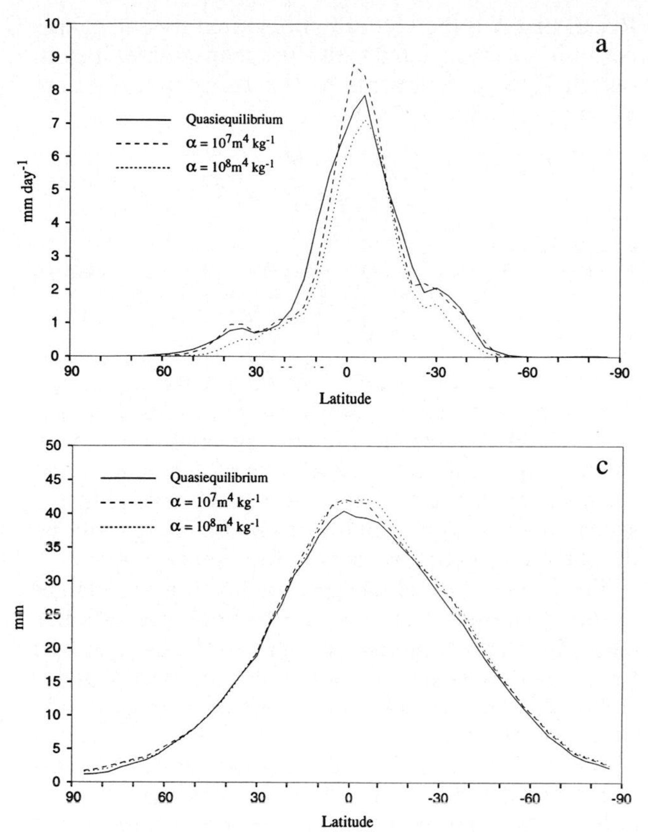

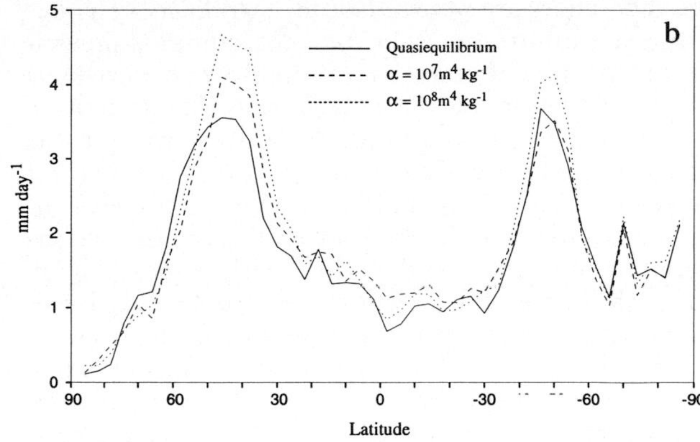

FIG. 11.2. The simulated zonally averaged January means of (a) cumulus precipitation rate; (b) large-scale precipitation rate; and (c) precipitable water. In each panel, the solid curve is obtained with the quasi-equilibrium closure, the dashed curve is based on the prognostic closure with $\alpha = 10^7$ m^4 kg^{-1}, and the dotted curve is based on the prognostic closure with $\alpha = 10^8$ m^4 kg^{-1}. All results were obtained with the CSU GCM.

confirm that the cumulus cloud-base mass flux is insensitive to the choice of τ_D, provided that (11.18)–(11.20) are satisfied. Figures 11.1b and 11.1c show that the cumulus kinetic energy and cloud work function depend strongly on the value of α used, again in agreement with our earlier analysis.

We conclude that, even though we currently have no physical theory to determine α or τ_D, we can nevertheless predict the cumulus cloud-base mass flux and the associated heating and drying rates by choosing sufficiently small values of α and τ_D. We cannot claim, however, to meaningfully predict the cumulus kinetic energy or the cloud work function, unless α and τ_D are known. Fortunately, the cumulus kinetic energy and the cloud work function are not of primary interest in modeling the large-scale circulation of the atmosphere.

As mentioned above, computational instability can occur for very small values of α, for which the adjustment time becomes shorter than the numerical time step. This sets a practical lower limit to α in actual applications of the prognostic closure. Our experience with the CSU GCM (discussed later) indicates that $\alpha = 10^7$ m^4 kg^{-1} is compatible with time steps commonly used in GCMs. Figure 11.1 indicates that this value of α is small enough so that reducing it further would not significantly alter the solution.

In summary, the prognostic closure will closely approximate quasi equilibrium with circulations of time scale τ_{LS} if we choose sufficiently small values of τ_D and α. For the limit in which both τ_D and α go to zero, we expect to recover pure quasi equilibrium. This expectation is examined more closely in the next section.

11.3. Results

In practice, we predict K by using (11.6) with an implicit time-differencing scheme and then find M_B by using (11.7). Of course, we predict T and q rather than A. An explicit scheme is currently being used for the convective terms of the T and q equations. A 15-min time step was used to produce the results presented below.

The prognostic closure was tested first in a variety of experiments with a one-dimensional (1D) version of the GCM, and then in full three-dimensional (3D) global simulations. Only a few results from the global simulations will be presented here; the 1D results and additional 3D results will be presented elsewhere. We have used both the prognostic and quasi-equilibrium closures in December–January simulations. In addition to one run with the quasi-equilibrium closure, referred to below as the "control" run, we have made two runs with the prognostic closure, using $\alpha = 10^8$ and 10^7 m^4 kg^{-1}. In all tests of the prognostic closure we used $\tau_D = 600$ s.

Figure 11.2 shows a few results from these tests. The zonally averaged cumulus precipitation rate, shown in Fig. 11.2a, is about the same in all three of the runs;

the differences are not statistically significant. There is some suggestion, however, that the cumulus precipitation maxima become more intense as α decreases. The precipitation rate due to large-scale supersaturation shown in Fig. 11.2b appears to be stronger with the prognostic closure than with the quasi-equilibrium closure, especially in middle latitudes. This may indicate that the two closures lead to different interactions between convection and stratiform cloudiness. Figure 11.2c shows that the prognostic closure leads to larger precipitable water amounts, especially in the tropics.

In summary, the differences in the results obtained with the prognostic and diagnostic closures are small but perhaps not negligible. Further study will be needed to quantify and evaluate these differences in more detail. It is not clear that the prognostic closure has converged to quasi equilibrium for $\alpha = 10^7$ m^4 kg^{-1}, but in order to test smaller values of α it will probably be necessary to reduce the time step.

The prognostic version of the parameterization is much faster than the quasi-equilibrium version; the time required for the cumulus parameterization is cut in half on a Cray Y-MP.

Additional results will be presented elsewhere.

11.4. What determines α?

Although the results presented above are promising, it is clear that a fully satisfactory version of the prognostic closure should include physically based methods to determine τ_D and, especially, α. We are currently studying this problem and report our progress to date in this and the following section.

The definition of α is given by (11.7), in terms of the convective mass flux and the CKE. An understanding of it can thus be sought by investigating these two quantities and the relationships between them. At each level, the convective mass flux satisfies

$$M_C \equiv M_B\eta = \rho\sigma(1-\sigma)(w_u - w_d), \qquad (11.21)$$

where η is the "normalized cloud mass flux" at any level (see A–S for a discussion of η), σ is the fractional area covered by rising motion, and subscripts u and d denote the upward and downward branches of the convective circulation, respectively (e.g., Randall et al. 1992). In the present context, the downward branch is the environment, which undergoes "compensating subsidence." We can apply (11.21) to each convective subensemble. The kinetic energy per unit volume associated with the vertical component of the convective motion, at a particular level, is

$$\frac{1}{2}\rho\overline{w'^2} = \frac{1}{2}\rho\sigma(1-\sigma)(w_u - w_d)^2; \qquad (11.22)$$

a derivation of (11.22) is given by Randall et al. (1992). Comparing (11.21) and (11.22), we see that

$$\frac{1}{2}\rho\overline{w'^2} = \frac{M_B^2\eta^2}{2\rho\sigma(1-\sigma)}. \qquad (11.23)$$

Recall that K is the vertically integrated kinetic energy per unit area associated with the subensemble, including all three components of the convective motion. Then (11.23) and (11.7) give

$$\alpha = \frac{1}{2\epsilon}\int_{z_S}^{z_T} \frac{\eta^2}{\rho\sigma(1-\sigma)}\,dz, \qquad (11.24)$$

where

$$\epsilon \equiv \frac{1}{K}\int_{z_S}^{z_T} \frac{1}{2}\rho\overline{w'^2}dz \qquad (11.25)$$

is the ratio of the vertically integrated CKE in the vertical component of the convective motion to the total vertically integrated kinetic energy contained in all three components, and z_S and z_T are the heights of the earth's surface and the cloud top, respectively. In the special case of equipartition of kinetic energy among the three convective velocity components, $\epsilon = 1/3$.

The form of (11.24) suggests that α should be larger for deeper clouds. Certainly the range of integration is larger for deeper clouds, and the vertically averaged value of η^2 may also be larger. We suspect that for clouds of a given depth, variations of α arise mainly from variations of ϵ.

Schubert (1973) showed that, for typical tropical soundings, the diagonal elements of the kernel matrix (which we interpret here as corresponding to our J) are about an order of magnitude smaller for shallow clouds than for deep clouds. If α were independent of cloud type, this would imply that the adjustment time $(\alpha/|J|)^{1/2}$ is longer for shallow clouds than for deep clouds—an implausible conclusion. This line of reasoning reinforces our tentative conclusion that α is smaller for shallower clouds.

Figure 22.15 of the paper by Xu (see chapter 22) shows the numerically simulated variations of α^{-1} in two different runs labeled as Q03 and Q02, respectively. These results[1] were produced with a cumulus ensemble model, as described by Xu. Both runs were performed with a prescribed, idealized, time-dependent large-scale advective process, which serves as the primary forcing for the cumulus convection. The run labeled Q02 has an imposed strong geostrophic shear of the zonal wind, while that labeled Q03 has no geostrophic shear. The run with shear has a maximum value of α close to 10^8 m^4 kg^{-1}, while that without shear has a maximum value near 3×10^7 m^4 kg^{-1}. Apparently shear increases α. From (11.26), we surmise that ϵ decreases in the presence of shear; that is, shear acts to increase the fraction of the CKE in the horizontal components of the motion and to decrease the fraction in the vertical component. That is exactly what Xu and Arakawa (1992a) found in their analysis of Q02 and Q03. In

[1] The values of α computed by Xu are "bulk" values, representative of all of the active convection in the simulations.

addition, they noted that the run with shear exhibited considerable mesoscale organization and that the enhanced horizontal disturbance kinetic energy was mostly associated with mesoscale circulations.

From (11.7), we find that

$$M_B = \left(\frac{K}{\alpha}\right)^{1/2}, \qquad (11.26)$$

which shows that large values of α, which are favored by shear, reduce the convective mass flux for a given value of K. Since M_B is the agency through which the convection modifies its environment, we conclude that shear inhibits the feedback of the convection on the mean flow. In particular, shear inhibits convective stabilization of the environment, allowing larger values of the cloud work function (more CAPE) to accumulate in response to the large-scale forcing. This is consistent with our equilibrium solution for A, given by (11.11), which shows that A increases with α. To the extent that shear inhibits convective stabilization, it will interfere with the adjustment to cloud work function quasi equilibrium. We thus expect that a diagnostic parameterization based on the quasi-equilibrium assumption will produce noticeable errors in predicting the evolution of strongly sheared convective systems with considerable mesoscale organization. This is consistent with the numerical results of Xu (1992) and Xu and Arakawa (1992a,b), who also showed that shear acts to prolong organized convection.

Of course, the convective mass flux produces a buoyancy flux that generates CKE, and so in a sense the mass flux is self-generating. That is why, as shown by (11.8), larger values of α favor slower growth of the convective mass flux when A is positive. This suggests that the convective response to large-scale forcing may be delayed by shear. Again, this is consistent with the numerical results of Xu and Arakawa (1992a,b).

Finally, we note that (11.26) gives a value of α for each subensemble. As discussed earlier, it is reasonable to expect that the various subensembles will require different values of α, although a single value was used in the calculations reported here.

11.5. What determines τ_D?

The dissipation term that appears in (11.6) needs to be more clearly defined. What does it represent? We assume for simplicity that A is defined so as to include the effects of liquid water and ice loading on the updraft. [This point is discussed by Cheng and Arakawa (1990).] In that case, the dissipation term of (11.6) represents only true kinetic energy dissipation. Lilly (1982) and Wu (1991) have argued that the kinetic energy dissipation rate of convective storms is minimized when the storms are strongly helical and that the helicity is controlled by the ambient shear. It thus appears that both α and τ_D depend on the environmental shear in which the convective clouds develop.

Perhaps the safest statement that can be made here is that our current understanding of cumulus dissipation is very rudimentary.

11.6. Concluding discussion

We have argued in section 11.2 that, because (11.12) and (11.3) are identical, the prognostic closure will give the same convective mass flux as the quasi-equilibrium closure and, in particular, that the value of M_B obtained with the prognostic closure will not depend significantly on the values of α and τ_D used, at least if they are sufficiently small. Some caution is needed, however, because the prognostic closure does not necessarily "feel" the same values of J and F as the quasi-equilibrium closure. As discussed in section 11.1, in order to apply the quasi-equilibrium hypothesis it is necessary to group the contributions to the time change of A into "convective" and "nonconvective" components. In practice, it is not always clear which processes are convective and which are not. For example, the upper-tropospheric stratiform clouds of a convective cloud system are for the most part produced by convective detrainment; their effects on the large-scale temperature and moisture soundings can hardly be called nonconvective. On the contrary, in many cases they are merely extensions of the cumulus effects. The precipitation and radiative processes associated with such stratiform clouds should, therefore, logically be included as part of the response of cumulus convection, rather than as part of the large-scale forcing. By treating the effects of convectively produced stratiform clouds as "nonconvective" processes, quasi-equilibrium closure incorporates some aspects of the convective feedback into the large-scale forcing. In so doing, it effectively alters both J and F. Since the quasi-equilibrium closure and the prognostic closure can produce different values of J and F, they may produce different values of M_B, even though both effectively use (11.3) to determine M_B. Further research is needed to explore this issue.

Notwithstanding these caveats, we have shown that, in the limit of small α and τ_D, the prognostic closure reduces, in principle, to the quasi-equilibrium closure proposed by A–S. It nevertheless has important advantages over quasi-equilibrium closure:

- The computational algorithm is drastically simplified. The kernel calculation is no longer necessary. The integral equation for the cloud-base mass flux disappears from the problem. The result is a much faster, shorter, and simpler computer code that can be more easily modified as our knowledge increases. The computer code based on the prognostic closure is also considerably more portable from one model to another.
- The mathematical difficulties that arise in applying the quasi-equilibrium closure with the restriction that all mass fluxes be nonnegative are neatly sidestepped.

• Because it simplifies the computation, the prognostic closure facilitates further improvements of the parameterization, such as enlarging the spectrum of clouds to include penetrative moist convective updrafts that originate above the planetary boundary layer and including the effects of convective-scale downdrafts.

• The distinction between the large-scale "forcing" and the convective "response" becomes moot, because neither has to be explicitly defined. In particular, it is no longer necessary to assume that the effects of the stratiform clouds associated with cumulus convection and the radiative cooling associated with cumulus anvil clouds are part of the large-scale forcing rather than the convective feedback. All components of a convective cloud system can thus be treated in a consistent way.

• Finally, the prognostic closure has the potential to allow simulation of the dynamical processes through which a cumulus ensemble maintains the near-neutral stability of its environment. So long as we use somewhat arbitrary prescribed values of τ_D and α, we cannot claim to be physically modeling the process by which the CKE dynamically adjusts to changes in the large-scale circulation. If a theory can be developed to determine τ_D and α, however, a true advance will have been achieved.

Our prognostic closure is related to the quasi-equilibrium closure in much the same way as a primitive equation (PE) model is related to a quasigeostrophic (QG) model. A PE model can produce QG motion by explicitly simulating the geostrophic adjustment process. A PE model is in some respects simpler than a QG model; for example, the QG model determines the vertical motion field through the inconvenient "ω equation," while PE models typically determine the vertical velocity from the much simpler continuity equation. The use of a PE model does not imply a rejection of the idea that the large-scale motions of the atmosphere are approximately geostrophic.

Similarly, our prognostic closure can produce a quasi equilibrium between the large-scale forcing and the convective response by explicitly simulating the conversion of CAPE into CKE. As explained above, the prognostic closure is simpler than quasi-equilibrium closure; for example, quasi-equilibrium closure entails a "kernel" calculation, while the prognostic closure does not. The use of our prognostic closure does not imply a rejection of the quasi-equilibrium hypothesis.

Just as the study of the geostrophic adjustment process has led to improved understanding of geostrophic motion, the study of our prognostic closure may yield better understanding of the physical basis of quasi equilibrium.

Acknowledgments. Professor Akio Arakawa and Dr. Kuan-Man Xu made numerous suggestions that led to improvements of the manuscript. This research has been supported by the National Science Foundation (NSF) under Grant ATM-8907414 to Colorado State University. Computing resources have been provided by the National Center for Atmospheric Research, which is sponsored by the NSF.

Chapter 12

The Kuo Cumulus Parameterization

DAVID J. RAYMOND

Physics Department and Geophysical Research Center, New Mexico Institute of Mining and Technology, Socorro, New Mexico

KERRY A. EMANUEL

Center for Meteorology and Physical Oceanography, Massachusetts Institute of Technology, Cambridge, Massachusetts

12.1. Introduction

The Kuo cumulus parameterization (Kuo 1965, 1974) is one of the earliest and most enduringly popular schemes for parameterizing cumulus convection. The purposes of this paper are to describe the scheme and some of the more notable variations on it and to make a critical evaluation of its suitability for use under various circumstances.

12.2. Derivation of Kuo scheme

The derivation presented here of the Kuo scheme leaves out many of the motivating arguments presented by Kuo in his two papers. We start from the large-scale equations in pressure (x, y, p) coordinates for potential temperature θ,

$$\frac{d\theta}{dt} = \frac{\partial\theta}{\partial t} + \nabla\cdot(\mathbf{v}\theta) + \frac{\partial}{\partial p}(\omega\theta) = \frac{LC + Q_r}{\pi} - \frac{1}{\pi}\frac{\partial F_k}{\partial p}, \quad (12.1)$$

and for water vapor mixing ratio q,

$$\frac{dq}{dt} = \frac{\partial q}{\partial t} + \nabla\cdot(\mathbf{v}q) + \frac{\partial}{\partial p}(\omega q) = -C - \frac{\partial F_q}{\partial p}, \quad (12.2)$$

where $(\mathbf{v}, \omega)$ is the velocity in pressure coordinates, F_q and F_k are the vertical molecular and eddy fluxes of water vapor and sensible heat (the horizontal fluxes being ignored), Q_r is the radiative heating rate, and $\pi = c_p(p/p_0)^\kappa = c_pT/\theta$ is the Exner function. In addition, L is the latent heat of condensation for water vapor, T is the temperature, p_0 is a reference pressure, c_p is the specific heat per unit mass at constant pressure for air, $\kappa = R/c_p$ where R is the gas constant for air, and C is the condensation minus evaporation rates for water vapor. Note that since we are working in pressure coordinates, upward fluxes correspond to negative values of F_k and F_q.

We now assume that

$$-\int_0^{p_s} C dp = (1-b)\left[\int_0^{p_s} \nabla\cdot(\mathbf{v}q)dp + F_{qs}\right] \equiv -(1-b)gM_t, \quad (12.3)$$

where F_{qs} is F_q evaluated at the surface pressure, p_s, and where b is a constant whose meaning is discussed below. The conventional bulk formula for the surface moisture flux transformed to pressure coordinates is

$$F_{qs} = -g\rho_s C_D|\mathbf{v}_s|(q_{ss} - q_s), \quad (12.4)$$

where C_D is the drag coefficient, $\mathbf{v}_s$ is the low-level wind, g is the acceleration of gravity, q_s is the low-level mixing ratio, ρ_s is the surface air density, and q_{ss} is the saturation mixing ratio at the sea surface temperature and pressure. The quantity

$$M_t = -\frac{1}{g}\left[\int_0^{p_s} \nabla\cdot(\mathbf{v}q)dp + F_{qs}\right] \quad (12.5)$$

is called the *moisture accession* by Kuo.

Integrating (12.2) in pressure and assuming that F_q vanishes at zero pressure and that ω is negligible at the surface results in

$$\int_0^{p_s} \frac{\partial q}{\partial t} dp = gbM_t. \quad (12.6)$$

Equation (12.6) shows that the parameter b is related to the rate at which the atmosphere locally moistens in a vertically integrated sense.

Defining the net cumulus heating as

$$Q_c = LC - \frac{\partial F_k}{\partial p}, \quad (12.7)$$

(12.1) can be written

$$\frac{d\theta}{dt} = \frac{Q_c + Q_r}{\pi}, \quad (12.8)$$

and from (12.3) we have that

$$\int_0^{p_s} Q_c dp = gL(1-b)M_t - F_{ks}, \quad (12.9)$$

where F_{ks} is the surface value of F_k.

In the traditional Kuo scheme F_{ks} is ignored and the vertical structure of Q_c is assumed to be in the form of a relaxation toward a moist adiabat θ_a; that is,

$$\frac{Q_c}{\pi} = \frac{\theta_a - \theta}{\tau}, \quad (12.10)$$

where τ is a relaxation time that is potentially a function of the horizontal space variables and time but not of pressure. Multiplying (12.10) by $\tau\pi$ and integrating in pressure leads to an expression for the relaxation time constant:

$$\tau = \frac{1}{gL(1-b)M_t}\int_0^{p_s} \pi(\theta_a - \theta)dp. \quad (12.11)$$

Thus, the time constant for relaxation to a moist adiabat decreases as the moisture accession M_t becomes larger. In regions where $M_t < 0$ or $\theta > \theta_a$, Q_c is set to zero. Note that the moisture accession contains contributions from both moisture convergence and surface moisture fluxes. In many circumstances the former is thought to greatly outweigh the latter. Thus, cumulus convection is thought to be driven primarily by the convergence of moisture.

The choice of moist adiabat in the relaxation equation, (12.10), needs to be specified. Kuo chose that adiabat arising from moist parcel ascent of boundary-layer air. This is in contrast to the usual choice in adjustment models (e.g., Manabe et al. 1965) of a profile that has the same integrated moist static energy ($c_pT + Lq + gz$) as the original sounding. For this reason, the subcloud-layer moist static energy is not directly affected by convection in the Kuo scheme, whereas in convective adjustment the subcloud-layer moist static energy is generally reduced by convection.

12.3. Extensions of Kuo scheme

One modification to the Kuo scheme has to do with the definition of moisture accession. The horizontal moisture divergence can be expanded to

$$\nabla \cdot (\mathbf{v}q) = q\nabla \cdot \mathbf{v} + \mathbf{v} \cdot \nabla q. \quad (12.12)$$

Often the second term on the right side of (12.12) is much less than the first and can therefore be dropped. From mass continuity, $\nabla \cdot \mathbf{v} = -\partial\omega/\partial p$. Furthermore, integration by parts demonstrates that

$$-\int_0^{p_s} \frac{\partial \omega}{\partial p} q dp = \int_0^{p_s} \omega \frac{\partial q}{\partial p} dp, \quad (12.13)$$

to a good approximation, since ω is zero at zero pressure and close to zero at the earth's surface. Therefore, an alternative way to write the moisture accession is

$$M_t = -\frac{1}{g}\left(\int_0^{p_s} \omega \frac{\partial q}{\partial p} dp + F_{qs}\right). \quad (12.14)$$

Krishnamurti et al. (1976) actually terminated the integration in (12.14) at cloud base rather than the surface and omitted the surface flux as well.

Anthes (1977) proposed that the moistening parameter, b, should be related to the mean relative humidity in the troposphere, RH, in the following way:

$$b = \left(\frac{1 - \mathrm{RH}}{1 - \mathrm{RH}_c}\right)^n, \quad (12.15)$$

with b not allowed to exceed one. Term RH_c is a critical value for the relative humidity below which no precipitation occurs, and n is a positive exponent of order unity. As the relative humidity increases, the moistening rate decreases, dropping to zero when relative humidity reaches 100%. In this way more of the moisture accession goes into moistening the atmosphere when the humidity is low.

Molinari (1982) proposed an alternate parameterization for b. By forcing the fractional relaxation rates of potential temperature and mixing ratio toward a saturated moist adiabat to be equal, we obtain a value of b. This value of b is a function of space and time and has the virtue of preventing the development of a saturated but moist-unstable environment. Since such a state is physically unlikely and causes problems in large-scale numerical models, this would seem to be a reasonable procedure. Molinari found that the above-mentioned scheme by Anthes allowed saturated unstable conditions to occur.

Various authors have replaced (12.10) with a more realistic assumption about the vertical heating profile. Anthes (1977) used a cloud model to predict heating profiles in the context of a Kuo-like model. Molinari (1985) proposed a generalization of the Kuo scheme that replaces the right sides of (12.1) and (12.2) with

$$\frac{d\theta}{dt} = \frac{1}{\pi}[gL(1-b)M_tQ_1 + Q_r] \quad (12.16)$$

and

$$\frac{dq}{dt} = -g(1-b)M_tQ_2, \quad (12.17)$$

where Q_1 and Q_2 are normalized heating and moistening profiles containing the effects of both condensation and vertical eddy fluxes, and subject to

$$\int_0^{p_s} Q_1 dp = \int_0^{p_s} Q_2 dp = 1. \quad (12.18)$$

The dependence of Q_1 and Q_2 with height can be obtained from diagnostic studies of convection (e.g., Yanai et al. 1973, etc.).

12.4. Critique

Cumulus parameterizations must be judged both on their fidelity to generally accepted physical principles and on how they actually perform in numerical models. This section analyzes the Kuo and related schemes with respect to the first criterion.

The premise underlying all physical parameterization is that some aspect of the microscale chaotic process is in statistical equilibrium with the macroscale system. In the kinetic theory of gases, slight deviations from the Maxwell–Boltzmann distribution are assumed to respond instantly to changes in the strain rate of the gas; the direct expression of this is the law of viscous stress. In theories of turbulence, the turbulent kinetic energy, or else its rate of production, is assumed to be in equilibrium with macroscopic fluid processes.

It is worth noting that these statistical equilibrium assumptions are rooted in causality; changes in the viscous stress are *caused* by changes in the strain rate, and turbulence is *caused* by instability of the macroscopic flow.

Like classical turbulence closures, the Arakawa–Schubert quasi-equilibrium assumption postulates a statistical equilibrium based on energy; convection consumes available potential energy at the rate at which the macrofluid system supplies it. This meets the causality requirement since convection is fundamentally caused by conditional instability, which is reflected by the degree of convective available potential energy in the macrosystem.

The Kuo scheme, as originally introduced, postulates a statistical equilibrium of water substance; convection is assumed to consume water (not energy) at the rate it is supplied by the macrofluid system. (In its later incarnation, this was modified to allow for a fractional consumption of water by convection.) As such, the Kuo closure fundamentally violates causality; convection is not caused by the macroscale water supply. The ill-posedness of this closure underlies many of the problems encountered with the Kuo scheme in practice. Since the degree of conditional instability is free to fluctuate, it does, leading to relatively large exchanges between available potential energy and kinetic energy in the macrosystem. Thus, the scheme makes the macrosystem "CISK-able," which reflects the aliasing of conditional instability up to the macrosystem. Other closures may not permit CISK (conditional instability of the second kind). It has been shown recently, for example, that the Betts–Miller scheme does not support CISK (Neelin and Yu 1993).

It is also apparent that the Kuo scheme fails to replicate solutions to simple, well-known problems. In radiative–convective equilibrium, evaporation from the ocean is exactly balanced by precipitation, but if b is nonzero, (12.6) shows that the atmosphere will continue to moisten. (With the Anthes or Molinari improvements, an equilibrium is also possible, but this equilibrium has a saturated atmosphere—a highly unrealistic feature.)

There are further technical problems with Kuo schemes. First, it is worth pointing out that the concept of moisture convergence as originally stated by Kuo is not Galilean invariant. This is evident from the expansion in (12.12), as the second term on the right side of this equation takes on different values for flow velocities that differ only by an additive constant. The modified form of the moisture accession, (12.14), is a clear improvement in this regard. However, the additional modifications of Krishnamurti et al. (1976) take the model quite far in spirit from Kuo's original ideas, since the amount of convection is no longer related to the total column-integrated moisture accession.

As noted, the original Kuo scheme ignores the contribution of the surface sensible heat flux, F_{ks}, to the convective heating. This is unlikely to be of significance over the ocean but may be important over land.

The Kuo scheme has also been criticized because the vertical profiles of convective heating do not agree with those observed in real situations. This is understandable, in that the original scheme derives this profile from the difference between the actual and moist-adiabatic potential temperature profiles. Molinari (1985) proposed replacing the original relaxation scheme with fixed profiles of heating and moistening. Real profiles are quite variable, however, and are determined in part by the large-scale forcing. Failing to allow the heating and moistening to evolve in response to changes in the surrounding atmosphere can result in serious inconsistencies.

The original Kuo scheme at least has the virtue of allowing the heating profile to adjust to changes in the environment. The exception is at and below cloud base, where the original scheme can never produce any heating or cooling, due to the choice of moist adiabat. This is unrealistic when there are moist downdrafts, which are important in the dynamics of mesoscale convective systems and which are also a crucial negative feedback mechanism in large-scale dynamics. The use of a moist static energy–conserving moist adiabat, as is generally done in adjustment schemes, might be a sensible alternative. This choice invariably produces cooling at low levels and therefore simulates the effects of downdrafts.

Acknowledgments. This work was supported by National Science Foundation Grant ATM-8914116.

PART III

Representation of Convection in Mesoscale Models

Chapter 13

A Hybrid Parameterization with Multiple Closures

WILLIAM M. FRANK

Department of Meteorology, The Pennsylvania State University, University Park, Pennsylvania

13.1. Introduction

Much of the current research using numerical models is focused on circulation systems that are most appropriately studied using models with 10–50-km grid spacings and time steps on the order of several minutes. These models can resolve mesoscale circulations reasonably well but are incapable of accurately resolving cumulus convection. Since convective-scale processes are often important, and are frequently dominant, on such scales it is necessary to employ cumulus parameterization techniques.

The relatively fine-grid meshes of mesoscale models present both problems and opportunities to the cumulus parameterizer. The major problem is that the cumulus clouds, which in nature have lifetimes of up to an hour or more, evolve on time and space scales that are not distinct from the resolved grid-scale circulations. Therefore, it should not be assumed that the convection can be parameterized as a series of processes that restore the grid-scale circulation to a known equilibrium state within a single time step. While individual convective clouds respond to their mesoscale environments, the environments respond to the clouds almost as rapidly.

The major advantage of a parameterization scheme operating in a mesoscale model is that the model can resolve the immediate environment of the cloud well enough to allow the use of relatively complex cloud models that interact with their surroundings. For example, thermodynamic processes resulting from mixing between clouds and their environments are important components of the evolution of both the cloud and the grid-scale circulations. Such processes cannot be estimated accurately if the properties of the cloud's environment are not well simulated by the model. Accurate resolution of the vertical stability in the lower levels and the local forced uplifting at cloud base are also crucial to accurate parameterization of convection. Neither can be estimated with confidence in a coarse-grid model.

Molinari and Dudek (1992) suggest that the best approach to cumulus parameterization in mesoscale models appears to be to use a scheme that operates simultaneously with and interacts directly with the explicit moisture scheme of the host model. They termed such schemes hybrid schemes, and that terminology is used here. The general approach is to use the mesoscale model to resolve the primary circulations of interest and the local processes that influence the development of the convective clouds. The parameterization makes use of a steady-state cloud model that interacts with the grid-scale fields and provides the host model with the net heating, drying, and condensate produced by the convection.

13.2. The Frank and Cohen parameterization

The first fully hybrid cumulus parameterization scheme was designed by Frank and Cohen (1987)—hereafter referred to as FC. [Zhang and Fritsch (1986) had earlier used simultaneous cumulus parameterization and explicit moist processes but their parameterization did not exchange condensate between the cloud model and the explicit moisture scheme.] Frank and Cohen use simple steady-state updraft and downdraft plume models with specified entrainment and detrainment rates to determine the convective processes. Convective heating and drying, as well as cloud water and precipitation detrained from the cloud model, are supplied to the grid scale. The mesoscale model's explicit moisture scheme is used simultaneously with the parameterization. The trigger function (defined as the method that determines whether or not the parameterization is activated) is based on low-level stability (an initial updraft velocity is specified, and the parameterization is called if the parcel reaches the level of free convection).

The FC closure specifies the cloud-base mass flux as a function of the three-dimensional mass flux convergence into the subcloud layer:

$$M_{cu} = \bar{M} - M_{cd} - \tilde{M}, \qquad (13.1)$$

where M_{cu} is the updraft mass flux at cloud base, and $\bar{M}$, M_{cd}, and $\tilde{M}$ are the mass fluxes of the grid scale, convective downdrafts, and convective environment at cloud base, respectively. This closure is based on the simple assumption of mass continuity but it requires estimates of the relative strengths of the downdraft and environmental mass fluxes at cloud base. These relationships are empirically tuned to observations of tropical convection.

The philosophy of the FC closure is that the rate of convective overturning is determined by the rate of forced lifting of updraft air at cloud base. If the air is convectively unstable, it will tend to overturn at a rate faster than that of the grid-scale uplifting. The convection is not constrained to a rate that would maintain quasi equilibrium with the grid-scale circulation, but approximate quasi equilibrium is nonetheless observed to occur in the model. The explanation of this apparent paradox is that the grid-scale circulation reacts so rapidly to the latent heat release by the clouds that only weak warming of the levels above cloud base is realized. The small excess of convective heating above the rate of cooling by the grid-scale uplifting results in rapid intensification of both the convection and the grid-scale vertical motion without significant temperature increases. Stabilization tends to occur in the model, as it generally does in the real world on small scales, primarily by cooling and drying of the boundary layer by downdrafts.

Simulations of tropical mesoscale convective systems with the FC scheme show that the large stratiform rain regions in the systems are strongly dependent upon water vapor and condensate detrained from the parameterized clouds. Inclusion of ice process in the explicit moisture scheme is also important.

Chen and Frank (1991) use a modified version of the FC scheme in the Pennsylvania State University–National Center for Atmospheric Research Mesoscale Model to study vortex formation in mesoscale convective complexes (MCCs). An example of their results is shown in Fig. 13.1. In the case shown, an initial short wave overran a weak low-level jet and moist tongue. A line of thunderstorms developed. The moist air and condensate detrained from the parameterized clouds, along with the mesoscale ascent forced by the short wave, produced a large, active stratiform rain area to the rear of the convective line. The latent heating within the stratiform region caused the formation of a strong mesoscale vortex that closely resembles the vortices frequently observed in MCCs. Both the parameterized convection and the explicit moist processes play crucial roles in this simulation, and the transfer of condensate from the parameterization to the explicit moisture scheme is important to the formation of the large stratiform rain area.

The mass flux cloud models used in the FC parameterization and others estimate convective feedbacks to the grid-scale temperature and moisture fields as the sums of the vertical advection by the compensating mass flux outside the clouds plus the effects of detrainment of cloud air and condensate [see Eqs. (12) and (13) of Frank and Cohen (1987)]. Therefore, the calculations depend heavily upon the vertical mass flux

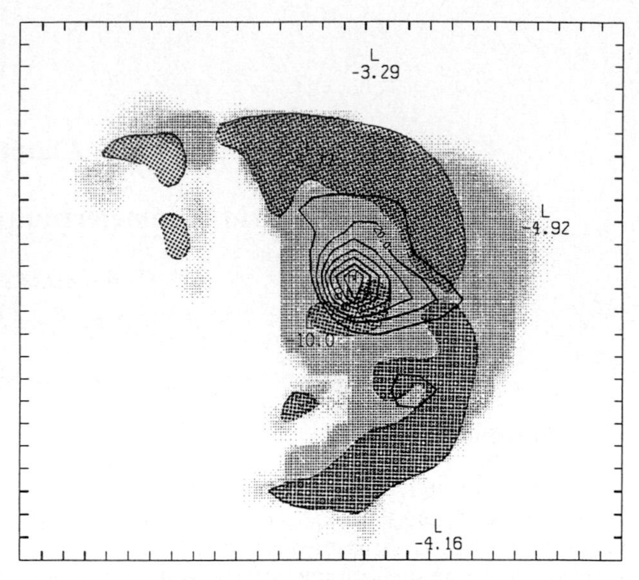

FIG. 13.1. Relative vorticity at 525 mb (10^{-5} s^{-1}, solid lines), stratiform rain areas (light shading), and deep convective areas (dark shading) at t = 12 h. The vortex has formed within the stratiform rain region (from Chen and Frank 1991).

profiles of the model clouds and upon the amount of cloud air detrained at each level. (The vertical mass flux of the updraft at each level is determined from the cloud-base mass flux plus the accumulated effects of entrainment and detrainment during parcel ascent, and the downdraft mass flux is essentially similar, though in the opposite direction.)

Unfortunately, entrainment and detrainment are poorly understood processes. They are thought to vary significantly with the thermodynamic properties of the cloud environment. As a cloud parcel rises, portions of it mix with various amounts of environmental air. Some of these mixtures will be positively buoyant and remain in the cloud. Environmental air in those mixtures is entrained. Conversely, mixtures of updraft air that have been sufficiently cooled by evaporation of cloud water (due to the subsaturation of the mixture) so that they become negatively buoyant will cease rising and be detrained. In general, entrainment is favored by extremely unstable conditions (where virtually all mixtures are still warm) and by moist environmental air (since less cloud water evaporates). Detrainment is favored by dry environmental air and near-neutral stabilities. (Entrainment and detrainment are also likely to be dependent upon shear and other dynamic processes.)

Kain and Fritsch (1990) have developed a buoyancy sorting technique that simulates mixing of updraft air with environmental air across a spectrum of mixing rates. They determine which of the mixtures are positively buoyant (and are entrained) and which are negatively buoyant (and are detrained). The goal of this technique is to provide a physical basis for estimating the entrainment and detrainment processes in mass flux cloud models. The Kain and Fritsch mixing technique has performed well in diagnostic tests and is described in more detail in chapter 16. It has recently been incorporated into the FC scheme.

13.3. The Frank, Chen, and Cohen hybrid scheme with multiple closures

One problem common to almost all cumulus parameterizations is that they are designed as complete packages. They tend to have three primary components:

1) a trigger function to activate the parameterization,
2) a cloud model to estimate the convective properties as functions of the grid-scale variables, and
3) a closure assumption, usually formulated to determine the total amount of convection.

These components are not fully separable in most parameterizations. (For example, the cloud model may be used to compute terms in the trigger function or closure.) While most parameterizations start with reasonably sound hypotheses, the problem of implementation is sufficiently complex that they end up including large numbers of interwoven assumptions affecting all three of the above components. Further, each of these components may be scale and model dependent. Comprehensive evaluations of most parameterizations are difficult because it is often impossible to isolate and compare their important assumptions and functions.

The FC parameterization has been restructured by the author and Dr. S. Chen to isolate the major components of the scheme, as discussed below. This is the version of the parameterization that they and their colleagues are currently using, and it is informally referred to as the Frank, Chen, and Cohen (FCC) hybrid scheme.

The FCC hybrid scheme is designed for use in mesoscale models. It has been structured such that the cloud model is completely within an isolated package of subroutines, so that the cloud model, trigger function, and closure can be easily isolated. This permits experimentation with various closures and trigger functions without changing the cloud model. Further, the parameterization cloud model is designed to be fully interactive with the explicit moisture scheme of the host model. Condensate from the host model is entrained into the cloud model, and both cloud water and precipitation produced by the parameterization are passed to the host model explicit moisture scheme. Figure 13.2 is a schematic diagram of the operation of the FCC scheme.

The structure of the FCC cloud model is similar to that described by Frank and Cohen (1987) except that a version of the Kain and Fritsch (1990) mixing technique for computing entrainment and detrainment has

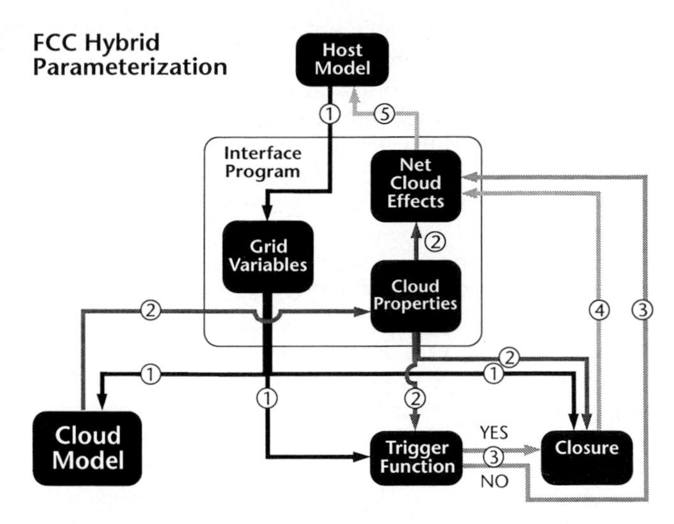

FIG. 13.2. Operational plan of the FCC parameterization. 1) The grid-scale variables (T, q, u, v, w, explicit condensate) are fed to the interface program, which then distributes them to the cloud model, closure, and trigger function subroutines. 2) The cloud model computes the net convective effects (per unit cloud-base mass flux) upon the temperature, moisture, momentum, and hydrometeor fields. These are distributed to the trigger function, closure, and net effects subroutines. 3) The trigger function determines whether to proceed. If so, the closure subroutine is activated. If not, the convective tendencies are set to zero. (Note—if the trigger function does not require output from of the cloud model, it is incorporated into step 1.) 4) The closure routine estimates the amount of cloud-base mass flux required to satisfy the desired closure hypothesis, computes in Eq. (13.2), and sends that value to the net convective effects routine. 5) The net convective effects routine determines the convective tendency terms [as in Eq. (13.1)] and feeds them back to the host model.

been added. The cloud model requires only the thermodynamic variables (temperature, mixing ratio, and condensate) from the host model sounding. It is run for a unit mass flux, and it computes the convective heating and drying at each level as tendency terms that are to be added to the host model forecast equations [Eqs. (1) and (2) of Cohen and Frank (1987)].

It is important to note that in this cloud model all of the convective tendencies are linear functions of the cloud-base mass flux, as in Eq. (13.2):

$$\left(\frac{\partial T}{\partial t}\right)_c = \alpha\left(\frac{\partial T}{\partial t}\right)_{\rm cm}, \qquad (13.2)$$

where $(\partial T/\partial t)_{\rm cm}$ is the temperature tendency per unit mass flux computed using the cloud model and α is the cloud mass flux at cloud base determined from the closure subroutine. Such linearity is an important feature, and parameterization cloud models should be designed to have this property where possible. For example, a typical closure assumption is to assume that the cumulus parameterization will provide sufficient heating to stabilize the grid column in a single time step (e.g., Arakawa and Schubert 1974). If the convective heating at every vertical level is a linear function of cloud-base mass flux, as in FCC, then a single run of the cloud model is sufficient to compute the rate of convective stabilization per unit mass flux, and hence α in Eq. (13.2). If the rate and vertical distribution of convective heating vary with cloud-base mass flux, then the cloud model must be run iteratively until the desired degree of stabilization is obtained.

The FCC hybrid scheme is currently being tested prognostically with a variety of closures based on stability, rate of stability change, moisture convergence, and low-level mass convergence. Several of these closures approximate, but are not exactly equivalent to, those of Arakawa and Schubert (1974), Kuo (1974), Fritsch and Chappell (1980), Frank and Cohen (1987), and others. They necessarily differ from the original closures to varying degrees since they use the FCC cloud model rather than those of the original schemes. A buoyancy sorting technique for downdrafts, designed to improve the distribution of convective precipitation between convective downdrafts and the explicit moisture scheme, has been developed and is currently being evaluated.

13.4. Summary

Recent numerical studies of mesoscale convective systems suggest that cumulus parameterization schemes intended for use in models with roughly 10–50-km grid meshes should utilize a hybrid scheme that allows direct interactions between the cloud model and the explicit moisture scheme of the host model. The parameterization should use a cloud model of sufficient sophistication to estimate the vertical distribution of convective properties, including heating, drying, and production of hydrometeors, based on physical interactions between the cloud model and the grid-column sounding. Further, it is highly desirable to construct the parameterization such that the cloud model is sufficiently separate from the trigger function and closure mechanism to allow the latter two components to be varied independently. Without such separation it is very difficult to isolate critical assumptions or to tune the parameterization to the host model and simulated weather conditions.

The FC and FCC parameterizations are both hybrid schemes. Although FCC was derived from the Frank and Cohen (1987) parameterization, it has been substantially restructured. One key element is an improved cloud model, designed to be fully interactive with explicit moisture schemes. The cloud model is now designed to be used interchangeably with a variety of closures and trigger functions. In a sense, it differs from most previous schemes in that it is intentionally not a closed, complete package but rather a set of parts and procedures. This seems appropriate since the current state of the art of cumulus parameterization is such that no schemes are dependable on any scale, let alone on several. The goal of this scheme is not to provide the ultimate parameterization but rather to provide a methodology and set of tools with which modelers can develop and test suitable procedures for their own applications.

Acknowledgments. This work was supported by National Science Foundation Grants ATM-8817915 and ATM-9115950.

Chapter 14

An Overview of Cumulus Parameterization in Mesoscale Models

JOHN MOLINARI

Department of Atmospheric Science, State University of New York at Albany, Albany, New York

14.1. Introduction

The problem of cumulus parameterization is nowhere more difficult than in mesoscale numerical weather prediction models. Cotton and Anthes (1989) describe the very concept of cumulus parameterization as "muddy" and "not well posed" for models with grid spacing below 50 km. Along these lines, some researchers have omitted cumulus parameterization entirely and instead directly simulated cumulus convection on the grid (Rosenthal 1978). Conversely, Fritsch and Chappell (1980) and Frank and Cohen (1987) have designed cumulus parameterizations for grid spacing as small as 10 or 20 km. The problem is further complicated by the need to simulate the development of mesoscale (and thus grid resolvable) precipitation structure in nature from initially unresolvable convection. This process blurs the distinction between convective and stratiform precipitation, and as a result blurs the meaning of parameterized and unparameterized condensation in numerical models.

This chapter will address the issues raised above and discuss the appropriate form for cumulus parameterization in mesoscale numerical models. No attempt will be made to evaluate closures or the conditions under which cumulus parameterization is invoked (the "triggering" problem), even though these are important as well. Emphasis will be placed on (i) the need for exchange of hydrometeors between the parameterized cumulus clouds and the grid scale, and (ii) the scale dependence of the optimum solution.

14.2. Definitions of current approaches

For the purposes of this discussion, "mesoscale model" will refer to hydrostatic models with grid spacing between 10 and 50 km. Unparameterized (or "explicit") condensation will refer to the practice of calculating condensation heating only upon grid-scale saturation. Parameterized ("implicit") condensation will refer to the practice of defining implicit subgrid scale convective clouds that produce grid-scale heating and precipitation, generally in the absence of grid-scale saturation. The parameterized approach requires closure conditions that relate the properties and transports of the implicit clouds to the grid-scale variables. A single characteristic unambiguously distinguishes the two methods: in implicit methods, the properties of the cloud(s) differ from those of the grid; in explicit methods, cloud and grid are synonymous.

A conceptual grouping of current approaches for incorporating cumulus convection into mesoscale numerical models is proposed as follows: (i) the *traditional* approach utilizes cumulus parameterization at convectively unstable points and explicit condensation at convectively stable points or upon supersaturation; (ii) the *fully explicit* approach uses only explicit methods regardless of stability; (iii) the *hybrid* approach parameterizes convective-scale updrafts and downdrafts at convectively unstable grid points but also "detrains" a fraction of the parameterized cloud and precipitation particles to their respective grid-scale forecast equations. This process (a) directly couples the parameterized and explicit condensation physics in the model and (b) allows the path and phase changes of previously implicit particles to be explicitly predicted over subsequent time steps.

In large-scale models, only the traditional approach has been used. In mesoscale models, all three of the above have been used, with a bewildering variety of results. A discussion of cumulus parameterization problems unique to mesoscale models, and a comparison of the various methods, will be presented in the sections that follow.

14.3. Special problems for mesoscale modelers

a. Scale separation

Ooyama (1971, 1982a), Arakawa and Schubert (1974), Frank (1983), and Arakawa and Chen (1987) have addressed the conceptual basis for cumulus parameterization. It requires in principle the existence of a spectral gap between the scales being parameterized and those being resolved on the grid. This ensures that the eddies have a time scale much smaller than the grid-scale motions, so that their integrated influence can be incorporated in a single time step. In nature, no spectral gap exists between the cloud scale and the mesoscale, and the above condition appears not to hold for mesoscale models. Ooyama (1982a) addressed the time-scale problem in terms of local rotational constraints. Under strong rotation, the local deformation radius can shrink enough to produce a long-lasting, inertially stable disturbance. In such cases, grid-scale divergent circulations are controlled by the slowly varying primary circulation. By this reasoning, the time-scale separation requirement is indeed met, even in mesoscale models, under sufficiently strong rotation. Ooyama noted that it was this characteristic that allowed the success of cumulus parameterization in numerical simulation of mature hurricanes using 10–20-km grid spacing.

In a mesoscale regional model, however, a range of dynamic stability regimes exists. At the large number of grid points without strong rotation, no clear separation of scale between the grid and the scale being parameterized will exist. The problem grows with increasing resolution. At a grid spacing of 10 km, the grid scale approaches the preferred scale for instability of convection in nature. In practice, this means that explicit convective clouds may form at a grid point while *essentially similar* clouds are simultaneously being parameterized. Calculation techniques are such that modelers do not allow "double counting": energy and moisture are conserved in such situations because grid-scale condensation is computed at the end of the time step after all other processes have acted (e.g., Kanamitsu 1975). Nevertheless, the presence of the same physical process in parameterized and unparameterized forms at the same grid point seems ambiguous at best.

It should be emphasized that the simultaneous presence of parameterized convective clouds and grid-scale saturation is not in itself objectionable. For instance, parameterized convection may saturate the upper troposphere through detrainment; there is no reason that subsequent parameterized convective clouds could not continue to create heat and moisture sources and sinks in the saturated layer at the same time that grid-scale condensation is occurring in the layer. Rather, the problem arises from the simultaneous presence of parameterized clouds and grid-scale saturated updrafts in convectively *unstable* layers. This most often develops at grid spacings of 25 km or less, because the cumulus parameterization has produced saturation of grid-scale layers in which convective instability still remains. This issue will be addressed further in section 14.6.

b. Mesoscale organization

Houze (1989) has described the process by which groups of cumulonimbus clouds often develop mesoscale structure over several hours. After convective elements have produced upper-tropospheric saturation, frozen hydrometeors ejected by active cumulus towers advect downstream and slowly descend, grow by vapor deposition (producing diabatic heating), and later melt and partly evaporate in the lower troposphere. Nearly half of the precipitation in the convective system typically reaches the ground in this manner as stratiform rain. The heating profile associated with the various phase changes noted above drives upper-level updrafts and lower-tropospheric downdrafts on a scale larger than that of the clouds, frequently extending 100 km or more from the convection. The area-averaged vertical heating profile differs dramatically from that of individual cumulonimbus clouds. Further details of this mesoscale organization of convection are provided by Houze (1989).

The existence of this process complicates the cumulus parameterization problem. In large-scale models, for which the entire mesoscale circulation remains subgrid scale, Arakawa and Chen (1987) provide convincing evidence that mesoscale effects are parameterizable in principle. Emanuel (1991) has proposed a specific parameterization of unsaturated mesoscale downdrafts for large-scale models. The mesoscale modeler, however, faces additional difficulties, because the final state can be resolved, yet a fraction of the resolvable stratiform rain condensed previously in convective updrafts. The mesoscale modeler must simulate (i) the vertical distributions of subgrid-scale convective sources of heat, moisture, and hydrometeors; (ii) their time variation over the life cycle of the system, as mesoscale circulations increase in areal coverage; and (iii) the explicit updraft–downdraft couplet, saturated aloft and unsaturated below, that is left behind. The essence of the problem can be seen in the following. Assume it is possible to define a "perfect" mesoscale cumulus parameterization that, if it persists long enough, produces the appropriate vertical profiles of stability, relative humidity, and vertical motion characteristic of mesoscale organization. The model could then in principle smoothly make the transition to explicit grid-scale microphysics. This scheme would fail, however, if the cumulus parameterization were not supplying the grid with hydrometeors, because the grid-scale equations would then have to generate cloud and precipitation particles. The spinup time for this process, plus the time for the particles to reach the ground, would produce a spurious gap in precipitation in the model. Alternatively, stratiform precipitation might never occur at the intensity observed (Molinari and Corsetti 1985).

The conceptual problem of scale separation and the practical problem of mesoscale organization associated with cumulus parameterization in mesoscale models have led researchers to experiment with the fully explicit approach. This approach will be discussed in the following section.

14.4. The fully explicit alternative

The fully explicit approach provides an alternative to the difficulties above. Such complex processes as generation, advection, and phase changes of hydrometeors can be addressed using direct prediction equations. No arbitrary closures are needed, and the interactions between the convective scale (to the extent it is resolved) and larger scales occur freely on the grid. An internal consistency occurs between vertical momentum fluxes and fluxes of other variables, because all occur explicitly in the model. The fully explicit approach has been used with apparent success in the simulation of fronts, tropical cyclones, and explosively growing middle-latitude cyclones.

Despite the successes noted above, the fully explicit approach has repeatedly failed in mesoscale models in the presence of large instability or weak grid-scale forcing (see Molinari and Dudek 1992 for details). In addition, mesoscale organization of convection has never been simulated using the fully explicit approach in hydrostatic mesoscale models. These failures arise as a result of the following.

(i) Convective precipitation is unrealistically delayed with the fully explicit approach because no condensation can occur until the grid scale saturates; in nature, convective precipitation often occurs in the absence of grid-scale saturation.

(ii) If the grid scale does saturate in a convectively unstable layer, the instability cannot be removed by eddies as in nature, because updrafts and downdrafts cannot occur simultaneously using the fully explicit approach. Instead, only the grid-scale vertical motion acts, and advects the high low-level moist static energy upward, producing a tropospheric deep layer of convective instability, similar to that shown by Molinari and Dudek (1986, 1992). This represents the reverse of what was found in nature by Riehl and Malkus (1958) and Betts (1974b), in which the area-averaged middle-tropospheric moist static energy minimum remained in place during convection.

(iii) This absolutely unstable column overturns on the scale of the grid, stabilizing only by "entraining" air from adjacent grid points, while producing extreme overprediction of rainfall in the process.

The fully explicit approach is likely to succeed in mesoscale models only when grid-scale forcing is large and instability relatively small. Large forcing ensures that vertical motions will be sufficient to minimize the delays in precipitation onset. Modest convective instability suggests the vertical eddy fluxes are sufficiently small to minimize errors in the vertical distribution of heating. Otherwise, the problems noted above make the fully explicit approach unsuitable for most applications. The difficulties with both the traditional and fully explicit approaches have led to the development of the hybrid approach discussed below.

14.5. Hybrid approach

The hybrid approach explicitly couples the convective parameterization to the grid scale in the following manner:

$$\frac{\partial \bar{q}_a}{\partial t} = \cdots + S^a_{\text{convec}},$$

where q is a mixing ratio, the bar represents a gridpoint average, S is the source term from the convective parameterization, and a represents cloud water, rain, cloud ice, or one or more forms of frozen precipitation. The dots refer to grid-scale advection, phase changes, and various particle interactions. The definition of the hybrid approach was proposed by Molinari and Dudek (1992). Kreitzberg and Perkey (1976) originated the idea of detraining a fraction of parameterized precipitation to the grid scale, while Frank and Cohen (1987) have developed the most complete hybrid approach.

In effect, the hybrid approach parameterizes *convective*-scale evaporation, condensation, and vertical eddy fluxes but does not parameterize heat and moisture sources that arise from the more slowly evolving process of detrained precipitation-sized particles falling between (or downwind of) convective clouds. This latter process evolves separately through the grid-scale equations. Arakawa and Chen (1987) argued that at least some part of mesoscale organization has to be parameterized due to the lack of a spectral gap between cloud and mesoscale. It is proposed that the hybrid approach, by separating out the forcing mechanism for the mesoscale component, removes the need to parameterize the mesoscale any further. This property may allow the hybrid approach to be used for smaller grid spacings than the traditional approach without encountering severe scale separation problems. These conceptual benefits of the hybrid approach have not been clearly stated by its originators. The major practical benefit of the hybrid approach lies in its potential for realistic simulation of the transition to mesoscale structure (see, for instance, Dudhia 1989).

In addition to its benefits, the hybrid approach has an inherent limitation: detrained particles are carried by grid-scale motions. In nature such particles fall between clouds and thus experience an environment that differs from the grid-scale average. The importance of this limitation varies with model resolution: for 40–50-km grid spacing, terminal velocity is usually much greater than either grid-area-averaged or between-cloud vertical velocity, so that particle evolution may not differ dramatically in the two cases. For a 10-km grid spacing, however, the grid-scale updraft may approach in strength a convective updraft in nature, and convective precipitation intended to fall between clouds may be unrealistically suspended. For such a fine grid spacing, microphysical processes may not be realistically simulated.

The one remaining problem of the hybrid approach is the lack of knowledge of actual precipitation detrainment profiles for deep convection. This problem, however, is solvable in principle.

In summary, the hybrid approach shares with cumulus parameterization the difficulties of defining a closure condition on the mesoscale and has potential problems with vertical advection of convectively generated particles. Both problems become significant as grid spacing reaches 10–15 km. Offsetting the weaknesses of the hybrid approach are the desirable separation of fast and slow processes into parameterized and explicit forms and the realistic simulation of mesoscale organization. Because mesoscale organization occurs so frequently and so strongly affects some fundamental inputs to the grid scale such as the vertical heating distribution, these characteristics of the hybrid approach offer significant potential advantages.

14.6. Recommendations

Figure 14.1 presents recommended solutions as a function of grid spacing. The optimum solution depends more upon local rotational constraints than upon grid spacing, but several dynamical regimes are likely to be present in a regional mesoscale model, so Fig. 14.1 represents the most general solution for each grid spacing.

For large-scale models (grid spacing greater than 50–60 km), the traditional approach provides the only solution. The hybrid approach cannot drive realistic mesoscale circulations in such models, because the mesoscale cannot be resolved. For models with grid spacing less than 2–3 km, the fully explicit approach is clearly superior to parameterized approaches, even though a 1-km grid spacing can simulate only the largest of convective clouds (Lilly 1990).

It is between these two extremes of resolution that the choices become more complex. For the hybrid approach to be effective, grid spacing must be small enough to resolve mesoscale organization but not so small that scale separation problems arise. It is suggested that the hybrid approach is the preferred choice for grid spacings from 20 or 25 to 50 km.

For grid spacings from about 3 km to 20–25 km, it remains uncertain whether a general solution exists. It may seem obvious to state that cumulus parameterization must not be used at grid spacings where growth of explicit cumulus clouds is favored. As noted earlier, however, the fully explicit alternative often fails at such grid spacing, because it can simulate neither the onset of convection nor the release of instability, which requires simultaneous updrafts and downdrafts. The hybrid approach, which contains a cumulus parameterization, suffers from the same problems on this scale as the traditional approach. Molinari and Dudek (1992) suggested that mesoscale modelers can solve this problem only by going either to much higher resolution (and explicitly resolve the convection) or to lower resolution where the hybrid approach may suffice.

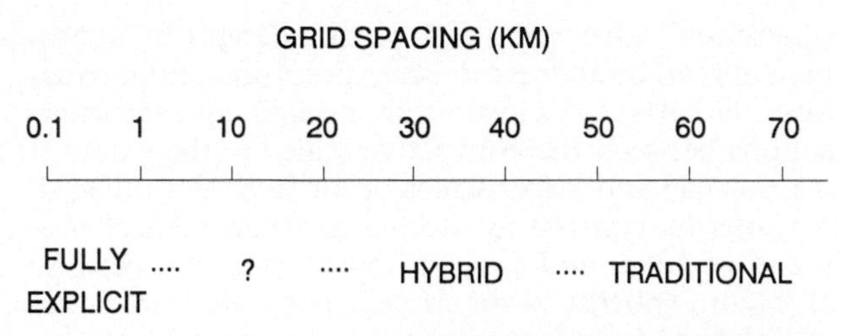

FIG. 14.1. Proposed form for cumulus parameterization in regional mesoscale models as a function of grid spacing. The scale is logarithmic below 10 km and linear above. The question mark indicates the lack of an obvious solution, and the dots represent transition regions between the choices. It is assumed the model covers a wide enough area that the approach must simulate convective effects over a range of thermodynamic and inertial stability regimes.

An alternative view has been proposed by Zhang and Fritsch (1987), Zhang and Gao (1989), and Zhang and Cho (1992a), who have produced strikingly accurate simulations with 20-km grid spacing. A key aspect of their simulations was the fairly rapid generation of grid-scale saturated updrafts that coexist with parameterized convection. The creation of grid-scale saturation has great practical significance in models, because once it occurs, the grid-scale vertical heating profile can be much different, and more strongly coupled to grid-scale dynamics, than for parameterized heating. The key to determining whether the Zhang–Fritsch approach will be consistently successful in mesoscale models lies in understanding (i) how ensembles of convective clouds produce larger-scale saturation in nature, (ii) how this process occurs in successful model integrations, and (iii) whether the latter represents the former. Understanding of both nature and numerical models should be greatly facilitated by cloud ensemble models, which are nonhydrostatic and able to simulate individual clouds. Xu (1991) has made a promising start in this direction.

Finally, it must be noted that in mesoscale models it is not possible to understand the *interaction* of the convective scale and mesoscale, because cumulus parameterization fixes that interaction a priori through closure conditions. Ooyama (1982a) noted that modelers must not "play the game with loaded dice." Even a perfect forecast on the mesoscale does not mean the process by which individual clouds produced a mesoscale disturbance has been understood. Additional detailed studies are needed of both the successes and failures of cumulus parameterization in high-resolution models.

Acknowledgments. This work was supported by National Science Foundation Grant ATM8902487.

Chapter 15

Convective Parameterization for Mesoscale Models: The Fritsch–Chappell Scheme

J. MICHAEL FRITSCH AND JOHN S. KAIN

Department of Meteorology, The Pennsylvania State University, University Park, Pennsylvania

15.1. Introduction

The Fritsch and Chappell (1980) (hereafter FC) convective parameterization scheme was developed to facilitate numerical simulations of mesoscale convective systems. Since simulation of these systems requires resolution of meso-β-scale features, the parameterization was designed for models with grid increments of about 10–30 km. Grid elements this small allow for a somewhat different approach to the convective parameterization problem from that used for simulations that employ much larger elements (e.g., the 100–500-km grid increments used in general circulation and global climate models). In particular, grid elements that span several hundred kilometers are large enough to contain entire mesoscale convective systems and therefore the effects of convectively generated stratiform clouds and mesoscale circulations must also be parameterized. In contrast, grid increments of about 20 km are sufficient to resolve the convectively generated mesoscale circulations and associated cloud regions so that the parameterization problem reduces to incorporating the effects of only the convective clouds. It is assumed that if the convective clouds are properly parameterized, the resolvable-scale governing system of equations will develop the appropriate mesoscale mass, heat, moisture, and momentum transports.

The following two sections outline the key assumptions, constraints, and cloud model utilized in the FC scheme. Section 15.4 briefly describes some of the applications of the FC scheme in simulations of mesoscale convective systems. The final section summarizes some of the deficiencies of the FC scheme and suggests areas for future improvements, especially to the scheme's cloud model.

15.2. Description of scheme

Assuming that changes produced by convection can be spread uniformly over the time period that convection is active in a numerical model grid element, the convectively produced changes can be expressed as

$$\left.\frac{\partial \bar{\chi}}{\partial t}\right|_{\text{conv}} = \frac{\hat{\chi} - \chi_0}{\tau_c}, \tag{15.1}$$

where χ_0 is the value of a grid element variable before convection, $\hat{\chi}$ the grid element value after convection, and τ_c the characteristic time period that convection is active in the grid element. Thus, very simply, the convective parameterization problem is to determine $\hat{\chi}$ and τ_c.

Formulation of the convective parameterization follows the same general approach as other techniques. Resolvable-scale quantities in the numerical model are utilized to establish constraints on the amount of convection, and a cloud model is used to estimate the vertical structure of the convective mass flux that satisfies the constraints.

Specification of the amount of convective activity originates with the concept of potential buoyant energy (PBE) and available buoyant energy (ABE). Consider the following. The vertical component of the equation of motion for an air parcel of unit mass in an updraft, ignoring pressure perturbations due to the updraft and frictional effects, is

$$\frac{dw}{dt} = g\left(\frac{T_u - T}{T}\right) = g\beta, \tag{15.2}$$

where T_u denotes the updraft virtual temperature and T is the environment (or gridpoint) virtual temperature. In an atmosphere containing a deep conditionally unstable layer and high relative humidity at low levels (as in Fig. 15.1), an air parcel rising from near the surface without mixing will achieve saturation at its lifting condensating level (LCL). Frequently, $\beta < 0$ at the LCL, but continued rise of the parcel in the conditionally unstable environment results in an increase of β. The parcel's level of free convection (LFC) is reached at the point where β becomes positive. Once the parcel reaches its LFC, it will continue to rise, eventually reaching its equilibrium temperature level (ETL), where β decreases to zero. The increase in kinetic energy of the parcel, associated with its vertical acceleration in rising between its LFC and its ETL, is given by

$$\left.\frac{1}{2}w^2\right|_{\text{ETL}} - \left.\frac{1}{2}w^2\right|_{\text{LFC}} = \int_{\text{LFC}}^{\text{ETL}} g\beta dz. \tag{15.3}$$

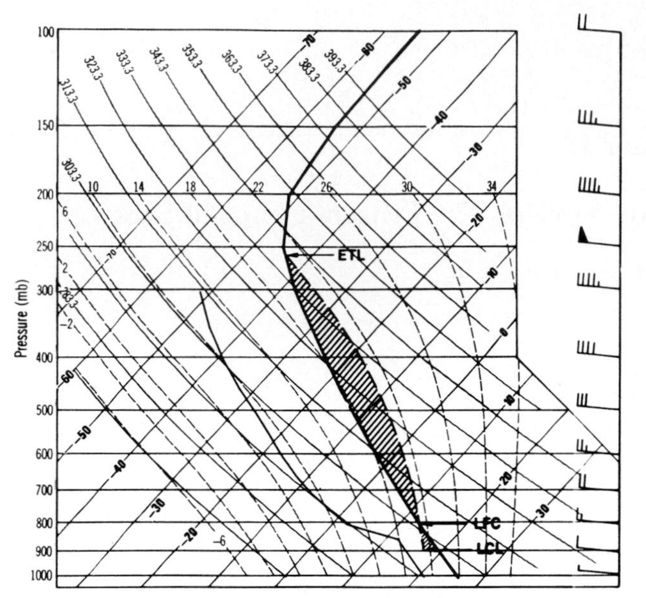

FIG. 15.1. Thermodynamic diagram (skew T–log$p$) of conditionally unstable sounding. Heavy and thin solid lines indicate environmental temperature and dewpoint, respectively. The heavy dashed line is a moist adiabat for a lifted parcel with mean thermodynamic characteristics of the lowest 100-mb layer. Shaded area to the left of the environmental temperature line defines the negatively buoyant region between the LCL and the LFC. Shaded area to the right of the environmental temperature line defines the positive buoyant area between the LFC and the ETL. Full wind barb—5 m s^{-1}; flag—25 m s^{-1}.

This is defined as buoyant energy, that is, the kinetic energy derived from buoyancy. The buoyant energy is proportional to the positive area in Fig. 15.1. Potential buoyant energy is the buoyant energy that would accrue to a parcel in rising from its LFC to its ETL. The PBE is identical to the buoyant energy for the sounding and thermodynamic trajectory of the parcel shown in Fig. 15.1. If there is no LFC for a given parcel, the PBE for that parcel is zero. Often a parcel must overcome an appreciable depth of negative buoyancy if it is to rise to its LFC. This is shown as the negative area in Fig. 15.1. This negative buoyancy may be reduced or eliminated by surface heating, differential horizontal temperature or moisture advection, or lifting of the conditionally unstable environment if this environment is also convectively unstable. The PBE is said to be "available" (PBE = ABE) if the negative buoyancy is eliminated or overcome and a parcel can reach its LFC. Therefore, only after the PBE within a grid column becomes available does convection begin. Once convection is triggered, it is assumed that the subsequent convective activity is sufficient to remove the ABE in a specified time period τ_c. The basis for this assumption follows from Fritsch et al. (1976), who observed that the large scale typically took many hours to generate potential buoyant energy, but once this energy became available, on the mesoscale the convection removed the energy in a small fraction of the time that it took to generate it.

The condition of complete removal of ABE by convection after time τ_c is

$$\widehat{\mathrm{ABE}} = \int_{\widehat{\mathrm{LFC}}}^{\widehat{\mathrm{ETL}}} g\left[\frac{\hat{T}_U(z) - \hat{T}(z)}{\hat{T}(z)}\right]dz = 0, \quad (15.4)$$

where the caret indicates the value of a parameter after adjustment for convection. In particular, $\hat{T}_U(z)$ is the vertical distribution of temperature in the updraft that results from lifting the convectively modified air, $\hat{T}(z)$. Since the areas of updrafts and downdrafts may be a substantial fraction of a mesoscale model grid element area, gridpoint temperatures are an area-weighted mean of updraft, downdraft, and environmental temperatures. Thus, the gridpoint temperature adjusted for convection, $\hat{T}(z)$, is defined by

$$\hat{T}(z) = A^{-1}[\hat{T}_E(z)A_E(z) + \hat{T}_U(z)A_U(z) + \hat{T}_D(z)A_D(z)], \quad (15.5)$$

where $A = A_E + A_U + A_D$, and the subscripts E, U, and D identify environment, updraft, and downdraft, respectively.

An additional simplification that results from using small grid elements is that all convective clouds in an element can be assumed to be alike. It is further assumed that the characteristics of the clouds that develop in a grid element will remain unchanged during the time period required for the clouds to transit the grid element, unless this time is longer than 1 h. The time for clouds to move through a grid element is estimated by dividing the grid length by the mean environmental wind speed over the cloud depth. If this time period (τ_c) is greater than 1 h, the computed rate of grid-scale stabilization by the convection is still applied for only 1 h. The 1-h limit is introduced because the mesoscale environment may undergo substantial changes over this time period, and it is unrealistic to assume that the cloud characteristics would remain the same. Therefore, at the end of the 1-h period, the cloud characteristics are determined again, but for the adjusted environment. A lower time limit for τ_c and for maintaining similar cloud characteristics is also imposed. This lower limit (30 min) is based upon the typical lifetime of a single cell; that is, once convection begins, it is assumed that the cloud continues through a normal life cycle.

With the assumption that all convective clouds in the mesoscale grid element are alike, there is a specific area (at cloud base) of updraft and corresponding areas of downdraft and environment that will produce a vertical distribution of $\hat{T}$ that satisfies (15.4) in the time period τ_c. To arrive at the proper set of areas, an iterative procedure is necessary. A first guess of the updraft area is taken to be 1% of the grid element area. For an initial unit of updraft air, the corresponding amount of downdraft air is calculated using the cloud model. The environmental temperature change produced by this single updraft–downdraft "cloud" unit is obtained from the time integral (over the period τ_c) of

$$\frac{\partial T_E(z)}{\partial t} = -w_E(z)[\Gamma - \gamma(z)] - \frac{L}{c_p} C^*, \quad (15.6)$$

where the compensating environmental vertical motion w_E is assumed to occur hydrostatically. The second term on the right-hand side of (15.6) is the environmental cooling rate associated with the evaporation of condensate C^* in the anvil. Compensating environmental vertical motion is obtained by invoking mass continuity between the cloud circulations and the environment within a grid element (see FC).

Normally the first guess at the updraft area A'_U will not remove all the ABE but will produce only a reduction, ΔABE. A new estimate for the number N of updraft–downdraft "cloud" units that will remove all the ABE is obtained from

$$N^{(m)}(z) = \frac{\text{ABE}}{\Delta\text{ABE}^{(m-1)}} N^{(m-1)}, \quad (15.7)$$

where m is the number of the iteration. Equations (15.4), (15.5), (15.6), and (15.7) are iterated until

$$\text{ABE} - \Delta\text{ABE} = 0 \pm 0.05\text{ABE}, \quad (15.8)$$

which is defined as an acceptable level of convergence. Usually only four or five iterations are necessary until (15.8) is satisfied. The total updraft cloud area is then NA'_U.

Finally, the equations governing changes in momentum and moisture follow the same mathematical development as the mass conservation. It is assumed that no significant change in horizontal momentum occurs during the brief period of time a parcel spends in the active updrafts and downdrafts. Major alterations to parcel momentum are permitted only to occur after a parcel reaches the anvil or boundary layers. As the parcel momentum is gradually incorporated into the mesoscale governing system, environmental pressure forces are then free to act on the new momentum fields.

15.3. Cloud model

A simple entraining-plume cloud model that contains both updrafts and moist downdrafts is utilized to help provide the magnitude and vertical distribution of convective cloud effects to the resolvable-scale governing system of equations. The complete cloud model is described in FC. Only a brief summary of some of the special characteristics of the cloud model is presented here.

a. Trigger function

The criteria used to determine when and where deep convection occurs in a numerical model (i.e., when PBE becomes ABE) are collectively termed the convective trigger function. For the original FC scheme, the trigger function is defined as follows.

Beginning with the lowest 100-mb layer, the LCL for a mixed-layer parcel with a mean-layer virtual temperature $\bar{T}(k)$ and mixing ratio $\bar{r}(k)$ is computed. At the LCL, the parcel is checked for buoyancy using

$$T_U^s - T + \Delta T \begin{cases} >0, & \text{buoyant} \\ \leqslant, & \text{stable,} \end{cases} \quad (15.9)$$

where T_U^s is the temperature of the saturated updraft and T is the grid element temperature provided by the mesoscale model. The temperature increment ΔT is defined by

$$\Delta T = c_1 w_G^{1/3}, \quad (15.10)$$

where c_1 (°C s$^{1/3}$ cm$^{-1/3}$) is a unit number, and w_G is the mesoscale model vertical motion (cm s^{-1}) at the LCL. The temperature increment ΔT crudely simulates local (subgrid scale) forcing as a function of the grid-scale vertical motion in the layer being lifted. For example, if mesoscale lifting is 1.0 cm s^{-1}, the thermal perturbation is 1.0°C; for 10 cm s^{-1}, it is 2.15°C. In support of this type of approach it is evident from Bean et al. (1972), Bean et al. (1975), and Chen and Orville (1980) that thermals are stronger and larger when low-level convergence is present.

If the mixed parcel for the lowest 100-mb layer is stable, then the 100-mb layer 50 mb higher than the previous layer is mixed, lifted, and checked for buoyancy. This procedure is repeated up to the 600–700-mb layer where, if the parcel is still stable, no convection is permitted to occur. If, however, a buoyant parcel is found, the scheme proceeds to calculate the thermodynamic path of the updraft to cloud top (CT).

b. Updraft

The equivalent potential temperature θ_e of the updraft is computed for the LCL and the parcel is lifted and mixed through 50-mb increments until the condition

$$\int_{\text{LCL}}^{\text{CT}} (T_U^s - T_E)\rho dz = 0 \quad (15.11)$$

is satisfied. Since the entrainment rate is a function of the cloud depth (see FC), (15.11) must be solved iteratively. The updraft parcel temperatures and vertical motions are then determined using the Clausius–Clapeyron equation and the parcel form of the vertical motion equation. Freezing of updraft condensate is introduced at $T_U = -25$°C. Once the updraft vertical motion is obtained, the updraft area can be determined from

$$A_U(k) = \frac{M_U(k)}{\rho_U(k) w_U(k)}, \quad (15.12)$$

where M_U is the vertical mass flux of the updraft and $\rho_U(k)$ is obtained from the equation of state assuming the pressure is the same in updraft and environment.

c. *Condensate lost in anvil*

It is assumed that all condensate produced above the ETL is detrained and subsequently evaporated (sublimated) into the environment. Air outside the cloud anvil in the highest model layer occupied by convective clouds is saturated first. If the condensate increment is not totally evaporated in the highest layer, then successively lower layers are saturated. This source of moisture is important since it contributes to the production of the mesoscale stratiform cloud (i.e., the "anvil cloud").

d. *Moist downdrafts*

Although there are two processes that typically initiate and maintain moist downdrafts, evaporational cooling and precipitation drag, only the negative buoyancy from evaporative cooling is included in the FC cloud model. Downdrafts begin at the level of free sink (LFS) and descend to the surface. The level of free sink is analogous to the LFC and occurs at the highest level (below the ETL) where the temperature of a saturated mixture of equal amounts of updraft and environmental air becomes less than the environmental temperature. The equivalent potential temperature at the LFS is found by mixing parcel and environment in equal amounts according to the technique developed by Foster (1958). Between the LFS and the surface, the thermodynamic structure of the downdraft is determined in much the same manner as the updraft, except that a constant relative humidity for each layer is specified instead of always requiring saturated conditions, and melting of condensate is introduced (at $T_D = 0°C$) instead of freezing. This approach crudely takes into account the fact that all downdraft air is not saturated nor does initially saturated air necessarily remain so during its entire downward path. Downdraft temperature, vertical motion, and area are determined in the same manner as for updrafts.

e. *Relationship of downdraft mass flux to updraft*

Recall from section 15.2 that sufficient convection must occur in the time period τ_c so that the ABE is eliminated. The removal of ABE is accomplished by warming of the environment through compensating subsidence forced by the updrafts, and by the replacement of subcloud-layer high-θ_e air by moist downdraft low-θ_e air. In order to determine the amount (area) of deep convective updrafts that will consume the ABE, it is necessary to determine the fractional amount N_D of moist downdraft air that occurs with each unit of updraft air. This fractional amount is given by

$$N_D = \frac{E_D}{R_D}, \tag{15.13}$$

where E_D is the rate that condensate is being made available from the updraft for evaporation in the moist downdraft and R_D is the rate of condensate loss (evaporation) over the entire depth of a unit downdraft. The rate at which condensate is being made available to the downdraft from the updraft is determined from an empirical relationship between the observed precipitation efficiency of convective clouds and the vertical wind shear of the environment (see Marwitz 1972; Foote and Fankhauser 1973; and FC).

15.4. Simulation of mesoscale convective systems

The first real-data application of the FC scheme was presented in Zhang and Fritsch (1986). They incorporated the scheme into the Pennsylvania State University (Penn State)–National Center for Atmospheric Research (NCAR) Mesoscale Model (Anthes et al. 1987) and produced an 18-h simulation of the meso-β-scale structure and evolution of the 1977 Johnstown flood. Figures 15.2a,b show mesoanalyses of the observed and model-simulated conditions 6 h into their simulation. It is evident that the model exhibited skill in reproducing some of the observed meso-β-scale features. Sensitivity experiments conducted on this case (Zhang and Fritsch 1988a) clearly showed that it was not possible to reproduce the observed meso-β-scale features without the convective parameterization, especially without the moist downdrafts. The experiments also showed that it is absolutely essential that the physics of the resolvable-scale governing system be "compatible" with the physics of the convective parameterization and with the primary physical processes that establish the structures and circulations of the mesoscale weather systems. For example, when a convective parameterization in a mesoscale model introduces midlevel warming and moistening into the resolvable-scale governing system, pronounced low-level and midlevel mesoscale convergence and ascent often result. In the "real world," the ascent is modulated by such processes as precipitation drag, melting, subcloud-layer evaporation, etc. Unless these processes are included in the resolvable-scale governing system, the mesoscale circulations tend to respond much too strongly to the forcing from the convective parameterization and thereby cause physically based model instabilities to develop (Zhang and Fritsch 1988a; Zhang 1989; Giorgi 1991).

Following the Johnstown flood simulation, the FC scheme was used in simulations of several other mesoscale convective systems. Among these were the simulation of an intense mesoscale convective complex that generated a long-lived inertially stable mesovortex (Zhang and Fritsch 1988b) and a squall-line event from PRE-STORM (Preliminary Regional Experiment for

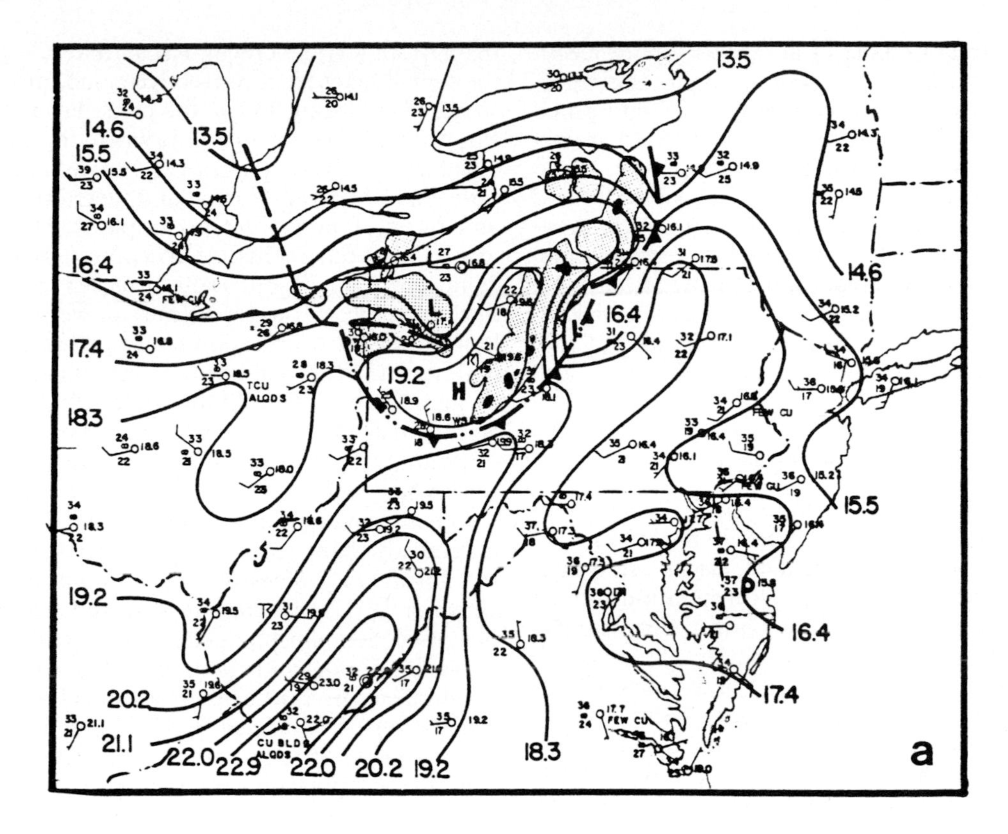

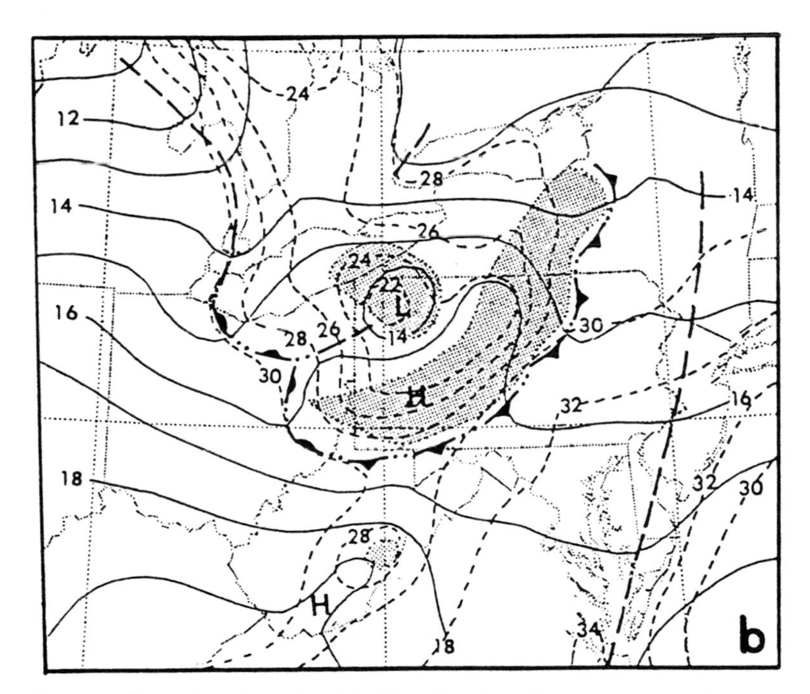

FIG. 15.2. (a) Mesoscale analysis for 1800 UTC 19 July 1977 (adapted from Hoxit et al. 1978). Heavy dashed lines indicate troughs. Cold- and warm-frontal symbols alternated with double dots indicate moist-downdraft outflow boundaries. The light shading denotes the level 1 radar echoes and the dark shading denotes level 3 (or greater) radar echoes. Reflectivity boundary is not defined if the echo contour is open. A full wind barb is 5 m s^{-1}. (b) Analysis of sea level pressure (mb, solid lines) and surface temperature (°C, dashed lines) for 6-h forecast verifying at 1800 UTC 19 July 1977. Heavy dashed lines indicate troughs. Cold- and warm-frontal symbols alternated with double dots indicate moist-downdraft outflow boundaries. Shading indicates area of active convection at verification time.

STORM-Central) (Zhang et al. 1989). More recent applications include a simulation of an upstream-propagating mesoscale convective system that formed over the High Plains and advanced southward against the flow at all levels (Stensrud and Fritsch 1991), and a simulation of the 14 July 1987 Montreal flash flood (Bélair et al. 1991). In addition to helping to demonstrate the potential for numerically predicting mesoscale convective systems, the FC scheme has been used in many other numerical studies as a vehicle to help understand the structure and dynamics of mesoscale phenomena (e.g., Gao et al. 1990; Zhang and Cho 1992b).

15.5. Opportunities for improvement

During and following the time the FC scheme was developed, a number of deficiencies in the scheme became evident. Some of these deficiencies are fundamental properties of the scheme's formulation and are very difficult, if not impossible, to correct. For example, the magnitude and vertical distribution of convective heating within a grid element are largely determined by the compressional warming from compensating subsidence. Because there is no a priori way to determine how the subsidence should be distributed on the mesoscale (i.e., into the grid elements surrounding the element containing the convection), all the subsidence is forced to occur in the element in which the convection develops. The resulting heating and moistening profiles are then fed into the mesoscale model, where it is assumed that the resolvable-scale governing equations will "more correctly" distribute the mesoscale effects of the convection. The dynamical consequences of this procedure are unknown.

An additional difficulty with the FC scheme arises as a result of the procedure for estimating the characteristic time period τ_c in which convection eliminates the ABE in a grid element. Recall that τ_c is defined as the time it takes for a convective cloud, moving at the speed of the mean flow in the cloud layer, to traverse the grid element. In situations where the winds in the cloud layer are very weak or very strong, τ_c can become unrealistically large or small and therefore upper (1 h) and lower (30 min) limits are imposed. The upper and lower limits can themselves become unrealistic, such as in the case of a hurricane or a fast-moving squall line where deep convection may move through a 20-km grid element in an 8–12-min period. In such situations, the FC scheme may systematically lag reality by maintaining effects of the convection too long in a given grid element. It remains to determine the severity of such errors in simulations of extremely strong or weak wind events.

Other recognized deficiencies in the FC scheme pertain mostly to the cloud model. In particular, the original formulation does not allow detrainment, except above the ETL. As shown by Kain and Fritsch (1990), this restriction can produce highly unrealistic vertical mass flux profiles in situations with small ABE and a dry midlevel environment. The vertical mass flux profiles are further distorted by FC's omission of precipitation drag and by their assumption that pressure gradient forces can be ignored during the ascent of a convective cloud parcel.

Some of the deficiencies listed above have already been addressed in an improved form of the FC scheme described in the next chapter of this monograph. Moreover, plans have been made to incorporate the Zhang and Cho (1991) cloud momentum parameterization formulation and to test the resultant version of the FC scheme in the quasi-operational version of the Penn State–NCAR model (Warner and Seaman 1990).

Acknowledgments. This work was supported by NSF Grants ATM-9222017 and ATM-9024434.

Chapter 16

Convective Parameterization for Mesoscale Models: The Kain–Fritsch Scheme

JOHN S. KAIN AND J. MICHAEL FRITSCH

Department of Meteorology, The Pennsylvania State University, University Park, Pennsylvania

16.1. Introduction

The Kain–Fritsch (KF) convective parameterization scheme (CPS) is based on the same fundamental closure assumption as the Fritsch–Chappell (FC) (1980) scheme—convective effects are assumed to remove convective available potential energy in a grid element within an advective time period. Its development was motivated by ongoing observational and numerical investigations of mesoscale convective systems that have revealed the potentially significant impact of certain physical processes that were not represented in the FC scheme. For example, in the FC scheme, detrainment from convective clouds to their environment occurs over a limited vertical depth near cloud top. Yet, it has become evident from diagnostic studies (e.g., Leary and Houze 1980; Gamache and Houze 1983) that midlevel detrainment of mass and moisture from deep convective clouds plays an important role in the development of some mesoscale convective systems.

Detrainment effects are more realistically distributed vertically in the KF scheme through the implementation of a new cloud model. This cloud model modulates the two-way exchange of mass between cloud and environment (i.e., entrainment and detrainment) as a function of the buoyancy characteristics of various mixtures of clear and cloudy air. In some environments, the vertical distribution of convective effects changes substantially with the addition of the new cloud model.

The KF scheme is also formulated to assure conservation of mass, thermal energy, total moisture, and momentum. Rigorous conservation of these quantities was not essential for most applications of the FC scheme, since they typically involved relatively short simulations (less than 24 h in duration) with regional-scale models. However, continuing advances in computing power have made it feasible to implement this type of scheme in larger-scale models and over longer time periods. For these and other more general applications, adherence to conservation principles can be critically important.

In our description of the KF scheme in the following subsection, we focus primarily on these two improvements to the original FC scheme. For a more detailed discussion on the KF scheme's closure and operating principles, the reader is referred to chapter 15 on the FC scheme. We follow with some preliminary diagnostic results. Considerations for future additions to the scheme are discussed in the last section.

16.2. Major components of the KF scheme

a. Mathematical formulation of the convective parameterization

Following Anthes (1977), the heating tendency due to subgrid-scale convective processes can be expressed as

$$\left.\frac{\partial\bar{\theta}}{\partial t}\right|_{\text{conv}} = \frac{L}{\pi}\frac{\overline{dq}}{dt} - \frac{\partial(\overline{\omega'\theta'})}{\partial p}, \qquad (16.1)$$

where θ is potential temperature (K), L is the latent heat released during phase change of a unit mass of water substance (J kg^{-1}), π is Exner's function, $c_p(p/p_0)^{R/c_p}$, where c_p is the heat capacity of dry air (J kg^{-1} K^{-1}), p is the pressure (hPa), p_0 = 1000 hPa, and R is the gas constant for dry air (J kg^{-1} K^{-1}); dq/dt is the rate of phase change of water substance (kg kg^{-1} s^{-1}), and ω is the vertical pressure velocity [(kg s^{-1})(m s^{-2})(m^{-2})]. The overbar denotes the grid-scale value in a numerical model while the primes indicate the subgrid-scale perturbations. In a manner similar to McBride (1981), the vertical velocity is normalized by the area of a model grid element so that the vertical pressure velocity ω is directly proportional to the convective mass flux. For example, ω_u, representing the updraft mass flux, is given by $\omega_u = -M_u g/A$, where M_u is the updraft mass flux (kg s^{-1}), g is the acceleration due to gravity (m s^{-2}), and A is the horizontal area occupied by a grid element (m^2).

The second term on the rhs of (16.1) can be approximated as the sum of the individual contributions from the updraft mass flux ω_u, the moist downdraft mass flux ω_d, and the compensating mass flux in the environment surrounding the convective drafts $\tilde{\omega}$; that is,

$$\frac{\partial(\overline{\omega'\theta'})}{\partial p} = \frac{\partial[(\omega_u - \bar{\omega})(\theta_u - \bar{\theta}) + (\omega_d - \bar{\omega}) \times (\theta_d - \bar{\theta}) + (\tilde{\omega} - \bar{\omega})(\tilde{\theta} - \bar{\theta})]}{\partial p}, \tag{16.2}$$

where $\omega_d > 0$. By definition, the thermodynamic environment of the updraft and downdraft is given by the resolvable-scale variables, so $\tilde{\theta} = \bar{\theta}$, and the third term on the rhs of (16.2) can be eliminated. Furthermore, if local compensation of convective mass fluxes is assumed—that is, $\bar{\omega} \approx \omega_u + \omega_d + \tilde{\omega} = 0$—(16.2) can be written as

$$\frac{\partial(\overline{\omega'\theta'})}{\partial p} = \frac{\partial[\omega_u\theta_u + \omega_d\theta_d - (\omega_u + \omega_d)\bar{\theta}]}{\partial p}. \tag{16.3}$$

For a given vertical layer in a numerical model, (16.3) can be written in finite-difference form as

$$\frac{\Delta(\overline{\omega'\theta'})}{\Delta p} = \frac{1}{\Delta p}\Big\{\overbrace{(\omega_{u2}\theta_{u2} - \omega_{u1}\theta_{u1})}^{\text{A}} + \overbrace{(\omega_{d2}\theta_{d2} - \omega_{d1}\theta_{d1})}^{\text{B}} - [(\omega_{u2} + \omega_{d2})\bar{\theta}_2 - (\omega_{u1} + \omega_{d1})\bar{\theta}_1]\Big\}, \tag{16.4}$$

where the subscript 2 denotes the top of a given model layer and the subscript 1 denotes the bottom.

Expressions for θ_{u2} and θ_{d1} can be written as functions of θ_{u1} and θ_{d2}, respectively. For example, in convective updrafts, as parcels rise from the bottom to the top of a model layer, their potential temperature changes as a function of mixing with the environment and latent heat release/absorption. Specifically, the term A in (16.3) can be written as

$$\omega_{u2}\theta_{u2} = \omega_{u1}\theta_{u1} - \epsilon_u\bar{\theta}_m + \delta_u\theta_{um} - \frac{L}{\pi}\omega_{u2}\Delta q_u, \tag{16.5}$$

where ϵ_u is the rate of entrainment of environmental mass into the updraft, $\bar{\theta}_m$ is the mean environmental potential temperature in the layer, δ_u is the rate of detrainment of updraft mass into the environment, θ_{um} is the mean updraft potential temperature in the layer, and Δq_u is the total mass of water substance, per unit mass of air, that changes phase (to a higher energy state, i.e., $\Delta q_u > 0$ for evaporation, melting, and sublimation) in updraft parcels as they rise through the layer. Entrainment and detrainment rates, ϵ and δ, respectively, are expressed in the same units as ω and are always positive. Similarly, the term B can be written in an analogous form for convective downdrafts as

$$\omega_{d1}\theta_{d1} = \omega_{d2}\theta_{d2} + \epsilon_d\bar{\theta}_m - \delta_d\theta_{dm} - \frac{L}{\pi}\omega_{d1}\Delta q_d. \tag{16.6}$$

Substitution of (16.5) and (16.6) into (16.4) yields

$$\frac{\Delta(\overline{\omega'\theta'})}{\Delta p} = \frac{1}{\Delta p}[(\omega_{u2} + \omega_{d2})\bar{\theta}_2 - (\omega_{u1} + \omega_{d1})\bar{\theta}_1 + (\epsilon_u + \epsilon_d)\bar{\theta}_m - \delta_u\theta_{um} - \delta_d\theta_{dm} - \frac{L}{\pi}(\omega_{u2}\delta q_u - \omega_{d1}\delta q_d)]. \tag{16.7}$$

Upon substitution of (16.7) into (16.1), the latent heating terms can be eliminated to yield

$$\left.\frac{\Delta\bar{\theta}}{\Delta t}\right|_{\text{conv}} = \frac{1}{\Delta p}[(\omega_{u2} + \omega_{d2})\bar{\theta}_2 - (\omega_{u1} + \omega_{d1})\bar{\theta}_1 + (\epsilon_u + \epsilon_d)\bar{\theta}_m - \delta_u\theta_{um} - \delta_d\theta_{dm}]. \tag{16.8}$$

From this expression, it can be seen that convective heating in a model layer is given by the sum of the fluxes of environmental potential temperature through the top and bottom of the layer (recall that $\tilde{\omega} \approx \omega_u + \omega_d$), minus the flux into convective drafts, plus the flux from the convective drafts into the environment, where the net mass flux into the layer is zero. In terms of total mass in a layer, any mass surplus or deficit created by entrainment into and/or detrainment out of convective drafts is exactly balanced by compensating fluxes through the top and bottom of the layer. In this way, the resolvable-scale grid in a numerical model "feels" updrafts and downdrafts within a grid element only indirectly. In contrast to FC, the grid-scale temperature in a layer is not in any way a function of the updraft or downdraft temperature in that layer unless there is active detrainment in the layer.

Following the same logic, one can arrive at an expression for the net tendency of specific humidity q_v due to subgrid-scale convection; that is,

$$\left.\frac{\Delta\bar{q}_v}{\Delta t}\right|_{\text{conv}} = \frac{1}{\Delta p}[(\omega_{u2} + \omega_{d2})\bar{q}_{v2} - (\omega_{u1} + \omega_{d1})\bar{q}_{v1} + (\epsilon_u + \epsilon_d)\bar{q}_{vm} - \delta_u q_{vum} - \delta_d q_{vdm}]. \tag{16.9}$$

In addition, liquid water detrainment from convective clouds supplies moisture to the resolvable scale. This process can be represented as a source of cloud water q_c, if used in a model with explicit prediction of cloud water,

$$\left.\frac{\Delta\bar{q}_c}{\Delta t}\right|_{\text{conv}} = -\frac{\delta_u q_{lum}}{\Delta p}, \tag{16.10}$$

where q_{lum} is the mean updraft liquid water mixing ratio in a layer. Alternatively, if cloud water is not an explicitly predicted variable on the resolvable scale, it is assumed that cloud water evaporates (sublimates) in the cloud environment, necessitating an additional source term in (16.9) and an evaporative (sublimative) cooling term in (16.8).

Finally, as in FC, momentum transport in convective clouds is crudely simulated by assuming conservation of momentum in convective drafts. This yields the corresponding momentum tendency equations

$$\left.\frac{\Delta\bar{u}}{\Delta t}\right|_{\text{conv}} = \frac{1}{\Delta p}[(\omega_{u2} + \omega_{d2})\bar{u}_2 - (\omega_{u1} + \omega_{d1})\bar{u}_1 + (\epsilon_u + \epsilon_d)\bar{u}_m - \delta_u u_{um} - \delta_d u_{dm}] \quad (16.11)$$

and

$$\left.\frac{\Delta\bar{v}}{\Delta t}\right|_{\text{conv}} = \frac{1}{\Delta p}[(\omega_{u2} + \omega_{d2})\bar{v}_2 - (\omega_{u1} + \omega_{d1})\bar{v}_1 + (\epsilon_u + \epsilon_d)\bar{v}_m - \delta_u v_{um} - \delta_d v_{dm}]. \quad (16.12)$$

Equations (16.8) and (16.9) are essentially equivalent to the flux form of the apparent heat source and moisture sink equations derived by Ooyama (1971), Yanai et al. (1973), Arakawa and Schubert (1974), and others, although they are derived under a slightly different set of assumptions. Use of the flux form is essential for conservation of advected quantities. Conservation of moisture and thermal energy also relies on an accurate representation of latent heating effects in convective updrafts and downdrafts, the formulations of which are described in the next section.

b. An entraining–detraining plume model of convective updrafts

Convective updrafts (and downdrafts) have been traditionally represented in CPSs by Lagrangian one-dimensional entraining plume (ODEP) models (e.g., Arakawa and Schubert 1974; Kreitzberg and Perkey 1976; Tiedtke 1989). These models are desirable because of their computational simplicity, but they are quite inflexible with regard to interactions between clouds and their environment. In particular, both entrainment and detrainment rates must be prespecified in an ODEP, which unrealistically restricts the vertical distributions of convective effects. Most significantly, the vertical heating profile and the vertical distribution of moisture detrainment are severely constrained by the prespecified parameters. Since numerous studies have shown that the impact of moist convection on larger-scale processes is extremely sensitive to these vertical distributions (e.g., Gyakum 1983; Hack and Schubert 1986; Kuo and Reed 1988), an updraft model that is responsive to variations in convective environments is desirable.

A new entraining–detraining plume model (ODEDP) that allows for more realistic cloud–environment interactions and thermodynamic processes, while introducing minimal additional computational requirements, is introduced with the KF scheme. The unique feature of the ODEDP is the mixing scheme that it uses to modulate updraft entrainment and detrainment rates. The scheme computes the buoyancy variations induced by turbulent mixing, in various proportions, between clear and cloudy air. It allows those mixtures that remain positively buoyant in each model layer to continue to rise with the updraft, while the mixtures that lose their positive buoyancy, through evaporative cooling effects, detrain into the environment. As discussed in detail in KF (1990), this scheme provides a realistic element of cloud–environment interaction so that vertical distributions of environmental entrainment, updraft detrainment, and net updraft mass flux can vary considerably as a function of the cloud-scale environment.

The new cloud model utilizes a more detailed representation of cloud microphysical processes than the original FC cloud model. Updraft thermodynamic processes are based on conservation of equivalent potential temperature θ_e (using Bolton's 1980 formula) and total water substance. Conversion of condensate to precipitation is simulated using a Kessler-type (Kessler 1969) autoconversion equation, as in Ogura and Cho (1973). Ice-phase thermodynamics are included, with a gradual transition between the liquid and ice phases occurring within a specified temperature interval. The transition to ice-phase thermodynamics requires an adjustment of θ_e values, as discussed in KF.

c. Convective downdrafts

The parameterization of convective downdrafts in the KF CPS contains a number of procedural differences from the downdraft formulation in FC. Most of these differences were implemented to assure conservation of all variables. Conceptually, however, the algorithms are very similar in KF and FC. Based on the empirical evidence of Foster (1958), downdrafts are initiated at the highest level below about 500 hPa at which a mixture containing equal parts of updraft and environmental air, when brought to saturation, becomes negatively buoyant with respect to the environment. Like the updraft, downdraft thermodynamics are based on conservation of θ_e (with an adjustment for melting effects), and evaporation of condensate is assumed to maintain a specified value of relative humidity at each level, typically 100% in the cloud layer and 90% below. Downdraft vertical velocity is computed using the buoyancy equation, and the downdraft is allowed to penetrate downward to the lowest layer where integrated buoyancy effects allow negative vertical velocity to be maintained.

The downdraft mass flux is related to the updraft mass flux through a precipitation efficiency relationship. The precipitation efficiency is determined by equally weighting estimates based on the vertical shear of the horizontal wind (as in FC) and cloud-base height (Fujita 1959; Zhang and Fritsch 1986). This efficiency estimate is applied to the total rate of precipitation generation in the updraft. The downdraft mass flux then corresponds to the maximum transport of mass that can be maintained at the specified relative humidity and over the estimated depth for the given availability of liquid water.

d. Convective environment

The KF scheme accommodates the influence of convective updrafts and downdrafts on their environment in a considerably different manner than the FC scheme. Detrainment of updraft temperature, moisture, and momentum are all calculated explicitly, as included in Eqs. (16.8)–(16.12). In the original FC scheme, the effects of updraft overshooting and detrainment on the temperature field are inferred from consideration of energy conservation above the equilibrium temperature level; the effects of updraft water vapor and momentum detrainment are not included. The total tendencies expressed by Eqs. (16.8)–(16.12) are held constant throughout the convective time period, rather than integrating forward in time as is done in the FC scheme, and there is no explicit area averaging as in FC.

16.3. Preliminary results

A primary motivation behind the development of the new cloud model and, to a lesser degree, the other modifications to the FC scheme was to expand the design of the original FC scheme to encompass a broader spectrum of convective environments. The efficacy of the FC scheme has been demonstrated in numerical simulations of continental mesoscale convective systems in the midlatitudes (e.g., Zhang and Fritsch 1986, 1988b; Zhang et al. 1989), but it has not been well tested in, for example, maritime tropical environments. However, preliminary results of diagnostic tests of the KF scheme in tropical environments are very encouraging. For example, consider the heating profiles diagnosed for the GATE [GARP (Global Atmospheric Research Program) Atlantic Tropical Experiment] environment by Frank and McBride (1989). These profiles are particularly well suited for comparison with heating profiles generated by the KF scheme because they are computed as a function of the stage of development of mesoscale convective systems (MCSs). Since the KF scheme parameterizes only the effects of deep convection (with the assumption that mesoscale stratiform components will be explicitly resolved), Frank and McBride's relatively high temporal resolution allows us to focus on the initial stages of MCSs, which were observed to be dominated by deep convection in the GATE environment (Houze and Betts 1981).

Figure 16.1a shows a composite of pre-MCS soundings for the GATE sounding array. Figure 16.1b shows the corresponding diagnosed apparent heat source for the initial stage (first 3 h) of MCSs in this environment. Figure 16.2a shows two heating profiles generated by the KF scheme with this input sounding. The dashed

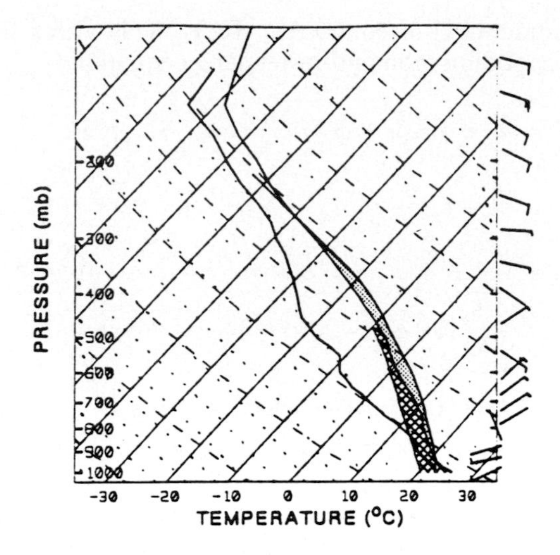

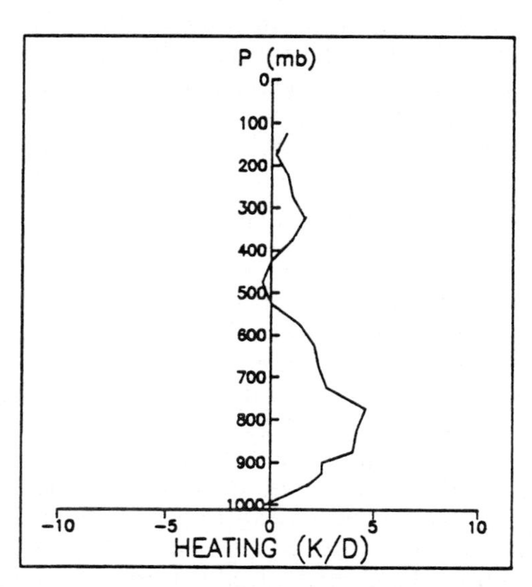

FIG. 16.1. Tropical (GATE composite) environmental sounding showing the positive area given by the new cloud model (stippled) and the downdraft negative area given by the downdraft plume model in the KF scheme (cross-hatched) (a) and diagnosed vertical distributions of diabatic heating from Frank and McBride (1989) for the initial stages of mesoscale convective systems in this environment (b).

profile is derived by turning off lateral detrainment (below cloud top) in the new cloud model so that it behaves like a simple entraining plume model. The solid profile is derived by allowing the new cloud model to execute normally. Clearly, the new cloud model has a substantial, and favorable, impact on the vertical heating profile generated by the KF scheme, with the parameterized vertical distribution matching the diagnosed distribution remarkably well. The magnitudes of the parameterized and diagnosed values differ considerably, but this is likely to be due to the disparity in

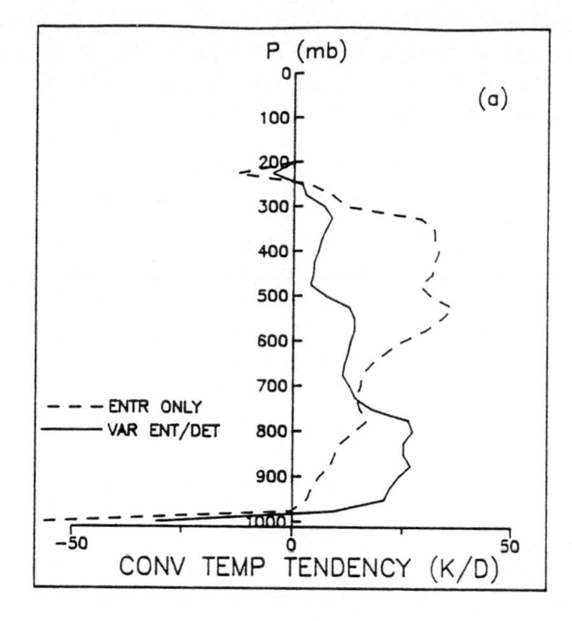

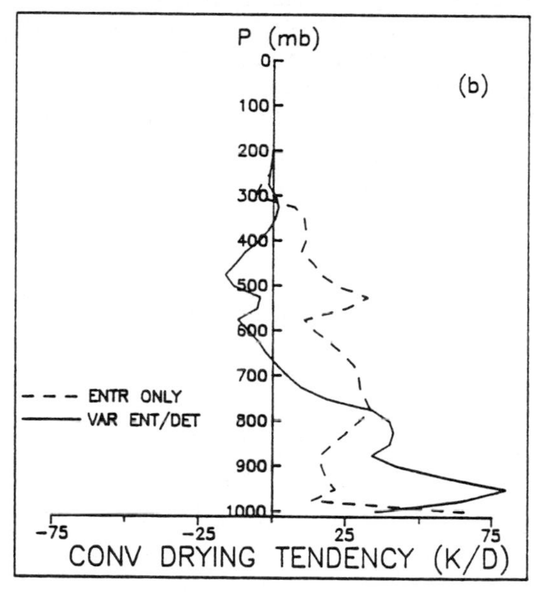

FIG. 16.2. Parameterized convective heating (a) and drying (b) profiles as a function of cloud model type for the GATE composite sounding shown in Fig. 16.1.

horizontal scale; the observational network has a horizontal scale of approximately 800 km, while the assumed horizontal scale of the parameterized convection is 25 km.

The parameterized convective drying profile derived using the new cloud model (Fig. 16.2b) is also consistent with the diagnosed distribution in this environment. The diagnosed drying maximizes at a lower level than the heating and becomes negative (moistening) in the mid- and upper troposphere during the early stages of the GATE systems (Frank, personal communication). The change in sign of the parameterized tendency is largely due to the detrainment of cloud hydrometeors, which is included as a moistening effect in the profiles shown Fig. 16.2b. The ability of the KF scheme to generate realistically this mid- and upper-tropospheric moisture source for resolvable-scale circulations is likely to be an important ingredient in the successful simulation of MCSs (Molinari and Dudek 1992). Numerous studies have indicated that deep convective clouds supply a significant fraction of the moisture that eventually falls as precipitation in the stratiform regions of MCSs (e.g., Leary and Houze 1980).

The KF scheme has been tested in a diagnostic mode in various other types of convective environments with equally encouraging results (KF 1990). A more thorough assessment of its performance is being carried out through testing in three-dimensional prognostic simulations (e.g., Kain and Fritsch 1992) and forecasts [in the "semioperational" version of the Pennsylvania State University (Penn State)–National Center for Atmospheric Research (NCAR) Mesoscale Model (Warner and Seaman 1990)] in various environments. In general, the preliminary results substantiate the basic hypothesis of FC (and Kreitzberg and Perkey 1976): a realistic parameterization of the intensity and vertical distribution of the effects of deep convection on the mesoscale can be achieved without regard to instantaneous larger-scale tendencies.

16.4. Considerations for further modifications

The continued rapid development of numerical models and methods must be accompanied by corresponding development of physical parameterizations if improvements in numerical weather prediction are to be expected. Listed below are two areas of change that are likely to be addressed in the KF scheme in the near future.

a. Feedback to the resolvable scale through convective mass sources and sinks

Derivation of Eqs. (16.8), (16.9), (16.11), and (16.12) requires an assumption that mass tends to be conserved in every model layer by vertical motions in the convective environment that exactly compensate for the vertical mass fluxes in convective drafts. The quantitative validity of this assumption becomes questionable as resolvable-scale grid lengths come down below the Rossby radius of deformation and approach the scale of individual convective clouds.

In general, a more realistic approach may be to solve for the compensating environmental motions on the resolvable scale by including convective mass source and sink terms in a resolvable-scale continuity equation. In hydrostatic models, however, this approach may still have serious drawbacks. For example, one could force compensating subsidence to occur on the resolvable-scale grid. Mathematically, this would be identical to the current approach for the first time step in a convective time period. In subsequent time steps, however, convectively induced subsidence would be vertically advecting quantities that are evolving with time at any given point. This would seem to be more realistic than the current approach, which feeds back the same values for convective tendencies at a given point at each time step during the convective time period. Yet, this approach would still force all of the compensation to occur locally, and in the vertical only.

Alternatively, one could introduce subgrid-scale mass sources and sinks in terms of horizontal pressure gradient forces. This could be implemented by adjusting the geopotential or pressure at each level and each time step to reflect the unresolvable vertical transports of mass into or out of a layer. This approach may be difficult to implement within the framework of current numerical models. Furthermore, it forces all of the compensating motions to occur through horizontal wind fields. Subsidence warming would presumably occur some distance away from the active convection, in contrast to both theory and observations (e.g., Lilly 1960; Fritsch 1975).

The mass source–sink type of feedback may be practicable only in nonhydrostatic models. Within the nonhydrostatic set of governing equations, mass sources and sinks can simultaneously induce responses in both the horizontal and vertical wind fields through the perturbation pressure field (Golding 1990). Current plans are to incorporate the KF scheme with this type of feedback in the nonhydrostatic version of the Penn State–NCAR model (Dudhia 1993).

b. Convective momentum transports and detrainment induced by horizontal momentum

Numerous studies have indicated that conservation of momentum in convective updrafts and downdrafts may be a poor assumption under some conditions (LeMone et al. 1984; Matejka and LeMone 1990b; Gallus and Johnson 1992). In particular, it appears that updraft and downdraft parcels can undergo substantial horizontal accelerations in response to local pressure gradient forces. A more sophisticated parameterization of convective momentum transport, such as that proposed by Zhang and Cho (1991), may be appropriate for the KF scheme.

A more realistic momentum parameterization may also allow for the implementation of a mass detrainment mechanism based on the differences in horizontal momentum between a cloud and its environment. Clearly, when convective updrafts rise through cloudy environments, some mechanism other than evaporatively induced negative buoyancy must be operative in the detrainment of updraft mass into the environment.

Acknowledgments. This work was supported by National Science Foundation Grants ATM-90-24434 and ATM-92-22017.

Chapter 17

A Method of Parameterizing Cumulus Transports in a Mesoscale Primitive Equation Model: The Sequential Plume Scheme

DONALD J. PERKEY AND CARL W. KREITZBERG

Department of Physics and Atmospheric Science, Drexel University, Philadelphia, Pennsylvania

17.1. Introduction

During the 1960s and early 1970s, parameterization of cumulus activity for use in cyclonic-scale forecast and extended-range climate numerical models was the topic of a great deal of research. Parameterization efforts for cyclonic-scale forecast models concentrated on the convective transport of temperature and moisture in the conditionally unstable atmosphere of the tropics (Lilly 1960), especially emphasizing the role of latent heat release in tropical cyclone development (Charney and Eliassen 1964; Kuo 1965; Ooyama 1969). In the global circulation and extended-range climate models, researchers focused their efforts on determining a quick, acceptable method of convective adjustment (Manabe and Strickler 1964; Arakawa et al. 1968; Oliger et al. 1970; Krishnamurti and Moxim 1971).

The above papers on tropical cyclone development stressed that the convection being parameterized was deep, organized, and undiluted. These assumptions may be acceptable for the mature and advanced stages of tropical cyclones but they have major drawbacks when dealing with the initial stages of convection in tropical disturbances and with the diversity of convection in extratropical cyclones. The most restricting assumptions are 1) the cloud is undiluted by entrainment; 2) all the condensed liquid water falls out as precipitation; and 3) the total convective cloud mass and, therefore, the rate of destruction of model-developed, larger-scale instability are controlled by the low-level, large-scale moisture convergence. This third assumption, in essence, equates the rate of destruction of instability by subgrid-scale, parameterized processes to the rate of creation of instability by resolvable-scale circulations.

In many cases, the assumption of undiluted transport leads to considerable overestimation of the cloud top. One such case is that of a moist layer capped by dry middle and upper layers, conditions that are commonly present in potentially unstable atmospheres. This undiluted temperature assumption was usually coupled with neglect of vertical momentum so that the cloud top was defined as the level where the cloud-base moist adiabat intersected the environmental lapse rate curve. However, in many cases the effects of entrainment and water loading kill the cloud buoyancy at much lower levels (Simpson 1971). Therefore, the undiluted ascent and zero vertical momentum assumptions led to errors in the estimate of which model levels were affected by cumulus transport. Because of this lack of accurate cloud-top definition, some authors stated at the time that only three-layer models were justified (Charney and Eliassen 1964; Ooyama 1969).

Because convective clouds through latent heat release are a principal source of energy in a conditionally unstable atmosphere and because the vertical scale of convection is the same order of magnitude as the vertical scale of cyclonic disturbances, the vertical distribution of heating caused by convective latent heat release must be properly parameterized. The resulting circulation of the large-scale cyclonic system is significantly different when influenced by low-level heating rather than by upper-level heating. For example, if heating occurred in the upper troposphere, the induced circulation would consist of strong outflow in a shallow upper layer and weak inflow of air throughout a deep lower layer. However, if the heating occurred in the lower troposphere only, strong inflow would be expected in a shallow low layer with weak outflow in a rather deep layer above.

Although both of these types of circulations lead to conversion of potential energy into kinetic energy, their roles in cyclone development are quite different. The strong inflow in lower layers associated with low- and midlevel cloud tops enhances boundary-layer convergence of mass and moisture that in turn enhances future convective activity and cyclone development. The deep convection associated with the mature stage of cyclone development provides the vertical energy transport to higher altitudes necessary for the maintenance of the cyclone kinetic energy. Thus, proper formulation of the cumulus cloud depth and vertical distribution of latent heat release is important in distinguishing different stages of weather system evolution and can have significant dynamic effects (e.g., see Gyakum 1983; Hack and Schubert 1986). With the arrival of larger and faster computers, higher resolution in both the horizontal and vertical was feasible, but prediction of convective bases and tops needed to be improved before the increased vertical resolution could be fully exploited.

The second assumption, complete condensate fallout, does not allow cumulus moisture transport in the liquid form. It is apparent that not all of the cloud condensate falls out but that some of it is advected by the cloud vertical velocity to higher levels where it evaporates and mixes into the environment (Houze 1977; Zipser 1977). Thus, the assumption of complete condensate removal by precipitation leads to underestimation of the cumulus moisture transport. In some cases, this can lead to the wrong sign on the convective induced upper-level temperature changes. With no liquid transport, the temperature changes caused by the convection are positive while the inclusion of liquid transport followed by evaporation reduces the warming and, in some cases, even causes cooling at upper levels where much of the cloud water is left behind.

Two concerns arise with the third assumption: 1) the rate of convective-scale destruction or release of instability may not be controlled by the rate of large-scale convergence of moisture or supply of large-scale instability, and 2) while it may be acceptable to assume that cumulus have their roots in the surface boundary layer in the tropics, cumulus in extratropical cyclones are not necessarily rooted in the boundary layer. Fritsch et al. (1976) observed that instability was often created over periods of days by large-scale processes and then released by convective-scale processes over periods of hours. While it may be true that over the long haul the creation and release of instability must balance, this is not necessarily the case over periods of hours. It is also true that because vertical instability in midlatitude systems is very sensitive to the vertical velocity, mesoscale circulations that have much stronger vertical velocities than large-scale circulations must be either resolved or parameterized correctly in order to properly determine the convective-scale latent heat release (Frank 1983).

Some of the concern with the assumption of low-level moisture convergence could be alleviated by assuming that the total cloud mass is proportional to the total three-dimensional moisture convergence in a vertical column, but this still does not account for mid-level-based cumulus that have their roots in a frontal zone (Kreitzberg and Brown 1970). Because the frontal surface may be considered as a "material" boundary, the convergence or divergence of moisture below the frontal surface does not affect the convective development above the surface or front. Thus, before moisture convergence relationships could be applied to extratropical cyclones, they at least had to be modified to include parameters that related to the height of the cumulus base and the possible existence of atmospheric layers that act as material boundaries.

As was stated earlier, the extended-range climate models were in need of a quick, acceptable method of convective adjustment. Through the early 1970s, researchers concentrated on developing adjustment schemes based on conservative properties. The quantity being conserved was either internal energy for dry convection (Manabe and Strickler 1964; Oliger et al. 1970) or static energy, which is defined as the sum of internal, potential, and latent energy for moist convection (Arakawa et al. 1968; Krishnamurti and Moxim 1971). Fundamentally this approach had the same restrictions as the tropical-cyclonic-scale adjustment schemes discussed above. In addition, most of these methods did not incorporate the effect of environmental subsidence about the active convection. In the case of unorganized tropical cumulus, the subsiding return flow induced by convection is an important modifying and energy transporting phenomenon. In extreme cases the dry return flow can actually dissipate the disturbance (Zipser 1970).

Possibly the most damaging assumption was the almost total neglect of mesoscale circulations. Mesoscale circulations contribute significantly to the atmospheric transport of heat, moisture, and momentum (for example, see Maddox 1980). In general, numerical models of this era were low-resolution (grid intervals of order 300 km) models with parameterization schemes based on the properties of individual clouds. Thus, the mesoscale was neither explicitly simulated nor explicitly included in parameterization schemes.

Since the early 1970s, many investigators have continued to work on the parameterization problem both by modifying schemes developed before that time and by developing new schemes. Examples that are discussed elsewhere in this volume include the Kuo parameterization scheme (Kuo 1965, 1974), the Arakawa–Schubert scheme (Arakawa and Schubert 1974), and the Fritsch–Chappell scheme (Fritsch and Chappell 1980). The Kuo parameterization scheme, partly because of its simplicity, has been modified and used by many investigators for many different meteorological conditions (e.g., Anthes 1977; Molinari 1982; Krishnamurti et al. 1983). As with the sequential plume scheme, the new schemes were developed with certain needs in mind; the Arakawa–Schubert scheme focused on tropical convection and aimed at global circulation models, while the Fritsch–Chappell scheme focused on improving the understanding of mesoscale weather systems.

The parameterization requirements for global-scale models are somewhat different from the requirements for mesoscale models, climate model requirements are different from those for forecast models, and the requirements for tropical models vary from those of midlatitude models. For example, global-scale models with their large grid areas must to some extent parameterize not only the convection that occurs on the cumulus scale but must include the convection on the mesoscale, that is, organized mesoscale systems. On the other hand, mesoscale forecast models wish to explicitly predict the mesoscale organization of the convective systems. Climate models are particularly sensitive to all three phases of water (vapor, liquid, and ice) transports by convection so that their radiation schemes can properly account for the radiatively active water substance. Because of their short forecast period, mesoscale forecast models are not generally thought to be as sensitive to radiation effects. Tropical convection is more likely to be surface rooted than midlatitude convection and, thus, may be somewhat more restricted in its variety.

It was with these historical and physical perspectives in mind that the sequential plume cumulus parameterization scheme was developed. This scheme was developed to be used in a regional-scale or mesoscale primitive equation model with a 35-km horizontal grid and 15 vertical levels between the surface and 16 km. This primitive equation model explicitly carried cloud water and rainwater as dependent variables. Note that, starting from large-scale initial conditions, this model, with its 35-km grid interval, can explicitly resolve the development of mesoscale transports; thus, the subgrid-scale convective parameterization scheme is relieved of this function. In this discussion we will attempt to provide some further insights into the sequential plume scheme. Section 17.2 will describe the sequential plume scheme, while section 17.3 will review results of simulations that used the sequential plume parameterization scheme.

17.2. Description of the sequential plume parameterization scheme

It is not the intent of this section to provide a complete explanation, either descriptive or mathematical, of the sequential plume model but simply to provide an overview of the scheme and then concentrate on describing the main areas where this scheme varies from other schemes and why and how this scheme attempts to address some of the concerns stated above. For a complete description of the scheme see Kreitzberg and Perkey (1976, 1977). Figure 17.1 shows a schematic flowchart of the sequential plume model.

At the heart of the sequential plume scheme is a one-dimensional Lagrangian updraft model with somewhat detailed dynamics and microphysics that consider condensation and freezing latent heat release, cloud and precipitating water, entrainment of environmental air (temperature, moisture, cloud and rainwater, and horizontal momentum) by the updraft, water loading, and vertical inertia of the updraft. One of the input quantities of the sequential plume scheme is the initial updraft radius (typically 3 km) that influences the updraft entrainment rate; that is, small radius updrafts entrain a larger percentage of their volume than larger radius updrafts. The entrainment rate affects the vertical development of the updraft and can introduce a limiting factor to the updraft depth.

Also included in the sequential plume model is modification of the updraft model cloud water and rainwater profiles by rainout. This updraft model with rainout is used to determine the depth of the cloud and the temperature, specific humidity, cloud water, and horizontal momentum properties of the convective cloud volume as well as the convective rainfall amount. Note that horizontal momentum effects were not included in Kreitzberg and Perkey (1976); horizontal momentum is currently included as a passive quantity.

To determine the cloud base with which to initiate the updraft model, the environmental temperature and humidity sounding (i.e., a single vertical column obtained from the primitive equation model) is searched for releasable instability, that is, a quantity related to the positive area between the environmental sounding and a simplified "cloud" sounding. In this case the simplified "cloud" sounding is the appropriate moist adiabat without the complexities of water loading, freezing, etc. Each level in the sounding is tested for possible releasable instability and the level with the most instability is selected as the Lagrangian updraft model's cloud base.

Upon completion of determining the cloud volume properties, the environment is modified by subsidence in such a manner as to conserve mass; that is, the amount of subsidence at a given level is calculated such that the subsiding mass of environmental air at a given level equals the mass of convective cloud air that rises through the level. Thus, the mass of cloud air moving up through any given layer is compensated by an equal amount of subsiding environmental air. In order to accomplish this step, the fractional area of cloud must be evaluated. The assumption upon which this evaluation is based will be discussed below. At this point, one has two soundings, one representing the cloud volume and the other representing the modified environmental air. These two volumes are mixed according to the calculated percentage of cloud air and environmental air.

One now has a new cloud-modified environmental sounding that is tested again to determine if further convection is possible. If further convection is possible, that is, the sounding still contains releasable instability, the above process is repeated; if no further releasable instability is present, that is, the releasable instability has been removed from the sounding, then the differences in temperature, moisture, cloud water, rainwater, and horizontal momentum between the initial environmental sounding and this resultant cloud-modified environmental sounding are calculated and converted into tendencies by dividing by a characteristic time scale of the convection (typically 40 min). The parameterization is not invoked every time step, but is typically

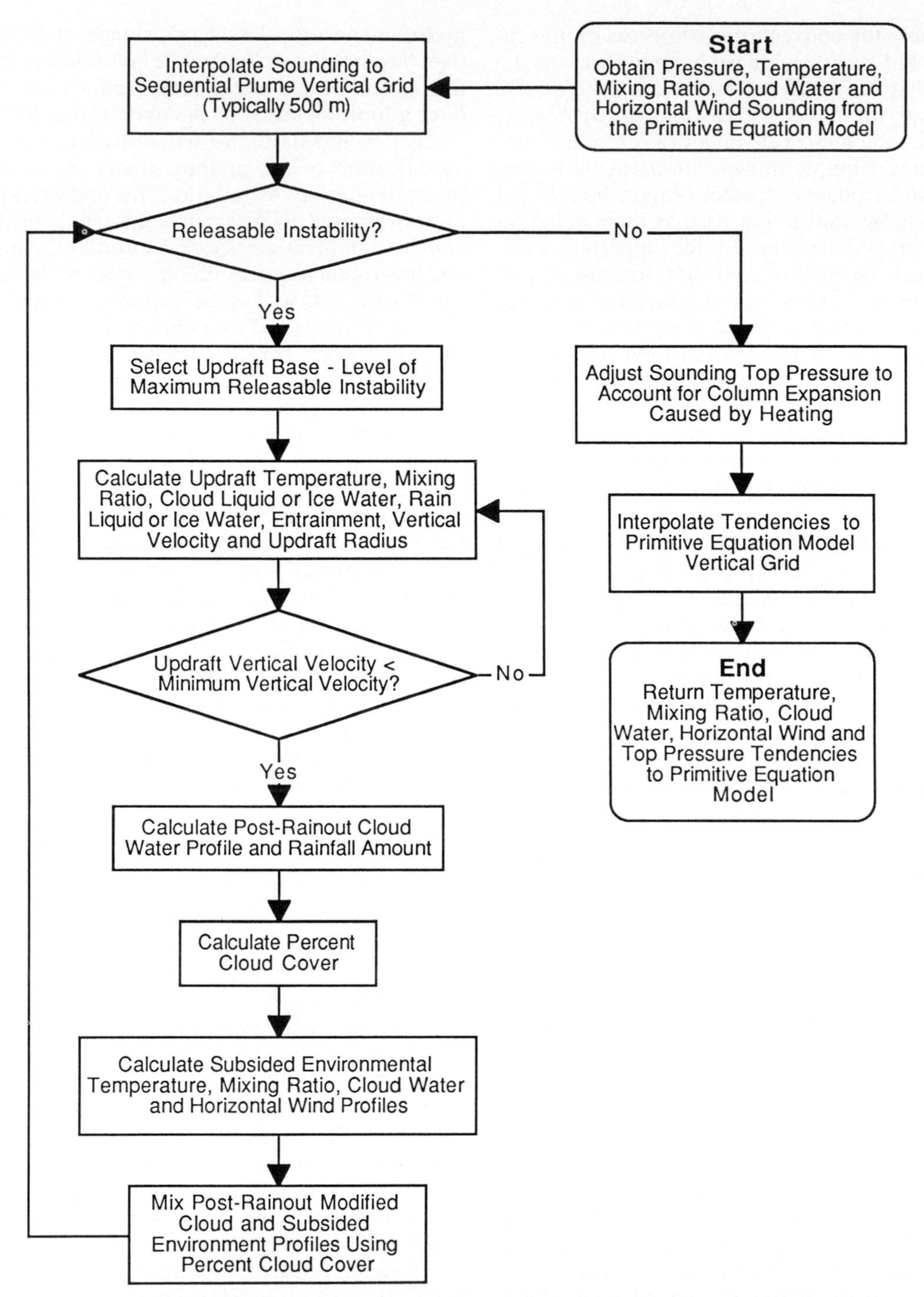

FIG. 17.1. Schematic diagram of the sequential plume parameterization scheme.

employed only every 20 min and the changes induced by convection are then averaged with those from the previous convective calculation. A 1–2–1 horizontal smoother is also applied in both the x and y directions. In other words, the convective tendencies at any given time step in the primitive equation model are a horizontally smoothed combination of convective parameterization calculations at two times.

Throughout the calculations involved in the sequential plume scheme, great care has been taken to assure conservation of mass, water substance, and energy. With this overview of the sequential plume scheme in mind, we will discuss some of the critical assumptions present in the model.

a. Updraft calculation

The updraft portion of the cloud model employed by the sequential plume model is based on the updraft models similar to those discussed by Simpson and Wiggert (1969) and Simpson (1971). These updraft models account for condensate- and freezing-derived thermal buoyancy, condensate drag, entrainment, aerodynamic drag, and vertical momentum. The top of the updraft occurs when the updraft vertical velocity decreases to near zero. Thus, although not perfect, these models yield a better estimate of the cloud top than is available by simply using a cloud-base moist adiabat. This, as stated above, becomes more critical as the primitive equation model's vertical resolution is increased.

b. Precipitation rainout

The microphysical processes of updraft conversion of cloud water to rainwater, collection of cloud water by rainwater, and evaporation of rainwater along with rainwater fall velocity are parameterized following Kessler (1969). Before the updraft air is mixed with the subsided environment, the cloud water profile is modified by allowing the cloud's rainwater to fall back through the cloud and collect cloud water. This process washes out much of the lower-level cloud water and increases the precipitation that reaches the ground. Again, this process follows the parameterization suggested by Kessler (1969). It is important to note that because less rain falls through the upper levels of the cloud than through the lower levels, more of the cloud water is left behind in the cloud volume at upper levels than at lower levels.

Although the mixing process was not one of the primary concerns mentioned above, the relative importance of this process has grown since the original formulation of the scheme. In particular, in recent years there has been increased awareness of the potential influence convective moisture sources may have on the model atmosphere's radiative budget and of the climatic impact of this interaction. There is no doubt that convection not only transports temperature and vapor but that it also transports liquid water. The sequential plume model as stated above mixes cloud and environmental cloud water, that is, the cloud updraft profile of cloud water that is left after rainwater rainout is mixed with subsided environmental air. The liquid water then either evaporates into the subsided air if the subsided air's relative humidity is less than 100% or remains as liquid water in the environment. Thus, this scheme often results in cooling and moistening of unsaturated environmental air or additions of hydrometeors to the environment, usually in the form of ice as the cloud top is frequently above the environmental freezing level. This process, of course, could not happen with parameterization schemes that assume complete condensate rainout.

c. The closure assumption

The closure assumption of any parameterization scheme is directly related to the demands placed on the scheme by the physical processes being parameterized. As stated above, in the convective areas observed in midlatitude occluded frontal zones (Kreitzberg and Brown 1970) the low-level moisture convergence does not control the total cloud mass and thus the rate of stabilization of the large-scale atmosphere. Midlatitude convective systems are not necessarily in balance with the large-scale convergence of moisture; convective "fuel" can build over hours and even days before it is released by convection rather suddenly (Fritsch et al. 1976). Thus, the sequential plume scheme employs a local closure assumption that is based on the assumption that, when the convection has completed its life cycle, the pressure of the modified environmental air should equal that under the cloud. More specifically, the hydrostatically integrated pressure from the top of the cloud to the ground through the cloud volume should be equal to that through the subsided environment. This is equivalent to saying that the hydrostatically integrated temperature from cloud top to ground in the cloud volume should equal that of the subsided environment; that is, the integrated cloud warming caused by condensation and freezing latent heating should equal the environmental warming caused by subsidence around the cloud.

17.3. Review of simulation results using the sequential plume parameterization scheme

This section presents a brief summary of a few simulations, using both idealized and real data, that used the sequential plume model as an integral part of a primitive equation model. The sequential plume model has been used to simulate many other cases, but the cases surveyed here were selected because parameterized convection was particularly important to the overall simulation.

a. Release of potential instability simulations

Kreitzberg and Perkey (1977) employed the sequential plume scheme within a two-dimensional primitive equation model to investigate the scale interaction between the large scale, mesoscale, and convective scale. Their results demonstrate how subgrid-scale parameterized convection can evolve into hydrostatic, mesoscale, resolvable-scale convection during a 12-h simulation. This evolution is driven by moistening and warming on the resolvable scale by the subgrid-scale parameterized transports. After several hours in which parameterized precipitation- and latent heat–driven transports dominated, a saturated, neutrally buoyant, mesoscale, resolvable-scale updraft developed (Kreitzberg and Perkey 1977, p. 1593).

The evolution of this updraft was driven by the following steps. 1) The sequential plume scheme created subgrid-scale precipitation in response to potential instability; 2) the resultant latent heating and subsidence warming (Kreitzberg and Perkey 1976, Fig. 4a, p. 471) caused the column to expand, thereby creating an area of high pressure at or near the top of the parameterized convection; 3) this tendency toward high pressure was transferred to the primitive equation model that in turn began to develop upper-level divergence; 4) as a result of the upper-level divergence, the surface pressure began to fall and low-level convergence was initiated; 5) this locally enhanced convergence increased the upward vertical velocity that in turn created resolvable-scale moistening by vertical advection. This resolvable-scale moistening more than countered the drying resulting from the subgrid-scale transport. In this case, the subgrid-scale transport was dominated by cloud-induced environmental subsidence drying rather than by the cloud updraft moistening (Kreitzberg and Perkey 1976, Fig. 4b, p. 471). Given continued large-scale, low-level convergence and upward vertical motion to provide moistening to destabilize the atmosphere and thus to maintain the subgrid-scale precipitation, in time the resolvable scale eventually became saturated throughout the middle levels and, therefore, became neutral to saturated vertical motion.

In light of the discussion in section 17.2b, it should be noted that early in the simulation the subgrid-scale transports created a cirrus layer, first by the moistening and cooling resulting from evaporation of subgrid-scale convective ice and then later by the addition to the already present resolvable-scale ice of subgrid-scale ice (Kreitzberg and Perkey 1976, p. 467, and Fig. 4c, p. 471). This resolvable-scale, upper-level cirrus layer developed before the saturation of the middle levels resulting from the processes described above.

Thus, in this example, large-scale, upward vertical motion initiated unorganized, subgrid-scale convection that in turn eventually developed into organized resolvable-scale convection through the release of potential instability.

b. Squall-line primitive equation simulations

Perkey (1976) demonstrated that the sequential plume scheme embedded within a primitive equation model successfully simulated a squall line propagating ahead of a cold front. In this particular case, a narrow band of enhanced moisture had to be included in the initial fields before the squall line could be simulated. Visual satellite pictures supported this enhancement. These pictures showed a narrow band of convectively active clouds between rawinsonde sites that was missed by the standard observations. Note the initial horizontal wind field, and thus, the initial convergence–divergence fields were unaltered. However, because the added moisture increased the potential instability and the releasable instability in the sequential plume scheme, convective-scale latent heat transports were activated and led to the simulation of the squall line.

Numerical simulations of the 6 May 1975 squall line suggested that an area of potentially unstable air in conjunction with low-level convergence initially fed this squall line (Chang et al. 1981). In addition, wet full-physics and dry (no latent heat; subgrid or resolvable scale) fine-mesh simulations "suggested that the rate of the cyclonic-scale occlusion processes was increased by convective latent heating and that these occlusion processes removed the low-level warm moist air which supported the convection . . ." (Chang et al. 1981, p. 1614). In mesomesh (35-km grid interval) simulations, Chang et al. observed that "convective heating was responsible for the strengthening and narrowing of the squall-band updraft. They continue by noting that as the squall band intensified, the subgrid-scale convection consumed low-level warm moist air and the resultant subgrid-scale convective heating grew into the upper troposphere. This upper-level heating reduced the model atmospheric lapse rate and thus stabilized the atmosphere. As a result of a more stable atmosphere, lower-tropospheric ascent decreased and the squall band subsequently dissipated. Chang et al. further speculated that "this sequence of events may explain the short life of squall lines in occlusions compared with those along cold fronts. Cold-frontal squall lines have a continuing supply of low-level warm moist air in the warm sector" (Chang et al. 1981, p. 1614). Because these events are clearly an example where the subgrid-scale convective stabilization and utilization of low-level fuel exceeded the rate of supply of fuel by the large scale, it is likely that this result could not have been derived using a parameterization scheme that required a balance between the low-level moisture convergence and the convective cloud mass.

c. Cyclogenesis simulations

Chen et al. (1983) compared full-physics simulations with dry three-dimensional primitive equation model simulations of an oceanic cyclone. They attributed the enhanced cyclogenesis observed in the full-physics simulation to the latent heat release (both subgrid-scale convection and resolvable-scale precipitation) enhanced vertical transports resulting from the upper-level high and low-level low pressure caused partially by subgrid-scale midlevel warming. This vertical communication or coupling through subgrid-scale convection and resolvable-scale vertical motions between the low-level and upper-level systems helped maintain the westward vertical tilt favorable for cyclogenesis.

Studying a midlatitude developing wave cyclone, Chang et al. (1982) discuss the effect of latent heat release (largely the result of subgrid-scale convection). Using the results of wet-minus-dry simulations, they indicated that the net effect of latent heat caused relatively cold temperatures and high geopotential heights at 700 mb, warm temperatures and low heights at 500 mb, and warm temperatures and high heights at 300 mb. Enhanced upward vertical motion was also present in the warm region at 500 mb. These patterns acted to increase the conversion of potential to kinetic energy especially in the lower to middle troposphere (Chang et al. 1984). The largely subgrid-scale convective latent heat release "stabilized the troposphere and reduced the large-scale temperature gradient." The result of this effect "decreased the intensity of the large-scale circulation" and "provided energy for the intensification of the regional-scale closed circulation" (Chang et al. 1982, p. 1570). Again we see the result of subgrid-scale latent heat release, primarily through its vertical transport of temperature, interact with and influence the mesoscale and large scale.

d. Mesoscale convective complex simulations

Maddox (1980) defined mesoscale convective complexes (MCCs) as highly organized, meso-α-scale convective systems. Numerical simulations of one of these systems using a primitive equation model with the sequential plume scheme indicated that subgrid-scale latent heat release created a relatively (wet minus dry simulations) warm 500-mb region with enhanced upward vertical motion. This region that was warm at midlevels exhibited increased 200-mb divergence and low-level convergence. With time, this subgrid-scale-driven system evolved into a resolvable-scale, moist-ascent, convective region with 200-mb jet-streak structure similar to those observed attending MCCs.

17.4. Summary

The sequential plume cumulus parameterization scheme was designed with specific requirements in mind. During the late 1960s and early 1970s, as computers began to increase in speed, memory, and storage size, increased model resolution, both horizontal and vertical, became feasible. With increased model vertical resolution, cumulus parameterization schemes needed to be able to predict the height of convection with increased accuracy. Therefore, more detailed cloud models, as compared to simply the moist adiabat, were required to predict the cloud top and cloud properties such as water loading, heat of fusion, and cloud water content. The sequential plume model incorporates a Lagrangian cloud model into the scheme to perform these calculations.

As the majority of primitive equation, real-data models of the time did not explicitly contain predictive equations for liquid water substance, both cloud-droplet size and precipitating-sized drops, parameterization schemes did not need to consider the interaction between resolvable and parameterized liquid and solid water substance. However, it was obvious that if improvements were to be made in increasing the accuracy of quantitative precipitation forecasts, then primitive equation models were going to have to incorporate more detailed cloud and precipitation microphysics. Therefore, the sequential plume scheme was the first cumulus parameterization scheme to explicitly require interaction between the primitive equation model's resolvable-scale cloud water and the parameterized cloud water. This interaction was accomplished through the detrainment of liquid or frozen cloud substance from the parameterization scheme.

Because of the Kreitzberg and Brown (1970) observations of occluded fronts, it was obvious that in order to be useful for midlatitude quantitative precipitation forecasts, cumulus parameterization schemes had to permit clouds that were not in balance with the large-scale, low-level moisture convergence. Also, as accurate quantitative precipitation forecasts required calculation of precipitation on a time scale of a couple of hours, time balance between the large scale and cumulus scale was not appropriate. Thus, the sequential plume scheme developed a local closure that centered on the integrated slice concept.

The specific cases of primary interest at the time the sequential plume scheme was developed did not emphasize moist cumulus downdrafts. However, because of the type of weather being studied, Fritsch and Chappell (1980) were among the first to emphasize the importance of the moist cumulus downdrafts in a parameterization scheme. As stated in Kreitzberg and Perkey (1976), the sequential plume model, because of its modularity, can be modified to include a parameterization component that includes the downdraft process. It is certainly true that for some types of convection the inclusion of moist downdrafts would be critical.

As of today, questions concerning the importance of convective vertical momentum transport remain largely unanswered. As stated above, the sequential plume scheme mixes horizontal momentum in the vertical as if this momentum were a passive property. However, it is evident from observations and cumulus cloud models that the momentum processes involved with convection should include not only transport but also creation of momentum.

Acknowledgments. Work by Dr. Kreitzberg was sponsored by the Office of Naval Research Grant N0014-86-K-0203. Work by Dr. Perkey was sponsored jointly by the Office of Naval Research Grant N0014-86-K-0203 and NASA–MSFC Grant NAG8-714.

PART IV

Representation of Convection in Climate Models

Chapter 18

Efficient Cumulus Parameterization for Long-Term Climate Studies: The GISS Scheme

ANTHONY D. DELGENIO AND MAO-SUNG YAO

NASA/Goddard Institute for Space Studies, New York, New York

18.1. Introduction

The Goddard Institute for Space Studies general circulation model (GISS GCM) differs from most other general circulation models in that it is designed for use exclusively on global climate change problems. Typical applications include assessments of the climatic response to increasing greenhouse gas concentrations and volcanic eruptions, evaluations of unforced variability of the atmosphere–ocean system, paleoclimatic simulations, and process studies of other planetary atmospheres.

The decadal time scales and subtle variations of the hydrologic cycle characteristic of such integrations place special demands on the physical parameterizations used in the model. For moist convection, the requirements are as follows. 1) Faithful representation of moisture and heat transports, sources and sinks, which determine water vapor, lapse rate, and cloud feedbacks in a changing climate. This necessitates a mass flux approach to parameterization, in which a calculated mass of subcloud air rises to its neutral buoyancy (detrainment) level and is compensated by downward-moving air that effects most of the heat and moisture transport. But in light of everything that has been learned about convection from field studies since the first mass flux schemes appeared in the late 1960s, it is imperative that we go beyond the traditional cumulus updraft–compensating subsidence rendering of cumulus dynamics. 2) Flexibility to respond to a changing climatic state. This argues against schemes that empirically adjust the vertical structure to specified current values of CAPE (convective available potential energy) or the cloud work function, as well as those that adjust the humidity to a fixed profile. 3) Computational efficiency to allow the model to be run on general purpose computers or workstations. This requires us to seek the simplest possible parameterization that can capture the essential physics of processes occurring in convective clusters. Given a physically viable framework, increasing detail can be incorporated as computational power inevitably increases, although such decisions necessarily involve trade-offs with the desire for increased fidelity in other physical parameterizations (e.g., boundary layer, ground hydrology, sea ice) and increased horizontal and vertical resolution.

18.2. The basic scheme: Model II

The operational version of the GISS GCM, known as Model II (Hansen et al. 1983), utilizes a simple mass flux cumulus parameterization. Moist convection is diagnosed to occur whenever a parcel of air lifted adiabatically from some level l is buoyant with respect to the environment at the next highest level. In terms of the moist static energy h, the instability criterion can be written

$$h_c = h_l > h^*_{l+1}, \qquad (18.1)$$

where the subscript c indicates the cloud parcel, and the asterisk the saturation value for the environment. Virtual temperature corrections to the buoyancy are neglected. The parcel rises to the highest consecutive level at which the buoyancy criterion is satisfied. Convection can be initiated from all levels below the tropopause. A fixed 50% of the mass of the cloud-base layer rises in each event. There is no entrainment into the rising plume, and thus one cloud top per cloud base.

Latent heat release serves only to maintain plume buoyancy; heating/cooling of the environment occurs by means of compensating environmental subsidence, detrainment of cloud air at the cloud top, and reevaporation of falling precipitation. The tendency of dry static energy s due to a single convective event can thus be written (in σ coordinates)

$$\frac{\partial s}{\partial t} \approx -M_c\frac{\partial s}{\partial \sigma} + (s_c - s)\frac{M_c}{\Delta\sigma}\delta(\sigma - \sigma_T) - LE, \quad (18.2)$$

where M_c is the cumulus mass flux in units appropriate to σ coordinates, σ_T the cloud-top level, $\Delta\sigma$ the layer thickness, L the latent heat of evaporation/sublimation, and E the evaporation rate. An analogous equation applies for the moisture tendency. Water condensed in the updraft at each level is not transported upward, but precipitation is allowed to reevaporate as it falls. Evaporation is calculated by making 25% of the mass of each (generally unsaturated) layer above cloud base available to the falling precipitation. All condensate is allowed to evaporate until either there is none left or the specified layer fraction saturates. This ensures that convective rain will not saturate any full GCM layer. The condensate remaining after that fraction of each layer saturates determines the convective precipitation that falls to the next level. Below cloud base, a larger fraction (50%) of the layer is made available for evaporation to indirectly account for the fact that some rain falls through convective clouds and cannot evaporate above cloud base. Precipitating ice melts immediately upon crossing the 0°C level.

The convective plume and subsiding environment transport grid-scale horizontal momentum. Convective cloud cover is assigned as proportional to the mean pressure thickness of all model layers up to cloud top; the visible optical thickness $\tau = 8$ per 100-mb depth of the cloud. All types of convection are predicted by the same criterion; differentiation between deep and shallow depends only on the cloud buoyancy.

18.3. Improvements for Model III

a. *Mass flux closure*

The quasi-equilibrium concept of cumulus interactions with the large-scale environment is well supported by available observations (Arakawa and Chen 1987). For climate models, though, the most appealing strategy is to develop a closure that produces quasi equilibrium as an output rather than specifying it as an input. This approach allows for the possibility of small yet potentially important climatic changes in CAPE or the cloud work function.

The fixed mass flux of Model II does not produce quasi equilibrium, because it reacts only as an "on–off" switch to variations in the rate of destabilization by large-scale processes. For Model III, currently under development, we have adopted a simple alternative. Since the scheme is triggered when a lifted parcel is buoyant with respect to the next highest layer, we transport enough mass to just neutralize the instability at cloud base, assuming an adjustment time equal to a physics time step (1 h); the required mass flux is obtained by iteration on an initial guess (Yao and DelGenio 1989). Reevaporation of falling precipitation is adjusted so that the available layer fraction above cloud base is half the fraction of the subcloud layer that rises in the updraft. Instability is judged by a modified version of requirement (18.1) that includes the effects of water vapor and condensate on buoyancy. Sensitivity tests show that doubling the adjustment time merely decreases the instantaneous mass flux while increasing the frequency of convective events, resulting in an almost identical monthly mean climate.

This closure constrains only the conditions at cloud base; the vertical structure is free to vary according to the model physics as represented by requirement (18.2). Thus, an appropriate test of the model is its ability to generate quasi-equilibrium variability. Arakawa and Chen (1987) have shown that in quasi equilibrium, variations in the degree of conditional instability of the troposphere (diagnosed, for example, by the vertical gradient of h^*) should be negatively correlated with variations in near-surface relative humidity (indicated by the boundary-layer value of $h - h^*$). Figure 18.1 demonstrates that the Model III closure does in fact exhibit such behavior at both tropical land and ocean locations.

Analyses of satellite data from ISCCP (International Satellite Cloud Climatology Project) show that, given the vertical resolution of current climate models, there are rarely more than two deep convective cloud-top levels present instantaneously within a GCM grid-box-sized area (DelGenio and Yao 1987). Thus, rather than admitting an entire spectrum of cumulus cloud tops at any moment, we allow for two cloud tops per cloud-base level in Model III (Yao and DelGenio 1989). The plumes are differentiated by entrainment rate; one is undilute, mimicking a convective core, while the mass flux in the other grows fractionally with height at a rate of $0.2\ \mathrm{km}^{-1}$. The nonentraining plume receives a fraction of the total cloud-base mass flux given by the large-scale convergence at cloud base; this allows in effect for a lower overall entrainment rate in situations conducive to organized cluster formation as opposed to isolated airmass thunderstorms. The two-plume configuration has several advantages: 1) it yields the observed bimodal spectrum of cumulus cloud tops in the time mean and the possibility of simultaneous deep and shallow convection without invoking a separate shallow convection parameterization; 2) it allows for some of the convection within a grid box to reach the tropopause, as observed, while producing the midtro-

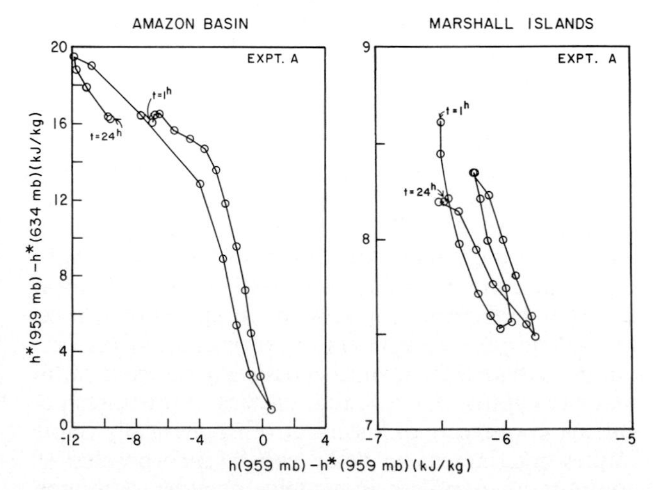

FIG. 18.1. Quasi-equilibrium behavior produced by a neutral buoyancy cumulus mass flux closure in the GISS GCM (Yao and DelGenio 1989).

posphere maximum in cumulus heating characteristic of heavily convecting regions in the tropics.

b. Convective downdrafts

GCMs with mass flux cumulus parameterizations commonly exhibit excessive cumulus heating and drying in the lower troposphere. This is a serious deficiency for models applied to climate problems, in which changes in low cloud cover and/or optical thickness can dominate cloud feedback. One possible reason for this problem is that the motions that compensate a cumulus updraft consist of more than just environmental subsidence. It is known, for example, that convective-scale downdrafts driven by precipitation loading and evaporation are a ubiquitous feature of precipitating convective systems. Downdraft air typically originates in the middle troposphere, with characteristics similar to those of air near the moist static energy minimum, and is carried to the surface. The downdraft is usually cool and dry and therefore fundamentally alters the convective stability of the planetary boundary layer (Barnes and Garstang 1982).

Downdrafts are the result of complex microphysics and mesoscale dynamics in convective clusters and are therefore difficult to represent both realistically and efficiently in GCMs. For GISS Model III we have implemented the first operational downdraft parameterization in any GCM (DelGenio and Yao 1988). For convective events penetrating more than two levels above cloud base, we simply test as the plume rises for the first level (if any) at which an evaporatively cooled equal mixture of cloud and environmental air is negatively buoyant. Such mixing is expected at midlevels of convective systems, where cloud-scale and mesoscale low pressure cause dynamic entrainment of air with low moist static energy from outside the cloud. For typical environmental conditions, roughly equal mixtures of cloudy and clear air are most likely to be negatively buoyant. If such a level is found, a downdraft forms there with the properties of the mixture. Once formed, the downdraft penetrates to cloud base, evaporating precipitation formed at lower levels to the extent necessary to remain as close to saturation as possible. The downdraft mass flux is specified to be one-third of the updraft mass flux, based on estimates from field studies (Johnson 1976) and mesoscale models (Simpson et al. 1982). Environmental subsidence below the downdraft formation level is thus reduced by a factor of 1/3, while M_c is reduced above the downdraft formation level by the mass of cloud air mixed into the downdraft.

Figure 18.2 (upper) illustrates that the inclusion of downdrafts significantly cools the planetary boundary layer in the GISS GCM; one consequence of this is an increase in the mass flux associated with shallow convection. Furthermore, even though downdrafts dry the boundary layer, they are moist relative to the environmental subsidence they replace. As a result, low-level relative humidity in the tropics increases when downdraft effects are included in the GCM (Fig. 18.2, lower). This has important possible ramifications for climate sensitivity. Most GCMs exhibit a positive component of cloud feedback due to decreasing low cloud cover in a warming climate (DelGenio 1993). This effect is probably associated with increased cumulus subsidence drying. To the extent that downdrafts offset subsidence drying, they may act to produce a more neutral cloud cover component of cloud feedback. In sea surface temperature perturbation experiments with the GISS GCM, for example, Model II cloud cover decreases by 1.3% in response to a 4°C warming, while a newer version of the GCM that includes downdrafts exhibits only a 0.4% decrease.

c. Mesoscale cirrus anvils

Cumulus clouds occupy a small fraction of the area of the tropics and therefore have a negligible radiative

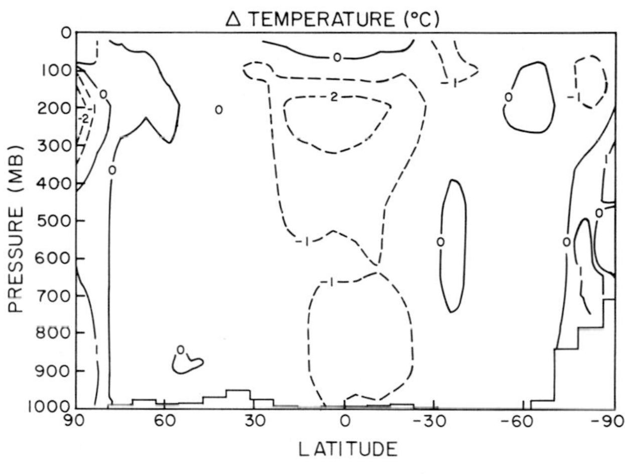

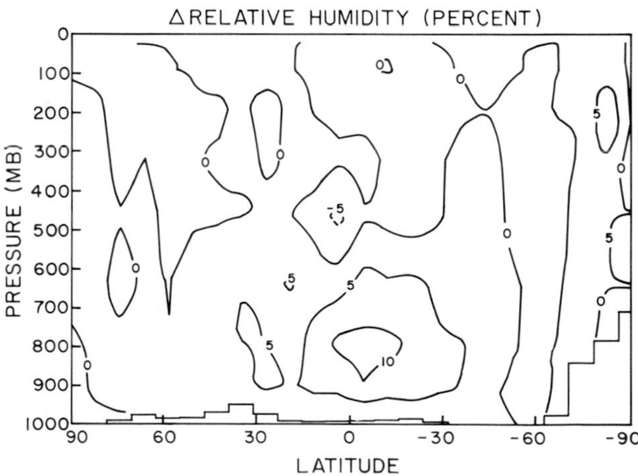

FIG. 18.2. Zonal mean distribution of the change in January temperature (upper) and relative humidity (lower) caused by the inclusion of a recent version of a convective downdraft parameterization in the GISS GCM.

impact. Convection is often organized, however, into mesoscale clusters capped by upper-troposphere anvils with long life cycles that are evident in satellite images (Fu et al. 1990). The detrainment term in mass flux convection schemes injects saturated vapor into the environment at cloud top and thereby triggers cirrus cloud formation. The clouds are typically too thin, though, and their frequency underpredicted.

Part of the problem is that anvils are dynamic; mesoscale updrafts within anvil clouds enhance condensation and produce significant stratiform precipitation (cf. Houze and Betts 1981). In addition, some of the ice found in anvils is not formed locally but instead is detrained from accompanying cumulus updrafts, at all altitudes above the freezing level. GISS Model III parameterizes these clouds with the aid of a prognostic cloud water budget for stratiform clouds (DelGenio and Yao 1990). In prognostic schemes, condensed water is carried as a predicted variable. Rather than forcing clouds to dissipate (evaporate or precipitate) in the same time step as they are formed, the scheme predicts the tendency of cloud water content due to microphysical sources (condensation) and sinks (autoconversion, accretion, evaporation). To simulate cumulus anvils, we add all convective condensate produced four model levels or more above cloud base (typically, from 550 mb to the tropopause) to the stratiform cloud water budget. The convective condensate thus injected into the stratiform anvil is treated as cloud water and evolves in time, precipitating only partially in a single time step but more efficiently as the mass of the anvil increases. This allows the anvils to persist for hours even if convection has ceased, and it permits us to utilize an interactive calculation of the anvil optical thickness, which has a dramatic effect on cloud feedback. Specifically, since the anvils at any level thicken as the climate warms, they reduce climate sensitivity by offsetting the Model II tendency for deeper convection to produce thinner cirrus (DelGenio 1993).

The GCM representation of anvil clouds is compared with in situ aircraft data acquired over Kwajalein in the tropical west Pacific (Heymsfield and Donner 1990) in Fig. 18.3. Despite its simplicity, the scheme produces a realistic temperature dependence of tropical ice water content. In part, this may be the result of radiative heating within the anvil that drives a grid-scale version of a mesoscale updraft in the upper troposphere. As a result, the stratiform fraction of tropical precipitation increases from less than 5% in Model II to a more realistic 15%–20% in the new version of the GCM. It is therefore not obvious that a separate mesoscale parameterization is required in climate GCMs.

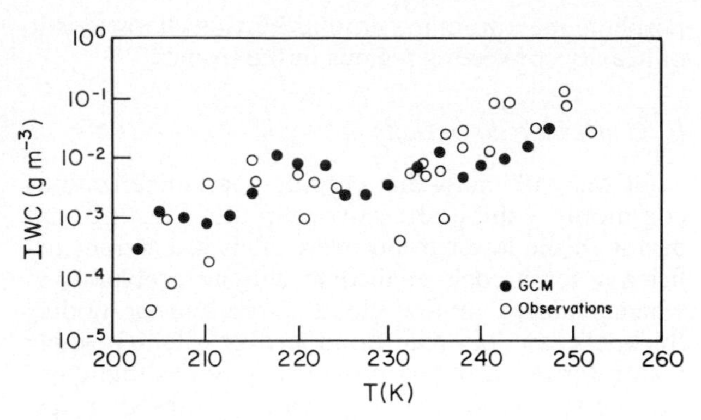

FIG. 18.3. Tropical ice water content versus temperature produced by the GISS cloud water budget parameterization (filled circles) compared with observations (open circles) from Heymsfield and Donner (1990).

We note that all the parameterization changes described in this paper have almost no effect on the combined contributions of water vapor and lapse rate feedbacks to climate sensitivity (DelGenio et al. 1991). Even more remarkable is the fact that *all* GCMs without exception produce similar water vapor–lapse rate feedbacks, regardless of whether they use drying mass flux schemes or moistening Kuo or convective adjustment schemes, and regardless of whether they have 2 or 20 vertical levels (Cess et al. 1990). This surprising result occurs because changing large-scale dynamical transports of heat and moisture always offset changes in cumulus heating and drying to produce almost constant relative humidity on climatic time scales (though not on shorter synoptic time scales). We therefore recommend that the emphasis in future cumulus parameterization research for climate models be directed toward quantifying the effects of moist convection on cloud feedback, which is much more sensitive than water vapor feedback to small errors in relative humidity. We also conclude that while one-dimensional models continue to be useful for physical process studies, three-dimensional models that incorporate the effects of large-scale dynamics explicitly will ultimately be the only reliable tool for inference of climate feedbacks.

Acknowledgments. We thank C. Shashkin for technical typing support. GCM physics development at GISS is supported by the NASA Climate Program.

Chapter 19

A Cumulus Representation Based on the Episodic Mixing Model: The Importance of Mixing and Microphysics in Predicting Humidity

KERRY A. EMANUEL

Center for Meteorology and Physical Oceanography, Massachusetts Institute of Technology, Cambridge, Massachusetts

19.1. Introduction

The observations of cumulus clouds presented in Part I of this monograph weigh strongly against the entraining plume model of clouds. In particular, aircraft observations show that the thermodynamic properties of clouds are extremely inhomogeneous, with a strong tendency for buoyancy sorting (chapter 2). This basic notion of mixing in clouds solves the paradox raised by Warner (1970) that while clouds entrain much mass from their environments they still ascend to the level predicted by undilute ascent from cloud base.

While it is clear that the entraining plume model is inadequate, it is not clear what to replace it with. One may consider that the entraining plume lies at one end of a spectrum of models, being based on the assumption that entrained air is immediately mixed with the cloudy air to form a completely homogeneous mixture at all levels. At the opposite end of the spectrum lies the model of Raymond and Blyth (1986), in which a uniform spectrum of mixtures is formed at each level of the cloud with each mixture then ascending or descending adiabatically to its new level of neutral buoyancy, where it is detrained.

The entraining plume model and the stochastic mixing model of Raymond and Blyth clearly constitute extreme assumptions about mixing in clouds; what actually happens must lie somewhere between these extremes. Attempts to use a more realistic model of mixing are at present seriously handicapped by lack of an observational basis for a more sophisticated scheme and by the fact that significantly more computation would be necessary in using such a scheme.

In view of these limitations, I have constructed a cumulus representation based on the episodic mixing model of Raymond and Blyth, which seems somewhat more realistic than the continuously entraining plume model. My model differs slightly from Raymond and Blyth's in that the sequence of mixing events is truncated differently, and precipitation may form and partially reevaporate in an unsaturated downdraft. The scheme discussed herein contains a number of small improvements over the one described in detail in Emanuel (1991).

19.2. The cloud model

The mixing tree for the cloud model is described in Fig. 19.1. As in the Raymond and Blyth scheme, air is assumed to ascend adiabatically from the subcloud layer to each level i between cloud base (level ICB) and cloud top (level INB). (The means of determining *how much* mass ascends to each level i will be described shortly.) Departing from Raymond and Blyth, who considered only nonprecipitating clouds, I assume that an amount of condensed water $\epsilon^i l_a^i$ is converted to precipitation. Here l_a^i is the adiabatic water content at level i and ϵ^i is the fraction of l_a^i converted to precipitation. This must be specified as a function of temperature, pressure, cloud depth, and/or the adiabatic water content itself; the set ϵ^i contains the principal parameters of this scheme.

After precipitation is removed, the remaining cloudy air is mixed with its (undisturbed) environment, forming a spectrum of mixtures. This spectrum of mixtures is formed by mixing in different fractions of environmental air, under the assumption that there is the same probability that a given mixing fraction will occur as any other mixing fraction. Each mixture then ascends or descends to its new level of neutral buoyancy, as in the Raymond and Blyth scheme. In that scheme, the mixed air is assumed to detrain to the environment. But if the mixture still contains condensed water, this will evaporate on detrainment, and the cooled mixture will descend. Thus, in principle, the mixing tree in Fig. 19.1 should be continued until all condensed water has been removed.

As a practical matter, it becomes computationally too expensive to continue much beyond the first or second mixing event, and it is necessary to truncate the mixing tree. I do so as follows. First, a spectrum of mixed parcels is formed by mixing the undilute cloudy air with its environment at level i, as just described. Then the mixed air ascends or descends according to its buoyancy. If it ascends, and if the amount of con-

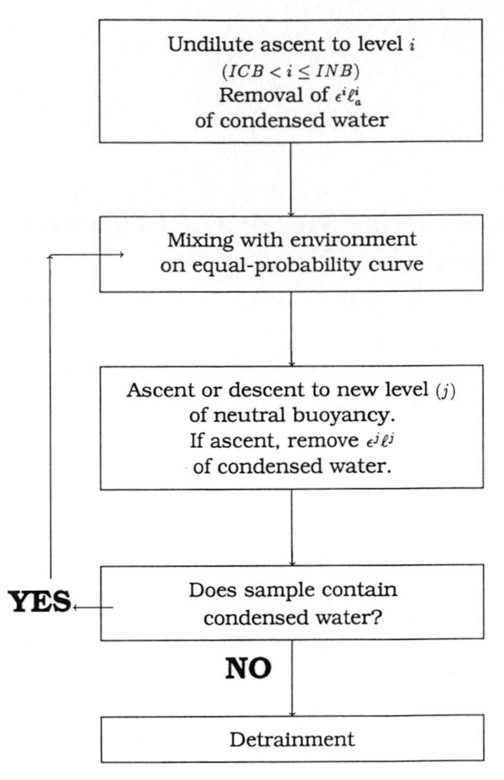

FIG. 19.1. Flowchart for episodic mixing model.

densed water in the mixture exceeds that of an undilute parcel lifted to some level *after* precipitation removal, then the excess condensed water is converted to precipitation. The amount of water converted to precipitation is then

$$P^{ij} = l^{ij} - \epsilon^j l^j_a, \tag{19.1}$$

where l^{ij} is the amount of condensed water formed in displacing the mixed air from level i to level j.

The detrainment level j is then chosen so that the detrained air has no buoyancy on mixing with its environment, assuming that all cloud water evaporates on mixing. The condition that is imposed is that

$$h^{ij} = [c_{pd}(1 - Q^{ij}) + c_{pv}Q^{ij}]T^j + gz^j, \tag{19.2}$$

where Q^{ij} is the total specific water content of the mixed air displaced from i to j after precipitation removal, and h^{ij} is the specific liquid water static energy of the mixture, also after precipitation removal. The liquid water static energy is

$$h = [c_{pd}(1 - Q) + c_{pv}Q]T + gz - Ll, \tag{19.3}$$

where Q is the total mass of water substance per unit mass of air, c_{pd} and c_{pv} are the heat capacities at constant pressure of dry air and water vapor, respectively; T is temperature, g is the gravitational acceleration, z is altitude, L is the (temperature dependent) latent heat of vaporization, and l is the mass of condensed water per unit mass of air. The liquid water static energy is conserved following slow displacements of a fluid parcel. It will also be unchanged following a more rapid displacement if the kinetic energy associated with the displacement is all converted back to heat. Note that the quantity $h - gz$ is the specific enthalpy. Also note that the condition (19.2), given the definition of h (19.3), does not imply that the temperature of the detrained air is exactly equal to that of the environment. It is easy to show that, by coincidence, this condition is much closer to (but not exactly the same as) the condition that the *virtual* temperatures of the detraining air and environmental air are equal, since

$$[c_{pd}(1 - q) + c_{pv}q]T \simeq c_{pd}T_v,$$

where T_v is the virtual temperature.

The requirement (19.2) determines the level j to which each mixture will ascend or descend. Virtually all of the descending mixtures will have no condensed water at j, so that (19.2) is almost equivalent to demanding that the mixture descends to its new level of neutral buoyancy. Mixtures that ascend to a level j for which $\epsilon^j = 1$ will also have no condensed water, so that (19.2) is also equivalent to a neutral buoyancy requirement for these mixtures. In intermediate cases, the air will detrain while still having positive buoyancy, but will lose it on mixing with the environment, due to evaporation of condensed water. While slightly unphysical in these cases, imposing (19.2) ensures that all detrained air (with one exception to be discussed presently) has no buoyancy and is thus "finished" convecting.

The one exception is that as there are a finite number of levels i, a particular mixture may not be able to ascend or descend all the way to the next level; it is then assumed to detrain at level i. This air will not necessarily satisfy (19.2). As the vertical resolution of the model decreases, this outcome becomes more likely. Essentially, a model with poor vertical resolution cannot distinguish between air that detrains at positive buoyancy with finite condensed water content and air that first mixes with its environment, then descends or ascends a distance less than a grid interval, and detrains at neutral buoyancy with no condensed water. Whatever the resolution, the appropriate integral constraints on heat and water will be obeyed, however.

A further consideration is that when $j > i$, there may be multiple levels j at which (19.2) is satisfied for a particular mixture. (This is because some condensed water is converted to precipitation, changing the value of h^{ij}.) In that case, the mixture detrains at the *highest* level for which (19.2) is satisfied.

The precipitation detrained at each level i for which $\epsilon^i > 0$ is added to a solitary precipitation shaft. A fraction σ^i_s, currently assumed constant, is assumed to fall outside of cloud into unsaturated air. [Cheng and Arakawa (1990) have developed a more sophisticated scheme for determining σ^i_s, accounting for effects of ambient shear.] This fraction evaporates at a rate E^i, depending on the precipitation content and the saturation deficit of the air. The evaporation cools the air and is assumed to drive a steady, unsaturated downdraft, which transports heat and water vapor downward. In this step, several additional parameters are introduced, including σ_s^i, as already discussed, the fall speed of the precipitation, and the fractional area σ_d occupied by the precipitation shaft. (If the evaporation depended linearly on the precipitation content, the parameterization would be independent of σ_d; I assume a square-root dependence of evaporation on precipitation content, thus introducing a square-root dependence of the result on σ_d. In practice, σ_d is chosen to give reasonable values of the precipitation content.)

The unsaturated downdraft proves to be a critical element of the scheme. It strongly cools and dries the subcloud layer, and moistens the air above, producing net moistening profiles quite different from the result of omitting such downdrafts.

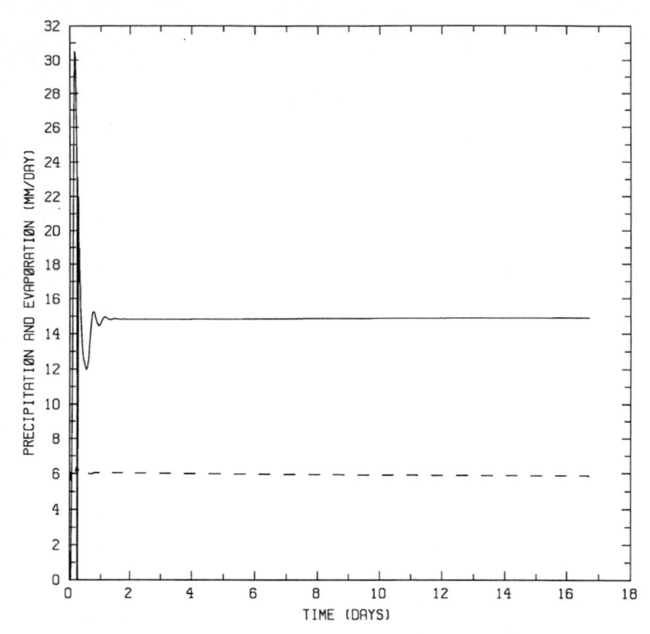

FIG. 19.2. Evolution with time (h) of the convective precipitation (solid) and surface evaporation (dashed), in millimeters per day.

19.3. The closure

With the above specifications, all that remains to be determined are the undilute mass fluxes M^i from the subcloud layer to each level i between cloud base and cloud top, including the latter. For computational reasons, I use a "soft quasi equilibrium" scheme that tends to drive these fluxes in the direction of quasi equilibrium without rigorously enforcing quasi equilibrium at each time step.

First, I represent each flux M^i as the product of a fractional area σ^i and an updraft mass flux per unit area:

$$M^i = \sigma^i \rho^i w^i, \qquad (19.4)$$

where ρ^i is the density at level i and w^i is the undilute updraft velocity at level i, calculated from the parcel's convective available potential energy (CAPE) for lifting to level i.

Now σ^i may be thought of as the fractional area occupied by the undilute updraft ascending to level i, or equivalently, as the frequency that such an updraft occurs over an interval of time. (It is actually the product of frequency and fractional area.) Suppose we were to fix σ^i at a constant value and run the parameterization within a model in which the large-scale forcing of convection (e.g., radiative cooling and surface heat fluxes) is fixed. After a short time, the mass fluxes M^i come into equilibrium so that the forcing is exactly balanced by the convective stabilization. They do this largely by allowing the CAPE to vary until the updraft velocities w^i yield the right mass fluxes M^i from (19.4). Obviously, as one makes σ^i smaller, the updraft velocities w^i and thus the CAPE will have to increase so as to compensate. On the other hand, something different happens if σ^i is increased. For small increases of σ^i, the updraft velocities w^i decrease to compensate. But beyond a critical value of σ^i, a further increase results *not* in a further diminution of w^i but in a decrease of the fraction of model time steps at which convection occurs. Large σ^i results in strong convective fluxes at infrequent intervals.

The strategy here is to drive the σ^i toward their critical values, so as to produce "smooth" convection (as opposed to violent bursts every so many time steps) while minimizing the system's CAPE. To accomplish this, we adjust σ^i according to

$$\frac{\partial \sigma^i}{\partial t} = \alpha(\Delta T_v)^i_{\min} - \beta \sigma^i, \qquad (19.5)$$

where α and β are constants, and

$$(\Delta T_v)^i_{\min} = \min_{\text{ICB}<j<i} (T^j_{vp} - T^j_v),$$

where ICB is the first model level above cloud base, T^j_{vp} is the virtual temperature of a parcel lifted adiabatically from the subcloud layer to level j, and T^j_v is the environmental virtual temperature at level j.

The effect of the first term on the right of (19.5) is to increase σ^i until the minimum buoyancy of a lifted parcel below level j approaches zero, at which point the convection would become episodic. The last term in (19.5) damps the evolution of σ^i. Integrations begin with $\sigma^i = 0$ and σ^i is reset to zero if there has been no convection for ten time steps (approximately 3 h).

The result of using (19.4) with (19.5) is a smoothly evolving convective forcing with minimal CAPE. Under conditions in which CAPE is stored up over a long time (i.e., there is a restraining inversion) and then

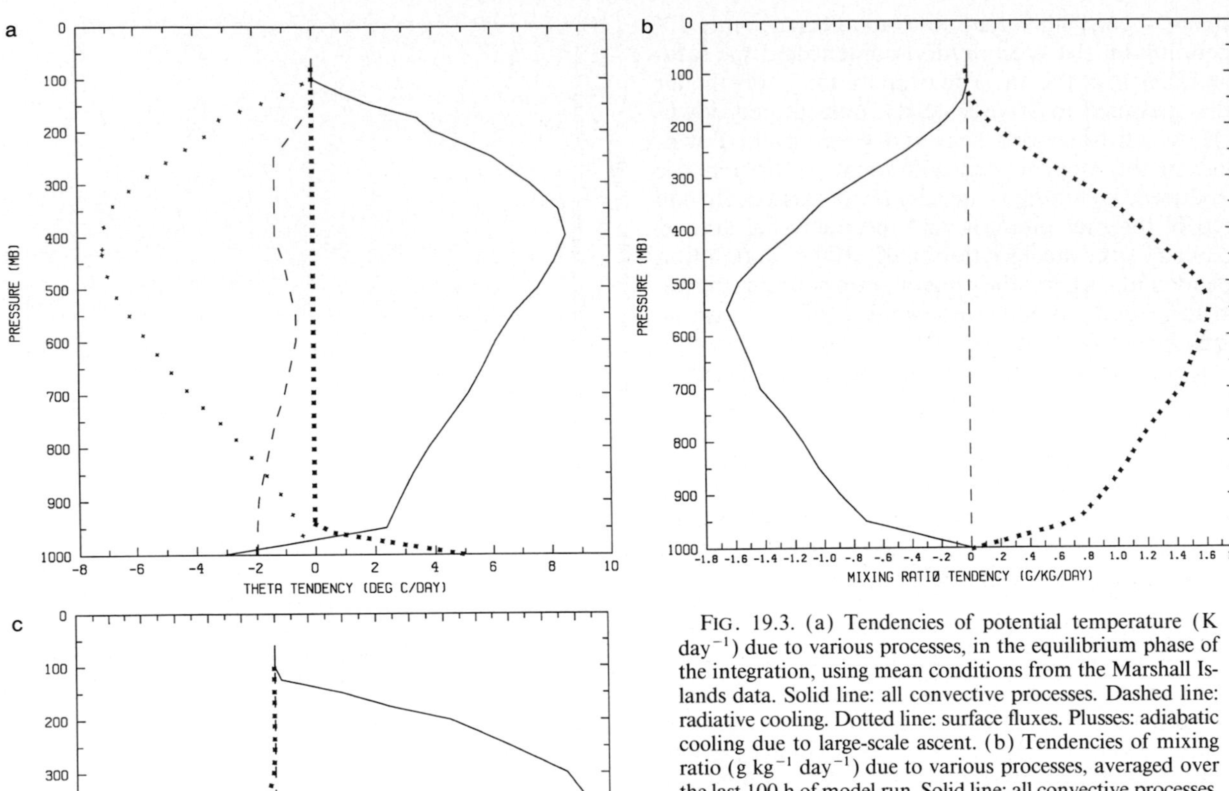

FIG. 19.3. (a) Tendencies of potential temperature (K day^{-1}) due to various processes, in the equilibrium phase of the integration, using mean conditions from the Marshall Islands data. Solid line: all convective processes. Dashed line: radiative cooling. Dotted line: surface fluxes. Plusses: adiabatic cooling due to large-scale ascent. (b) Tendencies of mixing ratio (g kg^{-1} day^{-1}) due to various processes, averaged over the last 100 h of model run. Solid line: all convective processes. Asterisks: moistening due to large-scale ascent. Convective drying and moistening by surface fluxes at the lowest level not plotted. (c) Net mass fluxes (10^{-3} kg m^{-2} s^{-1}) through each pressure level due to various convective drafts, averaged over the last 100 h of simulation. Solid line: all updraft mass fluxes, including entrained air flux. Dashed line: penetrative downdraft mass flux. Dotted line: mass flux in unsaturated downdraft. (d) Mass entrainment (dashed) and detrainment (solid) at each level. (e) The fraction of environmental air in a mixture entraining at level i (abscissa) and detraining at level j (ordinate) with the same liquid water static energy as environmental air at j. (f) The difference between virtual temperature (including condensate loading) of lifted subcloud-layer air and environmental virtual temperature, averaged over the last 100 h of the simulation. Units are kelvins. (g) As in (f) but for relative humidity.

released, (19.4) and (19.5) will produce a burst of strong convection over a time interval that depends on the value of α; otherwise, under quasi-equilibrium conditions, the results are quite insensitive to the values of α and β.

19.4. Some results

A simple (but incomplete) test of a cumulus scheme is to run it under conditions of constant forcing in a one-dimensional framework. This I do by specifying the radiative cooling and vertical velocity in the troposphere and by determining surface fluxes using the aerodynamic flux formulas with constant wind speed and sea surface temperature. The vertical velocity and radiative cooling profiles are specified from the Marshall Islands dataset as in Yanai et al. (1973), and with the sea surface temperature set at 27°C and the surface wind speed at 5 m s^{-1}. Note that the profiles calculated using the average forcing are not the same as the average of the profiles determined from the time-varying forcing. The scheme is run with a provisional specification of parameters. The fraction of rain falling into unsaturated air is 20%, the fractional area covered by the precipitation shaft is 1%, and ϵ^i is specified so that ϵ^i is 0 within 150 mb of cloud base, 1 above 500 mb above cloud base, and linearly varying between these two draft depths.

The evolution to equilibrium of precipitation and evaporation is shown in Fig. 19.2. Both quantities approach equilibrium over a period of about a day. The steady-state distributions of various quantities are

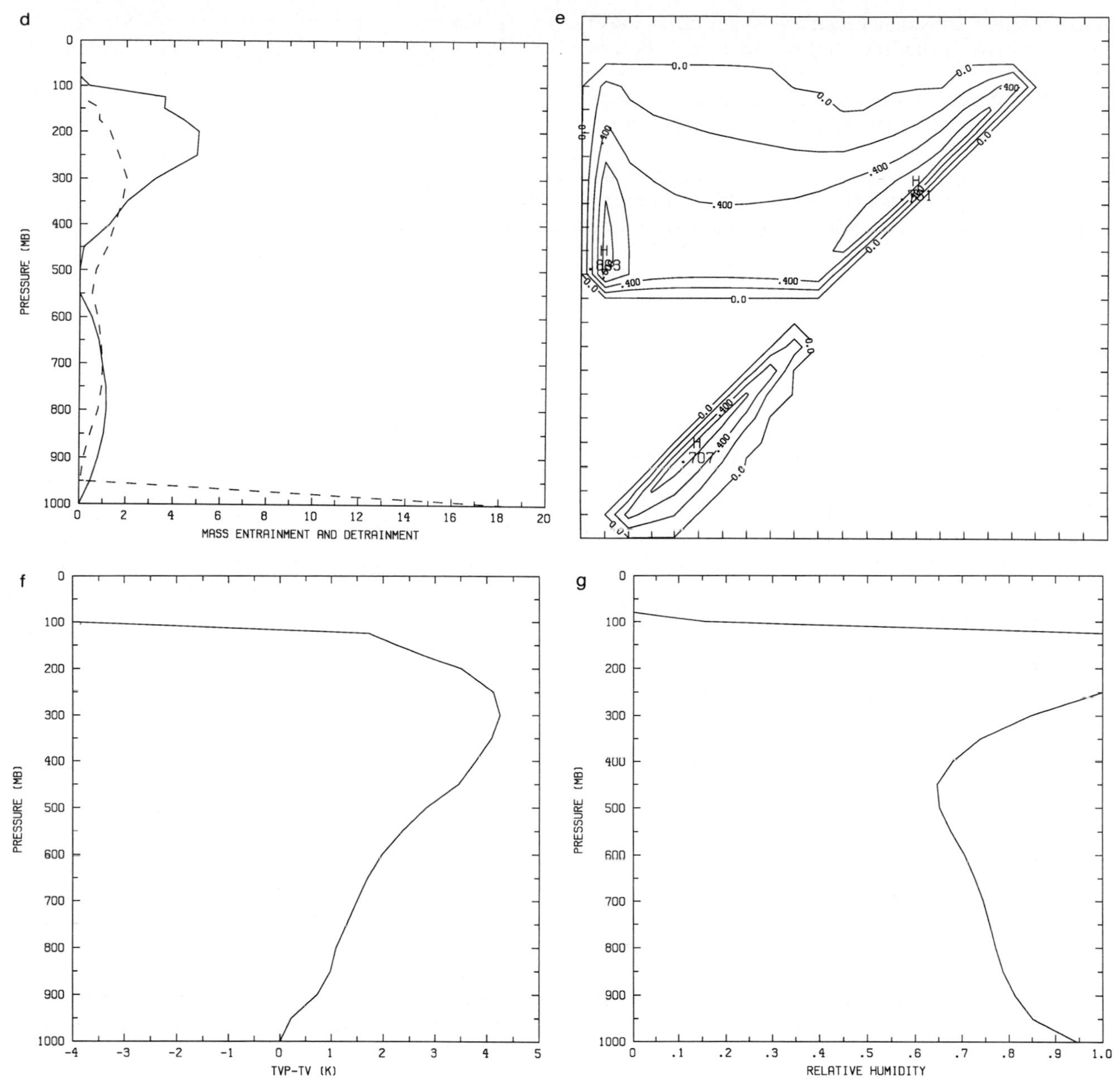

FIG. 19.3. (*Continued*)

shown in Fig. 19.3. Above the subcloud layer, radiative and adiabatic cooling are balanced by convective heating, while within the subcloud layer the convergence of sensible heat fluxes from the surface is an important contribution. (The negative contribution from convection results from the evaporation of rain.) The moistening of the atmosphere by large-scale ascent is balanced by convective drying. (Boundary-layer tendencies are not shown in Fig. 19.3b.) The mass fluxes show the substantial influence of entrainment and of saturated and, particularly, unsaturated downdrafts. The rate of entrainment and detrainment at each level is shown in Fig. 19.3d. The diagram in Fig. 19.3e shows the mixing fraction of environmental air in mixtures formed at level i and detrained at level j. A small amount of air entrained at low levels ascends to high levels; most of it descends in penetrative downdrafts. Higher up, all entrained air ascends in updrafts. The buoyancy of undilute air lifted from the subcloud layer is positive at all levels: small in the lower troposphere and increasing near the tropopause. The increase is necessary to allow for the net entrainment needed to increase the mass flux with height, which in this case is necessary to balance the radiative and advective forcing. The relative humidity in equilibrium has a minimum near 450 mb. The air is dry here because of environmental subsidence but moistens up at lower elevations due to shallow clouds and evaporation of precipitation.

The equilibrium relative humidity profile is of crucial importance in climate simulations, for which upper-tropospheric humidity has a strong effect on radiative transfer, both directly and through its influence on the formation of high, stratiform clouds. This equilibrium profile is, as one might expect, very sensitive to the microphysical parameters ϵ^i and σ_s^i. Figure 19.4 shows the equilibrium profile for an experiment identical to the control except that $\epsilon^i = 1.0$ everywhere, and no evaporation of rain is permitted. Since $\epsilon^i = 1.0$, there are no unsaturated downdrafts. This makes the scheme quite similar to that of Arakawa and Schubert (1974). Note that the troposphere is much drier, particularly at low levels. Thus, the equilibrium relative humidity of the atmosphere is very sensitive to assumptions about cloud microphysics.

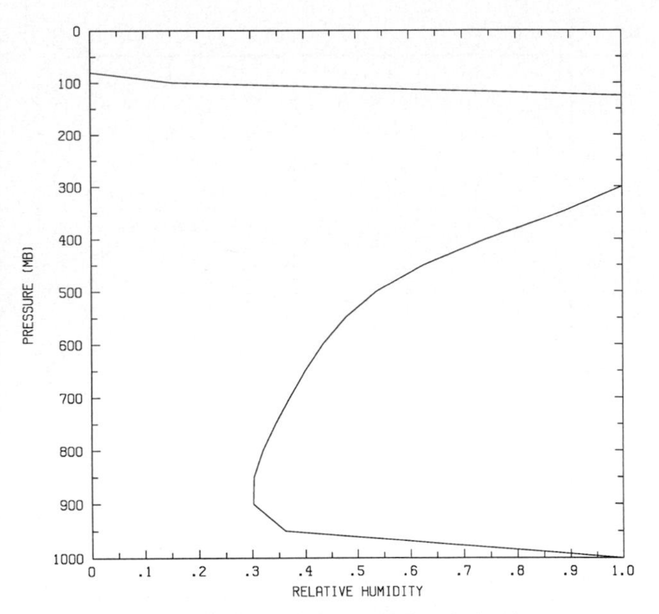

FIG. 19.4. Equilibrium relative humidity distribution when $\epsilon^i = 1.0$ for all i and no evaporation of rain is permitted.

Figure 19.5 shows what happens when large-scale subsidence is imposed and the mixing ratio is fixed above 600 mb. (Here the subsidence rate peaks at 48 mb day^{-1} at 600 mb.) A strong trade inversion forms, and the heat balance shows that in the inversion there is net cooling by convection; more mass is traveling downward in penetrative downdrafts than is going up in updrafts. The drying by the net subsidence is compensated by moistening by convection. The mean relative humidity is high, but not 100%, in the trade cumulus layer. Under the same forcing, the Arakawa–Schubert scheme would probably produce a saturated inversion layer, since finite amounts of condensed water are detrained at cloud top in that scheme. Here the cloud-top air mixes with its environment, then descends to neutral buoyancy and detrains. Thus, the influence of the various mixing assumptions is felt primarily in the water vapor distributions. (In the case of deep convection, however, detrained condensate may not be able to evaporate very much and penetrative downdrafts are much less important.)

19.5. Tuning convection schemes

We have seen that the relative humidity of convecting atmospheres is sensitive to detailed assumptions about cloud microphysics. This dependence is real and shows up in one form or another in all convection schemes. How shall the microphysical parameters be determined?

Naturally, one does one's best to determine such parameters directly from controlled laboratory experiments and from observations. The number of experiments that would need to be performed is large, however, and in practice the parameters will be uncertain within a fairly large range. Often, such parameters will be "tuned" within acceptable ranges (or even outside such ranges!) to give optimal performance in a weather forecast model. A drawback of this approach is that such tuning may turn out to be a compensation for other deficiencies of the large-scale model (e.g., turbulent diffusion).

Another approach to tuning convection schemes is to vary the parameters within acceptable limits so as to optimize the results of a semiprognostic test, in which tendencies produced by the scheme over a single time step are compared to the convective tendencies diagnosed as budget residuals from large-scale observations. This approach has the advantage that it is independent of a large-scale model but the drawback that it diagnoses instantaneous tendencies rather than equilibrium tendencies. (For the scheme described here, the tendencies at the first time step are always zero.) Very small differences between the diagnosed and observed heating can lead to large temperature changes over a relatively short time, so that "good" comparisons of diagnosed and predicted heating may be misleading.

I advocate another type of semiprognostic test, of a kind employed by Betts and Miller (1986). Using appropriate observational arrays {such as the GATE [Global Atmospheric Research Program (GARP) Atlantic Tropical Experiment] ship array}, calculate all the nonconvective processes, such as horizontal advection, vertical velocity, surface fluxes, and radiative transfer, over a period of several days. Using the convection scheme, calculate the total tendencies of water vapor and temperature and integrate these over the period of several days to predict the evolution of temperature and relative humidity. These can then be compared to observations. The relative humidity would, in particular, constitute a very sensitive test of the scheme, which would also have ample time to come

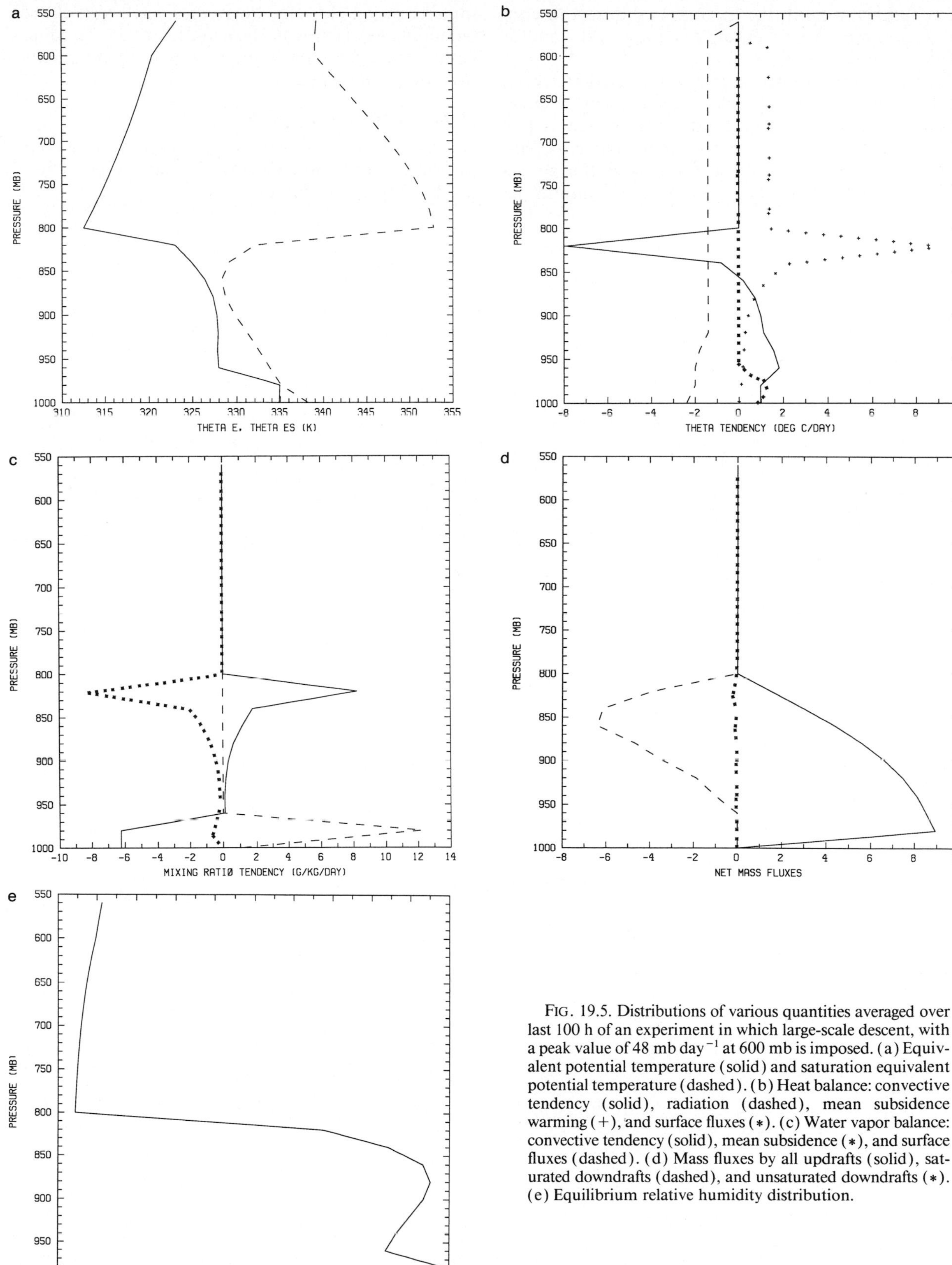

FIG. 19.5. Distributions of various quantities averaged over last 100 h of an experiment in which large-scale descent, with a peak value of 48 mb day^{-1} at 600 mb is imposed. (a) Equivalent potential temperature (solid) and saturation equivalent potential temperature (dashed). (b) Heat balance: convective tendency (solid), radiation (dashed), mean subsidence warming (+), and surface fluxes (*). (c) Water vapor balance: convective tendency (solid), mean subsidence (*), and surface fluxes (dashed). (d) Mass fluxes by all updrafts (solid), saturated downdrafts (dashed), and unsaturated downdrafts (*). (e) Equilibrium relative humidity distribution.

into equilibrium with the forcing. In contrast to classical semiprognostic schemes, this test would work for all convection schemes.

By varying the values of the various parameters to yield the best results of such a test, applied in many different circumstances, a model-independent set of optimal parameters can be obtained as well as a quantitative assessment of errors accruing from the parameterization. In view of the importance of cumulus parameterizations for climate studies, such tests should be conducted forthwith, bearing in mind that much better measurements of upper-tropospheric relative humidity will be necessary.

19.6. Software availability

The convection scheme described herein has been put in the form of a Fortran 77 subroutine. This is available from the author.

Part V

Representation of Slantwise Convection

Chapter 20

Parameterization of Slantwise Convection in Numerical Weather Prediction Models

THOR ERIK NORDENG

Norwegian Meteorological Institute, Oslo, Norway

20.1. Introduction

In two-dimensional motion (i.e., x–z direction), absolute momentum ($m = fx + v$) and potential temperature (equivalent potential temperature if the air is saturated) are conserved. It can be shown (e.g., Eliassen 1962) that the atmosphere may be stable to upright convection but unstable to perturbations in the direction of the sloping m surface provided that the slope of the m surface is less (more horizontal) than the slope of the potential temperature surface. Bennetts and Hoskins (1979) and Emanuel (1979, 1983a) suggested the importance of this instability in the formation of banded cloud and precipitation structures. It was named conditional symmetric instability (CSI). It manifests theoretically as roll vortices oriented along the shear vector (parallel to the thermal wind). Emanuel (1983b, 1988) presented synoptic evidence that suggests the presence of convection coexisting with atmospheric conditions that support the release of CSI. Shapiro (1982) and Emanuel (1983a) suggested that the effect of slantwise convection in a region where CSI is released is to adjust the atmosphere from a state of convective instability along an m surface to a state of neutrality along the m surface. Nordeng (1987) developed a scheme for parameterization of slantwise convection along these lines and tested it on real data in a mesoscale numerical weather prediction model (mesh size 50 km). A slightly different version of the scheme was implemented into the Pennsylvania State University (Penn State)–National Center for Atmospheric Research (NCAR) Mesoscale Model and tested on the case described in Emanuel (1983a) (Lindstrom and Nordeng 1992).

In this paper, the theory for CSI and slantwise convection is reviewed (section 20.2), while the Nordeng scheme is described in section 20.3. Examples of some results obtained with the scheme may be found in section 20.4.

20.2. Theoretical background

Shutts and Cullen (1987) showed that a diagnostic equation for the geopotential tendency for the *semi-geostrophic* (inviscid and adiabatic) equations (Eliassen 1949; Hoskins 1975) may be formed. This equation is elliptic provided that the matrix

$$\mathbf{Q} = \begin{pmatrix} m_x & m_y & m_z \\ n_x & n_y & n_z \\ \theta_x & \theta_y & \theta_z \end{pmatrix}$$

is positive definite, that is, it has only positive eigenvalues. Here, $m = fx + v_g$ and $n = fy - u_g$. Instability can be assumed when $\mathbf{Q}$ has negative eigenvalues. Shutts and Cullen (1987) discusses the property of the matrix $\mathbf{Q}$. The matrix is positive definite if (i) $m_x > 0$, (ii) $(\partial n/\partial y)_{m,z} > 0$, (iii) $(\partial\theta/\partial z)_{m,n} > 0$; (i) is the criterion for inertial stability. In particular, criterion (iii) shows that the instability criterion is that the potential temperature decreases with height along the intersection of the m and the n surfaces. In two-dimensional motion $\mathbf{Q}$ reduces to

$$\mathbf{Q} = \begin{pmatrix} m_x & m_z \\ \theta_x & \theta_z \end{pmatrix}$$

and the stability criterion is (i) $m_x > 0$: (inertial stability), (ii) $(\partial\theta/\partial z) > 0$: (vertical stability), (iii) $(\partial\theta/\partial z)_m > 0$: (the slope of the θ surface has to be less than the slope of the m surface).

CSI is usually described as a two-dimensional problem. However, the theory of Shutts and Cullen (1987) extends the CSI theory to three-dimensional motion. It should be stressed, however, that it is applicable to flow situations where the semigeostrophic approximation is justified.

So far we have been studying the dry problem. Introduction of moisture is straightforward in that equivalent potential temperature substitutes potential temperature in the analysis above. Potential temperature is used also in the following, but it is understood that it may be substituted by equivalent potential temperature when appropriate.

Hoskins and Draghici (1977) derived the diagnostic omega equation in geostrophic coordinates [$X = x + (v_g/f)$, $Y = y - (u_g/f)$] for the semigeostrophic equations. In these coordinates the omega equation looks exactly similar to the quasigeostrophic omega equation except that potential vorticity replaces static stability. The definitions of the geopotential and the ageostrophic wind are, however, slightly different than the common definition.

Iversen and Nordeng (1984) studied the various filtered (balanced) equations, of which the *balanced equation* (Charney 1962) is the best known. The equations (vorticity, balance, continuity, thermodynamic, and omega equations) are the natural extension of the semigeostrophic equations from an energetic and scaling point of view. Iversen and Nordeng showed that an approximate criterion for the ellipticity of the equations is that the potential vorticity is positive. In fact, the leading terms (second-order terms) of the generalized omega equation are (Iversen and Nordeng 1984)

$$S\nabla^2\omega + f\eta\omega_{pp} + \left(\alpha_x - f\frac{\partial v_\psi}{\partial p}\right)\omega_{px} + \left(\alpha_y + f\frac{\partial u_\psi}{\partial p}\right)\omega_{yp}, \quad (20.1)$$

where the wind is partitioned into a rotational and a divergent wind component

$$\mathbf{v} = \mathbf{v}_\psi + \mathbf{v}_\chi = \mathbf{k}\times\nabla\psi + \nabla\chi; \quad (20.2)$$

S is the static stability parameter $S = -(\alpha/\theta)(\partial\theta/\partial p)$; and η is the absolute vorticity ($\eta = v_x - u_y + f = \zeta + f$) where ζ is the relative vorticity.

Assuming that the rotational wind component obeys the thermal wind relation

$$\frac{\partial \mathbf{v}_\psi}{\partial p} = -\frac{1}{f}\mathbf{k}\times\nabla\alpha, \quad (20.3)$$

the criterion for an elliptic generalized omega equation is that $q > 0$, where q is the potential vorticity in pressure coordinates, but with only the rotational wind component retained:

$$q = -g\left(\eta\frac{\partial\theta}{\partial p} - \mathbf{k}\cdot\nabla\theta\times\frac{\partial\mathbf{v}_\psi}{\partial p}\right). \quad (20.4)$$

By changing to geostrophic coordinates $[X = x + (v_g/f),\ Y = y - (u_g/f),\ Z = p]$, a relation between the variation of potential temperature along the intersection between the m and n surfaces and the potential vorticity can be found:

$$\left[\eta_g + \frac{1}{f}\frac{\partial(u_g, v_g)}{\partial(x, y)}\right]\frac{\partial\theta}{\partial Z} = -\frac{q}{g} + \frac{1}{f}\frac{\partial(u_g, v_g, \theta)}{\partial(x, y, p)}, \quad (20.5)$$

where η_g is geostrophic absolute vorticity and q is defined as in (20.4) except that geostrophic values are used. The right-hand side of (20.5) may be written as $f^{-1}\nabla_3 m\times\nabla_3 n\cdot\nabla_3\theta = \zeta_g\cdot\nabla_3\theta$, while the left-hand side is $(\zeta_g\cdot\mathbf{k})\partial\theta/\partial Z$, where ζ_g is a generalized three-dimensional vorticity vector and it is parallel with the intersection line between the m and n surfaces. Here

$$\zeta_g = \left[-\frac{\partial v_g}{\partial p} + \frac{1}{f}\frac{\partial(u_g, v_g)}{\partial(y, p)},\ \frac{\partial u_g}{\partial p} + \frac{1}{f}\frac{\partial(u_g, v_g)}{\partial(p, x)},\ f + \frac{\partial v_g}{\partial x} - \frac{\partial u_g}{\partial y} + \frac{1}{f}\frac{\partial(u_g, v_g)}{\partial(x, y)}\right]; \quad (20.6)$$

$\zeta_g\cdot\nabla_3\theta$ is proportional to the semigeostrophic potential vorticity q_g, which is conserved for inviscid adiabatic motion in the geostrophic momentum approximation (Hoskins 1975):

$$q_g = -g(\zeta_g\cdot\nabla_3\theta). \quad (20.7)$$

The stability analysis of Shutts and Cullen (1987) implies that $\partial\theta/\partial Z < 0$ for stability [note that Z is defined as positive downward and that $\nabla_3 = (\partial/\partial x, \partial/\partial y, \partial/\partial p)$]. When the atmosphere is inertially stable, $\zeta_g\cdot\mathbf{k}$ is normally positive. The slantwise stability criterion is therefore that the isentropic absolute vorticity must be positive, which here is equivalent to saying that the tilt of the intersection line between the m and n surfaces is less (more upright) than the tilt of the potential temperature surface.

To summarize, Shutts and Cullen (1987) show that a necessary condition for the existence of a balanced state (stable) for the semigeostrophic approximation is that the potential temperature must increase with height along the intersection of the surfaces given by $m = fx + v_g$, $n = fy - u_g$. As demonstrated here, this is equivalent to requiring the (semigeostrophic) potential vorticity to be positive. This result is to be expected, since it was shown by Hoskins and Draghici (1977) that the semigeostrophic omega equation is elliptic provided that the potential vorticity is positive. Iversen and Nordeng (1984) showed that an approximate ellipticity criterion for the more general omega equation of the balanced model is that the potential vorticity [defined by Eq. (20.4)] is positive. We will therefore assume that an analogy exists between the vertical variation of potential temperature along the m, n direction and the potential vorticity in general; this will be used when constructing our parameterization scheme. Strictly speaking, however, such an analog has only been found for semigeostrophic flow. If we alternatively define m and n and $m = fx + v_\psi$, $n = fy - u_\psi$, Eq. (20.5) remains unchanged except that the rotational wind components replace the geostrophic wind components and q is given by (20.4).

The above results were derived by the assumption that the atmosphere is either dry, in which θ was used, or saturated, where θ_e should be used. Emanuel (1983b) introduced the concept of slantwise convective instability, $(\partial\theta_e/\partial z)_m < 0$, as an analogy to the criterion for vertical convective instability, $\partial\theta_e/\partial z < 0$, whether the air is saturated or not. This accounts for the possibility of air parcels lifted from the surface becoming unstable after reaching the lifting condensation level. He also introduced the slantwise convective available potential energy (SCAPE) as an analog to the more common convective available potential energy (CAPE) for vertical instability. SCAPE may be found by computing the difference between an air parcel lifted reversibly along an absolute momentum surface and its environment.

20.3. Parameterization of slantwise convection

From the discussion above, it is possible to decide how slantwise convection could be parameterized in numerical models of the atmosphere.

- We assume that a balanced state exists toward which the atmosphere tends to adjust if unstable.
- This balanced state is recognized as being slantwise convectively neutral; that is, $(\partial\theta_e/\partial z)_{m,n} = 0$.
- The analogy between the vertical variation of equivalent potential temperature along the m, n direction and the potential vorticity will be used to compute the vertical variation of equivalent potential temperature.

From these assumptions, Nordeng (1987) derived a scheme for parameterization of slantwise convection and tested it on some polar low cases in the Norwegian Sea. The scheme has also been tested on a case (Emanuel 1983b) where the appearance of slantwise convection was evident (Lindstrom and Nordeng 1992). The following closely follows Lindstrom and Nordeng (1992) but is reproduced here for the convenience of the reader.

By assuming that $\partial\theta_e/\partial Z = 0$ and neglecting the curvature term of (20.5), which is important only for strongly curved flow but zero for two-dimensional (including axisymmetric) flow, the vertical variation of equivalent potential temperature may be found from

$$\frac{\partial\theta_e}{\partial p} = \left(\frac{1}{\eta}\mathbf{k}\cdot\nabla\theta_e\times\frac{\partial\mathbf{v}}{\partial p}\right), \tag{20.8}$$

where, in light of the previous discussion, we have used the total wind instead of the geostrophic wind. One may argue that at least the rotational wind component should have been used, but we do not believe that this will affect the results significantly.

- We assume that the right-hand side of (20.8) can be determined from the model variables around the actual grid point and solve the left-hand side for equivalent potential temperature. We will call this profile θ_{ep}.
- We assume that the difference in temperature between a parcel lifted reversibly along an absolute momentum surface and its environment (at m, n constant) is the same as the difference between the temperature given by θ_{ep} and the environment (at x, y constant).

The last assumption eliminates horizontal interpolation to obtain the temperature difference (parcel minus environment) along the absolute momentum surface. This requires that the absolute vorticity does not vary too much in the horizontal direction [see Emanuel (1983b) or Nordeng (1987) for the derivation of this property]. Temperature and specific humidity (T_e, q_e) are derived from θ_{ep} by

$$T_c = T + \frac{\pi(\theta_{ep} - \theta_{es})}{c_p + \epsilon L^2 q_{\text{sat}}(T)/RT^2},$$
$$q_c = q_{\text{sat}}(T_c), \tag{20.9}$$

where $q_{\text{sat}}(T_c)$ is the saturation specific humidity of temperature T, and θ_{es} is the saturation equivalent potential temperature of the environment. Term π is the Exner function; (T_c, q_c) are used in the adjustment scheme.

The scheme may be looked at as a slanted extension of a Kuo-type scheme. The criterion for onset of slantwise convection is that there is a net moisture convergence into the grid column, that the atmosphere is convectively unstable along the intersection of the m and n surfaces, and that a lifting condensation level may be defined. To determine the amount of moisture convergence to be used for general moistening and the amount that is used for condensation (and precipitation fallout) the method of Geleyn (1985) is used. This approach has a self-regulating partitioning parameter between the two processes. The rainfall term $(\partial T/\partial t)_{\text{conv}}$ and the moistening term $(\partial q/\partial t)_{\text{conv}} + (\partial q/\partial t)_{\text{cvg}}$ become proportional to $T_c - T + (L/c_p)(q_c - q_w)$ and $q_w - q$, respectively. Here, q_w is the wet-bulb specific humidity obtained from

$$-L(q - q_w) = c_p(T - T_w),$$
$$q_w = q_{\text{sat}}(T_w, p), \tag{20.10}$$

so that

$$\left(\frac{\partial T}{\partial t}\right)_{\text{conv}} = I\frac{(T_c - T) + (L/c_p)(q_c - q_w)}{\int_{p_t}^{p_b}(q_c - q) + (c_p/L)(T_c - T)dp} \tag{20.11}$$

$$\left(\frac{\partial q}{\partial t}\right)_{\text{conv}} = -\left(\frac{\partial q}{\partial t}\right)_{\text{cvg}} + I\frac{(q_w - q)}{\int_{p_t}^{p_b}(q_c - q) + (c_p/L)(T_c - T)dp}, \tag{20.12}$$

where p_t is cloud top and p_b is cloud base (lifting condensation level) and

$$I = \int_{p_t}^{p_b}\left(\frac{\partial q}{\partial t}\right)_{\text{cvg}} dp. \tag{20.13}$$

Here $(\partial q/\partial t)_{\text{cvg}}$ is the moisture convergence. Heating and moistening are distributed along the intersection

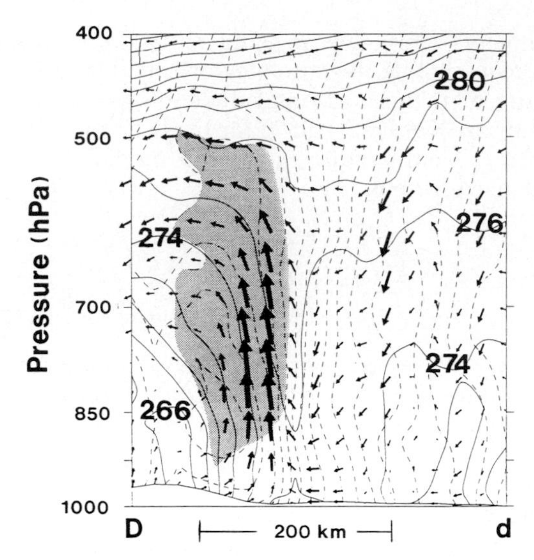

FIG. 20.1. Cross sections showing equivalent potential temperatures (full lines) at contour intervals of 2 K, absolute momentum (stippled lines) at contour intervals of 5 m s^{-1}, and tangential airflow (arrows) across a strong frontal zone in connection with a polar low development. The length of an arrow corresponds to a 15-min trajectory. The shaded regions have relative humidity above 90%, taken from Nordeng and Rasmussen (1992).

line between the constant m and n surfaces starting from the cloud base (p_b). Then,

$$m(p) = m(p_b) = v(p_b) = fx + v(p), \quad (20.14)$$

$$n(p) = n(p_b) = -u(p_b) = fy - u(p), \quad (20.15)$$

$$x = \frac{v(p_b) - v(p)}{f}, \quad y = \frac{u(p) - u(p_b)}{f}. \quad (20.16)$$

Here x and y are the relative coordinates as measured from the vertical axis of the point where the intersection line between the m and n surface crosses the pressure surface p. The heating and moistening at level p is distributed bilinearly between the four grid points surrounding (x, y). The scheme is used *together with* schemes for parameterization of upright convection and from time-scaling arguments it is used *only* when the atmosphere is stable to upright convection.

This approach mixes heat and moisture along the absolute momentum surfaces. It can be shown that the most unstable direction is *not* along the absolute momentum surfaces but rather along the equivalent potential temperature surface. In real cases, however, the difference between contours of θ_e and absolute momentum is probably small. Another approach would be to mix absolute momentum along the equivalent potential temperature surfaces (see chapter 21).

It should finally be mentioned that at least for two-dimensional motion, this parameterization is consistent with regard to conservation of momentum. Parameterization of vertical flux of horizontal momentum by convection is far from understood. The slantwise approach takes this flux implicitly into account. One may argue that upright convection is a special case of the more general slantwise convection (the absolute momentum surfaces are vertical for barotropic flow). Such an approach would eliminate the problem of how to describe vertical momentum transport due to convective motion. This approach has not been tested, however.

20.4. Discussion

One may ask if slantwise convection has to be parameterized at all. The horizontal scale of the phenomenon may be of the order of (several) hundreds of kilometers and should be resolved explicitly by the model equations. Model simulations have indeed confirmed the slantwise nature of rising motion (e.g., Kuo and Reed 1988). Figure 20.1, which is taken from Nordeng and Rasmussen (1992), shows explicitly resolved slanted rising motion and neutral stratification along absolute momentum surfaces in a model simulation with horizontal resolution 25 km. In this case the atmosphere is stable to vertical convection (Fig. 20.2). Shutts (1990) studied the "October storms" of 1987 with the U.K. Meteorological Office limited-area model (grid spacing 75 km) and found strong slantwise convective instability in the region of the storm develop-

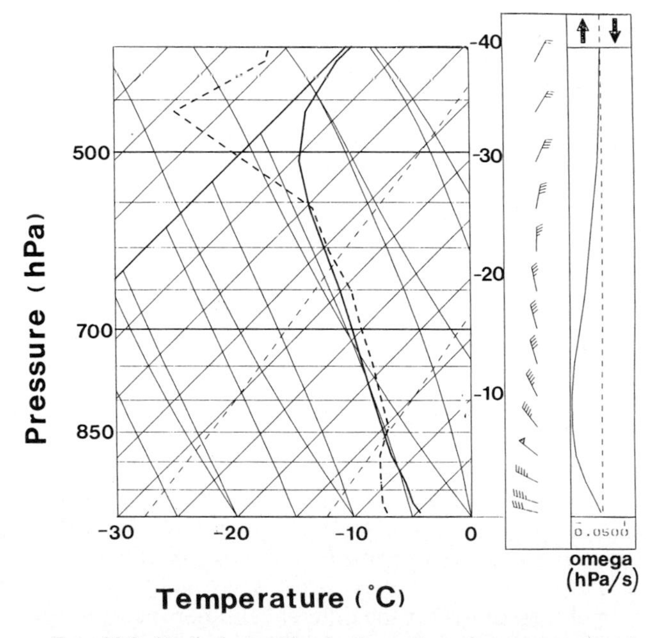

FIG. 20.2. Vertical sounding in the region of strongest vertical motion in Fig. 20.1. Temperature and dewpoint temperature are plotted in skew tephigrams with lines of constant temperature running upward to the right. Note that both cloud liquid water and water vapor specific humidity are used to compute dewpoint temperature. The arrows are wind direction (north is up) and speed (a flag is 25 m s^{-1}, a full barb is 5 m s^{-1}, and a half barb is 2.5 m s^{-1}). The vertical velocity ω (hPa s^{-1}) is visualized beside the winds, taken from Nordeng and Rasmussen (1992).

ment. It turned out, however, that the model seemed to be able to perform the right adjustment without parameterization. The simulation of the storm was almost perfect and the model atmosphere adjusted to a neutral state during the storm development.

Adjustment toward a neutral state, however, does not guarantee that the model performs in a dynamically correct way in general. In the case studied by Lindstrom and Nordeng (1992), it was demonstrated that the model atmosphere was slantwise unstable at the start of the integration. The instability gradually vanished during the integration. Since this happened over land, the rainfall pattern and rainfall intensity could be used to verify the simulation. Figure 20.3 shows total precipitation as predicted with the Penn State–NCAR model for resolutions 80 and 40 km in experiments with and without parameterization of slantwise convection. Note that *vertical* convection was parameterized (Anthes 1977) in both experiments. Observed precipitation is shown in Fig. 20.4. The precipitation was significantly better forecasted when slantwise convection was parameterized. The rainfall rate doubles at both resolutions and the area of maximum rainfall narrows (as observed). For this case, parameterization of slantwise convection has impacts not only on the precipitation pattern but also on the model dynamics. It strengthens the grid-scale updrafts and decreases the width of the grid-scale updraft and, therefore, the width of the grid-scale precipitation. In some areas the grid-scale precipitation is enhanced (in response to the more vigorous updrafts). In other areas, convective precipitation substitutes the grid-scale precipitation. Emanuel (1983a) predicted that slantwise convection should augment frontogenesis due to a stronger secondary circulation. The theoretical effect of a heat source in a slantwise unstable atmosphere is a dipole circulation consisting of ascending motion along the m surfaces and two branches of descending motion in both the cold and the warm air, giving an enhanced convergence (and frontogenesis) at the surface. There is evidence

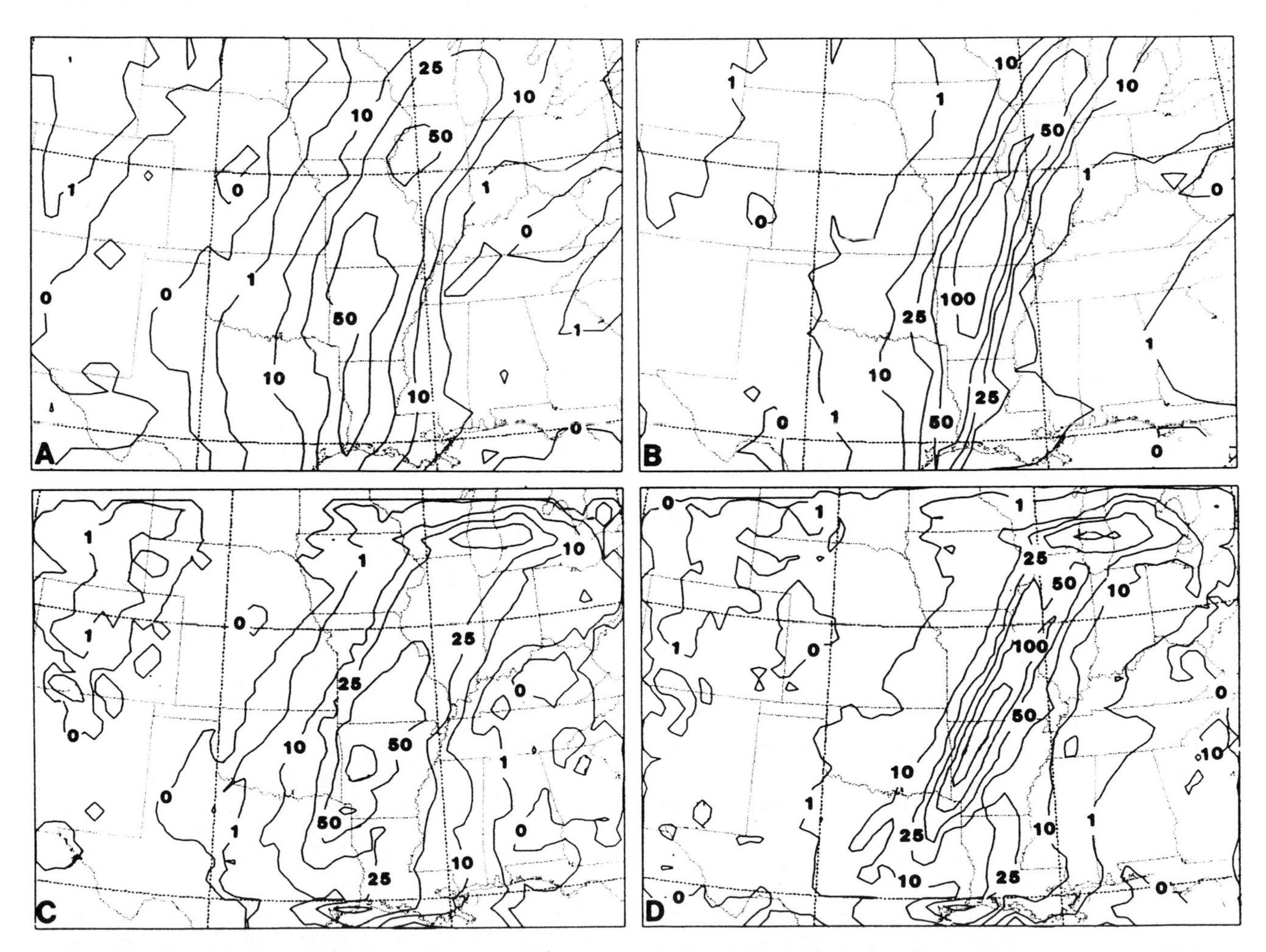

FIG. 20.3. Total 24-h precipitation from 80-km simulations [(a) and (b)] and 40-km simulations [(c) and (d)]. Panels (a) and (c) are with the Anthes–Kuo scheme for parameterization of vertical convection; (b) and (d) are with the Anthes–Kuo scheme for parameterization of vertical convection and Nordeng's scheme for parameterization of slantwise convection. Contour values of 0, 1, 10, 25, 50, 100, and 200 mm shown, taken from Lindstrom and Nordeng (1992).

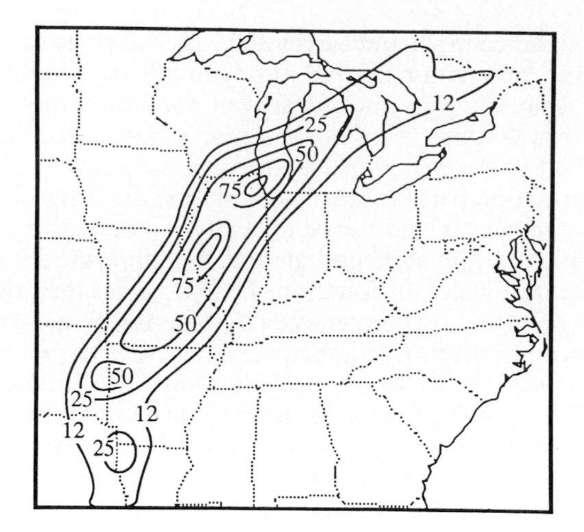

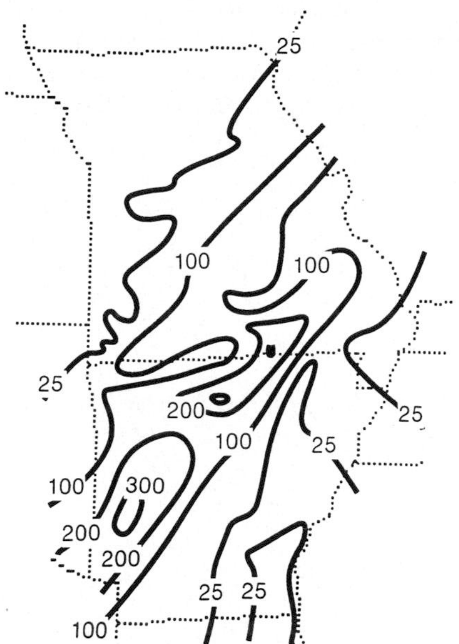

FIG. 20.4. (a) Synoptic-scale analysis of 24-h rainfall for the case studied by Lindstrom and Nordeng (Fig. 20.3). Contour values of 12, 25, 50, 75, and 100 mm. (b) Mesoscale analysis of 24-h rainfall for Arkansas and Missouri. Contour values of 25, 100, 200, and 300 mm, taken from Lindstrom and Nordeng (1992).

of this effect in the case studied by Lindstrom and Nordeng (1992). Figure 20.5 shows the absolute momentum after 12 h of simulation. There are clearly larger gradients in m ($\partial m/\partial x = f + \partial v/\partial x = f + \zeta$—absolute vorticity), suggesting a strong frontogenetic response in the run with parameterized slantwise convection. An interesting result is that the two simulations are more equal after 18 h of simulation than after 12 h of simulation, indicating that an effect of parameterized slantwise convection is to establish a secondary circulation more quickly than the model equations are able to do. This may easily be understood qualitatively. Since the slantwise convective scheme is activated as soon as the model atmosphere is *conditionally unstable* in the direction of the sloping absolute momentum surfaces, a circulation will be set up as a response to the diabatic heating. This may happen *before* the model equations explicitly set up such a circulation because the model atmosphere will have to saturate in response to the (weak) rising motion in that direction. (Remember that the convective scheme moistens the atmosphere in addition to releasing latent heat.) In reality, slantwise convection is expected to take place as individual updrafts and downdrafts, more similar to the motion found when the atmosphere is vertically unstable than the kind of rising motion found in connection with frontal upgliding.

Figure 20.6, which is taken from Nordeng (1987), shows a trough moving southward in the Norwegian Sea in connection to a major low over northern Scandinavia. A cross section through the trough (Fig. 20.7) shows that there is a geostrophic deformation corre-

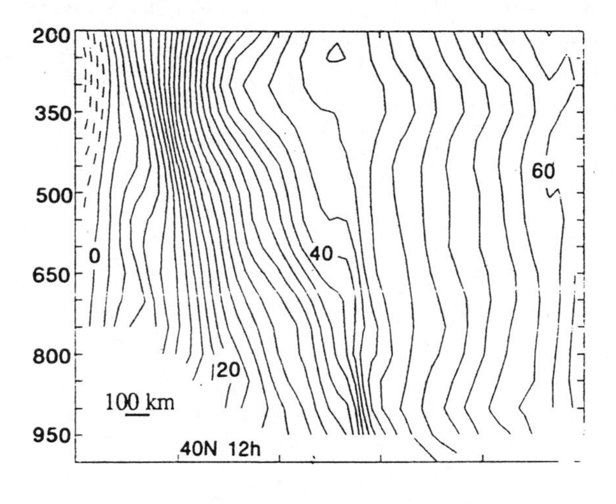

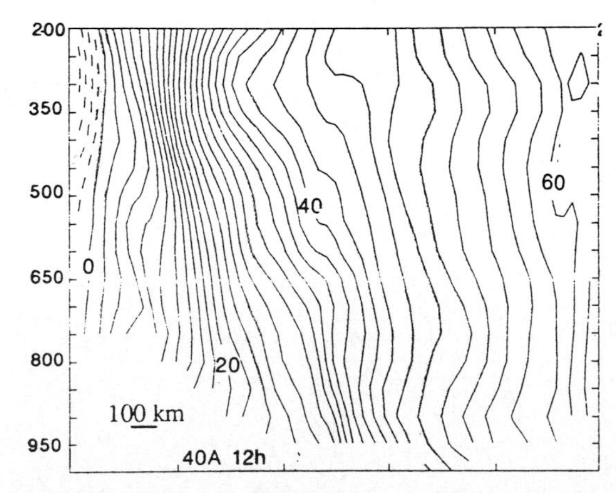

FIG. 20.5. Cross section of absolute momentum across the frontal zone in the case studied by Lindstrom and Nordeng. Model simulation with grid length 40 km—(a) with the Anthes–Kuo scheme for parameterization of vertical convection and Nordeng's scheme for parameterization of slantwise convection, and (b) with the Anthes–Kuo scheme for parameterization of vertical convection, taken from Lindstrom and Nordeng (1992).

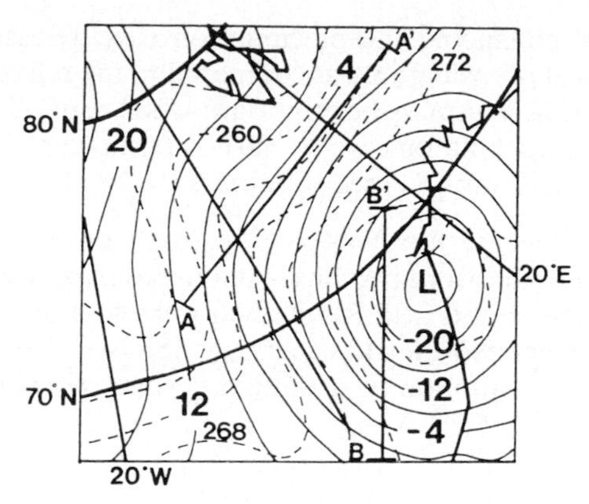

FIG. 20.6. Contours of 1000 hPa (solid lines) and (potential) temperature at 1000 hPa (dashed lines) for 1200 UTC 29 February 1984. Line BB' marks is cross section in Fig. 20.7, taken from Nordeng (1987).

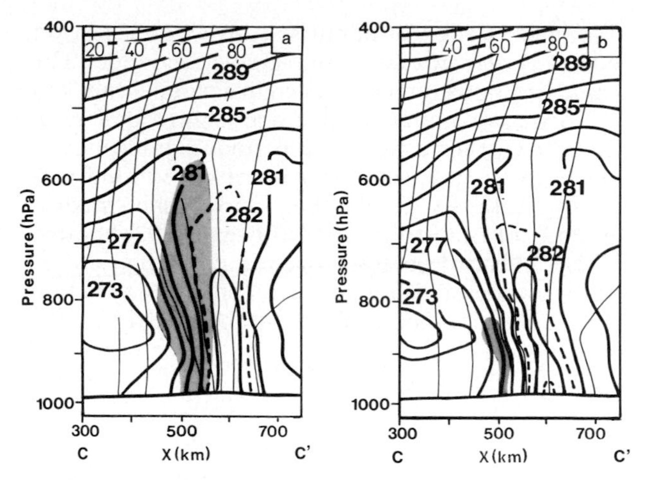

FIG. 20.8. Vertical cross section after 24 h from 1200 UTC 29 February 1984. Position of cross section is (61.5°N, 0.1°E)–(64.9°N, 5.3°E). Thick lines are contours of equivalent potential temperature. Thin lines are contours of absolute momentum. Shading denotes area of symmetric instability—(a) with parameterization of vertical convection only, and (b) with parameterization of slantwise convection in addition to parameterization of vertical convection.

sponding to a direct circulation. Thorpe and Emanuel (1985) studied an idealized flow with such a configuration and showed that the frontogenesis is enhanced by the existence of low stability to slantwise motion, that is, low potential vorticity. Some time later, a cross section through the trough reveals a wedge of warm air embedded in the cold air. The air is unstable for both vertical and slantwise convection (Fig. 20.8a). Parameterization of slantwise convection (Fig. 20.8b) adjusts the model atmosphere to be neutral to CSI. An interesting feature is that the effect of parameterization of slantwise convection is to set up a circulation as expected from the theory. Figure 20.9 shows the difference in vertical velocity from the run with parameterization of slantwise convection and the run with parameterization of vertical convection only. The imposed circulation due to the scheme is rising motion along the sloping absolute momentum surfaces with descending branches on either side.

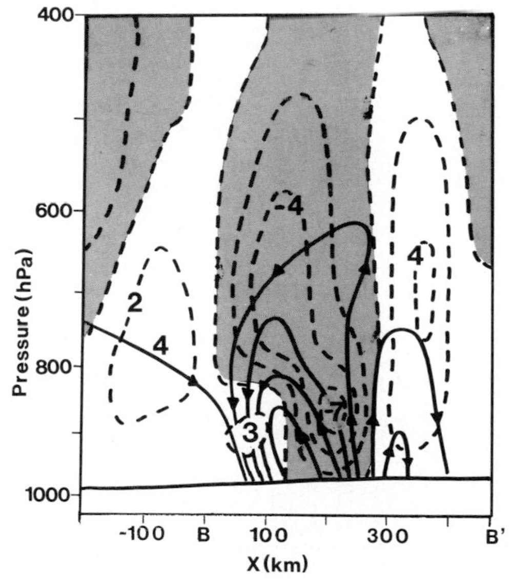

FIG. 20.7. Cross section BB' (see Fig. 20.6 for position of section) after 6 h of integration from 1200 UTC 29 February 1984. Dashed lines are contours of vertical velocity $\bar{\omega}$ at intervals of 1×10^{-3} hPa s^{-1}. Solid lines are contours of geostrophic forcing. Arrows on geostrophic forcing contours indicate direction of circulation. Areas of ascending motion are shaded, taken from Nordeng (1987).

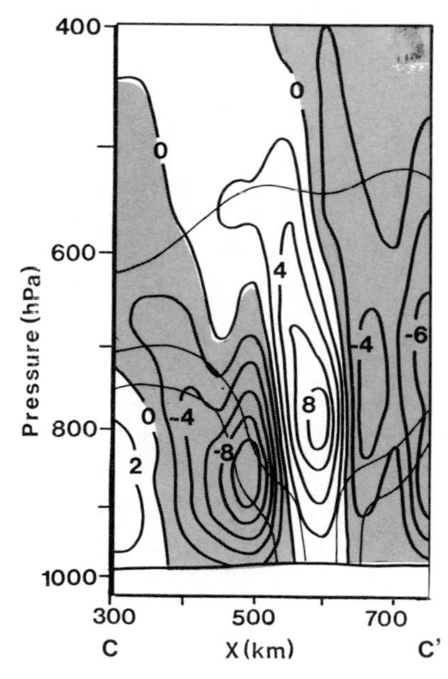

FIG. 20.9. Same cross section as Fig. 20.8 showing the difference in vertical velocity between the experiment with parameterization of slantwise convection and the experiment without this parameterization. Contour intervals are 2×10^{-3} hPa s^{-1}. Shading shows areas of ascending motion.

Parameterization of slantwise convection is not yet used in any *operational* weather prediction model. The scheme has, however, been tested in a parallel run with the Norwegian limited-area model (horizontal resolution 50 km) for a two-week period. Significant differences in the precipitation pattern were found in 8 out of 12 cases (T. E. Nordeng, unpublished study). The study of Lindstrom and Nordeng (1992) showed that one effect of parameterization of slantwise convection is to establish a strong secondary circulation before it is established explicitly in the model with the effect of enhancing the precipitation (i.e., release of latent heat) considerably as compared to the reference experiment. This may be important when running operational assimilation cycles, and in particular if diabatic initialization is used to initiate the model.

Acknowledgments. This paper is partly put together by using results from previously published papers. Scott S. Lindstrom and Erik Rasmussen were co-authors in those papers and they are acknowledged for their significant contribution in making this paper possible.

Chapter 21

A Parameterization Scheme for Symmetric Instability: Tests for an Idealized Flow

SIN CHAN CHOU AND ALAN J. THORPE

Department of Meteorology, University of Reading, Reading, England

21.1. Introduction

Current convective parameterization schemes make no explicit allowance for the different organization regimes into which clouds are known to fall. As far as cloud dynamics is concerned, there are distinct phenomena depending on the flow and thermodynamic structure in which the clouds develop. The following is probably not an exhaustive list:

convection:
isolated storms—supercell
short-lived showers
shallow (boundary layer) clouds
midlevel convection
squall lines—midlatitude
tropical
multicellular storms—traveling
stationary
mesoscale convective complexes

conditional symmetric instability:
frontal rainbands
axisymmetric bands in tropical cyclones.

There are special schemes to account separately for midlevel convection and shallow boundary-layer clouds but only in terms of their distinctive thermodynamic properties. To a cloud dynamicist, it seems surprising that separate account is not taken for all of the above cloud organization types. Nowhere is this lack of discrimination more acute than when it comes to the question of the momentum flux. But even the geometrical problem of the way in which a squall line, which may be 500 km long × 50 km wide, is different from a supercell, which may occupy an area of perhaps 50 km × 50 km, seems not to have been faced. The reasons have to do with the hypothesis that subgrid-scale clouds need to be statistically numerous for their overall effects to be representable. It is clear that most deep clouds occur in an organized fashion often well isolated from nearest neighbors.

Clearly the issue of mesoscale organization is an extremely difficult one, involving questions such as the scale separation of the clouds from the model grid spacing. It is the contention here that cloud bands that are essentially two-dimensional, such as frontal rain- or snowbands and squall lines, offer the most important challenge to conventional parameterization schemes. A further crucial difficulty for current schemes is that these phenomena can exist in a quasi- (or statistical) steady state for many hours and can, in that time, propagate over many grid squares. Here we consider in detail frontal rainbands that are due to conditional symmetric instability (CSI). These are particularly of interest as no operational forecast model as yet makes specific allowance for this instability. Is the space or time scale of these phenomena such that they need to be parameterized in current models? The growth rate of the instability might be a few hours, while such bands may be present over the complete life of a front of, say, 1–3 days. The spatial scales are highly anisotropic with cross-frontal scales being in the range of 20–200 km and alongfront scales from 200 to 2000 km. So these phenomena are uncomfortably at the spatial resolution of current global weather forecast models; they will be substantially subgrid scale for climate models for some time to come.

Nordeng (1987) included in an operational model with 50-km horizontal resolution a Kuo-type parameterization of CSI. The precipitation from this parameterization could be found in bands and complemented that from vertical convection and large-scale precipitation. Lindstrom and Nordeng (1992) showed that greater horizontal resolution can also affect the precipitation amount through stronger and narrower grid-scale updrafts. In Nordeng's scheme (chapter 20) the neutrality to symmetric instability is achieved by adjusting the equivalent potential temperature along the absolute momentum surfaces. Parcels of air are displaced along absolute momentum surfaces, the buoyancy is checked, and a modified Kuo scheme is applied on the temperature and moisture fields.

The formulation we will develop here is perhaps unique in cloud parameterization schemes in that we deliberately set out to confront, and answer, the problem of cloud momentum fluxes. We use the current European Centre for Medium-Range Weather Fore-

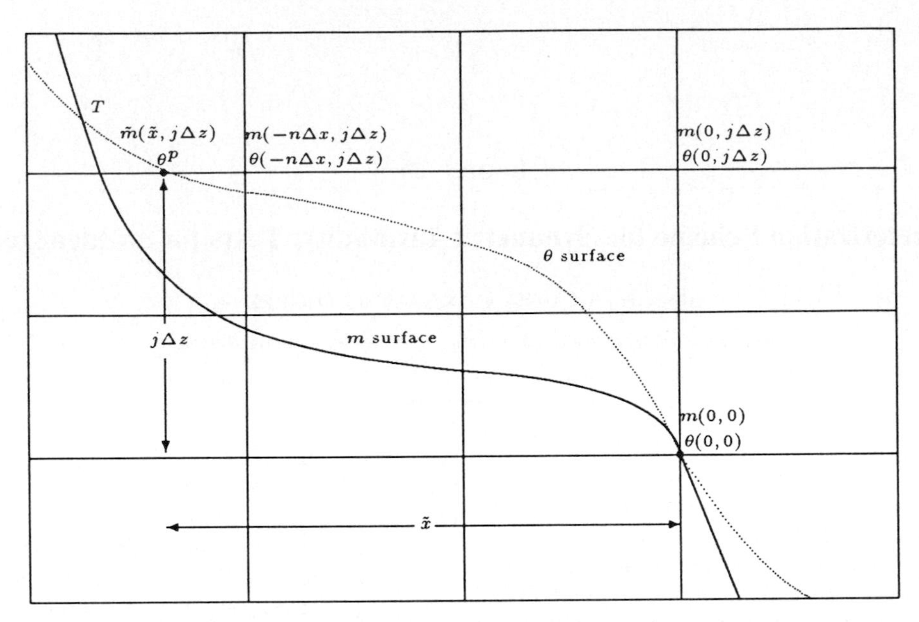

FIG. 21.1. Schematic cross section of an absolute momentum surface (solid line) and a potential temperature surface (dotted line). The parcel originating at the point (0, 0) ascends along the theta surface up to T; $\tilde{x}$ is the horizontal displacement at the height $j\Delta z$; Δx and Δz are the horizontal and vertical grid spacings.

casts (ECMWF) formulation for convection with the mass flux scheme but modify it to cope with CSI. To illustrate the scheme, we will use a dry analogy although the essential physics is not substantially different in concept for the full moist case. In the full implementation, there is, in addition to the momentum fluxes described here, latent heating due to condensation as parcels ascend along moist isentropic surfaces.[1] For the background to CSI the reader is referred to the following papers: basic dynamics—Bennetts and Hoskins (1979), Emanuel (1983a), Thorpe and Rotunno (1989), Jones and Thorpe (1992); observational evidence—Emanuel (1988), Sanders and Bosart (1985), and Thorpe and Clough (1991). For the background to the mass flux scheme we follow Tiedtke (1989).

21.2. The instability criterion

There are many ways to describe the criterion for symmetric instability. The linear model shows that the potential vorticity (PV) must be negative, which is equivalent for a geostrophic two-dimensional flow to a condition on the Richardson number. The linear model and nonlinear simulations, such as discussed in Thorpe and Rotunno (1989), show that the maximum growth rate is obtained for ascent close to the basic-state isentropes depending on the degree to which the motion satisfies the hydrostatic approximation. The parcel approach described by Emanuel (1983a) quantifies the forces acting on a displaced parcel. Thorpe et al. (1989) show that the neutral buoyancy displacement releases maximum parcel kinetic energy. For a parameterization scheme, the parcel model is useful because the schemes are usually based on lifting a parcel to calculate whether instability is present. All such calculations, even for convection, require the path of the lifted parcel to be prescribed. Bearing in mind the above results on the likely parcel path, we choose the neutral buoyancy trajectory. Results not shown here have been obtained for other assumed paths, but the mathematical formulation to be given now makes the neutral buoyancy assumption.

The instability criterion for a parcel lifted in a two-dimensional flow in a direction **x** orthogonal to the thermal wind but along a neutral buoyancy curve is

$$f\hat{m}\tilde{x} > 0, \qquad (21.1)$$

where $\hat{m}$ is the difference between the parcel absolute momentum ($m = v_g + fx$) and that of its current surroundings, and $\tilde{x}$ is its x displacement along a neutral buoyancy curve. The above criterion is general and applies to both hemispheres, either sign of baroclinity, and for ascending as well as descending parcels. For reference note that the criterion for convection arises from lifting a parcel vertically over a distance $\tilde{z}$: $g\hat{\theta}\tilde{z} > 0$, where $\hat{\theta}$ is the parcel's anomaly of potential temperature (or more accurately density). The scheme to be tested here will operate whenever convective instability, as just defined, is not present. There has been recent speculation about the use of a CSI scheme to account for both CSI and convection. While this is a

[1] This research forms part of the lead author's Ph.D. thesis available from the University of Reading.

possibility worth exploring, the practical difficulties are not discussed here.

In Fig. 21.1 a schematic of the disposition of θ and m surfaces is given along with the assumed parcel path. In the example of Fig. 21.1, taken from the Northern Hemisphere for a baroclinic flow with cold air lying to the west, the above general criterion reduces to $\hat{m} < 0$ as $\tilde{x} < 0$. This hypothetical undilute ascending parcel achieves neutral buoyancy at the point marked T at which $\hat{m} = 0$. The existence of such momentum anomalies has been found in nonlinear simulations—see Fig. 2 of Thorpe and Rotunno (1989)—and in frontal observations—see Thorpe and Clough (1991).

Here we deal with tests of the parameterization scheme in a two-dimensional flow in which $\partial/\partial y = 0$. In applying the scheme to an operational model, a local coordinate system must be calculated oriented along and across the thermal wind direction. This and the consequent interpolation will require the solution of the same practical implementation problems that are not faced here.

21.3. Mass flux scheme for CSI

Consider a dry atmosphere with a basic state in hydrostatic and geostrophic balance that is frictionless. The large-scale equations for heat and absolute momentum, $m = v_g + fx$, can be written as

$$\frac{\partial\bar{\theta}}{\partial t} + \bar{\mathbf{v}}\cdot\nabla\bar{\theta} + \bar{w}\frac{\partial\bar{\theta}}{\partial z} = -\frac{1}{\bar{\rho}}\nabla\cdot(\bar{\rho}\overline{\mathbf{v}'\theta'}) - \frac{1}{\bar{\rho}}\frac{\partial(\bar{\rho}\overline{w'\theta'})}{\partial z} + \frac{\bar{Q}_R}{Cp} \quad (21.2)$$

$$\frac{\partial\bar{m}}{\partial t} + \bar{\mathbf{v}}\cdot\nabla\bar{m} + \bar{w}\frac{\partial\bar{m}}{\partial z} = -\frac{1}{\bar{\rho}}\nabla\cdot(\bar{\rho}\overline{\mathbf{v}'m'}) - \frac{1}{\bar{\rho}}\frac{\partial(\bar{\rho}\overline{w'm'})}{\partial z}. \quad (21.3)$$

The overbar denotes average over a horizontal area that contains the ensemble of clouds. Terms $\overline{\mathbf{v}'\theta'}$, $\overline{\mathbf{v}'m'}$, $\overline{w'\theta'}$, and $\overline{w'm'}$ are eddy transports of heat and momentum in the horizontal and vertical, respectively, and $\bar{Q}_R/Cp$ is the radiative heating. The flow has been assumed to be two-dimensional; that is, $\partial/\partial y \approx 0$. The contributions to θ and m of the environment due to symmetric instability are

$$\frac{\partial\bar{\theta}}{\partial t} = -\frac{1}{\bar{\rho}}\frac{\partial(\bar{\rho}\overline{u'\theta'})_{\mathrm{SI}}}{\partial x} - \frac{1}{\bar{\rho}}\frac{\partial(\bar{\rho}\overline{w'\theta'})_{\mathrm{SI}}}{\partial z} \quad (21.4)$$

$$\frac{\partial\bar{m}}{\partial t} = -\frac{1}{\bar{\rho}}\frac{\partial(\bar{\rho}\overline{u'm'})_{\mathrm{SI}}}{\partial x} - \frac{1}{\bar{\rho}}\frac{\partial(\bar{\rho}\overline{w'm'})_{\mathrm{SI}}}{\partial z}. \quad (21.5)$$

The above expressions show that the adjustment toward a neutral state can be done through the distribution of heating and/or momentum in the atmosphere. Based on the linear theory, which shows the most unstable mode occurs in the direction parallel to the θ surfaces, a crucial assumption is made for the scheme: the unstable parcels move along θ surfaces in a neutral buoyant ascent.

The horizontal eddy flux term is normally neglected in upright convection when compared to the other large-scale terms. In CSI, the contribution of this term is not yet clear. A scale analysis suggests that its magnitude can be comparable to the vertical eddy flux term. Assume that the slope of the transversal circulation is parallel to the θ surfaces; that is,

$$\frac{u'}{w'} = -\frac{\partial\theta}{\partial z}\left(\frac{\partial\theta}{\partial x}\right)^{-1} = -\frac{N^2}{fV_z}, \quad (21.6)$$

where N^2 is the Brunt–Väisälä frequency, f is the Coriolis parameter, and V_z is the background horizontal wind shear.

The ratio between the horizontal and the vertical eddy flux terms can be scaled as

$$\frac{\partial(\bar{\rho}\overline{u'm'})}{\partial x}\left[\frac{\partial(\bar{\rho}\overline{w'm'})}{\partial z}\right]^{-1} \sim \frac{\Delta z}{\Delta x}\frac{N^2}{fV_z}. \quad (21.7)$$

Taking $\Delta z/\Delta x \sim 0.01$, $N^2 \sim 10^{-4}\,\mathrm{s}^{-2}$, $f \sim 10^{-4}\,\mathrm{s}^{-1}$, and $V_z \sim 10^{-2}\,\mathrm{s}^{-1}$, the ratio in (21.7) is of order 1 in higher horizontal resolution models, or 0.1 in a coarser horizontal grid. So, this term is kept in the equations.

Using (21.6), (21.5) can be rewritten as

$$\left(\frac{\partial\bar{m}}{\partial t}\right)_{\mathrm{SI}} = -\frac{1}{\bar{\rho}}\left[\frac{\partial}{\partial x}\left(-\bar{\rho}\frac{N^2}{fV_z}\overline{w'm'}\right) + \frac{\partial}{\partial z}(\bar{\rho}\overline{w'm'})\right]. \quad (21.8)$$

By doing a coordinate transformation such that X is perpendicular and Z parallel to the θ surfaces, the eddy flux evaluated along the θ surface is

$$\left(\frac{\partial}{\partial Z}\right)_\theta \bar{\rho}\overline{w'm'} = -\frac{N^2}{fV_z}\frac{\partial}{\partial x}\bar{\rho}\overline{w'm'} + \frac{\partial}{\partial z}\bar{\rho}\overline{w'm'}. \quad (21.9)$$

Assuming only small variations of N^2/fV_z, the lhs of (21.9) can be substituted into (21.8):

$$\left(\frac{\partial\bar{m}}{\partial t}\right)_{\mathrm{SI}} = -\frac{1}{\bar{\rho}}\frac{\partial}{\partial Z}(\bar{\rho}\overline{w'm'}), \quad (21.10)$$

where $\partial/\partial Z$ hereafter is assumed to mean the derivative along a θ surface. Note that in the limit of horizontal θ surfaces then $(X, Z) = (z, -x)$, whereas in the limit of vertical θ surfaces $(X, Z) = (x, z)$.

Essentially, the neutrality is achieved by mixing absolute momentum along the θ surfaces. The temperature field is not affected in this dry case, where potential temperature is conserved. However, in the moist version of the scheme, θ is no longer conserved, but θ_e, the equivalent potential temperature, is conserved. The temperature field would then be adjusted due to the condensation resulting from parcels ascending and becoming saturated.

The eddies can be parameterized in terms of mass fluxes, so the environment θ and m tendencies are expressed as

$$\left(\frac{\partial\theta}{\partial t}\right)_{\rm SI} = -\frac{1}{\bar{\rho}}\frac{\partial[M_u(\theta^p - \bar{\theta})]}{\partial Z} \qquad (21.11)$$

$$\left(\frac{\partial m}{\partial t}\right)_{\rm SI} = -\frac{1}{\bar{\rho}}\frac{\partial[M_u(m^p - \tilde{m})]}{\partial Z}, \qquad (21.12)$$

where θ^p is the parcel potential temperature, m^p is the parcel absolute momentum, and M_u is the CSI mass flux. Since $\theta^p - \bar{\theta} \approx 0$, the adjustment of the environment to a symmetrically neutral state reduces to (21.12). Due to the two-dimensional characteristic of the phenomenon, parcels conserve their absolute momentum. This turns into an advantage to the method because it is not necessary to make arbitrary estimates of the cloud absolute momentum; it is known exactly.

In CSI, the environment values are not the grid column values, as is the case for upright convection, but they are determined by the sloping path taken by the parcel. So, the environment absolute momentum, which will be denoted by $\tilde{m}$, must be calculated. In an operational model, such as that at ECMWF, one has access at a given grid column not only to the temperature and wind but also to their gradients. We use this fact here to obtain a computationally efficient method to estimate $\tilde{x}$ and $\tilde{m}$.

Therefore $\tilde{m}$ is estimated from a first-order approximation, using gridpoint values of m and $\partial m/\partial x$; that is,

$$\tilde{m} = m(0, z) + \frac{\partial m(0, z)}{\partial x}\tilde{x}, \qquad (21.13)$$

where $\tilde{x}$ is the horizontal distance between the parcel and the grid point, which is at $x = 0$. The θ surface where the parcel is positioned can be written similarly as

$$\tilde{\theta} = \theta^p = \theta(0, z) + \frac{\partial\theta(0, z)}{\partial x}\tilde{x}, \qquad (21.14)$$

where the neutral buoyant assumption has been used. Thus, the location of the parcel can be obtained by solving (21.14) for $\tilde{x}$:

$$\tilde{x} = [\theta^p - \theta(0, z)]\left[\frac{\partial\theta(0, z)}{\partial x}\right]^{-1}. \qquad (21.15)$$

The parcel path may span over the next grid columns; in this case an updated $\tilde{x}$ is estimated by taking the temperature and temperature gradient values from grid points nearest to the parcel. Here

$$\tilde{x}_{\rm upd} = n\Delta x + \Delta\tilde{x}, \qquad (21.16)$$

where Δx is the horizontal grid spacing, n is the number of grid columns over which the parcel path spanned, and

$$\Delta\tilde{x} = [\theta^p - \theta(n\Delta x, z)]\left[\frac{\partial\theta(n\Delta x, z)}{\partial x}\right]^{-1}. \qquad (21.17)$$

The recalculation of $\tilde{x}$ by taking nearer temperature gradients gives an improved $\tilde{m}$ estimate and consequently a better estimate of the cloud-top level.

Assuming no convective instability, $\partial\theta/\partial z \geqslant 0$, the numerator in (21.15), $\theta^p - \theta$, will always be negative or zero. The direction in which the θ surfaces slope will depend only on the sign of the horizontal temperature gradient $\partial\theta/\partial x$; consequently, the sign of $\tilde{x}$ will give the same information. Therefore, the instability criterion (21.1) basically compares the slope of the θ surfaces and the slope of m surfaces through the correlation of $\tilde{x}$ and $\hat{m}$ defined as the absolute momentum anomaly.

The instability criterion is tested from the lower levels upward. The first grid level that satisfies this criterion is taken as the cloud base. The parcel will continue to rise until $\hat{m}$ changes sign, that is, $\hat{m} = 0$ has occurred, and the parcel loses its lifting force. The cloud top is taken as the highest level before $\hat{m}$ changes sign.

Equation (21.12) can be rewritten as

$$\frac{\partial\tilde{m}}{\partial t} = -\frac{1}{\bar{\rho}}\frac{\partial[M_u(m^p - \tilde{m})]}{\partial Z}. \qquad (21.18)$$

The adjustments $\Delta\tilde{m}$, calculated from (21.18), should be applied to the position of the parcel; however, because the parcel at any level is rarely on a grid point, the tendencies are shared between the two nearest grid points, inversely proportional to their distances to the parcel. The velocity tendency $\Delta\tilde{v}$ is equal to $\Delta\tilde{m}$ as the adjustment is made at the location of the parcel.

It is interesting to note that from the definition of cloud base and cloud top, where $m^p - \tilde{m} = 0$, it follows that the restriction

$$\int_B^T \frac{\partial\tilde{m}}{\partial t}dZ = -\int_B^T \frac{1}{\bar{\rho}}\frac{\partial[M_u(m^p - \tilde{m})]}{\partial Z}dZ = 0 \qquad (21.19)$$

is immediately satisfied. Here B and T stand for cloud base and cloud top, respectively.

By way of simplification only updrafts are considered within the cloud. The bulk equations for mass, heat, and absolute momentum are

$$\frac{\partial M_u}{\partial Z} = E - D \qquad (21.20)$$

$$\frac{\partial(M_u m^p)}{\partial Z} = E\tilde{m} - Dm^p, \qquad (21.21)$$

where E and D are entrainment and detrainment of mass per unit length.

Turbulent entainment can occur within the cloud layer. Detrainment occurs at the cloud top. Overshooting and subcloud fluxes are included in some sensitivity tests. Overshooting is considered by adding detrainment to one more level above the cloud top, and subcloud fluxes are distributed in more than one layer.

The cloud-base mass flux is constrained by the CFL condition $M_u \leqslant \Delta z/\Delta t$. A value for $[M_u]_B$ chosen to be of the order of 0.5 kg m^{-2} s^{-1} satisfies this condition for $\Delta z = 750$ m and $\Delta t = 750$ s.

21.4. A potential vorticity view of the scheme

The instability criterion can be written in terms of the Ertel potential vorticity PV for a two-dimensional flow:

$$\bar{\rho}\mathrm{PV} = \frac{\partial m}{\partial x}\frac{\partial \theta}{\partial z} - \frac{\partial m}{\partial z}\frac{\partial \theta}{\partial x}. \quad (21.22)$$

Given that the presence of negative potential vorticity is one way to express the criteria for satisfying symmetric instability, it is interesting to analyze how this quantity can change following a parcel of air:

$$\frac{d\mathrm{PV}}{dt} = f\frac{g}{\theta_0}\mathbf{F}\cdot\nabla\theta + f\frac{g}{\theta_0}\boldsymbol{\zeta}\cdot\nabla Q. \quad (21.23)$$

On the rhs of (21.23), the first term is due to frictional forces $\mathbf{F}$ and the second due to diabatic processes Q such as radiative cooling, latent heat release, or sensible heat. Note that for a moist atmosphere, the instability criterion is in terms of the potential vorticity based on θ_e and that Eq. (21.23) then has an extra term due to the angle between θ_e and θ. In the present study, where the large- (resolved) scale atmosphere is assumed to be dry, two-dimensional, and frictionless, negative potential vorticity or symmetric instability cannot be produced in a initially stable environment. However, if we take into account the scale separation, potential vorticity can be defined as the scalar product of large- (resolved) scale absolute vorticity and large-scale potential temperature; that is, $\bar{\rho}\mathrm{PV} = \bar{\boldsymbol{\zeta}}\cdot\nabla\bar{\theta}$, where the overbar denotes large-scale area-averaged variables.

An equation for the large-scale potential vorticity with the above assumptions can be obtained by applying the dot product of $\nabla\bar{\theta}$ with the large-scale vorticity equation, which is

$$\frac{d\bar{\boldsymbol{\zeta}}}{dt} = \boldsymbol{\zeta}\cdot\nabla\bar{\mathbf{u}} - \nabla\times\frac{1}{\bar{\rho}}\nabla\cdot(\overline{\bar{\rho}\mathbf{u}'\mathbf{u}'}), \quad (21.24)$$

where on the rhs of (21.24) the first term is the vortex tilting term and the last term results from subgrid-scale effects, such as turbulent transports of relative vorticity. Note the horizontal transports of eddies are kept. Assuming the fluid is incompressible, the potential vorticity equation can be expressed in terms of large-scale and subgrid-scale variables, in the form

$$\frac{d}{dt}\left(\frac{\bar{\boldsymbol{\zeta}}\cdot\nabla\bar{\theta}}{\bar{\rho}}\right) = -\frac{\bar{\boldsymbol{\zeta}}}{\bar{\rho}}\cdot\nabla[\nabla\cdot(\overline{\mathbf{u}'\theta'})] - \frac{1}{\bar{\rho}}\left[\nabla\times\frac{1}{\bar{\rho}}\nabla\cdot(\overline{\bar{\rho}\mathbf{u}'\mathbf{u}'})\right]\cdot\nabla\bar{\theta}. \quad (21.25)$$

Thus, in two-dimensional atmosphere, with no resolved-scale frictional or diabatic effects, potential vorticity can still be changed following the trajectory of a parcel due to subgrid-scale frictional effects associated with turbulent mixing of heat and momentum.

In our idealized 2D model, where turbulent fluxes of heat are taken to be zero, this equation reduces to

$$\bar{\rho}\frac{d\mathrm{PV}}{dt} = -\frac{\partial}{\partial x}\left[\frac{1}{\bar{\rho}}\frac{\partial(\overline{\bar{\rho}v'u'})}{\partial x} + \frac{1}{\bar{\rho}}\frac{\partial(\overline{\bar{\rho}v'w'})}{\partial z}\right]\frac{\partial\bar{\theta}}{\partial z} + \frac{\partial}{\partial z}\left[\frac{1}{\bar{\rho}}\frac{\partial(\overline{\bar{\rho}v'u'})}{\partial x} + \frac{1}{\bar{\rho}}\frac{\partial(\overline{\bar{\rho}v'w'})}{\partial z}\right]\frac{\partial\bar{\theta}}{\partial x}. \quad (21.26)$$

Recalling Eq. (21.5) for absolute momentum, where $m' = v'$, Eq. (21.26) can be expressed as

$$\bar{\rho}\frac{d\mathrm{PV}}{dt} = \frac{\partial}{\partial x}\left(\frac{\partial\bar{m}}{\partial t}\right)_{\mathrm{SI}}\frac{\partial\bar{\theta}}{\partial z} - \frac{\partial}{\partial z}\left(\frac{\partial\bar{m}}{\partial t}\right)_{\mathrm{SI}}\frac{\partial\bar{\theta}}{\partial x}. \quad (21.27)$$

Thus,

$$\bar{\rho}\left(\frac{d\mathrm{PV}}{dt}\right)_{\mathrm{SI}} = \frac{\partial\dot{\bar{m}}}{\partial x}\frac{\partial\bar{\theta}}{\partial z} - \frac{\partial\dot{\bar{m}}}{\partial z}\frac{\partial\bar{\theta}}{\partial x}, \quad (21.28)$$

where $\dot{\bar{m}} = (\partial\bar{m}/\partial t)_{\mathrm{SI}}$.

The above expression gives the changes of potential vorticity of a parcel associated with the adjustments produced by the SI parameterization scheme. We will apply our SI scheme along θ surfaces, and in that case

$$\bar{\rho}\mathrm{PV} = -\left(\frac{\partial m}{\partial z}\right)_\theta\frac{\partial\theta}{\partial x}, \quad (21.29)$$

$$\bar{\rho}\left(\frac{d\mathrm{PV}}{dt}\right)_{\mathrm{SI}} = -\left(\frac{\partial\dot{\bar{m}}}{\partial z}\right)_\theta\frac{\partial\theta}{\partial x}. \quad (21.30)$$

In order to test the SI mass flux parameterization scheme and to understand its adjustment process, the scheme is applied initially in section 21.5 to an unstable environment. The scheme is assessed by the fraction of negative potential vorticity remaining in the domain. Further tests are applied in section 21.6 to the idealized flow, in which the initial conditions are stable to SI and a large-scale forcing is introduced so that a steady state can be examined where the forcing balances the parameterization tendencies.

21.5. Initially unstable flow

Two main groups of experiments were carried out: a single column cloud and a domain of grid columns.

a. Single column cloud

A profile of absolute momentum along a θ surface containing symmetric instability is constructed. It is assumed that horizontal and vertical gradients of θ and m are positive. So, the first level where $\partial\tilde{m}/\partial z > 0$ occurs, or that $\hat{m} = m^p - \tilde{m} < 0$, is taken as the cloud base.

The adjustment is done by applying Eq. (21.18), taking $M_u = 0.5$ kg m^{-2} s^{-1}, $\Delta z = 750$ m, and $\Delta t = 750$ s. To be representative of the ECMWF forecast model, an upstream difference scheme is used in discretizing (21.18) that becomes

$$\Delta\hat{m}(j) = -\frac{M_u\hat{m}(j+1) - M_u\hat{m}(j)}{\Delta z}\Delta t, \quad (21.31)$$

where height $= j\Delta z; j = 1, 2, 3 \cdots$.

The initial m profile along the θ surface shows a region in the center of the domain where the wind increases with height. This profile contains instability in the layer between 5.25 and 12.75 km, where the cloud base and cloud top are, respectively, considered to be located. (This is an unrealistically deep layer only being used here for illustrative purposes.) The vertical wind shear is of the order of 7×10^{-3} s^{-1} (Fig. 21.2a). The initial profiles of cloud mass flux M_u, absolute momentum anomaly $\hat{m}$, and the product M_u are shown in Fig. 21.2b. In this experiment no turbulent entrainment or detrainment occurred in the cloud layer, so the mass flux was maintained constant there. Detrainment occurs at the cloud top, where M_u vanishes. The subcloud fluxes reach zero in one layer just below the cloud base.

The instability is removed initially from upper levels downward. The cloud top descends more rapidly than the cloud base. This results in decreasing the depth of the unstable layer. The wind shear is at the same time reduced. The adjusted m profiles tend to stable rather than neutral profiles. The scheme also acts on regions outside the cloud layer, and hence the subcloud layer, which is initially very stable, has its stability reduced. All instability in the column was eliminated in about 4 h or less.

Other tests were done to reproduce overshooting and subcloud fluxes. For example, to include the effects of cloud overshooting, the detrained mass is distributed over two layers above cloud top. The subcloud-layer fluxes can be set to decrease to zero in one or two layers below the cloud base. The inclusion of these two processes, however, produced no significant changes to the adjusted profile. Very small differences can be noticed only near the cloud top and cloud base.

b. *Whole cloud domain*

1) INITIAL CONDITIONS

An idealized absolute momentum m and potential temperature θ field is constructed in a two-dimensional domain in x and z directions. The temperature field consists of a basic state and a known perturbation function:

$$\theta = \bar{\theta}(x) + \theta'(x, z), \quad (21.32)$$

where $\bar{\theta}(x) = f(\theta_0/g)\bar{V}_z x + 300$ K.

The perturbation field comes from the arbitrary function:

$$\frac{\partial\theta'}{\partial z} = \frac{\theta_0}{g}\left\{s_1 - s_0 \exp\left(\frac{-x^2}{a^2}\right)\sin[2\pi(z-d)]\right\}. \quad (21.33)$$

The values of the constants are

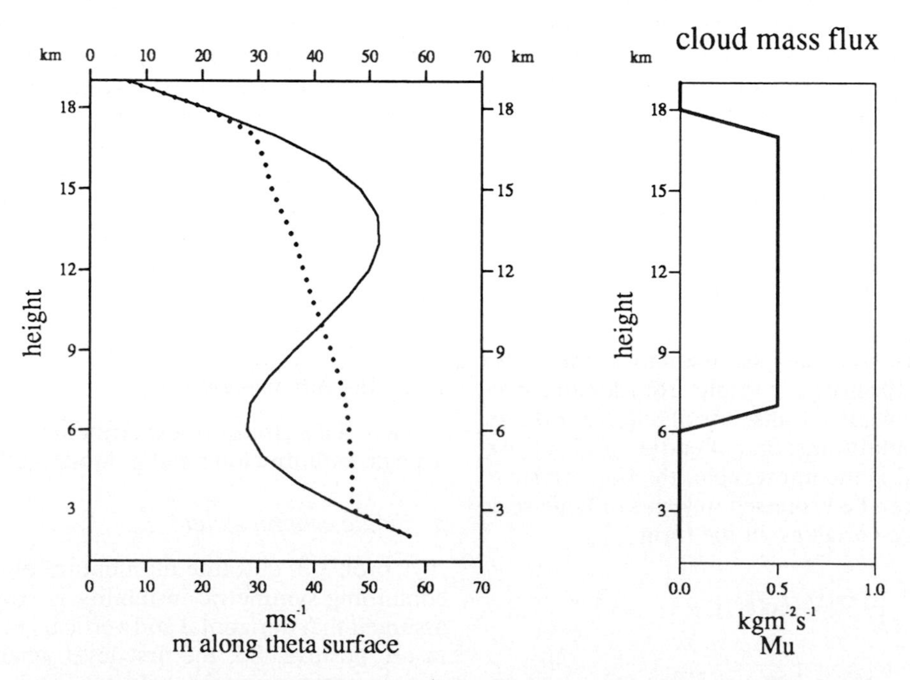

FIG. 21.2. (a) Absolute momentum (m s^{-1}) along the potential temperature surface. Initial (solid line) and adjusted (dotted line) profiles at $t = 4.2$ h. (b) Initial cloud mass flux profile.

$$\theta_0/g = 30 \text{ K m}^{-1} \text{ s}^2,$$

$$s_1 = 0.119 \times 10^{-3} \text{ s}^{-2},$$

$$s_0 = 0.575 \times 10^{-4} \text{ s}^{-2},$$

$$\bar{V}_z = 0.75 \times 10^{-2} \text{ s}^{-1},$$

$$f = 1.0 \times 10^{-4} \text{ s}^{-1}.$$

Here a and d are the two parameters that are changed to produce different instability patches, and z is nondimensional and varies between 0 and 1 from the bottom to the top boundary.

The temperature field can be constructed from $\theta(x, z) = \bar{\theta}(x) + \int_0^z (\partial\theta'/\partial z)dz$. The thermal wind relation is used to calculate $\partial v'/\partial z$, from the prescribed perturbation. The meridional wind is also assumed to be $v = \int_0^z [\bar{V}_z + (\partial v'/\partial z)]dz$. Finally, the absolute momentum is obtained from the meridional wind using $m = v + fx$.

The original full m–θ field was constructed from Eq. (21.33). It comprises 48 points in the horizontal, with a resolution of $\Delta x = 31.25$ km, and 24 levels in the vertical, with $\Delta z = 750$ m. This field is supposed to represent a "real" atmosphere. To simulate the coarse horizontal resolution of a large-scale model, another field was constructed by taking a lesser number of grid columns equally spaced. This new field has lower horizontal resolution than the original one, and we refer to these data as "the analytical form." In the interior of this domain, θ lines slope more than the m lines and potential vorticity is negative. This is the region unstable to SI.

2) EXPERIMENT CHARACTERISTICS

Several tests were carried out running the scheme over the whole domain. The parameters $\tilde{x}$ and $\hat{m}$ were estimated using Eqs. (21.15) and (21.13) as would be done in an operational version of this scheme (note that in the next subsection we use the known analytical values of $\tilde{x}$ and $\hat{m}$ rather than these estimates). Different instability regions were constructed. Some sensitivity tests were made to include processes such as entrainment, overshooting, and subcloud flux, as well as spatial resolution.

Entrainment is given by

$$E_u(z) = \epsilon_u M_u(z), \qquad (21.34)$$

where ϵ_u is the fractional turbulent entrainment rate, taken as $\epsilon_u = 1.0 \times 10^{-4} \text{ m}^{-1}$.

So, the entrainment rate in a layer $\Delta z = 750$ m can be written as $E(j) = 0.075 M_u(j-1)/\Delta z$.

Detrainment occurs at cloud top. To check the sensitivity of the scheme to cloud-top overshooting, detrainment is distributed over two layers and taken as

$$D(\text{top} + 1) = 0.75 M_u(\text{top})/\Delta z,$$

$$D(\text{top} + 2) = 0.25 M_u(\text{top})/\Delta z.$$

The cloud mass flux decreases to zero in the layer immediately below the cloud base. The sensitivity of the scheme to subcloud mass flux was also tested, with fluxes distributed as

$$M_u\hat{m}(\text{base} - 1) = 0.3 M_u\hat{m}(\text{base}),$$

$$M_u\hat{m}(\text{base} - 2) = 0.$$

We now assume that the forecast model to which this parameterization is to be applied has only a small number of grid columns in the domain. Experiments were done using six or eight grid columns.

Due to the fact that in SI every parcel rises following a sloping θ surface, the parcels originating from a given

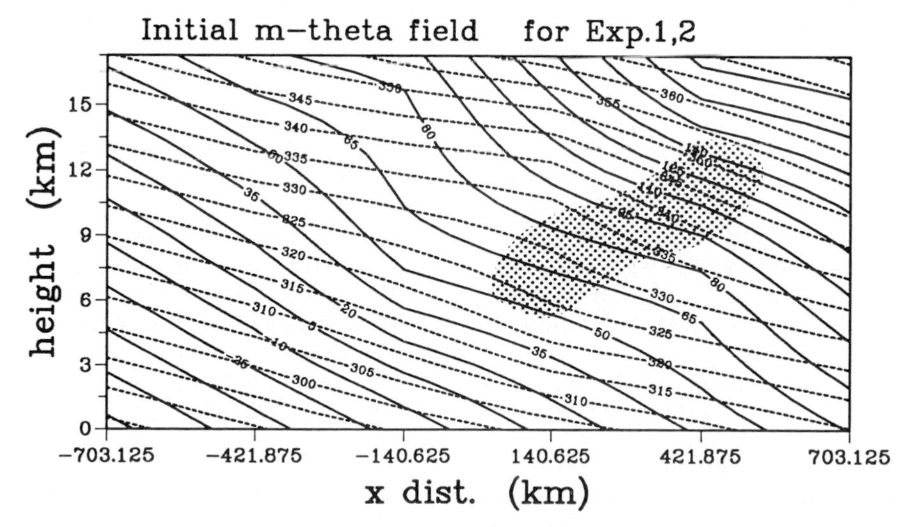

FIG. 21.3. Initial potential temperature (dashed lines, contours: 5 K) and absolute momentum (solid lines, contours: 15 m s^{-1}) fields for experiments 1 and 2. Region of negative potential vorticity is shaded.

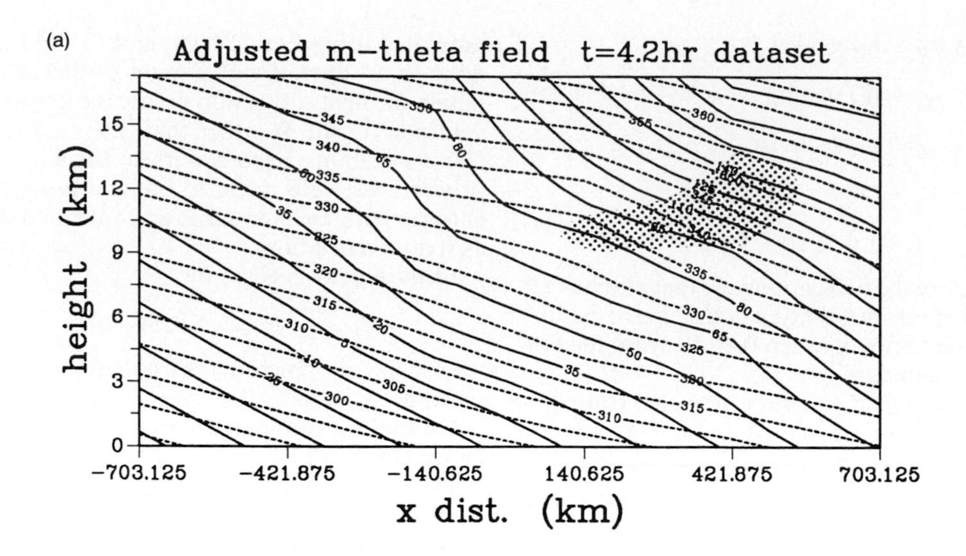

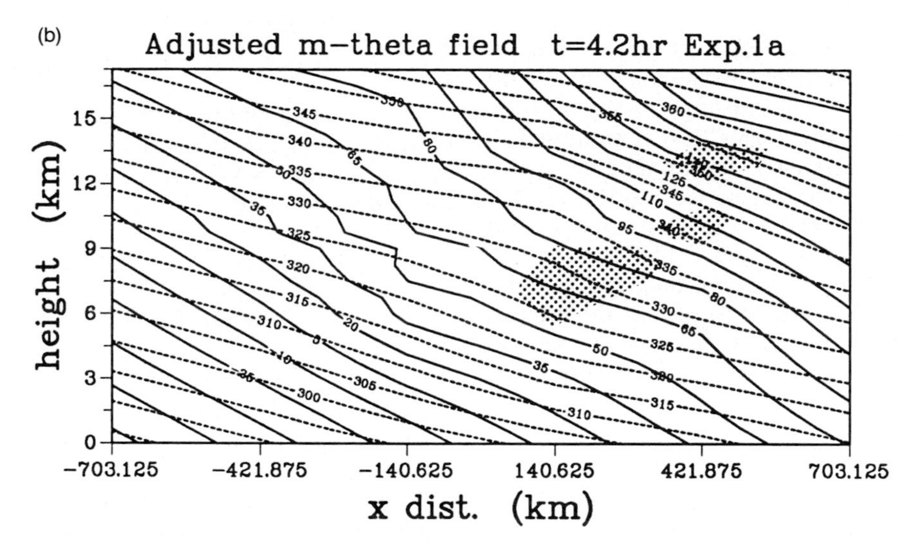

FIG. 21.4. Adjusted potential temperature (dashed lines, contours: 5 K) and absolute momentum (solid lines, contours: 15 m s^{-1}) fields. Region of negative potential vorticity is shaded. (a) Run using $\tilde{x}$ and $\tilde{m}$ from the initial dataset $t = 20\Delta t$; (b) experiment 1a—$t = 20\Delta t$; (c) experiment 1b—$t = 20\Delta t$; (d) experiment 2—$t = 10\Delta t$.

pressure level in a (vertical) grid column can be treated as essentially independent of those at other levels. This is unlike convection that is represented by the ascent of a single parcel lifted from near the surface. A test was included in which parcels from each pressure level in one grid column, unstable to SI, were separately lifted and the adjustments were applied simultaneously. This version of the scheme was called the "multibase cloud" scheme.

The experiments can be listed as the following: 1a—six grid columns; entrainment, overshooting, and subcloud fluxes included; 1b—eight grid columns; 2—six grid columns; multibase cloud scheme.

The horizontal resolution for the six-grid-column domain was Δx = 281.25 km and for the eight-grid column was Δx = 187.5 km. The vertical resolution was kept as Δz = 750 m.

The instability regions constructed for experiments 1 and 2 used the parameters a = 330 km and d = 0.20. The initial field of m–θ surfaces and PV surfaces is shown in Fig. 21.3.

Potential vorticity was integrated over the interior of the domain. The ratio between the area-integrated initial total PV and the area-integrated final total PV measures the conservation of PV by the scheme. The ratio between the initial integrated negative PV and the final integrated negative PV measures the fraction of the instability left in the domain by the scheme.

3) USING EXACT $\tilde{x}$ AND $\tilde{m}$ FROM ANALYTICAL FORM

To evaluate the performance of the scheme without recourse to any estimation of $\tilde{x}$ and $\tilde{m}$, the po-

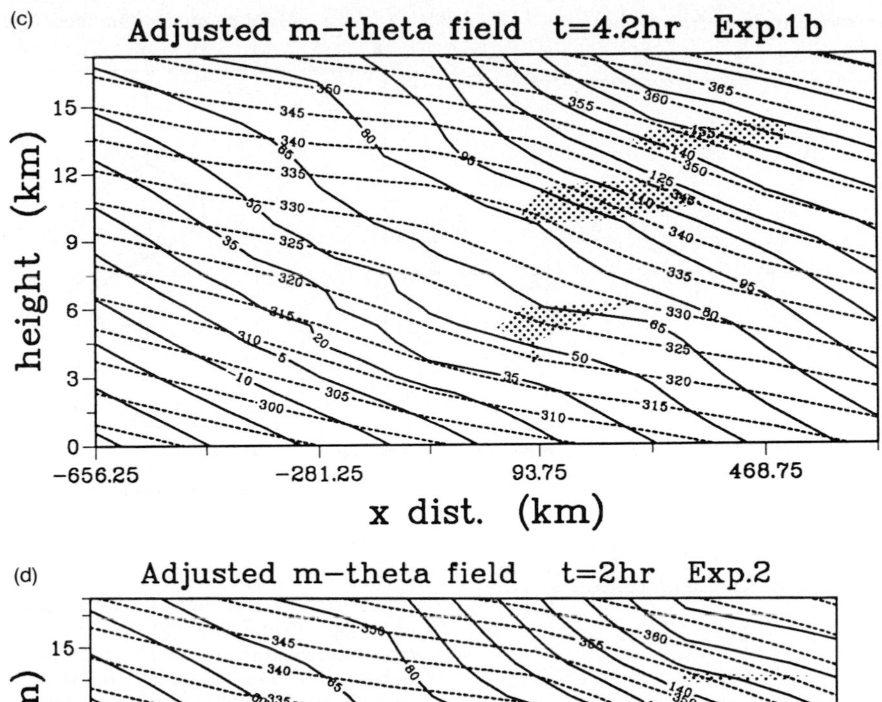

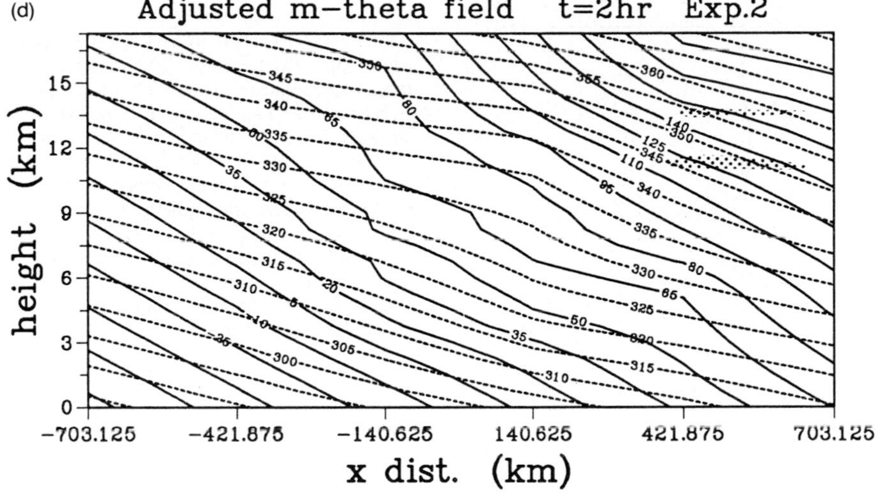

FIG. 21.4. (*Continued*)

sition of the θ surface, and its environment absolute momentum, were taken from the analytical formulas for the initial m–θ field rather than from the estimates of Eqs. (21.13) and (21.15). It was necessary, however, to apply a linear interpolation between neighboring gridpoint values to find the exact θ surface. The mass flux was kept constant in the cloud layer and experiment characteristics are otherwise as in experiment 1a.

The scheme removed 35% of the initial instability in 20 time steps, bringing the m surfaces almost parallel to the θ surfaces in the lower layers (Figure 21.4a). This run showed that the scheme succeeds in removing or reducing the instability in a domain

TABLE 21.1. Potential vorticity values for each experiment.

Exp no.	$-PV_{init}$ (PVU)	PV_{min} (PVU)	$-PV$	$+PV$	PV_{tot}	Comment
1a	0.0479	−0.085	0.42	0.98	1.004	$20\Delta t$, six grid columns, entrainment, overshooting and subcloud fluxes included
1b	0.0638	−0.127	0.16	0.95	1.008	$20\Delta t$, eight grid columns
2	0.0479	−0.085	0.04	0.98	1.010	$10\Delta t$, six grid columns, multibase cloud

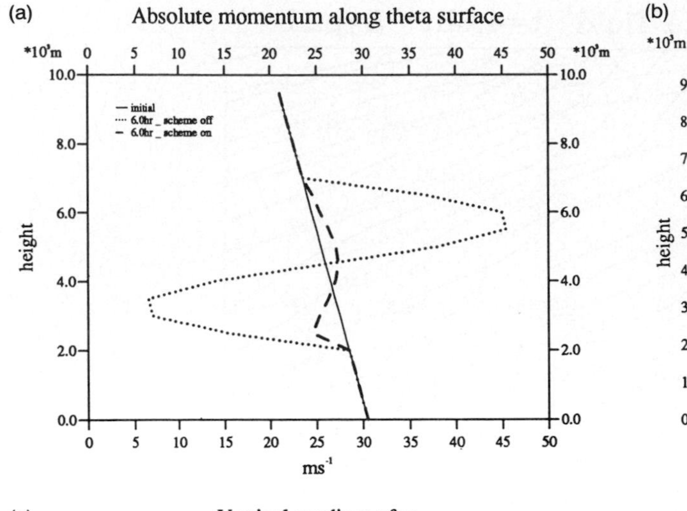

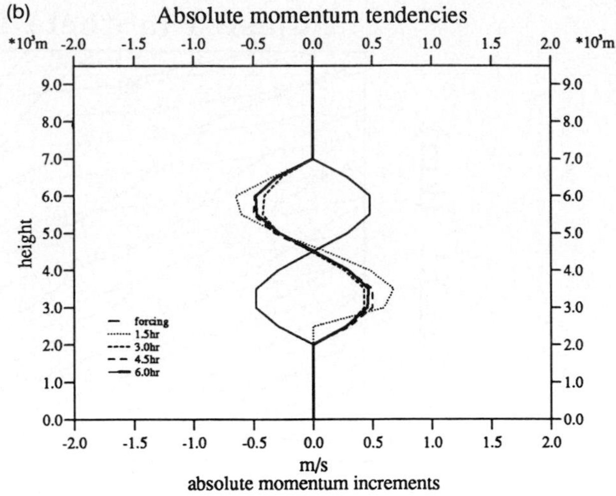

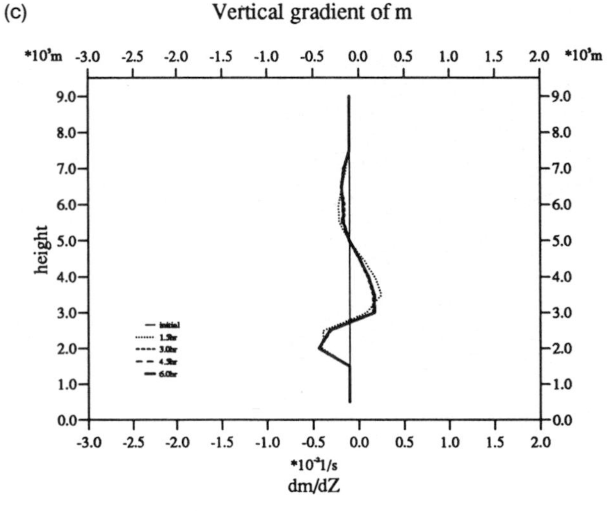

FIG. 21.5. Experiment with forcing applied to a single column cloud. (a) Profiles of absolute momentum along a potential temperature surface. The initial profile (solid line) is stable to SI and so has negative vertical wind shear along θ surfaces; it is taken from a cyclonic region with positive horizontal gradient of temperature. The dotted line shows the m profile at $t = 6$ h in the presence of the forcing but without the SI parameterization scheme acting. The dashed line shows the m profile for the case in which the scheme is included. (b) Profiles of the forcing and scheme tendencies on the absolute momentum (m s^{-1}), for every 1.5 h. (c) Profiles of the vertical gradient of absolute momentum ($\times 10^{-2}$ s^{-1}), for every 1.5 h. $M_u = 0.5$ kg m^{-2} s^{-1}, $A_c = 3.6$ m s^{-1} h^{-1}.

containing more than one grid column unstable to SI.

4) RESULTS OBTAINED BY ESTIMATION OF $\tilde{x}$ AND $\tilde{m}$

Initially in the six-grid-column domain the instability patch spanned over two grid columns, and in the eight-grid-column domain over four grid columns.

Figures 21.4b–d show the adjusted m–θ and PV fields for the tests after $20\Delta t$ and Table 21.1 summarizes some of these results. In Table 21.1, column 2 is the average of negative PV for the initial state (i.e., it is the sum of PV over all grid points at which PV < 0 divided by the number of unstable grid points). Column 3 gives the minimum value of PV found in the initial field. Negative PV, positive PV, and the total PV are, respectively, summed in the domain for every time integration. The ratios between the initial and the adjusted sums of $-$PV, $+$PV, and total PV after $20\Delta t$ are shown in columns 4–6.

Apart from the very small changes at the cloud-base and cloud-top regions, the effects of considering cloud-top overshooting and extended subcloud fluxes showed no significant changes or improvements to the scheme. Only for the cases of multibase cloud and for eight grid columns, the exclusion of those effects resulted in slightly more efficient reduction of instability in the domain.

The inclusion of entrainment of mass removed slightly more instability from this domain. Due to the mixing with the environment, the cloud was less "buoyant," and the cloud tops occurred a little lower. The increase of total PV after 20 time steps was equal to or less than 1%, indicating a relatively good conservation of total PV by the scheme. The relatively small decrease of $+$PV is due to the fact that the initial $+$PV is very large, and any small fraction is necessary to compensate the decrease of $-$PV to conserve the total PV in the domain.

The multibase cloud scheme reduces the instability in a much faster manner compared to the version of the scheme in which each time step lifts only the lowest unstable parcels. This scheme reduced instability in a time scale of about 7–10 h, whereas the multibase cloud scheme reduced it in about 2–4 h. The latter time scale agrees with that of the single column cloud tests.

In the eight-grid-column test, the scheme adjusted the m surfaces along the θ surfaces more smoothly, in spite of the larger initial $-$PV values. These results show that the spatial resolution may have an important effect on the performance of the scheme.

21.6. Forcing applied to an initially stable flow

The presence of a cumulus cloud affects the environment by cooling it through detrainment and warming it by cumulus-induced subsidence. Vertical shear of horizontal wind, if present, is reduced through mixing processes between cloud and environment. On the other hand, the existence of the cloud itself is a result of processes such as advection, radiative cooling, or boundary-layer fluxes. These are large-scale processes that produce an environment suitable for developing cumulus clouds.

In some prognostic cumulus parameterization studies observed data are taken as initial conditions, and the model is run allowing a single cloud to develop and then decay when finally the environment is adjusted toward a state that is neutral to convection. The period of time until the cloud is eliminated is called the adjustment time. Precipitation and heating distribution are compared with the observed data.

Assuming that the changes in the large-scale processes occur on time scales much longer than the time scale of the cloud effects, it is reasonable to have the large-scale forcing process applied during the time integration of the scheme. So, it is expected that clouds, or their parameterized effects, continuously counteract the action of the forcing in destabilizing the environment. This is the basis of the Arakawa–Schubert convective scheme (Arakawa and Schubert 1974) in which an ensemble of clouds follow quasi-equilibrium states with the large-scale forcing. The final adjusted state is not necessarily toward a neutral state, but toward one with clouds in equilibrium with the large-scale forcing. This is one of the advantages of including in the model a forcing term with a destabilizing effect. In cloud parameterization studies, a steady-state solution for the equations of motion can be sought. In the presence of forcing the adjustment time can also be determined. Another advantage of the use of forcing is the fact that it is a more realistic attempt to represent the atmosphere.

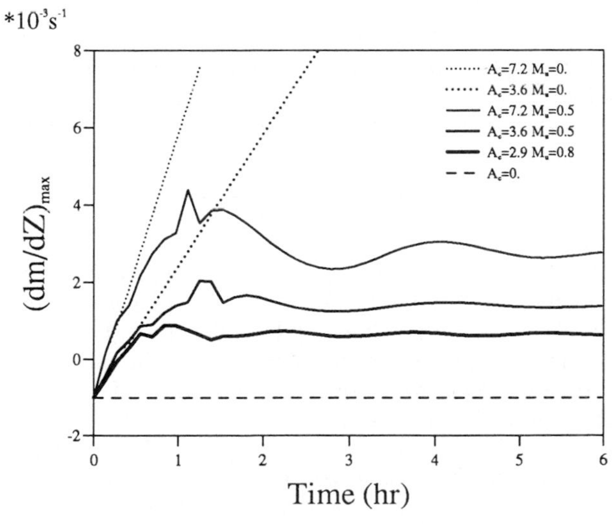

FIG. 21.6. Rate of change of the unstable layer shear, $(\partial m/\partial Z)_{\max}$, for different values of A_c (m s^{-1} h^{-1}) and M_u (kg m^{-2} s^{-1}) applied to a single-column cloud. Dashed and dotted lines lines refer to limiting cases when either no forcing or no scheme is acting. Solid line refers to experiments when both forcing and SI scheme are present.

In the set of following experiments, large-scale forcing is applied to the absolute momentum field, initially in the single column cloud, and afterward in the whole cloud domain. The forcing is maintained constant during the time integrations; thus, when the scheme is switched on, the quasi-equilibrium state is also a quasi-steady state of m–θ field.

The absolute momentum adjustments produced by both processes should be approximately balanced in the quasi-steady state; that is,

$$\frac{\partial \bar{m}}{\partial t} = \left(\frac{\partial \bar{m}}{\partial t}\right)_{\mathrm{LS}} + \left(\frac{\partial \bar{m}}{\partial t}\right)_{\mathrm{SI}} \approx 0. \qquad (21.35)$$

Similarly, the rate of change of negative or positive vorticity is approximately zero; that is,

$$\frac{\partial \mathrm{PV}}{\partial t} = \left(\frac{\partial \mathrm{PV}}{\partial t}\right)_{\mathrm{LS}} + \left(\frac{\partial \mathrm{PV}}{\partial t}\right)_{\mathrm{SI}} \approx 0. \qquad (21.36)$$

a. Single column cloud

The initial condition is defined by an absolute momentum profile of an environment that is symmetrically stable. Assuming the parcel trajectory is along the θ surface, this profile is represented by absolute momentum decreasing monotonically with height. In this idealized model, absolute momentum decreases linearly, so the $m(z)$ profile is given by $m(z) = m_0 + \alpha z$, where m_0 is the absolute momentum at the lowest model level, and α the shear of absolute momentum along the θ surface. A more negative value for α indicates an m profile less suitable for symmetric instability.

A simple time-independent forcing is designed with a sinusoidal function in height and applied at the midlevels of the profile. This forcing is given by

$$\left(\frac{\partial \bar{m}}{\partial t}\right)_{\mathrm{LS}} = \begin{cases} -A_c \sin \dfrac{2\pi(z - z_b)}{D_c}, & \text{if} \quad z_b < z < D_c + z_b \\ 0, & \text{otherwise} \end{cases} \qquad (21.37)$$

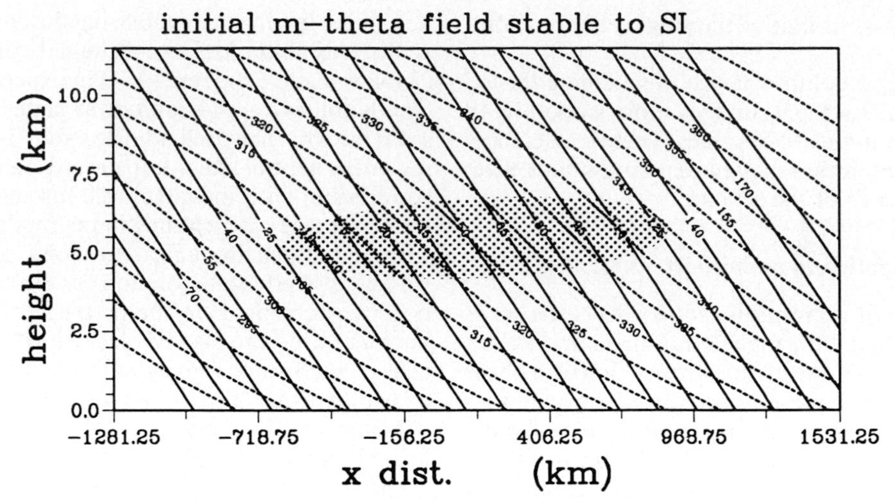

FIG. 21.7. Experiment with forcing applied to a whole cloud domain. Initial fields of potential temperature (dashed lines, contours: 5 K) and absolute momentum (solid lines, contours: 15 m s^{-1}). Region of *minimum* but still positive potential vorticity is shaded.

where z_b is the height at cloud base, D_c is the depth of the cloud layer, and A_c is a constant forcing amplitude. The advantage of this simple forcing is to keep the column integral absolute momentum constant.

The vertical resolution is taken as $\Delta z = 500$ m with 20 height levels, and the time step of integration as $\Delta t = 500$ s. The cloud mass flux M_u has to be smaller than 1 kg m^{-2} s^{-1} to satisfy the CFL condition.

When the scheme is switched on, to counteract the destabilization mechanism of the forcing, it is important to quantify the instability present in the column. Potential vorticity, measured by gradients along the θ surface, indicates how much instability there is in the column and how it evolves with time from Eqs. (21.29) and (21.30).

A comparison between the magnitudes of the large-scale forcing tendencies and the adjustments produced by the scheme is from the ratio:

$$\frac{(\partial m/\partial t)_{\mathrm{LS}}}{(\partial m/\partial t)_{\mathrm{SI}}} = \frac{-A_c \sin[2\pi(z - z_b)/D_c]}{-\bar{\rho}^{-1}(\partial/\partial x)[M_u(m^p - m)]}. \quad (21.38)$$

Assuming no mass entrainment through the lateral boundaries of the cloud, the cloud mass flux and the parcel momentum is maintained constant in the cloud layer, so the relation (21.38) can be rewritten as

$$\frac{(\partial m/\partial t)_{\mathrm{LS}}}{(\partial m/\partial t)_{\mathrm{SI}}} = \frac{-A_c \sin[2\pi(z - z_b)/D_c]}{\bar{\rho}^{-1}M_u(\partial m/\partial z)}. \quad (21.39)$$

In the steady state, the scheme should be approximately balancing the large-scale forcing; in this case, this relation should be approximately -1.

Thus, the steady-state profile of absolute momentum can actually be predicted by taking (21.39) equal to -1, and integrating m from the cloud base z_b to a height z within the cloud layer, thus giving

$$m(z) = m^p + \frac{\rho A_c D_c}{M_u 2\pi}\left[1 - \cos\frac{2\pi(z - z_b)}{D_c}\right]. \quad (21.40)$$

At the midlayer height, $z_h = z_b + D_c/2$, we have the following relation:

$$\frac{\Delta m}{\Delta z} = \frac{m(z_h) - m^p}{D_c/2} = \frac{2\rho A_c}{\pi M_u}. \quad (21.41)$$

Whenever the forcing is present, $A_c > 0$, there will always exist a "cloud" in the column, $\partial m/\partial z > 0$; however, this instability can be kept to a minimum by having the ratio $\bar{\rho}A_c/M_u$ small. Equation (21.41) shows the magnitude of the unstable shear that remains in the layer in the steady state.

A set of single column cloud experiments was carried out. In these experiments the initial profile, the amplitude of the forcing, and the cloud mass flux varied and the scheme was assessed under large-scale forcing conditions.

Results. Figures 21.5a–c show the results of the 6-h integration for the single-column cloud with forcing and scheme on. The forcing has an amplitude, $A_c = 3.6$ m s^{-1} h^{-1}, applied at every time step. The cloud mass flux was fixed at 0.5 kg m^{-2} s^{-1}, giving a ratio of $\bar{\rho}A_c/M_u = 2.0$ m s^{-1} km^{-1}, where density is taken as 1 kg m^{-3}. Figure 21.5a shows the initial and adjusted m profiles along the θ surface. The initial shear was taken as $\alpha = -1.0$ m s^{-1} km^{-1}.

At $t = 3.0$ h, the adjusted m profile has already reached a steady state. The steady "cloud" has its base at 2500 m and top at 7000 m. This is the layer in which the forcing is acting. The magnitude of the shear of the steady m profile is dictated by the forcing amplitude. At this time, the profile of the absolute momentum adjustments (Fig. 21.5b) due to the scheme is approximately balancing the absolute momentum tendencies

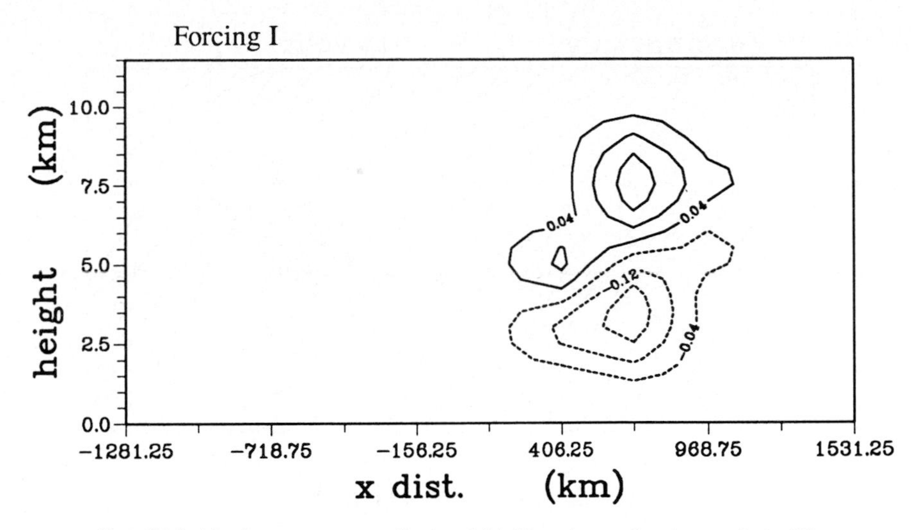

FIG. 21.8. Absolute momentum forcing field. Negative tendencies are dashed lines and positive tendencies are solid lines (contours: 0.08 m s^{-1} per time step).

produced by the forcing. It can be seen clearly from the profile of the vertical gradient of absolute momentum along the θ surface (Fig. 21.5c) that at $t = 3.0$ h a steady state has been established. The interpretation of the $(\partial m/\partial z)_\theta$ profile is that layers with negative gradients are stable to SI, while layers with positive gradients are unstable to SI. Positive values of $(\partial m/\partial z)_\theta$ occurred in a shallow layer between 3000 and 4000 m. A very stable subcloud layer was produced as a result of the negative tendencies of the forcing applied immediately above the subcloud layer. The shear is kept of order 10^{-3} s^{-1}, whereas in the absence of this parameterization scheme the shear is of the order of 10^{-2} s^{-1} after 3 h.

The performance of the scheme can be evaluated by comparing the dotted curves with the solid curves in Fig. 21.6. These curves show the maximum wind shear in the column and give a measure of the instability in the cloud layer and its progression in time. In the absence of the scheme, the forcing changes the initially stable shear to produce unstable shear that increases linearly with time. The rate of increase is controlled by the magnitude of the forcing. The presence of the counteracting effects of the parameterization scheme prevents the instability from growing steadily. For the initial two hours, the scheme together with the forcing results in rather noisy curves (in time) of unstable shear. After these initial hours, both the scheme and the forcing tendencies start balancing each other as the variations in the curves become smaller. A steady state can also be identified in these curves as the unstable shear converges to a constant value given by the relation (21.41). The different solid curves show the dependence of the remaining unstable shear to the ratio $\bar{\rho}A_c/M_u$. An adjustment time scale of about 3.5 h can be assumed, as after this time the variations are much smaller than they were initially.

Different initial profiles were tested: stable to SI ($\alpha = -1.0$ m s^{-1} km^{-1}), strongly stable to SI ($\alpha = -2.0$ m s^{-1} km^{-1}), and unstable to SI (figures not shown). After some time integration, the model reaches an equilibrium state that seems to be independent of the initial conditions.

b. *Whole cloud domain*

A two-dimensional domain with absolute momentum and potential temperature fields was constructed using Eq. (21.33). Some parameters were changed to produce an initial stable field. The modified parameters are $a = 1000$ km; $s_1 = 0.15 \times 10^{-3}$ s^{-2}; $s_0 = 0.57 \times 10^{-4}$ s^{-2}; and $\bar{V}_z = 0.8 \times 10^{-2}$ s^{-1}. The vertical resolution was taken as $\Delta z = 500$ m and the horizontal resolution was kept at $\Delta x = 281.25$ km. The time step of integration was taken as $\Delta t = 500$ s. The dimensions of the domain were 12.0 km × 2812.5 km, with 24 height levels and 11 grid columns. In this stable domain, absolute momentum surfaces slope more than the potential temperature surfaces. A region of minimum potential vorticity is located in the center (Fig. 21.7).

The forcing is maintained constant during the time integrations. It is given as

$$F_d(x, z) = A_d \exp\left[\frac{-(x - x_0)^2}{a_d^2}\right] \exp\left[\frac{-(z - z_0)^2}{b_d^2}\right] \times \sin\left[\frac{2\pi(z - z_0)}{D_d(x)}\right], \quad (21.42)$$

where x_0 and z_0 are the coordinate points chosen to be the origin of the forcing function. The exponential dependence in x and z produces a damping in the amplitude of the forcing toward its borders. The simple sinusoidal function in z produces a change of sign of

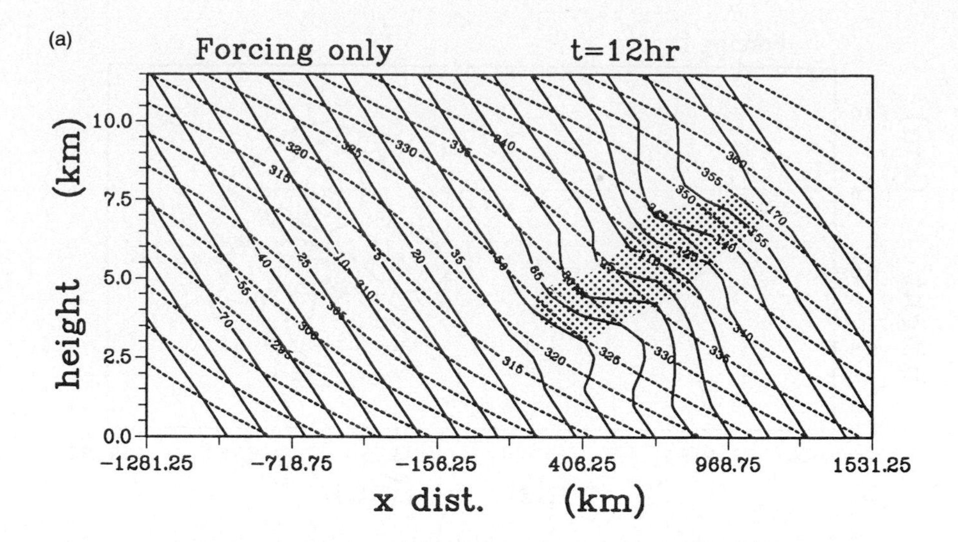

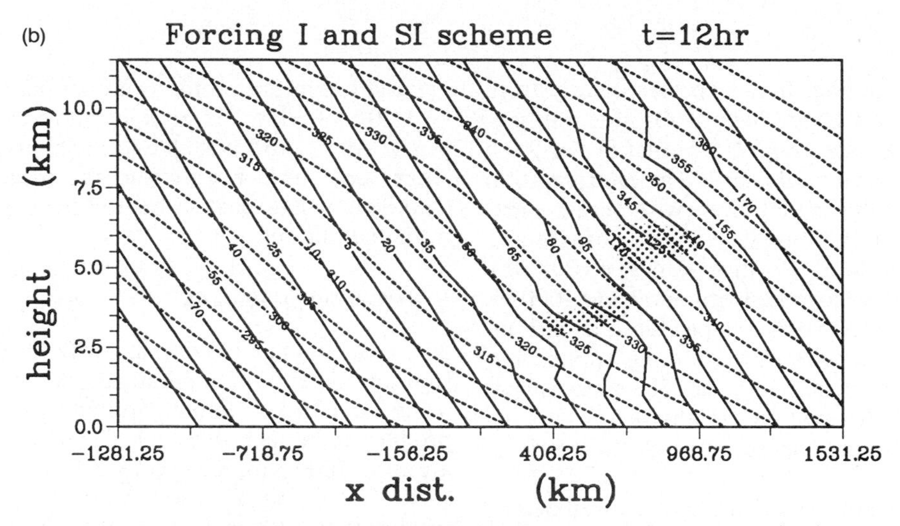

FIG. 21.9. Whole cloud domain experiment. (a) Forcing only (A_d = 2.2 m s^{-1} h^{-1}, tilt—1/225), t = 12 h. Adjusted potential temperature (dashed lines, contours: 5 K) and absolute momentum (solid lines, contours: 15 m s^{-1}) fields. (b) Forcing and SI scheme. Adjusted potential temperature (dashed lines, contours: 5 K) and absolute momentum (solid lines, contours: 15 m s^{-1}) fields. Negative potential vorticity is shaded. (M_u = 0.5 kg m^{-2} s^{-1}), t = 12 h.

the forcing tendencies in the domain and maintains as constant the volume integral absolute momentum after every time integration. The depth of the forcing, $D_d(x)$ varies for different grid columns producing a population of clouds with different cloud-top heights. The cloud base can also be different by tilting the forcing patch.

The multibase cloud version of the scheme was used in the next set of experiments with the whole cloud domain and large-scale forcing. These experiments consisted of varying the amplitude of the forcing and the magnitude of the cloud mass flux M_u and the shape of the forcing patch.

Results. A forcing of amplitude A_d = 2.2 m s^{-1} h^{-1} with its long axis sloping at a rate 1/225 is applied on the right half of the domain (Fig. 21.8). At t = 12 h a large instability patch is fully developed and spans over three grid columns (Fig. 21.9a). Strong stable regions develop above and below the unstable patch. Taking $\dot{m}_F$ as the absolute momentum tendencies applied by the forcing, a negative potential vorticity region will be produced wherever $\partial\dot{m}_F/\partial z > 0$ and $\partial\dot{m}_F/\partial x < 0$; according to the forcing characteristics this results in an elongated and tilted patch.

When the parameterization scheme is included with a cloud mass flux M_u equal to 0.5 kg m^{-2} s^{-1}, the neg-

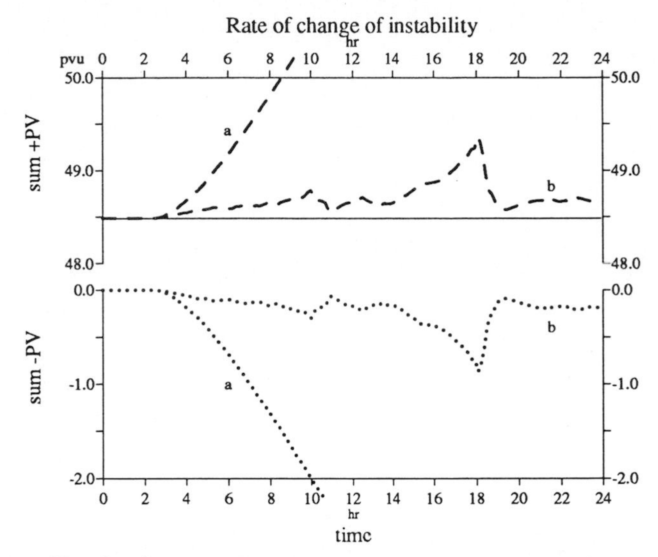

FIG. 21.10. Whole cloud domain experiment. Rate of change of potential vorticity summed in the interior of domain. (a) Forcing only. (b) Forcing and SI scheme ($M_u = 0.5$ kg m^{-2} s^{-1}). Total summation in solid, negative in dotted, and positive in dashed lines.

ative PV values are kept very small at $t = 12$ h; only the zero contour line shows (Fig. 21.9b). The regions of negative PV at this time are very similar to those at $t = 9$ h. When the forcing alone is modifying the absolute momentum field, θ surfaces slope much more than the m surfaces in the center of the domain. In the experiments in which the scheme is switched on, these surfaces are approximately parallel, or closer to a neutral state, as Fig. 21.9b shows.

Experiments with different values of $\bar{\rho}A_d/M_u$ showed that a steady state is established in which the unstable shear has a magnitude given by this ratio in accordance with Eq. (21.41). Although less instability is produced with weak forcing, the region where it acts tends almost to neutrality with m–θ surfaces parallel and smoother. The overstabilization above and below the cloud layer is minimized.

Figures 21.10a,b show the rate of change of the summation of the potential vorticity in the interior of the domain. The total potential vorticity presents a good conservation during the whole time integration. The amount of symmetric instability reduced by the scheme can be assessed by comparing the curves from the experiment where the forcing acts alone with the curves from the experiment with the scheme counteracting the forcing. This curve suggests an adjustment time of about 5 h.

A different forcing was constructed (figure not shown). The amplitude of the forcing was kept the same, but the tilt changed to approximately $-1/170$, oriented almost parallel to the θ surfaces. After 12 h of integration, this forcing produced smaller and weaker instability compared to the previous forcing. At $t = 12$ h, the potential vorticity field showed scattered and localized regions of negative PV, with reduced amplitude compared with the case in which the forcing is acting alone. The momentum surfaces show more structure.

21.7. Concluding remarks

The results given here indicate that a simple extension to the mass flux scheme for convection can operate to remove symmetric instability. The efficiency of the scheme was shown working under initially unstable conditions and initially stable conditions with forcing applied, either in a single column cloud or in the whole cloud domain. The rate at which the scheme effects a stabilization of the atmosphere depends on the choice of M_u. The closure of the scheme by relating M_u to the resolved vertical velocity or moisture convergence has not been considered here.

The adjustment time for the single column cloud occurs in about 3–4 h. The whole cloud domain requires a longer adjustment time due to the interactive effects of the population of clouds.

The forcing represents the large-scale flow's tendency to increase the shear locally to produce symmetric instability. As this is equivalent to producing a sink of PV locally, it can occur only by means of (unresolved) irreversible processes such as turbulence.

One conclusion from this study is that due to the sloping nature of parcel trajectories in SI there is a requirement to simultaneously perform adjustments due to the lifting of parcels from *each* vertical level in a grid column. This is different from convective instability, which is taken as a vertical adjustment. This simultaneous lifting of several parcels in a given grid column we have called the "multibase cloud" scheme.

The scheme can also be easily extended for the full moist atmosphere. Questions such as whether current operational models need to include such a scheme have not been addressed herein.

Acknowledgments. The authors would like to acknowledge the help of Dr. Martin Miller in formulating this scheme. Comments from Dr. Thor Erik Nordeng have improved the paper.

Part VI

Use of Explicit Simulation in Formulating and Testing Cumulus Representations

Chapter 22

Cumulus Ensemble Simulation

KUAN-MAN XU

Department of Atmospheric Science, Colorado State University, Fort Collins, Colorado

22.1. Introduction

The cumulus parameterization problem is the formulation of the collective effects of subgrid-scale cloud processes in terms of the prognostic variables of resolvable scale. A quantitative understanding of the basic physical processes for individual clouds and their mesoscale organization is essential in order to address this problem. The understanding is, however, still not adequate due to (i) the lack of direct observations of the quantities associated with an ensemble of cumulus clouds (e.g., the cumulus mass flux), (ii) the difficulties of measuring temperature and mixing ratio in and around clouds, and (iii) the existence of large sampling errors.

In the past decade, cumulus ensemble models (CEMs) have emerged as a useful tool in cumulus parameterization (e.g., Yamasaki 1975; Soong and Tao 1980; Tripoli and Cotton 1982; Krueger 1985; Lipps and Hemler 1986; Nakajima and Matsuno 1988; Gregory and Miller 1989). A CEM resolves individual clouds but covers a large horizontal domain. Such a model is able to simulate a variety of cloud regimes under different large-scale (synoptic scale) and underlying surface conditions. It is a valuable tool because it can provide high-resolution data for increasing the understanding of the dynamics of cumulus ensembles, the interaction between convective-scale and large-scale processes, and the role of mesoscale organization in the interaction.

Another reason for using CEM simulations in cumulus parameterization is a philosophical one, because the cumulus parameterization problem is analogous to the climate dynamics problem in many aspects (Arakawa and Chen 1987) although the actual physical mechanisms involved are quite different. Difficulties in developing a cumulus parameterization can be compared with those in developing a simplified climate model that includes a formulation of the collective effects of synoptic-scale eddies. Not surprisingly, almost the same technical methods are used in both problems: that is, statistics, diagnosis, and numerical simulation. Simulations of cumulus ensembles, which are the counterpart of simulations by general circulation models (GCMs) in the climate dynamics problem, should be performed more extensively in the future. The synergism among these approaches will lead to better understanding of the scientific issues involved in the cumulus parameterization problem and eventually to improvement of cumulus parameterizations.

The present study intends to give a concise review of existing CEM investigations. It also describes the utilization of a CEM to better understand the macroscopic behavior of cumulus ensembles.

This study addresses one of the most basic questions raised by Arakawa and Chen (1987): To what extent can cumulus convection be parameterized deterministically and diagnostically? Most existing cumulus parameterizations are *diagnostic,* in the sense that no prognostic equations for cloud-scale variables are introduced in the parameterization and *deterministic,* in the sense that the collective effects of cumulus convection are uniquely determined for a given large-scale condition. Another related question to be addressed is, Is it possible to parameterize cumulus clouds despite the influence of mesoscale organization? The answers to such questions are not necessarily unique; in particular, they are very likely dependent on the grid size of the numerical model to be used.

The rest of the paper is organized as follows. Section 22.2 describes the characteristics of existing CEMs. Section 22.3 briefly reviews the results obtained from existing CEM investigations. Section 22.4 presents some recent results from an analysis of the macroscopic behavior of cumulus convection and an evaluation of the Arakawa–Schubert (1974) cumulus parameterization using simulated data. Conclusions and discussion are given in section 22.5.

22.2. Characteristics of CEMs

CEMs differ from the more familiar isolated cloud models in that a CEM allows several clouds of various sizes to develop simultaneously (and randomly) inside the model domain. Two-dimensional CEMs are widely used to study tropical cumulus clouds and their mesoscale organization because they can cover a large horizontal domain in the range of 30–1000 km (e.g.,

Hill 1974; Yamasaki 1975; Soong and Ogura 1980; Chen and Orville 1980; Krueger 1985, 1988; Lipps and Hemler 1986; Nakajima and Matsuno 1988; Gregory and Miller 1989; Tao and Simpson 1989; Xu et al. 1992). Three-dimensional models were also developed with horizontal domains of 32 km × 32 km or less (Yau and Michaud 1982; Lipps and Hemler 1986; Tao and Soong 1986). More recently, Tao and Simpson (1989) used a horizontal domain of 96 km × 96 km to simulate a tropical squall line, but only a portion of the squall line could be covered. Currently it is not feasible to run a 3D model long enough to simulate the life cycle of mesoscale organization.

Most CEMs are designed to simulate the formation of an ensemble of cumulus clouds under a given large-scale condition as if the CEM was situated within a grid box of a large-scale numerical model. This requires that information on large-scale destabilizing and moistening rates be imposed on every CEM grid point. The CEM provides detailed information about cumulus ensembles such as the cloud mass flux, the in-cloud temperature and mixing ratio, and the rates of condensation and evaporation. All of these are not extensively observable (see chapter 4). Table 22.1 summarizes the major characteristics of various CEMs along with the domain sizes and lengths of the simulations. The detailed characteristics of CEMs are explained below.

The commonly adopted dynamic framework of most CEMs is the anelastic system. In addition, one of the most important components of a CEM is its parameterization of cloud microphysical processes. Most earlier CEM studies (e.g., Soong and Tao 1980; Krueger 1985, 1988; Lipps and Hemler 1986) used the Kessler-type (1969) two-category liquid water (cloud water and rainwater) microphysics. The significance of including ice-phase (cloud ice, graupel/hail, and snow) microphysical processes has been revealed by observations of tropical convective systems (e.g., Leary and Houze 1979) and modeling studies of tropical cyclones (Lord et al. 1984). Tao and Simpson (1989) and Xu and Krueger (1991) used three-category ice-phase bulk microphysics. These ice-phase parameterizations basically followed Rutledge and Hobbs (1984) and Lin et al. (1983), respectively.

Another important component of CEMs is their turbulence parameterizations in the subcloud layer (SCL) and within clouds. The most sophisticated turbulence parameterization is a third-moment closure (Krueger 1985, 1988). The primary motivation for adopting a higher-order turbulence closure is to improve the simulation of boundary-layer turbulence as well as in-cloud turbulence. Some CEMs used the simple K-type turbulence closure (Nakajima and Matsuno 1988; Tripoli and Cotton 1982) or determined the coefficient K from the turbulence kinetic energy (TKE) equation either diagnostically (e.g., Lipps and Hemler 1986) or prognostically (e.g., Soong and Ogura 1980; Tao and Soong 1986).

In CEMs the role of the SCL is emphasized through the use of a vertically stretched coordinate except in a few earlier studies (Soong and Ogura 1980; Soong and Tao 1980; Lipps and Hemler 1986). This is essential since most clouds originate in the SCL and have important feedbacks on the SCL by means of precipitation and downdrafts. The grid interval near the surface is 100 m in Krueger (1985, 1988), which is the smallest among the CEMs.

Initiation of clouds in CEMs has not yet become unique. Hill (1977) first used random surface heating as a mechanism to initiate clouds. Crook and Moncrieff (1988) imposed a momentum forcing generated by inertial–gravity waves to initiate cumulus convection. Most of the more familiar isolated cloud models used an initial warm bubble with a maximum perturbation temperature of up to 2 K. The horizontal extent of the initial bubble is several kilometers and its depth is about 2 km (e.g., Fovell and Ogura 1988; Rotunno et al. 1988). This method, however, was generally not adopted in CEMs, because the location of clouds is predetermined and the initial bubble does not resemble thermals in the SCL. In CEMs, therefore, only small random perturbations of temperature and/or moisture are commonly used (e.g., Tao and Soong 1986; Krueger 1988). The other purpose for imposing random perturbations is to keep the SCL horizontally inho-

TABLE 22.1. Summary of the major characteristics of various existing CEMs.

Study	Model	Microphysics	Turbulence	Domain	Integration (h)
Gregory and Miller	2D	Kessler	Prescribed fluxes	256 km	9
Krueger	2D	Kessler	third moment	30 km	2
Lipps and Helmer	2D	Kessler	Diagnostic TKE	32–64 km	4
	3D			24 km × 16 km	
Nakajima and Matsuno	2D	Kessler	K type	512 km	50
Tao and Simpson	2D	Kessler + ice	Prognostic TKE	512 km	12
	3D			96 km × 96 km	
Tao and Soong	3D	Kessler	Prognostic TKE	30 km × 30 km	6
Tao, Simpson, and Soong	2D	Kessler	Progsnotic TKE	64 km	6
	3D			32 km × 32 km	
Xu and Krueger	2D	Kessler + ice	third moment	512 km	120

mogeneous. Tao and Soong (1986) found that the collective effects of clouds are basically independent of the pattern of the random perturbations in simulations with identical large-scale conditions.

As the horizontal domain of CEMs increases, the Coriolis force becomes a more significant dynamical component of CEMs. In Xu and Krueger (1991) the Coriolis effect is incorporated into the 2D model developed by Krueger (1985, 1988) by prescribing the large-scale y-component pressure gradient in the v-momentum equation.

22.3. Statistical properties of cumulus ensembles

Most CEM investigations have focused on analyzing statistical characteristics of cumulus ensembles. The reality of simulated cumulus convection was generally examined through the comparison of simulated cloud-draft statistics with those obtained from aircraft measurements during the Global Atmospheric Research Program (GARP) Atlantic Tropical Experiment (GATE) (LeMone and Zipser 1980). Note that this is not the only method for examining the reality of simulated cumulus convection. Section 22.4a presents an alternative method.

Figure 22.1 shows a comparison of the mean vertical velocity and the mean diameter for both upward and downward cores from Lipps and Hemler's (1986) 3D simulation. The agreement of 3D (Fig. 22.1) and 2D (not shown) simulations with the LeMone–Zipser observations is quite remarkable except that the downdraft strength is slightly weaker. The cloud-draft statistics of Tao et al. (1987) agree with observations even better (not shown).

Comparisons between 2D and 3D CEMs suggest that 2D models are able to produce essentially the same statistical properties and budgets of cumulus ensembles as 3D models, given an identical large-scale condition (Lipps and Hemler 1986; Tao and Soong 1986; Tao et al. 1987). By comparing 2D with 3D simulations, Lipps and Hemler (1986) found that the heat and moisture budgets were very similar although the cloud mass fluxes were slightly larger in the 2D case especially in the lower troposphere (Fig. 22.2). The environmental subsidence, which is the main contributor to the heat and moisture budgets (Yanai et al. 1973), is almost identical between 2D and 3D simulations. Tao and Soong (1986) and Tao et al. (1987) reached similar conclusions from their CEM simulations (not shown).

More detailed analyses of simulated data have focused on individual components of the thermodynamic budget equations such as the rates of condensation and evaporation, the cumulus-scale (and turbulent) sensible heat flux, and moisture fluxes (e.g., Tao et al. 1987, 1991; Tao et al. 1993; Krueger 1988; Gregory and Miller 1989). As mentioned in section 22.2, these

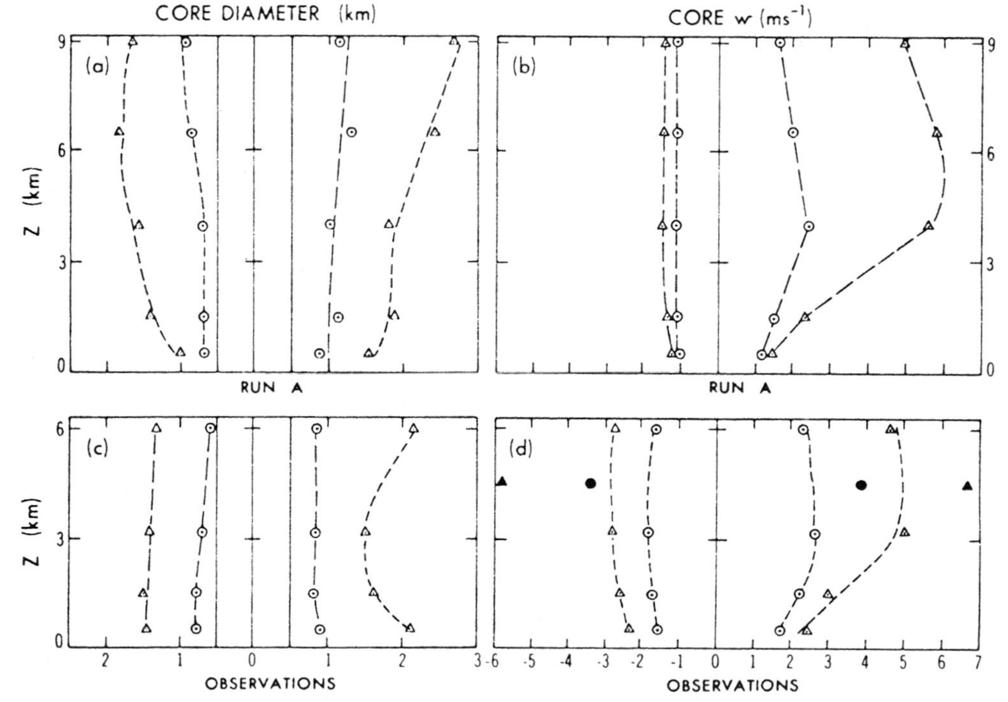

FIG. 22.1. Variation with height of median (50%) and 10% level (10% of cores stronger than value) for updraft and downdraft cores: (a) core diameter and (b) mean vertical velocity. Corresponding values in (c) and (d) are given for the observed data of LeMone and Zipser (1980). Solid symbols represent data of Gray (1965) at the 4.5-km level (from Lipps and Hemler 1986).

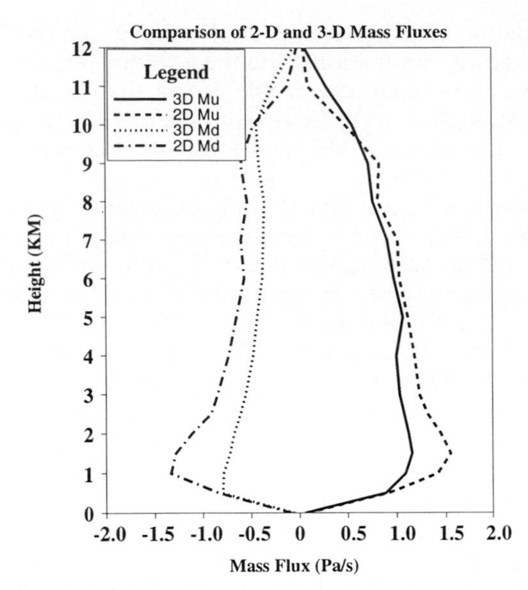

FIG. 22.2. Vertical profile of updraft and downdraft mass fluxes from 2D and 3D simulations (reproduced from Lipps and Hemler 1986).

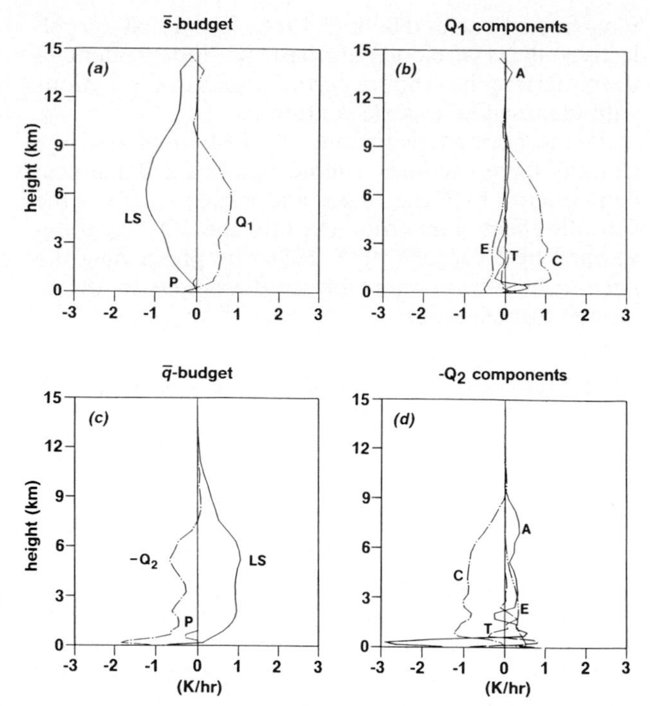

FIG. 22.3. Vertical profile of (a) dry static energy budget, (c) moisture budget [solid: due to large-scale vertical advection; dashed: due to cumulus convection (Q_1 and Q_2)], (b) Q_1 component, and (d) Q_2 component [solid: cumulus transport; dashed: turbulent transport; dash–dot: net condensation of cloud droplets; dash–double dot: raindrop evaporation] (courtesy of Steven K. Krueger).

individual components are not directly observable. (See chapter 4 for diagnosis method of obtaining these components from large-scale observations.) The aforementioned analyses revealed that condensation in the cloud layer and evaporation in the lower troposphere are the dominant terms in the budget equations. Their magnitudes are much larger than those of the vertical cumulus-scale (eddy) flux divergence. The role of the vertical eddy flux divergence is larger in cumulus drying than in cumulus heating. Figure 22.3 shows an example of such analyses from Krueger's (1988) simulations. Krueger (1988) further examined those components in the SCL (Fig. 22.4), where the turbulent and eddy flux divergences are important along with the rainwater evaporation.

Individual components of budget equations were further examined according to the intensities of updrafts and downdrafts (Tao et al. 1987; Krueger 1988). This type of analysis cannot be performed with any diagnosis method using large-scale data. Figure 22.5 shows an analysis of CEM data by Tao et al. (1987). The main conclusions are as follows. Condensation occurs only in the updraft areas and mainly takes place in the active updrafts with vertical velocities greater than 1 m s^{-1} (Fig. 22.5a). Evaporation mainly takes place in the downdraft areas (Fig. 22.5b). Inactive downdrafts ($w > -1$ m s^{-1}) and active downdrafts contribute equally to evaporation (Fig. 22.5b). Active updrafts contribute to the entire sensible heat flux in the cloud layer but updrafts and downdrafts have an equal contribution in the SCL (Fig. 22.5c). Inactive drafts contribute more to the moisture flux than to the sensible heat flux (Fig. 22.5d). Soong and Tao (1980) and Tao et al. (1987) found that both updraft mass flux and downdraft mass flux are dominated by active drafts (not shown) although their fractional areas are small (<4%). All classes of updrafts are warmer and more humid than the mean environment. The averaged downdrafts are slightly warmer in the cloud layer and slightly more humid above 850 mb but much drier below 850 mb than the mean environment [see Tao et al. (1987) for details].

Mesoscale circulations were simulated by Tao and Simpson (1989), Tao et al. (1991), and Tao et al. (1993) using their 2D CEM with a large domain size (≥512 km) and ice-phase cloud microphysics. They

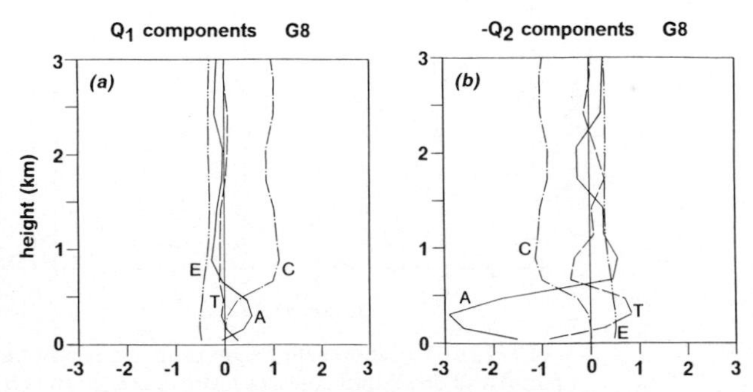

FIG. 22.4. Same as Figs. 22.3b,d except for z = 0–3 km (courtesy of Steven K. Krueger).

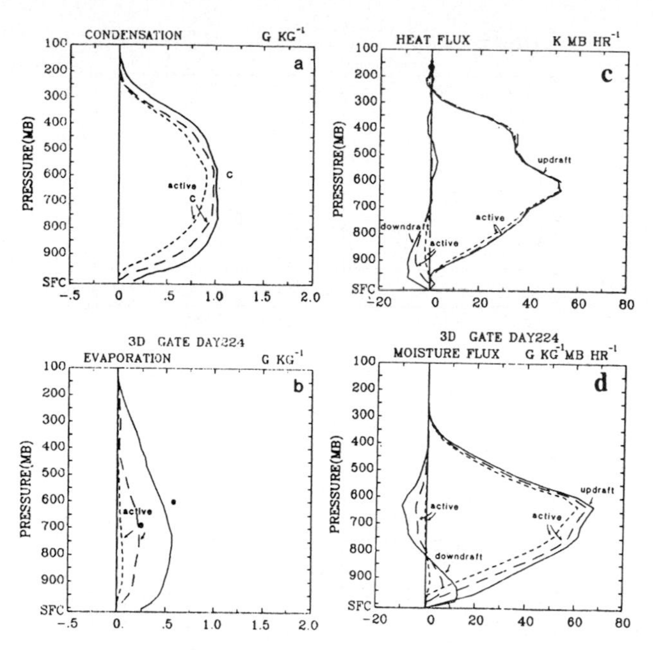

FIG. 22.5. Vertical profile of (a) condensation rates of updraft areas and active updraft areas ($|w|$ exceeds 1 or 2 m s^{-1}), (b) evaporation rates in cloud downdraft areas, (c) sensible heat fluxes, and (d) moisture fluxes inside cloud updraft and downdraft areas (from Tao et al. 1987).

separated the contributions to cumulus heating and vertical mass transport by convective-scale and mesoscale circulations. Their results qualitatively resemble diagnostic results based on observations during GATE (e.g., Houze 1982).

Soong and Tao (1984) studied the vertical transport of cumulus momentum in a tropical rainband using a 2D CEM. Tao and Soong (1986) repeated the study using a 3D CEM with a smaller domain. They found that the cumulus momentum transport can be either countergradient or downgradient, depending on the location of new cloud cell development relative to the environment wind shear. The generation of horizontal momentum by the pressure gradient force is the primary mechanism involved.

22.4. Simulations with the UCLA CEM

The University of California, Los Angeles (UCLA) CEM is a 2D model (Krueger 1985, 1988; Xu and Krueger 1991). It includes a third-moment turbulence closure, a three-phase cloud microphysics parameterization, and the Coriolis force. Radiation heating rate is currently prescribed.

The CEM has been used to perform numerical simulations with prescribed large-scale conditions (i.e., horizontally uniform large-scale vertical velocity or destabilizing and moistening rates, as well as large-scale y-component pressure gradient) and underlying surface conditions (Xu and Krueger 1991; Xu et al. 1992). The horizontal domain is usually 512 km wide with a 2-km grid size.

The depth of domain is usually 19 km with a stretched coordinate and 33 layers. Near the surface the grid interval is 100 m, while near the model top the grid interval is 1000 m. The upper and lower boundaries are rigid. The lateral boundary condition is cyclic. The initial thermodynamic conditions in all simulations are horizontally uniform. Each simulation was integrated for five days or longer. See Xu et al. (1992) for further details.

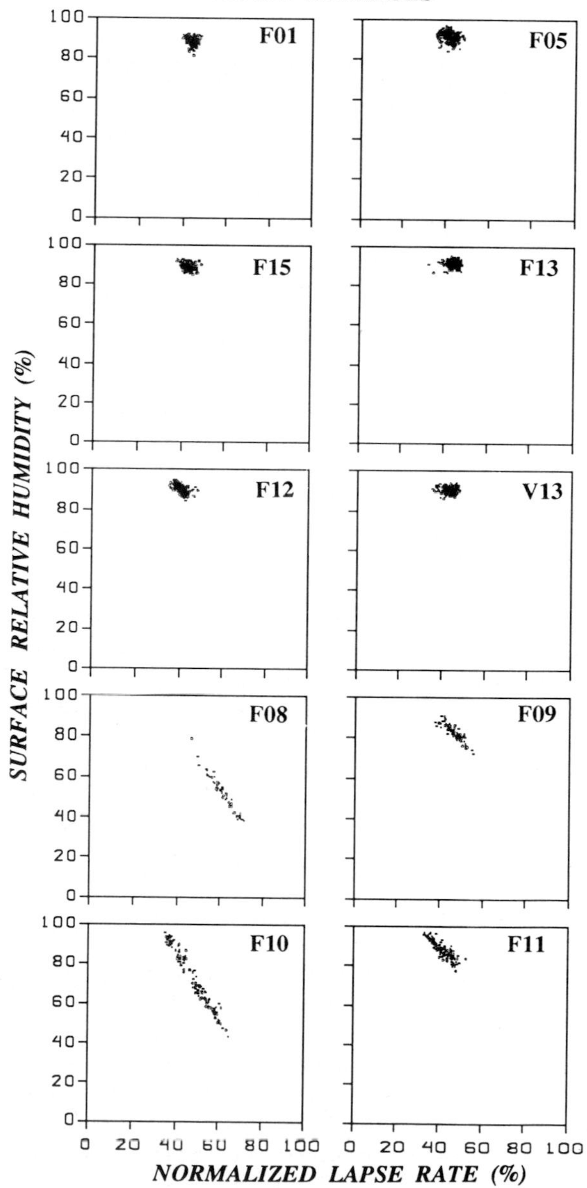

FIG. 22.6. Scatterplots of RH_S vs Γ_N for the 128-km averaging distance of ten different simulation experiments.

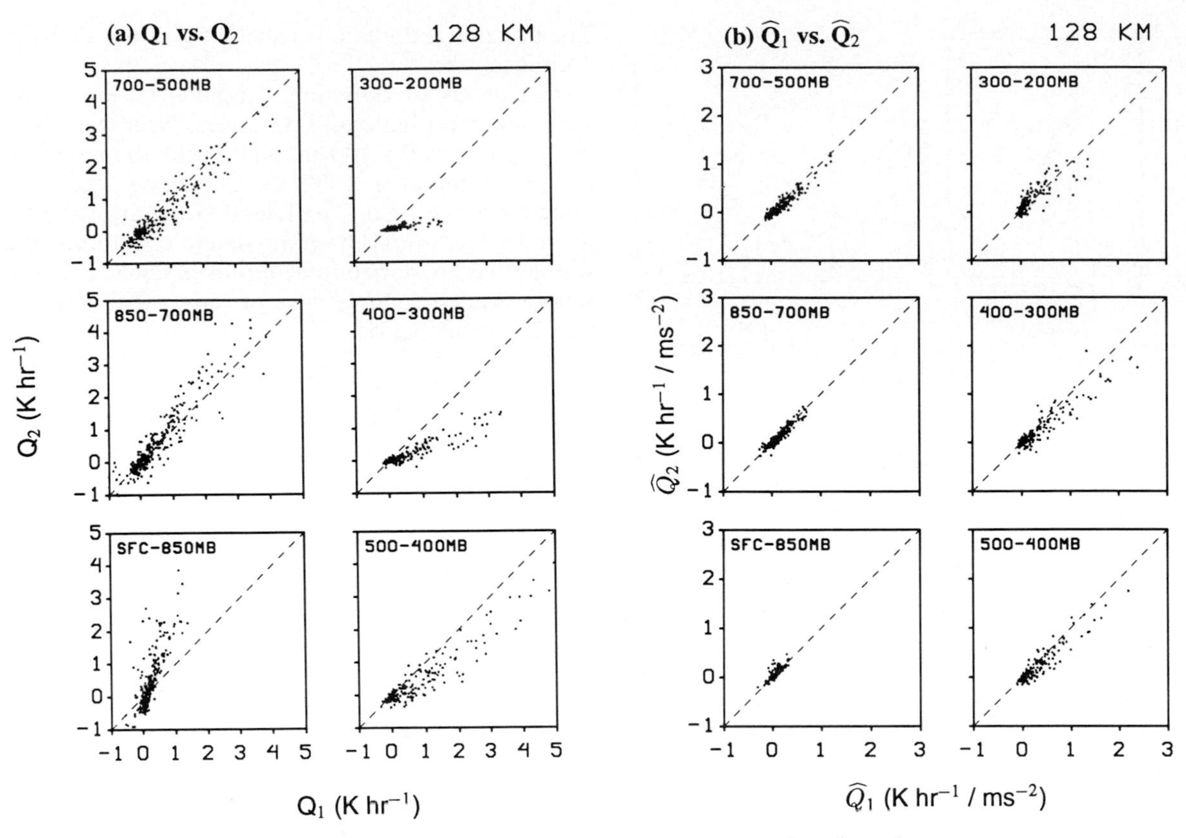

FIG. 22.7. Scatterplots of (a) $Q_1 - Q_R$ vs Q_2 and (b) $\hat{Q}_1 - \hat{Q}_R$ vs $\hat{Q}_2$ in selected layers for the 128-km averaging distance of Q02.

a. Comparison with observations

The statistical behavior of simulated cumulus convection is compared with that observed by Arakawa and Chen (1987) and Chen (1989). (Their main results can also be found in chapter 1.) This comparison is made in view of type I and type II closure assumptions in existing cumulus parameterization (Arakawa and Chen 1987). Type I closure constrains the coupling between *net* warming $\partial T/\partial t$ and *net* moistening $\partial q_v/\partial t$, that is, the coupling of temperature and humidity profiles. Type II closure constrains the processes that are responsible for Q_1 (apparent heat source) and Q_2 (apparent moisture sink). See chapter 4 for the definition of Q_1 and Q_2.

A highly simplified representation of type I closure is a *negative* correlation between the surface relative humidity RH_S and the normalized lapse rate Γ_N if deep cumulus clouds exist (Arakawa and Chen 1987). The latter is defined as $(\Gamma - \Gamma_{mS})/(\Gamma_d - \Gamma_{mS})$, where Γ is the mean lapse rate between the surface and 500 mb and the subscripts mS and d denote moist (at surface) and dry adiabats, respectively. This two-parameter space representation may give some limited indication for the existence of type I coupling in a statistical sense.

Figure 22.6 shows the scatter diagrams of RH_S versus Γ_N for ten different simulations under a variety of large-scale and underlying surface conditions. [See Xu et al. (1992) for a detailed description of these simulations.] Each data point in Figs. 22.6 and 22.7 represents an average over an hour in time and over 128 km in horizontal extent. For simulations with an underlying ocean surface (F01, F05, F12, F13, F15, and V13), the scatter is confined to a small segment in the high-humidity, conditionally unstable region of the RH_S versus Γ_N space because RH_S over the ocean changes little and the lapse rate is very close to the moist adiabat. The scatter is thus similar to that observed during GATE by Arakawa and Chen (1987, their Fig. 13). For simulations with an underlying land surface (F08, F09, F10, and F11), however, the scatter spreads over a much longer range in the two-parameter space. A negative correlation between RH_S and Γ_N is much more clearly evident in these simulations. Again, this is similar to that observed over the Asian continent by Arakawa and Chen (1987, their Figs. 10–12). Thus, the type I coupling of simulated cumulus convection is similar to that observed.

Figure 22.7a shows scatterplots of $Q_1 - Q_R$ versus Q_2 and Fig. 22.7b shows those of $\hat{Q}_1 - \hat{Q}_R$ (normalized apparent heat source) versus $\hat{Q}_2$ (normalized apparent moisture sink) for each of the six layers in the vertical from a simulation described later, where Q_R is the ra-

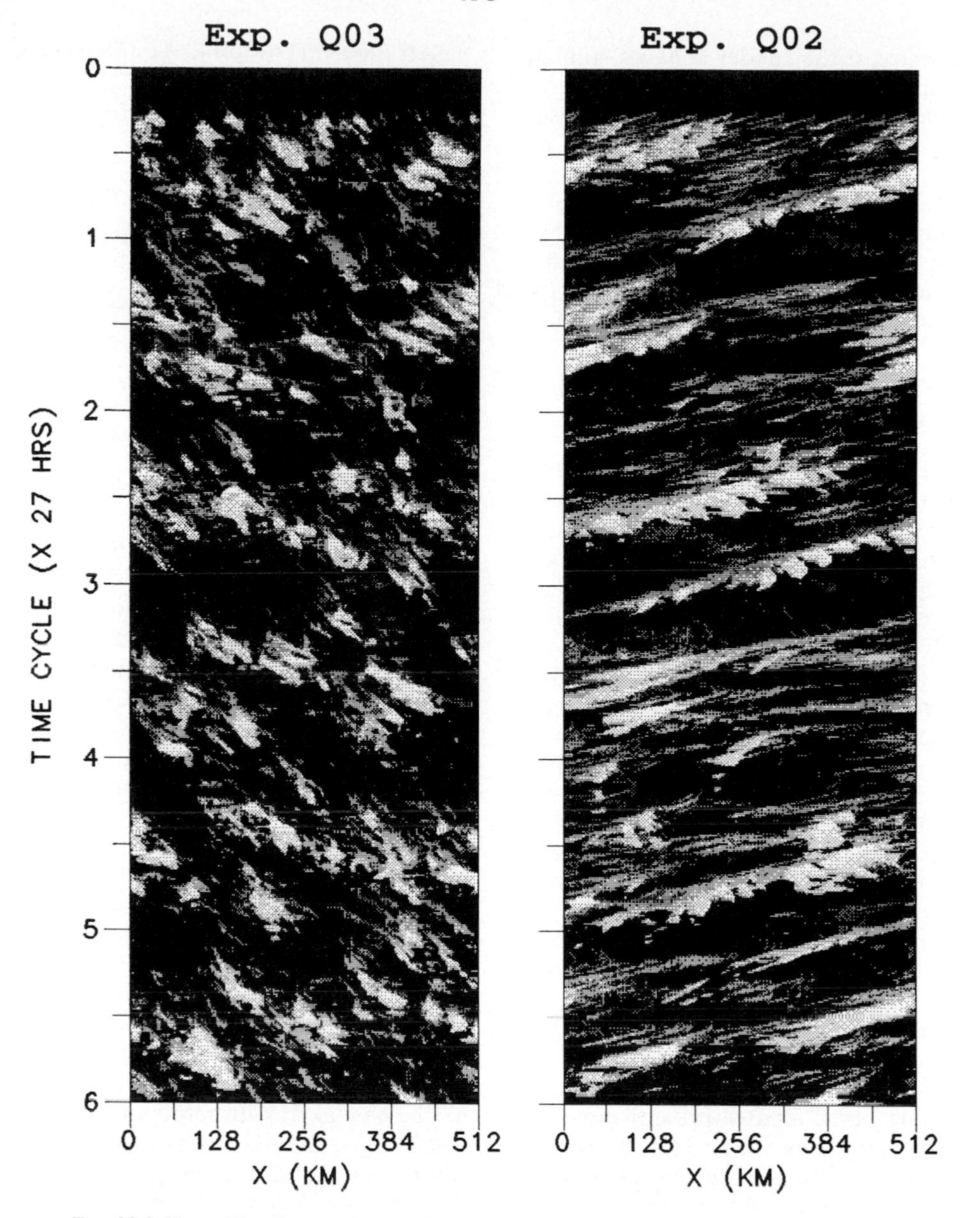

FIG. 22.8. Hovmöller diagrams (x–t sections) of cloud-top temperature for Q02 and Q03, based on 15-min averages. Cloud-top temperature is indicated by a linear gray scale: white—200 K; black—300 K (from Xu et al. 1992).

diative heating rate. As described by Arakawa and Chen (1987; see also chapter 1), if $\hat{Q}_1 - \hat{Q}_R = \hat{Q}_2$, then it represents a perfect coupling between $Q_1 - Q_R$ and Q_2 for nondetraining cumulus clouds. In other words, the subsidence between clouds is the only process contributing to Q_1 and Q_2. When $\hat{Q}_1 - \hat{Q}_R \neq \hat{Q}_2$, the detrainment of cloudy air and/or large-scale condensation–evaporation processes are also involved.

The apparent best fit in Fig. 22.7a deviates from the diagonal line and rotates clockwise with height. This feature is similar to that observed from GATE by Arakawa and Chen (1987, their Fig. 14). The regression line of the normalized quantities (i.e., the diagonal line) in Fig. 22.7b reflects the primary role of the subsidence between clouds in contributing to Q_1 and Q_2. The scatter around the diagonal line indicates that processes other than subsidence are also operating. The physical interpretation for the scatter was provided by Arakawa and Chen (1987); that is, the detrainment effect is responsible for the scatter on the left side of the diagonal line and the large-scale condensation/evaporation for the rest of scatter in Fig. 22.7b. Therefore, we can see that the detrainment effect increases with height and the large-scale condensation–evaporation effect exists at all levels but less frequently in the upper troposphere. This result is consistent with one's physical intuition.

It should be expected if the model behaves similarly to the real atmosphere.

From the above, we have confirmed the conclusion obtained from observed datasets by Arakawa and Chen (1987) that both type I and type II coupling are reasonably strong. In other words, the statistical behavior of simulated cumulus convection is similar to that observed, from the point of view of type I and type II closures. These conclusions were also confirmed (see Xu 1991) using a sophisticated statistical method.

b. *Modulation of cumulus activity by time-varying large-scale processes*

As noted above, a number of simulations have been performed with the UCLA CEM. In the rest of the paper, the concentration will be on two simulations, Q02 and Q03. These simulations were performed with *identical* large-scale advective effects (i.e., destabilizing and moistening rates) and with an underlying ocean surface. The advective effects typical of the GATE mean condition (Fig. 1 in Xu et al. 1992) vary with time with the period of 27 h. The primary motivation for specifying the time dependence of the large-scale advective effects is to study the response of cumulus activity to time-varying large-scale processes. Simulations Q02 and Q03 differ in the prescribed x-component geostrophic wind profiles: one with strong wind shear (Q02) and the other without (Q03). These wind profiles are used as the initial conditions and also represent the large-scale y-component pressure gradient. Both simulations were integrated for 11 days of physical time.

Figure 22.8 shows the spatial distribution and time evolution of cloud-top temperature for Q02 and Q03. In Fig. 22.8 cirrus anvils associated with cumulonimbi appear white, while black areas represent cloud-free ocean surface. The short tick in the vertical axis corresponds to the maximum large-scale advective effects. One feature is that the simulated cumulus convection is in fact modulated by the time-varying large-scale processes. But the modulation is not completely deterministic. Another feature to note is that the duration, horizontal extent, intensity, and propagation speed of cloud systems are quite different between the simulations. Cumulus convection in Q02 is more organized under strong wind shear than in Q03 (no shear). There are several long-lived mesoscale systems in Q02.

Figure 22.9 shows the domain-averaged, hourly precipitation rates of Q02 and Q03. The precipitation rate is rather strongly modulated by the imposed time-varying large-scale advective effect (dashed line), but it also undergoes high-frequency fluctuations. The degree of modulation differs between the sheared (Q02) and nonsheared (Q03) simulations as shown by the differences in phase delays and magnitude of high-frequency fluctuations. In addition, Fig. 22.9 shows that the modulation is not completely deterministic.

An averaging over nine cycles, starting from 21 to 263 h, was taken with respect to the phase of the imposed large-scale processes. We call such an averaging in time "ensemble averaging." Figure 22.10 shows the ensemble mean and standard deviation of the precip-

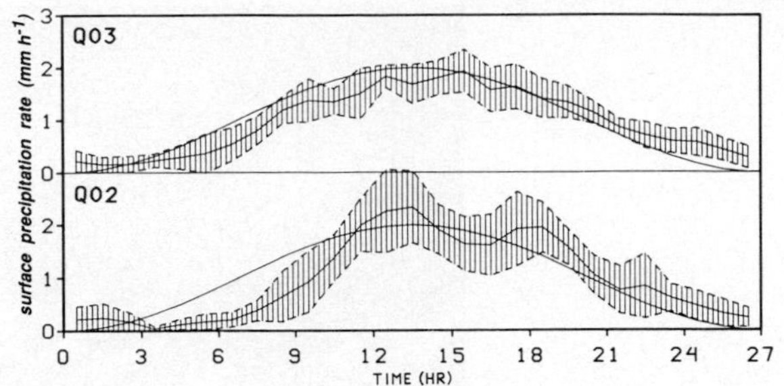

FIG. 22.10. The ensemble mean (thick solid line) and associated standard deviation (error bars) of surface precipitation rate for Q03 (top) and Q02 (bottom). The abscissa is the phase of the imposed large-scale advective processes. The thin solid line corresponds to the dashed line in Fig. 22.9 (from Xu et al. 1992).

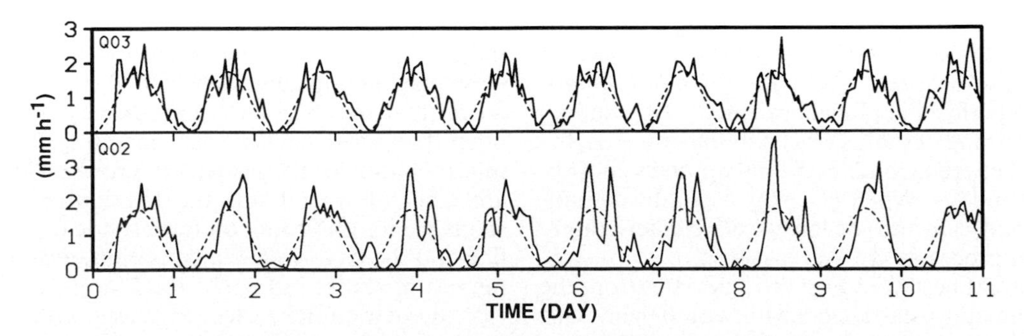

FIG. 22.9. Time sequences of the hourly rate of entire domain-averaged surface precipitation (solid) for Q02 and Q03. As a reference, time sequence of the imposed large-scale advective effect is shown with an arbitrarily chosen amplitude by the dashed line (from Xu et al. 1992).

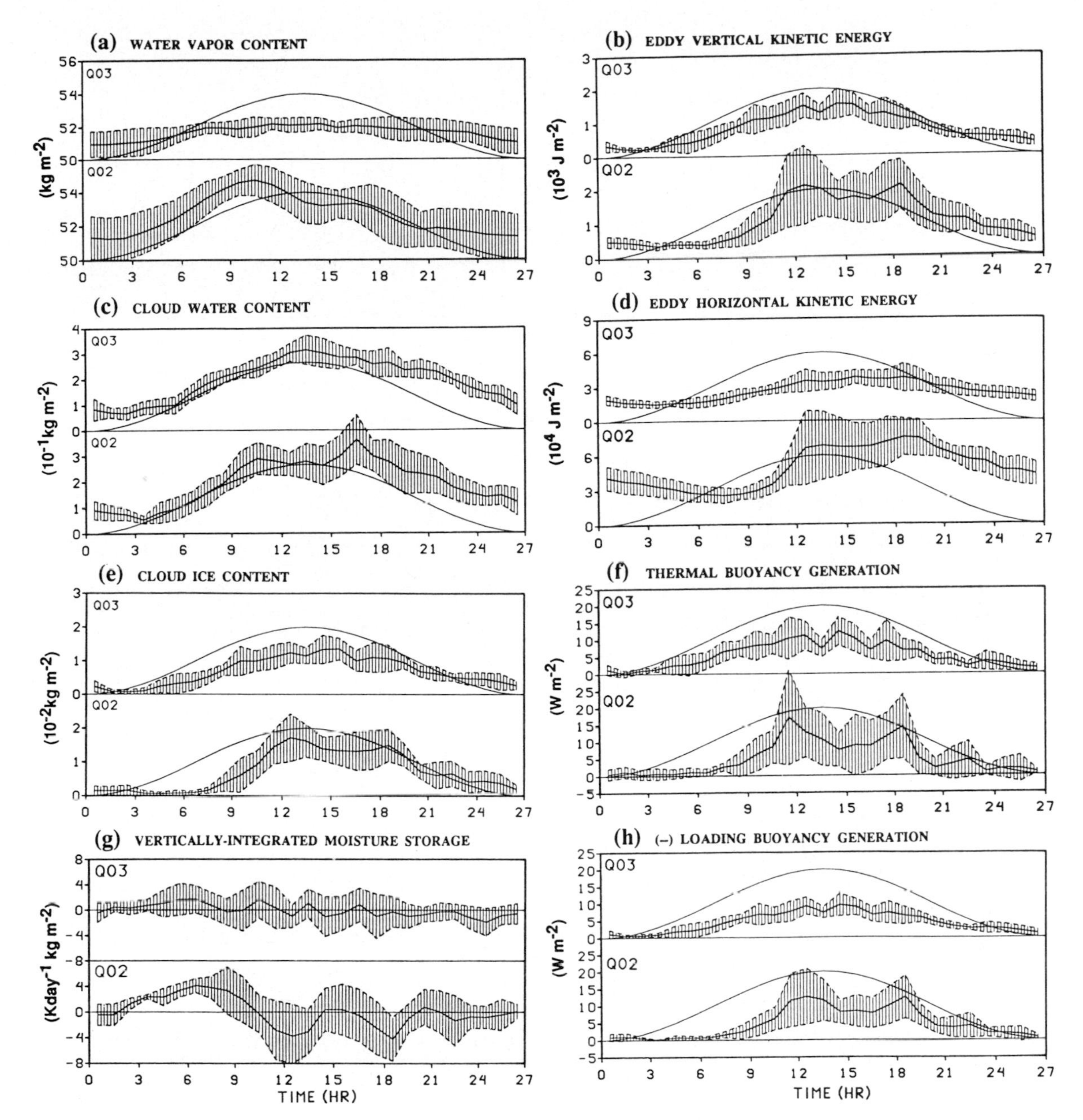

FIG. 22.11. Same as Fig. 22.10 except for some bulk properties of simulated cumulus convection.

itation rate shown in Fig. 22.9. Some of the features seen in Fig. 22.9 also appear in Fig. 22.10. Differences between Q02 and Q03 appear mainly in the amount of systematic phase delay. Simulation Q02 shows a more pronounced phase delay, which is caused by the existence of mesoscale organization (Xu et al. 1992).

To further illustrate the differences between the simulations, the ensemble mean and standard deviation of bulk properties of cumulus ensembles are shown in Fig. 22.11. These bulk properties are the *vertically integrated* quantities of water vapor, cloud water, cloud ice, moisture storage (time derivative of water vapor mixing ratio), vertical and horizontal eddy kinetic energy, and thermal and loading buoyancy generation rates. The last two quantities are defined as the eddy correlations between vertical velocity and buoyancy (Xu et al. 1992).

Features to note in Fig. 22.11 are that (i) some of bulk properties, such as cloud water content, cloud ice content, vertical eddy kinetic energy (EKE), and thermal and loading buoyancy generation rates, are highly modulated by the prescribed large-scale processes; (ii) the systematic phase delay in Q02 is more distinct than in Q03 especially in the horizontal EKE (Fig. 22.11d); and (iii) the nonsystematic fluctuations in Q02 are generally larger than in Q03. Also note that the ensem-

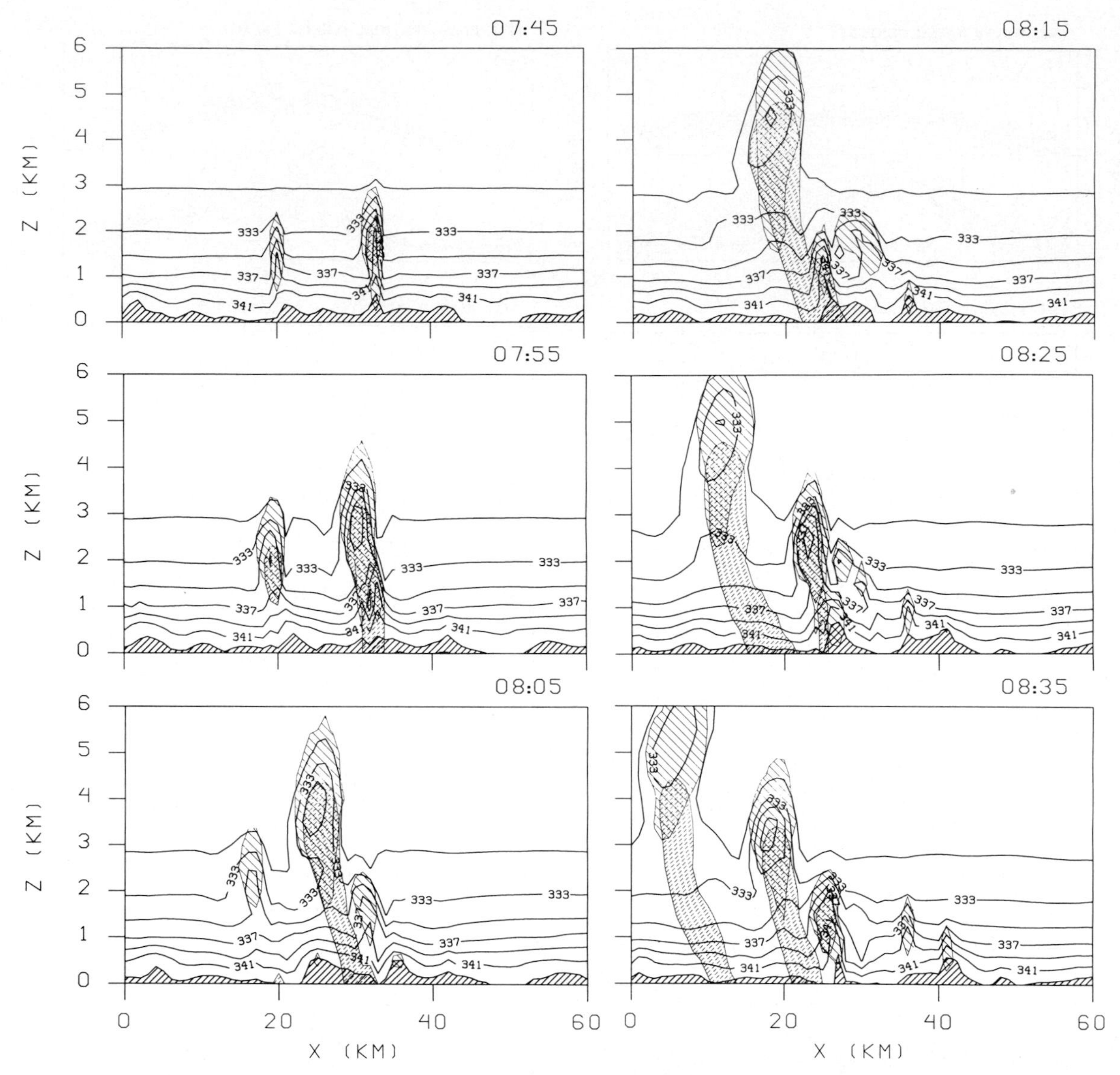

FIG. 22.12. Cross section (x–z) of moist static energy (thick solid line), cloudy areas (thin solid hatch), and precipitating areas (thin dashed hatch) from 0745 to 0835 of a high-resolution simulation with a domain size of 512 km. The thick solid hatch represents high moist static energy that exceeds 343 kJ kg^{-1}.

ble means of bulk properties except the moisture storage in Q03 follow the time variation of large-scale processes with a negligible phase delay.

In Q02 the phase relations among different bulk properties are much more complicated. The primary peak occurs from 10 to 14 h in different bulk properties. It results from the mature stage of the mesoscale convective system where active updrafts are the main contributors to the precipitation. This primary peak in precipitation is associated with the secondary peaks in the cloud water content at 10 h (Fig. 22.11c) and horizontal EKE at 12 h (Fig. 22.11d). It is also associated with the primary peak in the cloud ice content (an indicator of deep convection) at 13 h (Fig. 22.11e). The secondary peak in precipitation at 19 h is associated with the increase of stratiform precipitation. It is related to the primary peaks in the cloud water content at 16 h (Fig. 22.11c) and in the horizontal EKE at 19 h (Fig. 22.11d). The pronounced phase delay in the horizontal EKE of Q02 is associated with the slow evolution of mesoscale organization after its initial stage.

Another important feature to note is that the net (thermal + loading) buoyancy generation rate (Figs. 22.11f,h) is almost zero, indicating the quasiequilibrium of EKE. Nevertheless, a negligible net generation rate exists to compensate for dissipation of EKE. The moisture storage (Fig. 22.11g) in Q03 is about zero, but it is large between 2 and 10 h in Q02 as a consequence of the inhibition of deep cloud growth by shear (see also Fig. 22.11a).

In summary, cumulus activity is rather strongly modulated by large-scale processes, in spite of the existence of some nonmodulated features. This suggests that cumulus convection is basically parameterizable in the presence of mesoscale organization. The existence of mesoscale organization mainly increases the phase delay of the modulation.

c. *Interaction of subcloud layer with cumulus convection*

Budget component analyses (see Figs. 22.3 and 22.4) suggest that the SCL is important in the eddy transport of heat and moisture (Krueger 1988). The downdraft is an important component of the interaction of SCL with cumulus convection. Krueger (1988) studied this interaction by examining the downdraft and updraft initiation processes. These processes can be better seen from Fig. 22.12, which is obtained from a more recent simulation with a larger domain size (512 km with a grid size of 1 km) than that (30 km) in Krueger (1988). Figure 22.12 shows the x–z sections of moist static energy, cloudy regions (thin solid hatch), and precipitation regions (dashed hatch) at every 5 min. The moist static energy is approximately conserved for moist convection.

Figure 22.12 shows that downdrafts are formed below the active updrafts in rainshafts. They bring down air with low moist static energy (e.g., x = 32 km at 0755 and x = 25 km at 0825). The downdraft initially contains updraft air (e.g., the larger cloud cell at 0755) that originates within the SCL in the region with very high moist static energy (x = 33 km at 0745). As the larger cloud cell grows, the precipitation area is gradually separated from the cloudy region, indicating the separation of the downdraft from the updraft. The lower moist static energy near the cloud edges (e.g., at 0755) suggests that the updraft air is mixed with environmental air slightly above the originating level (at approximately 500 m). As the downdraft develops, midlevel air with low moist static energy flows into it and is eventually carried down into the SCL. On the other hand, penetrative downdrafts and evaporative cooling by rain in the SCL create gust fronts (e.g., x = 31 km at 0805). The SCL circulation induced by downdrafts leads to convergence and new cloud formation (e.g., x = 25 km at 0815) in the moistest SCL regions (solid hatched regions). The SCL sensible and latent heat fluxes (not shown) are also modulated by the initiation of updrafts and downdraft-induced circulations.

Figure 22.13 shows the time evolution of near-surface water vapor mixing ratio (at 47 m) difference δq from the initial state for both Q02 and Q03. During the first 6–12 h, δq increases primarily due to surface turbulence fluxes. Once cumulus convection begins, δq patterns are largely determined by downdrafts associated with deep convection that bring drier air into the boundary layer (Fig. 22.12). Especially in Q02, δq is large ahead of long-lived convective lines (Fig. 22.8) and sharply decreases behind them in mesoscale "wakes." It usually takes several hours for those wakes to recover due to the increase of surface turbulent fluxes and the absence of cumulus convection. Then cumulus convection restarts at some locations after δq becomes high enough. In Q03 (no shear) similar variations are less visible due to weak, short-lived, and more frequently occurring cloud organization (Fig. 22.8). The values of δq are generally lower because there are only a few hours for δq to increase before cumulus convection restarts. In Q02 deep convection is inhibited despite high δq in some locations, due to the low-level wind shear; only strong forcing at gust fronts can initiate deep convection.

Thus, it is shown from the above that the interaction of the SCL with cumulus convection is strongly related to the downdraft activity.

d. *Evaluation of Arakawa–Schubert cumulus parameterization*

The simulated data have also been used to evaluate the Arakawa–Schubert (1974, hereafter A–S) cumulus parameterization (Xu and Arakawa 1992). The details of the A–S parameterization can be found in chapter 10. This section gives a brief discussion of the conclusions from such an evaluation.

Two different semiprognostic tests were performed following a procedure similar to that used by Lord (1982). One uses the standard A–S parameterization with the cloud work function (CWF) quasi equilibrium (hereafter, control test) and the other with the CWF nonequilibrium based on the simulated time change of the CWF (hereafter, nonequilibrium test). The CWF is defined as the rate of generation of cloud-scale kinetic energy due to work done by the buoyancy force per unit cloud-base mass flux.

The semiprognostic tests were performed for input data (T, q_v profiles and their large-scale and turbulence tendencies) averaged over various subdomains of the CEM with the widths of 512, 256, 128, and 64 km, as well as over 1 h in time. These tests correspond to situations in which the parameterization is applied to different horizontal resolutions of large-scale models. The mesoscale processes are partly resolved in the range of high resolutions (e.g., 64 km).

Figure 22.14 shows the lag correlation between the apparent heat sources of simulation and parameterization for the 512-, 256-, 128-, and 64-km subdomains of Q02. The time lag from −5 h to +5 h is chosen. A positive lag time indicates that the parameterized variable precedes the simulated one.

Figure 22.14 indicates that (i) the standard A–S parameterization (i.e., the control test) performs reasonably well for all averaging distances and (ii) the impact

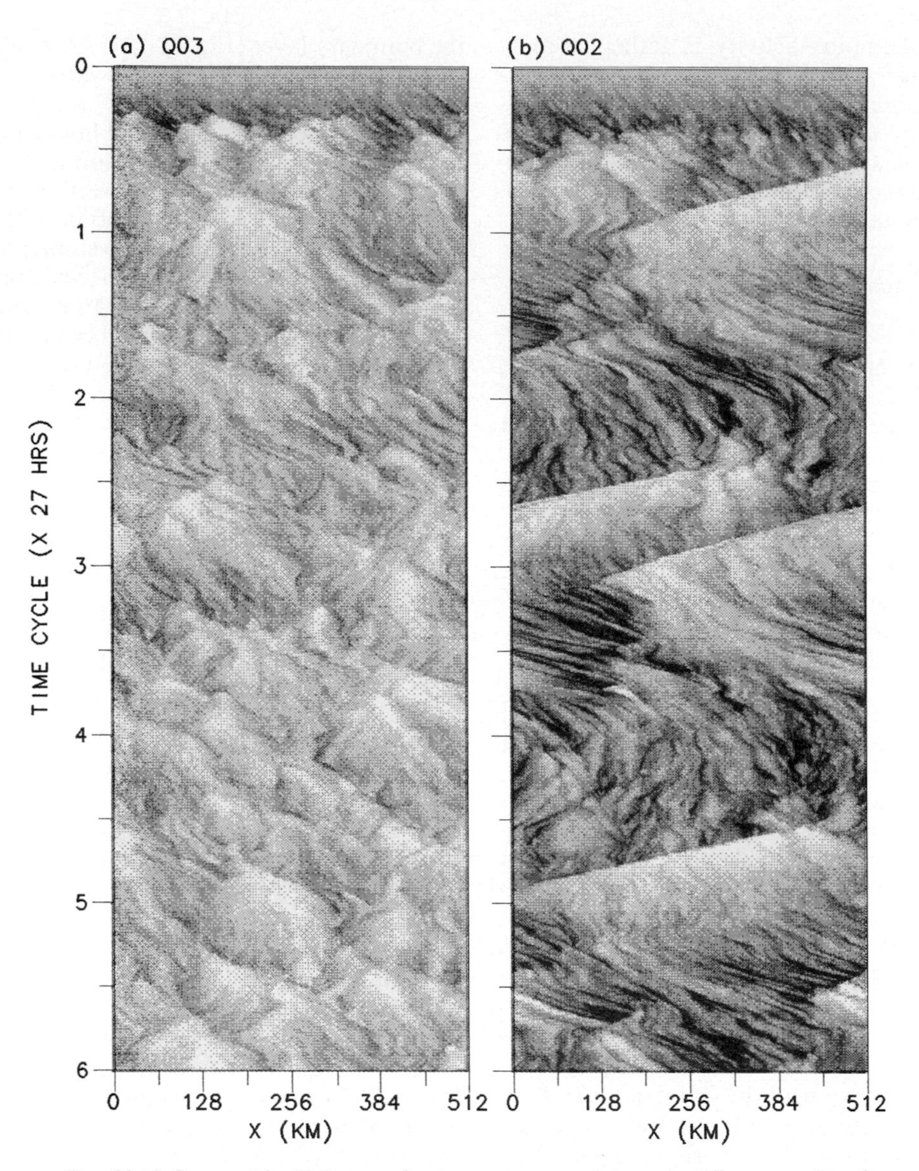

FIG. 22.13. Same as Fig. 22.8 except for the water vapor mixing ratio difference at 47 m from the initial state. The mixing ratio difference is indicated by a linear gray scale: white—< −2 g kg^{-1}; black—>2 g kg^{-1}.

of the CWF quasi-equilibrium assumption is not insignificant for the large subdomains. These suggest that the assumption is slightly worse when mesoscale processes are not even partially resolved; that is, the nondiagnostic effect of cumulus convection is significant for the large subdomains. The explanation for this result is as follows. The large-scale forcing as defined by Arakawa and Schubert (1974; see also Lord 1982), which is mainly due to advective processes, is likely to become more dominant for smaller scales. On the other hand, the CWF, which is completely determined by a sounding, is not significantly scale dependent. Therefore, the ratio of the time-change rate of CWF to the large-scale forcing decreases as the averaging distance decreases (Xu and Arakawa 1992).

On the other hand, Fig. 22.14 shows that the maximum correlation coefficients tend to decrease as the subdomain size decreases. Apparently the CWF quasi-equilibrium assumption is not responsible for this decrease. It is possible that the inherent nondeterministic nature of the parameterization problem becomes relevant for small subdomains because smaller-scale phenomena become resolvable. In other words, we expect that nondeterministic errors are the largest for the smallest resolvable scale and those for a *fixed larger scale* are not sensitive to the grid size used.

The above results also appear in other variables such as the surface precipitation rate, apparent moisture sink, and upward and downward mass fluxes (Xu 1991; Xu and Arakawa 1992).

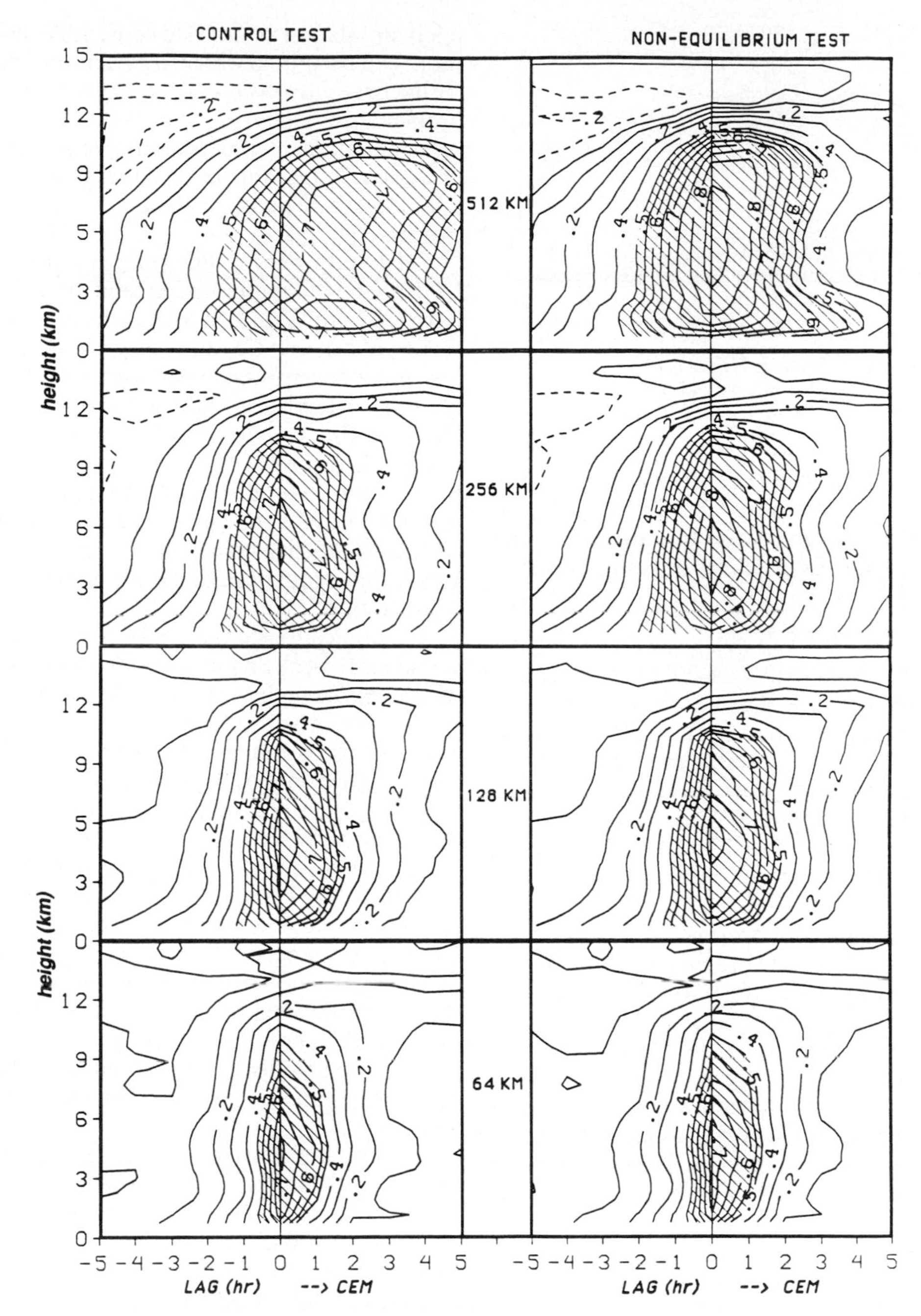

FIG. 22.14. Lag correlation coefficient between the apparent heat sources of simulation and parameterization as a function of height for the 512-, 256-, 128-, and 64-km subdomains of Q02 from the control test (left column) and the nonequilibrium test (right column). Shaded areas correspond to correlation coefficients greater than 0.5 with a contour interval of 0.05. The contour interval is 0.1 outside the shaded areas (from Xu and Arakawa 1992).

e. Bulk nondiagnostic aspects of cumulus ensemble

To include the nondiagnostic effects in a future cumulus parameterization, it is important to further understand the nonmodulated behavior of cumulus convection. As a starting point, the original A–S cumulus parameterization is modified to include the time change of the CWF and cumulus kinetic energy K for an ensemble of cumulus clouds (Xu 1991). The essence of the modified A–S parameterization can then be captured by

$$\frac{d^2M}{dt^2} + \frac{c}{2}\frac{dM}{dt} + \frac{k}{2\alpha}M = \frac{1}{2\alpha}F, \qquad (22.1)$$

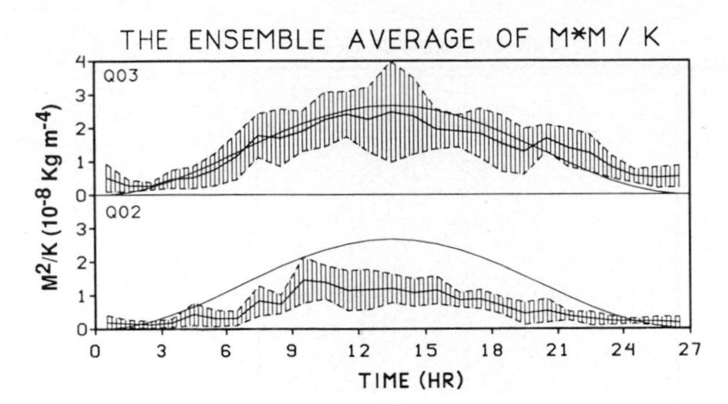

FIG. 22.15. Same as Fig. 22.10 except for the time sequence of α^{-1}.

where M is the bulk mass flux, c the dissipation rate of EKE, $-kM$ the adjustment of the CWF by the cumulus ensemble, F the large-scale forcing, and α the proportionality constant in the closure assumption; that is,

$$K = \alpha M^2. \tag{22.2}$$

From (22.1), the significance of the time-derivative terms is determined by the magnitude of α. When α is large, it is necessary to include the time-derivative terms in (22.1); that is, the nondiagnostic effects are important. The large α also corresponds to a large K for a given M. The EKE is usually larger in the presence of mesoscale organization (Fig. 22.11d). Therefore, α may be related to the strength of mesoscale convective organization.

From simulated data, the proportionality constant α can be calculated (Xu 1991). Figure 22.15 shows the ensemble average of α^{-1} for Q03 and Q02. The time variation of α^{-1} is not very large, especially during the convectively active period (9–18 h). This suggests that α is a "quasi constant" for a given convective cloud regime. The most important feature is that the magnitude of α^{-1} in Q03 is about twice as large as that in Q02; that is, α is larger when mesoscale convective organization is present (Q02). Therefore, α is a measure of the strength of mesoscale convective organization.

The result above suggests that nondiagnostic aspects of cumulus ensembles should be considered in a future cumulus parameterization when mesoscale organization is present. See chapters 1 and 11 for further discussion.

22.5. Conclusions and discussion

Numerical cumulus ensemble models (CEMs) have been used to study the macroscopic behavior of cumulus convection under a variety of imposed large-scale and underlying surface conditions. These models are able to produce a variety of convective cloud regimes observed in the real atmosphere: for example, scattered, squall type, nonsquall type, and diurnally modulated. It is emphasized in this study that the explicit simulation of cumulus ensemble should be used more extensively in order to better understand the scientific issues involved in the cumulus parameterization problem. By analyzing simulated data, one can gain more quantitative understanding of subgrid-scale cloud processes despite the lack of direct observations of the quantities associated with an ensemble of cumulus clouds.

This study mainly describes the utilization of the UCLA CEM to better understand the macroscopic behavior of cumulus convection in addition to a brief review of existing CEM investigations. The most basic question concerning the parameterizability of cumulus convection has been addressed in this study. It is found that cumulus activity is rather strongly modulated by large-scale processes but also undergoes high-frequency fluctuations. The degree of modulation is somewhat dependent upon the large-scale conditions. When the vertical wind shear is strong, there are some phase delays in the modulation due to the existence of mesoscale organization.

An implication of these results is that cumulus convection is basically parameterizable despite the presence of mesoscale organization. The phase delay in the modulation due to the existence of mesoscale organization suggests the need for a nondiagnostic cumulus parameterization. The existence of nonmodulated high-frequency fluctuations suggests that some nondeterministic behavior of cumulus ensembles is present.

The conclusions above are also confirmed by testing the sensitivity of a diagnostic and deterministic cumulus parameterization to the horizontal resolution of large-scale models. The nondiagnostic aspects of cumulus convection are more significant for coarse horizontal resolutions due to unresolved mesoscale organization. The nondeterministic aspects are most likely to be more serious for small grid sizes due to smaller differences in scales between individual clouds and grid size.

It is also emphasized that simulated data can provide a guide for incorporating some nondiagnostic effects and possibly some nondeterministic effects into a future parameterization. This will require a higher-order closure than the quasi-equilibrium assumption. This problem is similar to a higher-moment turbulence closure. In practice, however, it is more difficult to incorporate the higher-order closure into a cumulus parameterization since the nonmodulated behavior of cumulus convection is not well understood. Further investigations will be very useful in this regard.

The above conclusions are obtained from a 2D CEM and should be viewed with caution. On the other hand, the usefulness of 2D models has been clearly demonstrated in the present study. Ideally, 3D models with comparable domain sizes should be used; however, due to computer limitations at the present time, a great deal may still be learned from such 2D CEMs. Many questions still remain and can be addressed with simulated data. For example, How can the nondiagnostic and nondeterministic aspects be incorporated into a future parameterization? To what extent can the influence of mesoscale organization on cumulus convection be formulated by explicitly predicting some bulk properties of cumulus convection? How serious is the nondeterministic aspect of the modulation of cumulus convection by large-scale processes? What is the influence of downdrafts on the subcloud layer, and can this be incorporated into a cumulus parameterization? To what degree does radiation impact cumulus convection, especially under the presence of mesoscale organization? These and many other questions must be addressed in the future in order to improve cumulus parameterization.

Acknowledgments. The author would like to thank Professors Akio Arakawa and David A. Randall for their advice and encouragement, and Professor Steven K. Krueger for his help in using the cumulus ensemble model during the course of this work. The author also thanks Professor Steven K. Krueger, Ms. Tammy M. Weckwerth, and Dr. Wei-Kuo Tao for their comments on an earlier draft of the paper.

The work performed at UCLA under the supervision of Professor Akio Arakawa was supported jointly by NSF under Grant ATM-8910564 and NASA under Grant NAG 5-789. The computations were performed at the NCAR SCD and the computing facilities of UCLA. NCAR is sponsored by the NSF. The work was partially supported by the U.S. Department of Energy under Grant DE-FG02-92ER61363 to CSU (Randall, PI).

REFERENCES

Ackerman, T. P., K.-N. Liou, F. P. J. Valero, and L. Pfister, 1988: Heating rates in tropical anvils. *J. Atmos. Sci.*, **45**, 1606–1623.

Agee, E., 1982: An introduction to shallow convective systems. *Cloud Dynamics*, E. Agee and T. Asai, Eds., D. Reidel, 3–30.

——, and K. E. Dowell, 1974: Observational studies of mesoscale cellular convection. *J. Appl. Meteor.*, **13**, 46–53.

Albrecht, B., 1981: Parameterization of trade-cumulus amounts. *J. Atmos. Sci.*, **38**, 97–105.

——, V. Ramanathan, and B. Boville, 1986: The effects of cumulus moisture transports on the simulation of climate with a general circulation model. *J. Atmos. Sci.*, **43**, 2443–2462.

Anthes, R. A., 1977: A cumulus parameterization scheme utilizing a one-dimensional cloud model. *Mon. Wea. Rev.*, **105**, 270–286.

——, E.-Y. Hsie, and Y.-H. Kuo, 1987: Description of the Penn State/NCAR Mesoscale Model Version 4 (MM4), NCAR/TN-282 + STR, National Center for Atmospheric Research, Boulder, CO, 66 pp.

Arakawa, A., 1969: Parameterization of cumulus convection. *Proc. WMO/IUGG Symp. Numerical Weather Prediction*, Tokyo, Japan Meteor. Agency, 1–6.

——, 1972: Design of the UCLA general circulation model. Numerical Simulation of Weather and Climate. Tech. Rep. No. 7, Department of Meteorology, University of California, Los Angeles, 116 pp.

——, 1975: Modelling clouds and cloud processes for use in climate model. The Physical Basis of Climate and Climate Modelling, GARP Publication Series No. 16, 183–197.

——, and W. H. Schubert, 1974: Interaction of a cumulus cloud ensemble with the large-scale environment. Part I. *J. Atmos. Sci.*, **31**, 674–701.

——, and J.-M. Chen, 1987: Closure assumptions in the cumulus parameterization problem. *Short- and medium-range Numerical Prediction*, Collection of papers presented at the WMO/IUGG NWP Symposium, Tokyo, T. Matsuno, Ed., Meteor. Soc. Japan, 107–131.

——, and K.-M. Xu, 1990: The macroscopic behavior of simulated cumulus convection and semi-prognostic tests of the Arakawa–Schubert parameterization. *Proc. of the Indo–U.S. Seminar on Parameterization of Sub-Grid Scale Processes in Dynamical Models of Medium-Range Prediction and Global Climate*, Pune, India.

——, and ——, 1992: The macroscopic behavior of simulated cumulus convection and semiprognostic tests of the Arakawa–Schubert cumulus parameterization. *Physical Processes in Atmospheric Models*, Collection of papers presented at the Indo-U.S. Seminar on Parameterization of Sub-Grid Scale Processes in Dynamical Models of Medium Range Prediction and Global Climate, Pune, India, O. R. Sikka and S. S. Singh, Eds., Wiley Eastern Limited, 3–18.

——, A. Katayama, and Y. Mintz, 1968: Numerical simulation of the atmosphere. *Proc. WMO/IUGG Symp. on Numerical Weather Prediction*, Tokyo, Meteor. Soc. Japan, 8–12.

Arkell, R., and M. Hudlow, 1977: *GATE International Meteorological Radar Atlas.* Environmental Data Service, NOAA, 222 pp. [Available from the Superintendent of Documents, U.S. Govt. Printing Office, Washington, D.C., 20402; Stock No. 003-019-00038-1.]

Asai, T., 1970: Three-dimensional features of thermal convection in a plane Couette flow. *J. Meteor. Soc. Japan*, **48**, 18–29.

Augstein, E., H. Schmidt, and F. Ostapoff, 1974: The vertical structure of the atmospheric planetary boundary layer in undisturbed trade winds over the Atlantic Ocean. *Bound.-Layer Meteor.*, **6**, 129–150.

Baik, J. J., M. DeMaria, and S. Raman, 1990: Tropical cyclone simulations with the Betts convective adjustment scheme. Part II: Sensitivity experiments. *Mon. Wea. Rev.*, **118**, 529–541.

Baker, M. B., R. Corbin, and J. Latham, 1980: The influence of entrainment on the evolution of cloud droplet spectra. I: A model of inhomogeneous mixing. *Quart. J. Roy. Meteor. Soc.*, **106**, 581.

Barnes, G. M., and M. Garstang, 1982: Subcloud layer energetics of precipitating convection. *Mon. Wea. Rev.*, **110**, 102–117.

——, and K. Sieckman, 1984: Mass inflow normal to fast and slow tropical mesoscale convective cloud lines. *Mon. Wea. Rev.*, **112**, 1782–1794.

Bean, B. R., R. Gilmer, R. L. Grossman, and R. McGavin, 1972: An analysis of airborne measurements of vertical water vapor flux during BOMEX. *J. Atmos. Sci.*, **29**, 860–869.

——, C. B. Emmanuel, R. O. Gilmer, and R. E. McGavin, 1975: The spatial and temporal variations of the turbulent fluxes of heat, momentum and water vapor over Lake Ontario. *J. Phys. Oceanogr.*, **5**, 532–540.

Bélair, S., D.-L. Zhang, and J. Mailhot, 1991: Numerical simulation of the Montreal flash flood of 14 July 1987. Preprints, *Ninth Conf. on Numerical Weather Prediction*, Denver, Amer. Meteor. Soc., 790–793.

Bengtsson, L., M. Kanamitsu, P. Kållberg, and S. Uppala, 1982: FGGE 4-dimensional data assimilation at ECMWF. *Bull. Amer. Meteor. Soc.*, **63**, 29–43.

Beniston, M., 1985: Organization of convection in a numerical mesoscale model as a function of initial and lower boundary conditions. *Beitr. Phys. Atmos.*, **58**, 31–52.

Bennetts, D. A., and B. J. Hoskins, 1979: Conditional symmetric instability—A possible explanation for frontal rainbands. *Quart. J. Roy. Meteor. Soc.*, **105**, 945–962.

Betts, A. K., 1973a: A composite mesoscale cumulonimbus budget. *J. Atmos. Sci.*, **30**, 597–610.

——, 1973b: Non-precipitating cumulus convection and its parameterization. *Quart. J. Roy. Meteor. Soc.*, **99**, 178–196.

——, 1974: The scientific basis and objectives of the U.S. convection subprogram for the GATE. *Bull. Amer. Meteor. Soc.*, **55**, 304–313.

——, 1974: Thermodynamic classification of tropical convective soundings. *Mon. Wea. Rev.*, **102**, 760–764.

——, 1975: Parametric interpretation of trade-wind cumulus budget studies. *J. Atmos. Sci.*, **32**, 1934–1945.

——, 1976: The thermodynamic transformation of the tropical subcloud layer by precipitation and downdrafts. *J. Atmos. Sci.*, **33**, 1008–1020.

——, 1982: Saturation point analysis of moist convective overturning. *J. Atmos. Sci.*, **39**, 1484–1505.

——, 1983: Thermodynamics of mixed stratocumulus layers: Saturation point budgets. *J. Atmos. Sci.*, **40**, 2655–2670.

——, 1986: A new convective adjustment scheme. Part I: Observational and theoretical basis. *Quart. J. Roy. Meteor. Soc.*, **112**, 677–691.

——, and M. J. Miller, 1986: A new convective adjustment scheme. Part II. Single column tests using GATE wave, BOMEX, ATEX and arctic air-mass data sets. *Quart. J. Roy. Meteor. Soc.*, **112**, 693–709.

——, and B. Albrecht, 1987: Conserved variable analysis of the convective boundary layer thermodynamic structure over the tropical oceans. *J. Atmos. Sci.*, **44**, 83–99.

——, and J. Bartlo, 1991: The density temperature and the dry and wet virtual adiabats. *Mon. Wea. Rev.,* **119,** 169–175.
——, R. W. Grover, and M. W. Moncrieff, 1976: Structure and motion of tropical squall lines over Venezuela. *Quart. J. Roy. Meteor. Soc.,* **102,** 395–404.
Biggerstaff, M. I., and R. A. Houze, Jr., 1991: Kinematic and precipitation structure of the 10–11 June 1985 squall line. *Mon. Wea. Rev.,* **119,** 3034–3065.
——, and ——, 1993: Kinematics and microphysics of the transition zone of the 10–11 June 1985 squall line. *J. Atmos. Sci.,* **50,** 3091–3110.
Binder, P., 1990: On the parametric representation of the tropospheric thermodynamic structure for midlatitude convective situations. *Quart. J. Roy. Meteor. Soc.,* **116,** 1349–1358.
Blanchard, D. O., 1990: Mesoscale convective patterns of the southern High Plains. *Bull. Amer. Meteor. Soc.,* **71,** 994–1005.
Bluestein, H. B., and M. H. Jain, 1985: Formation of mesoscale lines of precipitation: Severe squall lines in Oklahoma during the spring. *J. Atmos. Sci.,* **42,** 1711–1732.
——, G. T. Marx, and M. H. Jain, 1987: Formation of mesoscale lines of precipitation: Non-severe squall lines in Oklahoma during the spring. *Mon. Wea. Rev.,* **115,** 2719–2727.
Blyth, A. M., W. A. Cooper, and J. B. Jensen, 1988: A study of the source of entrained air in Montana cumuli. *J. Atmos. Sci.,* **45,** 3944–3964.
Bolton, D., 1980: The computation of equivalent potential temperature. *Mon. Wea. Rev.,* **108,** 1046–1053.
Bower, K., T. Choularton, J. Latham, M. Baker, J. Nelson, A. Blyth, and J. Jensen, 1992: Warm cloud microphysics: measurements and discussion. *Proc. Int. Conf. Cloud Physics,* Montreal.
Bretherton, C. S., 1987: A mathematical model of moist convection between two parallel plates. Part I: "Linear" theory and cloud structure. *J. Atmos. Sci.,* **44,** 1809–1827.
——, 1988: Group velocity and the linear response of stratified fluids to internal heat or mass sources. *J. Atmos. Sci.,* **45,** 81–93.
——, and P. K. Smolarkiewicz, 1989: Gravity waves, compensating subsidence and detrainment around cumulus clouds. *J. Atmos. Sci.,* **46,** 740–759.
Brown, E., and R. Braham, Jr., 1959: Precipitation-particle measurements in trade-wind cumuli. *J. Meteor.,* **16,** 609–616.
Brümmer, B., 1978: Mass and energy budgets of a 1 km high atmospheric box over the GATE C-scale triangle during undisturbed and disturbed weather conditions. *J. Atmos. Sci.,* **35,** 997–1011.
Byers, H. R., and R. R. Braham, 1949: *The Thunderstorm.* U.S. Govt. Printing Office, Washington, D.C., 287 pp. [NTIS PB-234-515.]
——, and E. C. Hull, 1949: Inflow patterns of thunderstorms as shown by winds aloft. *Bull. Amer. Meteor. Soc.,* **30,** 90–96.
——, and R. Hall, 1954: A census of cumulus cloud height vs. precipitation in the vicinity of Puerto Rico during the winter and spring of 1953–54. *J. Meteor.,* **12,** 176–178.
Cess, R. D., and coauthors, 1990: Intercomparison and interpretation of climate feedback processes in 19 atmospheric general circulation models. *J. Geophys. Res.,* **95,** 16 601–16 615.
Chang, C.-B., D. J. Perkey, and C. W. Kreitzberg, 1981: A numerical case study of the squall line of 6 May 1975. *J. Atmos. Sci.,* **38,** 1601–1615.
——, ——, and ——, 1982: A numerical case study of the effects of latent heating on a developing wave cyclone. *J. Atmos. Sci.,* **39,** 1555–1570.
——, ——, and ——, 1984: Latent heat induced energy transformations during cyclogenesis. *Mon. Wea. Rev.,* **112,** 357–367.
Charney, J. G., 1955: The use of primitive equations of motion in numerical predictions. *Tellus,* **7,** 22–26.
——, and A. Eliassen, 1964: On the growth of the hurricane depression. *J. Atmos. Sci.,* **21,** 68–75.
Chen, C., and H. D. Orville, 1980: Effects of mesoscale convergence on cloud convection. *J. Appl. Meteor.,* **19,** 256–274.
Chen, J. M., 1989: Observational study of the macroscopic behavior of moist-convective processes. Ph.D. thesis, University of California, Los Angeles, 264 pp.
Chen, S. S., and W. M. Frank, 1992: A numerical study of the genesis of extratropical convective mesovortices. Part I: Evolution and dynamics. *J. Atmos. Sci.,* **49,** 2401–2426.
Chen, T.-C., C.-B. Chang, and D. J. Perkey, 1983: Numerical study of an AMTEX '75 oceanic cyclone. *Mon. Wea. Rev.,* **111,** 1818–1829.
Cheng, C.-P., and R. A. Houze, Jr., 1979: The distribution of convective and mesoscale precipitation in GATE radar echo patterns. *Mon. Wea. Rev.,* **107,** 1370–1381.
Cheng, M.-D., 1989a: Effects of downdrafts and mesoscale convective organization on the heat and moisture budgets of tropical cloud clusters. Part I: A diagnostic cumulus ensemble model. *J. Atmos. Sci.,* **46,** 1517–1538.
——, 1989b: Effects of downdrafts and mesoscale convective organization on the heat and moisture budgets of tropical cloud clusters. Part II: Effects of convective-scale downdrafts. *J. Atmos. Sci.,* **46,** 1540–1564.
——, and M. Yanai, 1989: Effects of downdrafts and mesoscale convective organization on the heat and moisture budgets of tropical cloud clusters. Part III: Effects of mesoscale convective organization. *J. Atmos. Sci.,* **46,** 1566–1588.
——, and A. Arakawa, 1990: Inclusion of convective down drafts in the Arakawa–Schubert cumulus parameterization. Tech. Rep., Department of Atmospheric Sciences, University of California, Los Angeles, 69 pp.
——, and ——, 1991a: Inclusion of convective downdrafts in the Arakawa–Schubert cumulus parameterization. *Extended Abstracts, 19th Conf. on Hurricane and Tropical Meteorology,* Miami, FL, Amer. Meteor. Soc., 295–300.
——, and ——, 1991b: Inclusion of rainwater budgets and convective downdrafts in a cumulus parameterization. *Ninth Conf. on Numerical Weather Prediction,* Denver, Amer. Meteor. Soc., 4 pp.
Cho, H.-R., and Y. Ogura, 1974: A relationship between cloud activity and low-level convergence as observed in Reed–Recker's composite easterly waves. *J. Atmos. Sci.,* **31,** 2058–2065.
——, and L. Cheng, 1980: Parameterization of horizontal transport of vorticity by cumulus convection. *J. Atmos. Sci.,* **37,** 812–826.
——, and T. L. Clark, 1981: A numerical investigation of the structure of vorticity fields associated with a deep convective cloud. *Mon. Wea. Rev.,* **109,** 1654–1670.
Chu, J.-H., M. Yanai, and C.-H. Sui, 1981: Effects of cumulus convection on the vorticity field in the tropics. Part I: The large-scale budget. *J. Meteor. Soc. Japan,* **59,** 1981.
Churchill, D. D., and R. A. Houze, Jr., 1984: Development and structure of winter monsoon cloud clusters on 10 December 1978. *J. Atmos. Sci.,* **41,** 933–960.
Chisholm, A. J., and J. H. Renick, 1972: The kinematics of multicell and supercell Alberta hailstorms. Alberta Hail Studies 1072, Research Council of Alberta, Hail Studies Report No. 72-2, 24–31.
Chong, M., and D. Hauser, 1990: A tropical squall line observed during the COPT 81 experiment in West Africa. Part III: Heat and moisture budgets. *Mon. Wea. Rev.,* **118,** 1696–1706.
——, P. Amayenc, G. Scialom, and F. Testud, 1987: A tropical squall line observed during the COPT 81 experiment in West Africa. Part I: Kinematic structure inferred from dual-Doppler data. *Mon. Wea. Rev.,* **115,** 670–694.
Clark, T. L., T. Hauf, and J. P. Kuettner, 1986: Convectively forced internal gravity waves: Results from two-dimensional numerical experiments. *Quart. J. Roy. Meteor. Soc.,* **112,** 899–925.
Cohen, C., 1989: Numerical experiments showing the response of parameterized convection to large-scale forcing. *J. Atmos. Sci.,* **46,** 132–149.
——, and W. M. Frank, 1989: A numerical study of lapse-rate adjustment in the tropical atmosphere. *Mon. Wea. Rev.,* **117,** 1891–1905.
Cooley, W. W., and P. R. Lohnes, 1971: *Multivariate Data Analysis.* Wiley, 364 pp.

Cotton, W. R., 1986: Averaging and the parameterization of physical processes in mesoscale models. *Mesoscale Meteorology and Forecasting,* P. S. Ray, Ed., Amer. Meteor. Soc., 614–635.

——, and R. A. Anthes, 1989: *Storm and Cloud Dynamics.* Academic Press, 883 pp.

Cox, S. K., and K. T. Griffith, 1979: Estimates of radiative divergence during Phase III of the GARP Atlantic Tropical Experiment: Part II. Analysis of Phase III results. *J. Atmos. Sci.,* **36,** 586–601.

Crook, N. A., and M. W. Moncrieff, 1988: The effect of large-scale convergence on the generation and maintenance of deep moist convection. *J. Atmos. Sci.,* **45,** 3606–3624.

Crum, F. X., and D. E. Stevens, 1983: A comparison of two cumulus parameterization schemes in a linear model of wave–CISK. *J. Atmos. Sci.,* **40,** 2671–2688.

Deardorff, J. W., 1980: Cloud-top entrainment instability. *J. Atmos. Sci.,* **37,** 131–147.

DelGenio, A. D., 1991: Convective and large-scale cloud processes in global climate models. Energy and Water Cycles in the Climate System. *Proc. NATO Advanced Study Institute,* Springer-Verlag.

——, and M.-S. Yao, 1987: Properties of deep convective clouds in the ISCCP pilot data set. Preprints, *17th Conf. on Hurricanes and Tropical Meteorology,* Miami, Amer. Meteor. Soc., 133–136.

——, and ——, 1988: Sensitivity of a global climate model to the specification of convective updraft and downdraft mass fluxes. *J. Atmos. Sci.,* **45,** 2641–2668.

——, and ——, 1990: Predicting cloud water variations in the GISS GCM. Preprints, *Conf. on Cloud Physics,* San Francisco, Amer. Meteor. Soc., 497–504.

——, A. A. Lacis, and R. A. Ruedy, 1991: Simulations of the effect of a warmer climate on atmospheric humidity. *Nature,* **351,** 382–385.

Dopplick, T. G., 1972: Radiative heating of the global atmosphere. *J. Atmos. Sci.,* **29,** 1278–1294.

Dudhia, J., 1989: Numerical study of convection observed during the winter monsoon experiment using a mesoscale, two-dimensional model. *J. Atmos. Sci.,* **46,** 3077–3107.

——, 1993: A nonhydrostatic version of the Penn State–NCAR mesoscale model. *Mon. Wea. Rev.,* **121,** 1493–1513.

Echternacht, K. L., and M. Garstang, 1976: Changes in the structure of the tropical sub-cloud layer from the undisturbed to disturbed states. *Mon. Wea. Rev.,* **104,** 407–417.

Eliassen, A., 1949: The quasistatic equations of motion with pressure as independent variable. *Geophys. Norv.,* **24,** 229–239.

——, 1962: On the vertical circulation in frontal zones. *Geophys. Norv.,* **24,** 147–160.

Emanuel, K. A., 1979: Inertial instability and mesoscale convective systems. Part I: Linear theory of inertial instability in rotating viscous fluids. *J. Atmos. Sci.,* **36,** 2425–2449.

——, 1983a: The Lagrangian parcel dynamics of moist symmetric instability. *J. Atmos. Sci.,* **40,** 2368–2376.

——, 1983b: On assessing local conditional symmetric instability from atmospheric soundings. *Mon. Wea. Rev.,* **111,** 2016–2033.

——, 1986: Some dynamical aspects of precipitating convection. *J. Atmos. Sci.,* **43,** 2183–2198.

——, 1988: Observational evidence of slantwise convective adjustment. *Mon. Wea. Rev.,* **116,** 1805–1816.

——, 1991: A scheme for representing cumulus convection in large-scale models. *J. Atmos. Sci.,* **48,** 2313–2335.

Emmitt, G. D., 1978: Tropical cumulus interaction with and modification of the sub-cloud region. *J. Atmos. Sci.,* **35,** 1485–1502.

Esbensen, S., 1975: An analysis of subcloud-layer heat and moisture budgets in the western Atlantic trades. *J. Atmos. Sci.,* **32,** 1921–1933.

——, 1978: Bulk thermodynamic effects and properties of small tropical cumuli. *J. Atmos. Sci.,* **35,** 826–837.

——, L. J. Shapiro, and E. I. Tollerud, 1987: The consistent parameterization of the effects of cumulus clouds on the large-scale momentum and vorticity fields. *Mon. Wea. Rev.,* **115,** 664–669.

——, J.-T. Wang, and E. I. Tollerud, 1988: A composite life cycle of nonsquall mesoscale convective systems over the tropical ocean. Part II: Heat and moisture budgets. *J. Atmos. Sci.,* **45,** 537–548.

Fankhauser, J. C., 1969: Convective processes resolved by a mesoscale rawinsonde network. *J. Appl. Meteor.,* **8,** 778–798.

——, 1971: Thunderstorm–environment interaction determined from aircraft and radar observations. *Mon. Wea. Rev.,* **99,** 171–192.

——, 1974: The derivation of consistent fields of wind and geopotential height from mesoscale rawinsonde data. *J. Appl. Meteor.,* **13,** 637–646.

——, G. M. Barnes, and M. A. LeMone, 1991: Structure of a midlatitude squall line formed in strong unidirectional shear. *Mon. Wea. Rev.,* **120,** 237–260.

Fiedler, B. H., 1984: The mesoscale stability of entrainment into cloud-topped mixed layers. *J. Atmos. Sci.,* **41,** 92–101.

——, 1985: Mesoscale cellular convection: Is it convection? *Tellus,* **37A,** 163–175.

Fitzjarrald, D. R., and M. Garstang, 1981a: Vertical structure of the tropical boundary layer. *Mon. Wea. Rev.,* **109,** 1512–1526.

——, and ——, 1981b: Boundary-layer growth over the tropical ocean. *Mon. Wea. Rev.,* **109,** 1762–1772.

Foote, G. B., and J. C. Fankhauser, 1973: Airflow and moisture budget beneath a northeast Colorado hailstorm. *J. Appl. Meteor.,* **12,** 1330–1353.

Foster, D. S., 1958: Thunderstorm gusts compared with computed downdraft speeds. *Mon. Wea. Rev.,* **86,** 91–94.

Fovell, R. G., and Y. Ogura, 1988: Numerical simulation of a midlatitude squall line in two dimensions. *J. Atmos. Sci.,* **45,** 3846–3879.

Frank, H. W., and G. B. Foote, 1982: The 22 July 1976 case study: Storm air-flow, updraft structure, and mass flux from triple-Doppler measurements. *Hail Storms of the Central High Plains, Volume II,* Colorado Associated University Press, 131–162.

Frank, W. M., 1977: The structure and energetics of the tropical cyclone. Part I: Storm structure. *Mon. Wea. Rev.,* **105,** 1119–1135.

——, 1983: The cumulus parameterization problem. *Mon. Wea. Rev.,* **111,** 1859–1871.

——, and C. Cohen, 1987: Simulation of tropical convective systems. Part I: A cumulus parameterization. *J. Atmos. Sci.,* **44,** 3787–3799.

——, and J. L. McBride, 1989: The vertical distribution of heating in AMEX and GATE cloud clusters. *J. Atmos. Sci.,* **46,** 3464–3478.

Fritsch, J. M., 1975: Cumulus dynamics: Local compensating subsidence and its implications for cumulus parameterization. *Pure Appl. Geophys.,* **113,** 851–867.

——, and C. F. Chappell, 1980: Numerical prediction of convectively driven mesoscale pressure systems. Part I: Convective parameterization. *J. Atmos. Sci.,* **37,** 1722–1733.

——, ——, and L. R. Hoxit, 1976: The use of large scale budgets for convective parameterization. *Mon. Wea. Rev.,* **104,** 1408–1418.

Fu, R., A. D. DelGenio, and W. B. Rossow, 1990: Behavior of deep convective clouds in the tropical Pacific deduced from ISCCP radiances. *J. Climate,* **3,** 1129–1152.

Fujita, T., 1959: Precipitation and cold-air production in mesoscale thunderstorm systems. *J. Meteor.,* **16,** 454–466.

Gallus, W. A., Jr., and R. H. Johnson, 1991: Heat and moisture budgets of an intense midlatitude squall line. *J. Atmos. Sci.,* **48,** 122–146.

——, and ——, 1992: The momentum budget of an intense midlatitude squall line. *J. Atmos. Sci.,* **49,** 422–450.

Gamache, J. F., and R. A. Houze, Jr., 1982: Mesoscale air motions associated with a tropical squall line. *Mon. Wea. Rev.,* **110,** 118–135.

——, and ——, 1983: Water budget of a mesoscale convective system in the tropics. *J. Atmos. Sci.,* **40,** 1835–1850.

Gao, K., D.-L. Zhang, M. W. Moncrieff, and H.-R. Cho, 1990: Mesoscale momentum budget in a midlatitude squall line: A numerical case study. *Mon. Wea. Rev.*, **118**, 1011–1028.

Gaynor, J. E., and E. F. Ropelewski, 1979: Analysis of convectively modified GATE boundary layer using in situ and acoustic sounder data. *Mon. Wea. Rev.*, **107**, 985–993.

Geleyn, J.-F., 1985: On a simple, parameter-free partition between moistening and precipitation in the Kuo scheme. *Mon. Wea. Rev.*, **113**, 405–407.

Gill, A. E., 1980: Some simple solutions for heat-induced tropical circulation. *Quart. J. Roy. Meteor. Soc.*, **106**, 447–462.

Giorgi, F., 1991: Sensitivity of simulated summertime precipitation over the western United States to different physics parameterizations. *Mon. Wea. Rev.*, **119**, 2870–2888.

Golding, B. W., 1990: The meteorological office mesoscale model. *Meteor. Mag.*, **119**, 81–96.

Gray, W. M., 1965: Calculation of cumulus vertical draft velocities in hurricanes from aircraft observations. *J. Appl. Meteor.*, **4**, 47–53.

——, 1973: Cumulus convection and larger scale circulations. I: Broadscale and mesoscale considerations. *Mon. Wea. Rev.*, **101**, 839–855.

Gregory, D., and M. J. Miller, 1989: A numerical study of the parameterization of deep tropical convection. *Quart. J. Roy. Meteor. Soc.*, **115**, 1209–1241.

——, and P. R. Rountree, 1990: A mass flux convection scheme with representation of cloud ensemble characteristics and stability-dependent closure. *Mon. Wea. Rev.*, **118**, 1483–1506.

Grell, G., Y.-H. Kuo, and R. J. Pasch, 1991: Semiprognostic tests of three cumulus parameterization schemes in middle latitudes. *Mon. Wea. Rev.*, **119**, 5–31.

Grossman, R. L., and D. R. Durran, 1984: Interaction of low level flow with the Western Ghat Mountains and off-shore convection in the summer monsoon. *Mon. Wea. Rev.*, **112**, 652–672.

——, and O. Garcia, 1990: The distribution of deep convection over ocean and land during the Asian summer monsoon. *J. Climate*, **3**, 1032–1044.

Gyakum, J. R., 1983: On the evolution of the *QE II* storm. II: Dynamic and thermodynamic structure. *Mon. Wea. Rev.*, **111**, 1156–1173.

Hack, J. J., and W. H. Schubert, 1986: Nonlinear response of atmospheric vortices to heating by organized cumulus convection. *J. Atmos. Sci.*, **43**, 1559–1573.

——, W. H. Schubert, and P. L. Silva Dias, 1984: A spectral cumulus parameterization for use in numerical models of the tropical atmosphere. *Mon. Wea. Rev.*, **112**, 704–716.

Hansen, J., G. Russell, D. Rind, P. Stone, A. Lacis, S. Lebedeff, R. Ruedy, and L. Travis, 1983: Efficient three-dimensional global models for climate studies: Models I and II. *Mon. Wea. Rev.*, **111**, 609–662.

Harris, C. W., and H. F. Kaiser, 1964: Oblique factor analytic solutions by orthogonal transformations, *Psychometrica*, **29**, 347.

Hauser, D., and P. Amayenc, 1986: Retrieval of cloud water and water vapor contents from Doppler radar data in a tropical squall line. *J. Atmos. Sci.*, **43**, 823–838.

Hayashi, Y., 1970: A theory of large-scale equatorial waves generated by condensation heat and accelerating the zonal wind. *J. Meteor. Soc. Japan*, **48**, 140–160.

Haynes, P. H., and M. E. McIntyre, 1987: On the evolution of vorticity and potential vorticity in the presence of diabatic heating and frictional or other forces. *J. Atmos. Sci.*, **44**, 828–841.

He, H., J. W. McGinnis, Z. Song, and M. Yanai, 1987: Onset of the Asian summer monsoon in 1979 and the effect of the Tibetan Plateau. *Mon. Wea. Rev.*, **115**, 1966–1995.

Heckley, W. A., M. J. Miller, and A. K. Betts, 1987: An example of hurricane tracking and forecasting with a global analysis–forecasting system. *Bull. Amer. Meteor. Soc.*, **68**, 226–229.

Heymsfield, A. J., and L. J. Donner, 1990: A scheme for parameterizing ice-cloud water content in general circulation models. *J. Atmos. Sci.*, **47**, 1865–1877.

——, P. N. Johnson, and J. E. Dye, 1978: Observations of moist adiabatic ascent in northeast Colorado cumulus congestus clouds. *J. Atmos. Sci.*, **35**, 1689–1703.

Hill, G. E., 1974: Factors controlling the size and spacing of cumulus clouds as revealed by numerical experiments. *J. Atmos. Sci.*, **31**, 646–673.

——, 1977: Initiation mechanisms and development of cumulus convection. *J. Atmos. Sci.*, **34**, 1934–1941.

Holland, J., and E. Rasmussen, 1973: Measurements of atmospheric mass, energy and momentum budgets over a 500-kilometer square of tropical oceans. *Mon. Wea. Rev.*, **101**, 11–55.

Hoskins, B. J., 1975: The geostrophic momentum approximation and the semigeostrophic equations. *J. Atmos. Sci.*, **32**, 233–242.

——, and I. Draghici, 1978: The forcing of ageostrophic motion to the semigeostrophic equations and in an isentropic coordinate model. *J. Atmos. Sci.*, **34**, 1859–1867.

Houze, R. A., 1977: Structure and dynamic of a tropical squall-line system. *Mon. Wea. Rev.*, **105**, 1540–1567.

——, 1982: Cloud clusters and large-scale vertical motions in the tropics. *J. Meteor. Soc. Japan*, **60**, 396–410.

——, 1989: Observed structure of mesoscale convective systems and implications for large-scale heating. *Quart. J. Roy. Meteor. Soc.*, **115**, 425–462.

——, and A. K. Betts, 1981: Convection in GATE. *Rev. Geophys. Space Phys.*, **19**, 541–576.

——, and C.-P. Cheng, 1981: Inclusion of mesoscale updrafts and downdrafts in computations of vertical fluxes by ensembles of tropical clouds. *J. Atmos. Sci.*, **38**, 1751–1770.

——, and P. V. Hobbs, 1982: Organization and structure of precipitating cloud systems. *Advances in Geophysics*, Academic Press, Vol. 24, 225–315.

——, and E. N. Rappaport, 1984: Air motions and precipitation structure of an early summer squall line over the eastern tropical Atlantic. *J. Atmos. Sci.*, **41**, 553–574.

——, C.-P. Cheng, C. A. Leary, and J. F. Gamache, 1980: Diagnosis of cloud mass and heat fluxes from radar and synoptic data. *J. Atmos. Sci.*, **37**, 754–773.

——, S. G. Geotis, F. D. Marks, and A. K. West, 1981: Winter monsoon convection in the vicinity of North Borneo. Part I: Structure and time variation of the clouds and precipitation. *Mon. Wea. Rev.*, **109**, 1595–1614.

——, S. A. Rutledge, M. I. Biggerstaff, and B. F. Smull, 1989: Interpretation of Doppler weather radar displays of midlatitude mesoscale convective systems. *Bull. Amer. Meteor. Soc.*, **70**, 607–619.

——, B. F. Smull, and P. Dodge, 1992: Mesoscale organization of springtime rainstorms in Oklahoma. *Mon. Wea. Rev.*, **118**, 613–654.

Hoxit, L. R., C. F. Chappell, and J. M. Fritsch, 1976: Formation of mesolows or pressure troughs in advance of cumulonimbus clouds. *Mon. Wea. Rev.*, **104**, 1419–1428.

——, R. A. Maddox, C. F. Chappell, F. L. Zuckerberg, H. M. Mogil, I. Jones, D. R. Greene, R. E. Saffle, and R. A. Scofield, 1978: Meteorological analysis of the Johnstown, Pennsylvania flash flood, 19–20 July 1977. NOAA Tech. Rep. ERL 401-APCL43, 71 pp.

Hsu, W.-R., and W.-Y. Sun, 1991: Numerical study of mesoscale cellular convection. *Bound.-Layer Meteor.*, **57**, 167–186.

Iversen, T., and T. E. Nordeng, 1984: A hierarchy of filtered models—Numerical solutions. *Mon. Wea. Rev.*, **112**, 2048–2059.

Janjic, Z. I., 1990: The step-mountain coordinate: Physical package. *Mon. Wea. Rev.*, **118**, 1429–1443.

Johnson, R. H., 1976: The role of convective-scale precipitation downdrafts in cumulus and synoptic-scale interactions. *J. Atmos. Sci.*, **33**, 1890–1910.

——, 1977: The effects of cloud detrainment on the diagnosed properties of cumulus populations. *J. Atmos. Sci.*, **34**, 359–366.

——, 1980: Diagnosis of convective and mesoscale motions during Phase III of GATE. *J. Atmos. Sci.*, **37**, 733–753.

——, 1981: Large-scale effects of deep convection on the GATE tropical boundary layer. *J. Atmos. Sci.*, **38**, 2399–2413.

——, 1982: Vertical motion in near-equatorial winter monsoon convection. *J. Meteor. Soc. Japan,* **60,** 682–689.
——, 1984: Partitioning tropical heat and moisture budgets into cumulus and mesoscale components: Implication for cumulus parameterization. *Mon. Wea. Rev.,* **112,** 1590–1601.
——, and M. E. Nicholls, 1983: A composite analysis of the boundary layer accompanying a tropical squall line. *Mon. Wea. Rev.,* **111,** 308–319.
——, and G. S. Young, 1983: Heat and moisture budgets of tropical mesoscale anvil clouds. *J. Atmos. Sci.,* **40,** 2138–2147.
——, and P. J. Hamilton, 1988: The relationship of surface pressure features to the precipitation and air flow structure of an intense midlatitude squall line. *Mon. Wea. Rev.,* **116,** 1444–1472.
——, W. A. Gallus, Jr., and M. D. Vescio, 1990: Near-tropopause vertical motion within the trailing stratiform region of a midlatitude squall line. *J. Atmos. Sci.,* **47,** 2200–2210.
Jones, S. C., and A. J. Thorpe, 1992: The three dimensional nature of "symmetric instability." *Quart. J. Roy. Meteor. Soc.,* **118,** 227–258.
Jordan, C. L., 1958: Semiprognostic tests of three cumulus parameterization schemes in middle latitudes. *Mon. Wea. Rev.,* **15,** 91–97.
Jorgensen, D. P., and M. A. LeMone, 1989: Vertical velocity characteristics of oceanic convection. *J. Atmos. Sci.,* **46,** 621–640.
——, E. J. Zipser, and M. A. LeMone, 1985: Vertical motions in intense hurricanes. *J. Atmos. Sci.,* **42,** 839–856.
——, M. A. LeMone, and B. J.-D. Jou, 1991: Precipitation and kinematic structure of an oceanic mesoscale convective system. Part I: Convective line structure. *Mon. Wea. Rev.,* **119,** 2608–2637.
Kain, J. S., and J. M. Fritsch, 1990: A one-dimensional entraining/detraining plume model and its application in convective parameterization. *J. Atmos. Sci.,* **47,** 2784–2802.
——, and ——, 1992: The role of the convective "trigger function" in numerical forecasts of mesoscale convective systems. *Meteor. Atmos. Phys.*
Kampé de Fériet, 1951: Averaging processes and Reynolds equations in atmospheric turbulence. *J. Meteor.,* **8,** 358–361.
Kanamitsu, M., 1975: On numerical prediction over a global tropical belt. Ph.D. thesis, The Florida State University, 281 pp.
Kao, C.-Y. J., and Y. Ogura, 1987: Response of cumulus clouds to large-scale forcing using the Arakawa–Schubert cumulus parameterization. *J. Atmos. Sci.,* **44,** 2437–2458.
Kessler, E., 1969: On the distribution and continuity of water substance in atmospheric circulations. *Meteor. Monogr.,* No. 32, Amer. Meteor. Soc., 84 pp.
Kloessel, K. A., and B. A. Albrecht, 1979: Low level inversions over the tropical Pacific—Thermodynamic structure of the boundary layer and the above-inversion moisture structure. *Mon. Wea. Rev.,* **117,** 87–101.
Knight, C. A., and P. Squires, 1982: *Hailstorms of the Central High Plains.* Vol. II. Colorado Associated University Press, 245 pp.
Knupp, K. R., and W. R. Cotton, 1985: Convective cloud downdraft structure: An interpretive survey. *Rev. Geophys.,* **23,** 183–215.
König, W., and E. Ruprecht, 1989: Effects of convective clouds on the large-scale vorticity budget. *Meteor. Atmos. Phys.,* **41,** 214–229.
Kreitzberg, C. W., and H. A. Brown, 1970: Mesoscale weather systems within an occlusion. *J. Appl. Meteor.,* **9,** 417–432.
——, and D. Perkey, 1976: Release of potential instability. Part I: A sequential plume model within a hydrostatic primitive equation model. *J. Atmos. Sci.,* **33,** 456–475.
——, and ——, 1977: Release of potential instability. Part II: The mechanism of convective/mesoscale interaction. *J. Atmos. Sci.,* **34,** 1569–1595.
Krishnamurti, R., 1975: On cellular cloud patterns—Part I: Mathematical model. *J. Atmos. Sci.,* **32,** 1353–1363.
Krishnamurti, T. N., and W. J. Moxim, 1971: On parameterization of convective and nonconvective latent heat release. *J. Appl. Meteor.,* **10,** 3–13.
——, and H. S. Bedi, 1988: Cumulus parameterization and rainfall rates III. *Mon. Wea. Rev.,* **116,** 583–599.
——, M. Kanamitsu, R. Godbole, C. B. Chang, F. Carr, and J. Chow, 1976: Study of a monsoon depression (ii), dynamical structure. *J. Meteor. Soc. Japan,* **54,** 208–225.
——, Y. Ramanathan, H.-L. Pan, R. Pasch, and J. Molinari, 1980: Cumulus parameterization and rainfall rates I. *Mon. Wea. Rev.,* **108,** 465–472.
——, S.-L. Lam, and R. Pasch, 1983: Cumulus parameterization and rainfall rates II. *Mon. Wea. Rev.,* **111,** 815–828.
Krueger, S. K., 1985: Numerical simulation of tropical cumulus clouds and their interaction with the subcloud layer. Ph.D. dissertation, University of California, Los Angeles, 205 pp.
——, 1988: Numerical simulation of tropical cumulus clouds and their interaction with the subcloud layer. *J. Atmos. Sci.,* **45,** 2221–2250.
Kuettner, J., and S. Soules, 1966: Organized convection as seen from space. *Bull. Amer. Meteor. Soc.,* **47,** 364–370.
Kuo, H. L., 1965: On formation and intensification of tropical cyclones through latent heat release by cumulus convection. *J. Atmos. Sci.,* **22,** 40–63.
——, 1974: Further studies of the parameterization of the influence of cumulus convection on large-scale flow. *J. Atmos. Sci.,* **31,** 1232–1240.
——, 1975: Instability theory of large-scale disturbances in the tropics. *J. Atmos. Sci.,* **32,** 2229–2245.
Kuo, Y.-H., and R. A. Anthes, 1984a: Semiprognostic tests of Kuo-type parameterization schemes in an extratropical convective system. *Mon. Wea. Rev.,* **112,** 1498–1509.
——, and ——, 1984b: Accuracy of diagnostic heat and moisture budgets using SESAME-79 field data as revealed by observing system simulation experiments. *Mon. Wea. Rev.,* **112,** 1465–1481.
——, and ——, 1984c: Mesoscale budgets of heat and moisture in a convective system over the central United States. *Mon. Wea. Rev.,* **112,** 1482–1497.
——, and R. J. Reed, 1988: Numerical simulation of an explosively deepening cyclone in the eastern Pacific. *Mon. Wea. Rev.,* **116,** 2081–2105.
Kurihara, Y., 1973: A scheme of moist convective adjustment. *Mon. Wea. Rev.,* **101,** 547–553.
Lafore, J.-P., and M. W. Moncrieff, 1989: A numerical investigation of organization and interaction of the convective and stratiform regions of tropical squall lines. *J. Atmos. Sci.,* **46,** 521–544.
——, J.-L. Redelsperger, and G. Jaubert, 1988: Comparison between a three-dimensional simulation and Doppler radar data of a tropical squall line: Transports of mass, momentum, heat, and moisture. *J. Atmos. Sci.,* **45,** 3483–3500.
Lau, K.-M., and L. Peng, 1987: Origin of low-frequency (intraseasonal) oscillations in the tropical atmosphere. *J. Atmos. Sci.,* **44,** 950–972.
Laursen, L., and E. Eliasen, 1989: On the effects of the damping mechanisms in an atmospheric general circulation model. *Tellus,* **41A,** 385–400.
Leary, C. A., and R. A. Houze, Jr., 1979: The structure and evolution of convection in a tropical cloud cluster. *J. Atmos. Sci.,* **36,** 437–457.
——, and ——, 1980: The contribution of mesoscale motions to the mass and heat fluxes of an intense tropical convective system. *J. Atmos. Sci.,* **37,** 784–796.
LeMone, M. A., 1983: Momentum transport by a line of cumulonimbus. *J. Atmos. Sci.,* **40,** 1815–1834.
——, and W. T. Pennell, 1976: The relationship of trade wind cumulus distribution to subcloud layer fluxes and structure. *Mon. Wea. Rev.,* **104,** 524–539.
——, and E. J. Zipser, 1980: Cumulonimbus vertical velocity events in GATE. Part I: Diameter, intensity and mass flux. *J. Atmos. Sci.,* **37,** 2444–2457.
——, and J. Jensen, 1990: Some aspects of the structure and dynamics of Hawaiian cloud bands. Preprints, *Conf. on Cloud Physics,* San Francisco, Amer. Meteor. Soc., 744–749.

——, and D. P. Jorgensen, 1991: Precipitation and kinematic structure of an oceanic mesoscale convective system. Part II: Momentum transport and generation. *Mon. Wea. Rev.*, **119**, 2638–2653.

——, G. M. Barnes, and E. J. Zipser, 1984: Momentum flux by lines of cumulonimbus over the tropical oceans. *J. Atmos. Sci.*, **41**, 1914–1932.

Lewis, J. M., 1975: Test of the Ogura–Cho model on prefrontal squall line case. *Mon. Wea. Rev.*, **103**, 764–778.

Lilly, D. K., 1960: On the theory of disturbances in a conditionally unstable atmosphere. *Mon. Wea. Rev.*, **88**, 1–17.

——, 1982: The development and maintenance of rotation in convective storms. *Topics in Atmospheric and Oceanographic Sciences: Intense Atmospheric Vortices.* L. Bengtsson and J. Lighthill, Eds., Springer-Verlag, 149–182.

——, 1990: Numerical prediction of thunderstorms—Has its time come? *Quart. J. Roy. Meteor. Soc.*, **116**, 779–798.

——, and B. F. Jewett, 1990: Momentum and kinetic energy budgets of simulated supercell thunderstorms. *J. Atmos. Sci.*, **47**, 707–726.

Lin, Y.-J., T.-C. C. Wang, R. W. Pasken, H. Shen, and Z.-S. Deng, 1990: Characteristics of a subtropical squall line determined from TAMEX and dual-Doppler data. Part II: Dynamic and thermodynamic structures and momentum budgets. *J. Atmos. Sci.*, **47**, 2382–2399.

Lin, Y. L., 1986: Calculation of airflow over an isolated heat source with application to the dynamics of V-shaped clouds. *J. Atmos. Sci.*, **43**, 2736–2751.

——, 1987: Two-dimensional response of a stably stratified shear flow to diabatic heating. *J. Atmos. Sci.*, **44**, 1375–1393.

——, and R. B. Smith, 1986: Transient dynamics of airflow near a local hear source. *J. Atmos. Sci.*, **43**, 40–49.

——, and S. Li, 1988: Three-dimensional response of a stably stratified shear flow to diabatic heating. *J. Atmos. Sci.*, **45**, 2987–3002.

——, R. D. Farley, and H. D. Orville, 1983: Bulk parameterization of the snow field in a cloud model. *J. Climate Appl. Meteor.*, **22**, 1065–1092.

Lindstrom, S. S., and T. E. Nordeng, 1992: Parameterized slantwise convection in a numerical model. *Mon. Wea. Rev.*, **120**, 742–756.

Lindzen, R. S., 1974: Wave-CISK in the tropics. *J. Atmos. Sci.*, **31**, 156–179.

Lipps, F. B., and R. S. Hemler, 1986: Numerical simulation of deep tropical convection associated with large-scale convergence. *J. Atmos. Sci.*, **43**, 1796–1816.

——, and ——, 1991: Numerical modeling of a midlatitude squall line: Features of the convection and vertical momentum flux. *J. Atmos. Sci.*, **48**, 1909–1929.

Lord, S. J., 1982: Interaction of a cumulus cloud ensemble with the large-scale environment. Part III: Semiprognostic test of the Arakawa–Schubert cumulus parameterization. *J. Atmos. Sci.*, **39**, 88–103.

——, and A. Arakawa, 1980: Interaction of a cumulus cloud ensemble with the large-scale environment. Part II. *J. Atmos. Sci.*, **37**, 2677–2692.

——, and ——, 1982: Interaction of a cumulus cloud ensemble with the large-scale environment. Part III: Semiprognostic test of the Arakawa–Schubert cumulus parameterization. *J. Atmos. Sci.*, **39**, 88–103.

——, W. C. Chao, and A. Arakawa, 1982: Interaction of a cumulus cloud ensemble with the large-scale environment. Part IV: The discrete model. *J. Atmos. Sci.*, **39**, 104–113.

——, H. E. Willoughby, and J. M. Piotrowicz, 1984: Role of a parameterized ice-phase microphysics in an axisymmetric, nonhydrologic tropical cyclone model. *J. Atmos. Sci.*, **41**, 2836–2848.

Luo, H., and M. Yanai, 1984: The large-scale circulation and heat sources over the Tibetan Plateau and surrounding areas during the early summer of 1979. Part II: Heat and moisture budgets. *Mon. Wea. Rev.*, **112**, 966–989.

Maddox, R. A., 1980: Mesoscale convective complexes. *Bull. Amer. Meteor. Soc.*, **61**, 1374–1387.

——, D. J. Perkey, and J. M. Fritsch, 1981: Evolution of upper tropospheric features during the development of a mesoscale convective complex. *J. Atmos. Sci.*, **38**, 1664–1674.

Malkus, J. S., 1958: On the structure of the trade wind moist layer. Papers. *Phys. Oceanogr. Meteor.*, **13**, No. 2, 47 pp.

——, C. Ronne, and M. Chaffee, 1961: Cloud patterns in hurricane Daisy, 1958. *Tellus*, **13**, 8–30.

Manabe, S., and R. Strickler, 1964: Thermal equilibrium of the atmosphere with a convective adjustment. *J. Atmos. Sci.*, **21**, 361–385.

——, J. Smagorinsky, and R. F. Strickler, 1965: Simulated climatology of a general circulation model with a hydrological cycle. *Mon. Wea. Rev.*, **93**, 769–798.

Marwitz, J. O., 1972: Precipitation efficiency of thunderstorms on the High Plains. *J. Rech. Atmos.*, **6**, 367–370.

Matejka, T., and M. A. LeMone, 1990a: The generation and redistribution of momentum in a squall line. *Proc. Fourth Conf. on Mesoscale Processes*, Boulder, CO, Amer. Meteor. Soc., 196–199.

——, and ——, 1990b: The generation and redistribution of momentum in a squall line. Preprints, *Fourth Conf. on Mesoscale Processes*, Boulder, CO, Amer. Meteor. Soc., 196–197.

McBride, J. L., 1981: An analysis of diagnostic cloud mass flux models. *J. Atmos. Sci.*, **38**, 1977–1990.

——, B. W. Gunn, G. J. Holland, T. D. Keenan, N. E. Davidson, and W. M. Frank, 1989: Time series of total heating and moistening over the Gulf of Carpentaria radiosonde array during AMEX. *Mon. Wea. Rev.*, **117**, 2701–2713.

McNab, A. L., and A. K. Betts, 1978: A mesoscale budget study of cumulus convection. *Mon. Wea. Rev.*, **106**, 1317–1331.

Miller, B. L., and D. G. Vincent, 1987: Convective heating and precipitation estimates for the tropical South Pacific during FGGE, 10–18 January 1979. *Quart. J. Roy. Meteor. Soc.*, **113**, 189–212.

Miller, L. J., F. I. Harris, and J. C. Fankhauser, 1982: The 22 June case study: Structure and evolution of internal airflow. *Hail Storms of the Central High Plains, Volume II*, Colorado Associated University Press, 35–59.

Miyakoda, K., J. Smagorinsky, R. F. Strickler, and G. D. Hembree, 1969: Experimental extended predictions with a nine-level hemispheric model. *Mon. Wea. Rev.*, **97**, 1–76.

Molinari, J., 1982: A method for calculating the effects of deep cumulus convection in numerical models. *Mon. Wea. Rev.*, **110**, 1527–1534.

——, 1985: A general form of Kuo's cumulus parameterization. *Mon. Wea. Rev.*, **113**, 1411–1416.

——, and T. Corsetti, 1985: Incorporation of cloud-scale and mesoscale downdrafts into a cumulus parameterization: Results of one- and three-dimensional integrations. *Mon. Wea. Rev.*, **113**, 485–501.

——, and M. Dudek, 1986: Implicit versus explicit convective heating in numerical weather prediction models. *Mon. Wea. Rev.*, **114**, 1822–1831.

——, and ——, 1992: Parameterization of convective precipitation in mesoscale numerical models: A critical review. *Mon. Wea. Rev.*, **120**, 326–344.

Moncrieff, M. W., 1978: The dynamical structure of two-dimensional steady convection in a constant vertical shear. *Quart. J. Roy. Meteor. Soc.*, **104**, 543–567.

——, 1981: A theory of organized steady convection and its transport properties. *Quart. J. Roy. Meteor. Soc.*, **107**, 29–50.

——, 1992: Organized mesoscale convective systems: Archetypical models, mass and momentum flux theories. *Quart. J. Roy. Meteor. Soc.*, **118**, in press.

——, and J. S. A. Green, 1972: The propagation and transfer properties of steady convective overturning in shear. *Quart. J. Roy. Meteor. Soc.*, **98**, 336–352.

——, and M. J. Miller, 1976: The dynamics and simulation of tropical cumulonimbus and squall lines. *Quart. J. Roy. Meteor. Soc.,* **102,** 373–394.

Monin, A. S., and A. M. Yaglom, 1971: *Statistical Fluid Mechanics: Mechanics of Turbulence.* The MIT Press, 769 pp. (English edition, J. L. Lumley, Ed.)

Moorthi, S., and M. J. Suarez, 1992: Relaxed Arakawa–Schubert: A parameterization of moist convection for general circulation models. *Mon. Wea. Rev.,* **120,** 978–1002.

Mourad, P. D., and R. A. Brown, 1990: Multiscale large eddy states in weakly stratified planetary boundary layers. *J. Atmos. Sci.,* **47,** 414–438.

Nakajima, K., and T. Matsuno, 1988: Numerical experiments concerning the origin of cloud cluster in the tropical atmosphere. *J. Meteor. Soc. Japan,* **66,** 309–329.

Neelin, J. D., and J.-Y. Yu, 1993: Modes of tropical variability under convective adjustment and the Madden-Julian oscillation. Part I: Analytical theory. *J. Atmos. Sci.,* **50,** submitted.

Nehrkorn, T., 1986: Wave-CISK in a baroclinic basic state. *J. Atmos. Sci.,* **43,** 2773–2791.

Newton, C. W., 1950: Structure and mechanism of the prefrontal squall line. *J. Meteor.,* **7,** 210–222.

Nicholls, M. E., 1987: A comparison of the results of a two-dimensional numerical simulation of a tropical squall line with observations. *Mon. Wea. Rev.,* **115,** 3055–3077.

——, R. A. Pielke, and W. R. Cotton, 1991a: Thermally forced gravity waves in an atmosphere at rest. *J. Atmos. Sci.,* **48,** 1869–1884.

——, ——, and ——, 1991b: A two-dimensional numerical investigation of the interaction between sea breezes and deep convection over the Florida peninsula. *Mon. Wea. Rev.,* **119,** 298–323.

Nicholls, S., and M. LeMone, 1980: The fair weather boundary layer in GATE: The relationship of subcloud fluxes and structure to the distribution and enhancement of cumulus clouds. *J. Atmos. Sci.,* **37,** 2051–2067.

Ninomiya, K., 1968: Heat and water budget over the Japan Sea and the Japan Islands in winter season—With special emphasis on the relation among the supply from sea surface, the convective transfer and the heavy snowfall. *J. Meteor. Soc. Japan,* **46,** 343–372.

——, 1971: Dynamical analysis of outflow from tornado-producing thunderstorms as revealed by ATS III pictures. *J. Appl. Meteor.,* **10,** 275–294.

Nitta, T., 1972: Energy budget of wave disturbances over the Marshall Islands during the years of 1956 and 1958. *J. Meteor. Soc. Japan,* **50,** 71–84.

——, 1975: Observational determination of cloud mass flux distributions. *J. Atmos. Sci.,* **32,** 73–91.

——, 1977: Response of cumulus updraft and downdraft to GATE A/B-scale motion systems. *J. Atmos. Sci.,* **34,** 1163–1186.

——, 1978: A diagnostic study of interaction of cumulus updrafts and downdrafts with large-scale motions in GATE. *J. Meteor. Soc. Japan,* **56,** 232–242.

——, 1983: Observational study of heat sources over the eastern Tibetan Plateau during the summer monsoon. *J. Meteor. Soc. Japan,* **61,** 590–605.

——, and S. Esbensen, 1974: Heat and moisture budget analyses using BOMEX data. *Mon. Wea. Rev.,* **102,** 17–28.

Nordeng, T. E., 1987: The effect of vertical and slantwise convection on the simulation of polar lows. *Tellus,* **39A,** 354–375.

——, and E. Rasmussen, 1992: A most beautiful polar low. A case study of a polar low development in the Bear Island region. *Tellus,*

Normand, C. W. B., 1946: Energy in the atmosphere. *Quart. J. Roy. Meteor. Soc.,* **72,** 145–167.

Ogura, Y., and H.-R. Cho, 1973: Diagnostic determination of cumulus populations from large-scale variables. *J. Atmos. Sci.,* **30,** 1276–1286.

——, and Y.-L. Chen, 1977: A life history of an intense mesoscale convective storm in Oklahoma. *J. Atmos. Sci.,* **34,** 1458–1476.

——, and M.-T. Liou, 1980: The structure of a midlatitude squall line: A case study. *J. Atmos. Sci.,* **37,** 553–567.

——, and C.-Y. J. Kao, 1987: Numerical simulation of a tropical mesoscale convective system using the Arakawa–Schubert parameterization. *J. Atmos. Sci.,* **44,** 2459–2476.

Oliger, J. E., R. E. Wellck, A. Kasahara, and W. M. Washington, 1970: Description of NCAR Global Circulation Model. Publication of Laboratory of Atmospheric Science, NCAR, Boulder, CO, 94 pp.

Ooyama, V. K., 1969: Numerical simulation of the life cycle of tropical cyclones. *J. Atmos. Sci.,* **26,** 3–40.

——, 1971: A theory on parameterization of cumulus convection. *J. Meteor. Soc. Japan,* **49,** 744–756.

——, 1982a: Conceptual evolution of the theory and modeling of the tropical cyclone. *J. Meteor. Soc. Japan,* **60,** 369–379.

——, 1982b: Schemes and Closures. *Workshop on the Impact of Cumulus Parameterization on Large-Scale Numerical Weather Prediction,* Tallahassee, FL. [Available from the author, HRD/AOML/NOAA, 4301 Rickenbacker Causeway, Miami, FL.]

——, 1987: Scale-controlled objective analysis. *Mon. Wea. Rev.,* **115,** 2479–2506.

Paluch, I. R., 1979: The entrainment mechanism in Colorado cumuli. *J. Atmos. Sci.,* **36,** 2467–2478.

Pan, V., and D. A. Randall, 1991: Arakawa-Schubert cumulus parameterization with a prognostic cumulus kinetic energy. *Ninth Conf. on Numerical Weather Prediction,* Denver, CO, Amer. Meteor. Soc., 4 pp.

Pandya, R., D. Durran, and C. S. Bretherton, 1993: Comments on "Thermally forced gravity waves in an atmosphere at rest." *J. Atmos. Sci.,* **50,** 4097–4101.

Payne, S. W., 1981: The inclusion of moist downdraft effects in the Arakawa–Schubert cumulus parameterization. Preprints, *Fifth Conf. on Numerical Weather Prediction,* Monterey, CA, Amer. Meteor. Soc., 277–284.

Pearce, R. P., and H. Riehl, 1969: Parameterization of convective heat and momentum transfer suggested by analysis of Caribbean data. *Proc. 1968 WMO/IUGG Symp. Numerical Weather Prediction,* Tokyo, Japan Meteor. Agency, 75–84.

Pedigo, C. B., and D. G. Vincent, 1990: Tropical precipitation rates during SOP 1, FGGE, estimated from heat and moisture budgets. *Mon. Wea. Rev.,* **118,** 542–557.

Perkey, D. J., 1976: A description and preliminary results from a fine-mesh model for forecasting quantitative precipitation. *Mon. Wea. Rev.,* **104,** 1513–1526.

Pruppacher, H. J., and J. Klett, 1978: *Microphysics of Clouds and Precipitation.* D. Reidel, 714 pp.

Puri, K., and M. J. Miller, 1990: Sensitivity of ECMWF analysis–forecasts of tropical cyclones to cumulus parameterization. *Mon. Wea. Rev.,* **118,** 1709–1741.

——, and P. Lonnberg, 1991: Use of high-resolution structure functions, and modified quality control in the analysis of tropical cyclones. *Mon. Wea. Rev.,* **119,** 1151–1167.

Raga, G., 1989: Characteristics of cumulus band clouds off the east coast of Hawaii. Ph.D. dissertation, University of Washington, 150 pp.

——, J. B. Jensen, and M. B. Baker, 1990: Characteristics of cumulus band clouds off the coast of Hawaii. *J. Atmos. Sci.,* **47,** 338–355.

Rand, H. A., and C. S. Bretherton, 1993: The relevance of mesoscale entrainment instability to the marine cloud-topped atmospheric boundary layer. *J. Atmos. Sci.,* **50,** 1152–1157.

Randall, D. A., 1980: Conditional instability of the first kind upside-down. *J. Atmos. Sci.,* **37,** 125–130.

——, 1989: Cloud parameterization for climate models: Status and prospects. *Atmos. Res.,* **23,** 345–362.

——, and G. J. Huffman, 1980: A stochastic model of cumulus clumping. *J. Atmos. Sci.,* **37,** 2068–2078.

——, and J. Wang, 1992: The moist available energy of a conditionally unstable atmosphere. *J. Atmos. Sci.,* **49,** 240–255.

——, Harshvardan, and D. A. Dazlich, 1991: Diurnal variability of the hydrological cycle in a general circulation model. *J. Atmos. Sci.*, **48**, 40–62.

——, Q. Shao, and C.-H. Moeng, 1992: A second-order bulk boundary-layer model. *J. Atmos. Sci.*, **49**, 1903–1923.

Rao, G. V., and T.-H. Hor, 1991: Observed momentum transport in monsoon convective cloud bands. *Mon. Wea. Rev.*, **119**, 1075–1087.

Rauber, R., K. Beard, and B. Andrews, 1991: A mechanism for giant raindrop formation in warm, shallow convective clouds. *J. Atmos. Sci.*, **48**, 1791–1797.

Raymond, D. J., 1983: Wave-CISK in mass-flux form. *J. Atmos. Sci.*, **40**, 2561–2572.

——, and A. M. Blyth, 1986: A stochastic mixing model for nonprecipitating cumulus clouds. *J. Atmos. Sci.*, **43**, 2708–2718.

——, and M. H. Wilkening, 1982: Flow and mixing in New Mexico mountain cumuli. *J. Atmos. Sci.*, **39**, 2211–2228.

——, and ——, 1985: Characteristics of mountain-induced thunderstorms and cumulus congestus clouds from budget measurements. *J. Atmos. Sci.*, **42**, 773–783.

——, and A. M. Blyth, 1986: A stochastic model for nonprecipitating cumulus clouds. *J. Atmos. Sci.*, **43**, 2708–2718.

——, and ——, 1989: Precipitation development in a New Mexico thunderstorm. *Quart. J. Roy. Meteor. Soc.*, **115**, 1397–1423.

——, R. Solomon, and A. M. Blyth, 1991: Mass fluxes in New Mexico mountain thunderstorms from radar and aircraft measurements. *Quart. J. Roy. Meteor. Soc.*, **117**, 587–621.

Redelsperger, J.-L., and J.-P. Lafore, 1988: A three-dimensional simulation of a tropical squall line: Convective organization and thermodynamic vertical transport. *J. Atmos. Sci.*, **45**, 1334–1356.

Reed, R. J., and E. E. Recker, 1971: Structure and properties of synoptic-scale waves disturbances in the equatorial western Pacific. *J. Atmos. Sci.*, **28**, 1117–1133.

Richman, M. B., and P. J. Lamb, 1985: Climatic pattern analysis of 3- and 7-day summer rainfall in the central United States: Some methodological considerations and regionalization. *J. Appl. Meteor.*, **24**, 1325–1343.

Riehl, H., 1979: *Climate and Weather in the Tropics.* Academic Press, 609 pp.

——, and J. S. Malkus, 1958: On the heat balance in the equatorial trough zone. *Geophysica,* **6**, 503–538.

——, and ——, 1961: Some aspects of hurricane Daisy, 1958. *Tellus,* **13**, 181–213.

——, T. Yeh, J. Malkus, and N. la Seur, 1951: The north-east trade of the Pacific Ocean. *Quart. J. Roy. Meteor. Soc.*, **77**, 598–626.

Rosenthal, S. L., 1978: Numerical simulation of tropical cyclone development with latent heat release by the resolvable scales I. Model description and preliminary results. *J. Atmos. Sci.*, **35**, 258–271.

Rotunno, R., J. B. Klemp, and M. L. Weisman, 1988: A theory for strong, long-lived squall lines. *J. Atmos. Sci.*, **45**, 463–485.

Roux, F., 1985: The retrieval of thermodynamic fields from multiple Doppler radar data using the equation of motion and the thermodynamic equation. *Mon. Wea. Rev.*, **113**, 2142–2157.

——, 1988: The West African squall line on 23 June during COPT 81: Kinematics and thermodynamics of the convective region. *J. Atmos. Sci.*, **45**, 406–426.

——, J. Testud, M. Payen, and B. Pinty, 1984: West African squall-line thermodynamic structure retrieved from dual-Doppler radar observations. *J. Atmos. Sci.*, **41**, 3104–3121.

Rutledge, S. A., and P. V. Hobbs, 1984: The mesoscale and microscale structure and organization of clouds and precipitation in midlatitude cyclones. XII: A diagnostic modeling study of precipitation development in narrow cold-frontal rainbands. *J. Atmos. Sci.*, **41**, 2949–2972.

——, and R. A. Houze, Jr., 1987: A diagnostic modeling study of the trailing stratiform region of a midlatitude squall line. *J. Atmos. Sci.*, **44**, 2640–2656.

Sanders, F., and L. F. Bosart, 1985: Mesoscale structure in the megalopolitan snowstorm of 11–12 February 1983. *J. Atmos. Sci.*, **42**, 1052–1061.

Sardeshmukh, P. D., and B. J. Hoskins, 1985: Vorticity balances in the tropics during the 1982–83 El Niño–Southern Oscillation event. *Quart. J. Roy. Meteor. Soc.*, **111**, 261–278.

Schneider, E. K., and R. S. Lindzen, 1976: A discussion of the parameterization of momentum exchange by cumulus convection. *J. Geophys. Res.*, **81**, 3158–3160.

Schubert, W. H., 1973: The interaction of a cumulus cloud ensemble with the large-scale environment. Ph.D. thesis, University of California, Los Angeles, 168 pp.

Seitter, K. L., and H.-L. Kuo, 1983: The dynamical structure of squall-line type thunderstorms. *J. Atmos. Sci.*, **40**, 2832–2854.

Shapiro, L. J., 1978: The vorticity budget of a composite African tropical wave disturbance. *Mon. Wea. Rev.*, **106**, 806–817.

——, and D. E. Stevens, 1980: Parameterization of convective effects on the momentum and vorticity budgets of synoptic-scale Atlantic tropical waves. *Mon. Wea. Rev.*, **108**, 1816–1826.

Shapiro, M. A., 1982: Mesoscale weather systems of the central United States. Cooperative Institute for Research in Environmental Studies (CIRES). National Oceanic and Atmospheric Administration (NOAA), Boulder, CO, 78 pp.

Shutts, G. J., and M. J. P. Cullen, 1987: Parcel stability and its relation to semigeostrophic theory. *J. Atmos. Sci.*, **44**, 1318–1330.

Simpson, J. S., 1971: On cumulus entrainment and one-dimensional models. *J. Atmos. Sci.*, **28**, 449–455.

——, 1972: Reply. *J. Atmos. Sci.*, **29**, 220–225.

——, 1983a: Cumulus clouds: Early aircraft observations and entrainment hypotheses. *Mesoscale Meteorology: Theories, Observations and Models,* Lilly and T. Gal-Chen, Eds., 355–373.

——, 1983b: Cumulus clouds: Interactions between laboratory experiments and observations as foundations for models. *Mesoscale Meteorology: Theories, Observations and Models,* Lilly and T. Gal-Chen, Eds., 399–412.

——, 1983c: Cumulus clouds: Numerical models, observations and entrainment. *Mesoscale Meteorology: Theories, Observations and Models,* Lilly and T. Gal-Chen, Eds., 413–445.

——, and V. Wiggert, 1969: Models of precipitating cumulus towers. *Mon. Wea. Rev.*, **97**, 471–489.

——, and ——, 1971: 1968 Florida cumulus seeding experiment: Numerical model results. *Mon. Wea. Rev.*, **99**, 87–118.

——, and G. van Helvoirt, 1980: GATE cloud–sub cloud layer interaction examined using a three dimensional cumulus model. *Beitr. Phys. Atmos.*, **53**, 106–134.

——, R. H. Simpson, D. A. Andrews, and M. A. Eaton, 1965: Experimental cumulus dynamics. *Rev. Geophys.*, **3**, 387–431.

——, G. Van Helvoirt, and M. McCumber, 1982: Three-dimensional simulations of cumulus congestus clouds on GATE day 261. *J. Atmos. Sci.*, **39**, 126–145.

Smith, R. B., and Y. L. Lin, 1982: The addition of heat to a stratified airstream with application to the dynamics of orographic rain. *Quart. J. Roy. Meteor. Soc.*, **108**, 353–378.

Smull, B. F., and R. A. Houze, Jr., 1985: A midlatitude squall line with a trailing region of stratiform rain: Radar and satellite observations. *Mon. Wea. Rev.*, **113**, 117–133.

——, and ——, 1987a: Dual-Doppler radar analysis of a midlatitude squall line with a trailing region of stratiform rain. *J. Atmos. Sci.*, **44**, 2128–2148.

——, and ——, 1987b: Rear inflow in squall lines with trailing stratiform precipitation. *Mon. Wea. Rev.*, **115**, 2869–2889.

Song, J.-L., and W. M. Frank, 1983: Relationship between deep convection and large-scale processes during GATE. *Mon. Wea. Rev.*, **111**, 2145–2160.

Soong, S.-T., and Y. Ogura, 1973: A comparison between axisymmetric and slabsymmetric cumulus cloud models. *J. Atmos. Sci.*, **30**, 879–893.

——, and ——, 1980: Response of tradewind cumuli to large-scale processes. *J. Atmos. Sci.*, **37**, 2035–2050.

——, and W.-K. Tao, 1980: Response of deep tropical cumulus clouds to mesoscale processes. *J. Atmos. Sci.*, **37**, 2016–2034.

——, and ——, 1984: A numerical study of the vertical transport of momentum in a tropical rainband. *J. Atmos. Sci.,* **41,** 1049–1061.

Squires, P., 1958a: The microstructure and colloidal stability of warm clouds. *Tellus,* **10,** 256–261.

——, 1958b: Penetrative downdraughts in cumuli. *Tellus,* **10,** 381–389.

——, 1958c: The spatial variation of liquid water and droplet concentration in cumuli. *Tellus,* **10,** 372–380.

——, and J. S. Turner, 1962: An entraining jet model for cumulonimbus updraughts. *Tellus,* **14,** 422–434.

Stark, T. E., 1976: Wave–CISK and cumulus parameterization. *J. Atmos. Sci.,* **33,** 2383–2391.

Stensrud, D. J., and J. M. Fritsch, 1991: Incorporating mesoscale convective outflows in mesoscale model initial conditions. Preprints, *Ninth Conf. on Numerical Weather Prediction,* Denver, Amer. Meteor. Soc., 798–801.

Stevens, D. E., 1979: Vorticity, momentum and divergence budgets of synoptic-scale wave disturbances in the tropical eastern Atlantic. *Mon. Wea. Rev.,* **107,** 535–550.

Stommel, H., 1947: Entrainment of air into a cumulus cloud. *J. Meteor.,* **4,** 91–94.

Suarez, M. J., A. Arakawa, and D. A. Randall, 1983: The parameterization of the planetary boundary layer in the UCLA general circulation model: Formulation and results. *Mon. Wea. Rev.,* **111,** 2224–2243.

Sui, C.-H., and M. Yanai, 1986: Cumulus ensemble effects on the large-scale vorticity and momentum fields of GATE. Part I: Observational evidence. *J. Atmos. Sci.,* **43,** 1618–1642. (See also Corrigendum, 1989: *J. Atmos. Sci.,* **46,** 1630.)

——, M.-D. Cheng, X. Wu, and M. Yanai, 1989: Cumulus ensemble effects on the large-scale vorticity and momentum fields of GATE. Part II: Parameterization. *J. Atmos. Sci.,* **46,** 1609–1629.

Takahashi, T., 1977: A study of Hawaiian warm rain showers based on aircraft observations. *J. Atmos. Sci.,* **34,** 1773–1790.

——, 1981: Warm rain study in Hawaii—Rain initiation. *J. Atmos. Sci.,* **38,** 347–369.

——, K. Yoheyama, and Y. Tsubota, 1989: Rain duration in Hawaiian trade-wind rainbands—Aircraft observation. *J. Atmos. Sci.,* **46,** 937–955.

Tao, W.-K., and S.-T. Soong, 1986: A study of the response of deep tropical clouds to mesoscale processes: Three-dimensional numerical experiments. *J. Atmos. Sci.,* **43,** 2653–2676.

——, and J. Simpson, 1989: Modeling study of a tropical squall-type convective line. *J. Atmos. Sci.,* **46,** 177–202.

——, ——, and S.-T. Soong, 1987: Statistical properties of a cloud ensemble: A numerical study. *J. Atmos. Sci.,* **44,** 3175–3187.

——, ——, and S.-T. Soong, 1991: Numerical simulation of a subtropical squall line over the Taiwan Strait. *Mon. Wea. Rev.,* **119,** 2699–2723.

——, ——, C.-H. Sui, S. Lang, J. Scala, B. Ferrier, M.-D. Chou, and K. Pickering, 1993: Heating, moisture and water budgets of tropical and midlatitude squall lines: Comparisons and sensitivity to longwave radiation. *J. Atmos. Sci.,* **50,** 673–690.

Telford, J. W., 1975: Turbulence, entrainment, and mixing in cloud dynamics. *Pure Appl. Geophys.,* **113,** 1067–1084.

——, and P. B. Wagner, 1974: The measurement of horizontal air motion near clouds from aircraft. *J. Atmos. Sci.,* **31,** 2066–2080.

Tennekes, H. 1973: A model for the dynamics of the inversion above a convective boundary layer. *J. Atmos. Sci.,* **30,** 558–567.

Testud, J., F. Roux, M. Chong, and G. Scialom, 1984: Comparison of the dynamical structure of two tropical squall lines observed during COPT-81 experiment. *Proc. 22d Conf. Radar Meteor.,* Zurich, Amer. Meteor. Soc., 37–42.

Thompson, R. M., Jr., S. W. Payne, E. E. Recker, and R. J. Reed, 1979: Structure and properties of synoptic-scale wave disturbances in the intertropical convergence zone of the eastern Atlantic. *J. Atmos. Sci.,* **36,** 53–72.

Thorpe, A. J., and K. A. Emanuel, 1985: Frontogenesis in the presence of small stability to slantwise convection. *J. Atmos. Sci.,* **42,** 1809–1824.

——, and R. Rotunno, 1989: Nonlinear aspects of symmetric instability. *J. Atmos. Sci.,* **46,** 1285–1299.

——, and S. A. Clough, 1991: Mesoscale dynamics of cold fronts-structure described by dropsoundings in FRONTS 87. *Quart. J. Roy. Meteor. Soc.,* **117,** 903–941.

——, M. J. Miller, and M. W. Moncrieff, 1982: Two-dimensional convection in non-constant shear: A model of midlatitude squall lines. *Quart. J. Roy. Meteor. Soc.,* **108,** 739–762.

——, B. J. Hoskins, and V. Innocentini, 1989: The parcel method in a baroclinic atmosphere. *J. Atmos. Sci.,* **40,** 1274–1284.

Tiedtke, M., 1989: A comprehensive mass flux scheme for cumulus parameterization in large-scale models. *Mon. Wea. Rev.,* **117,** 1779–1800.

——, W. A. Heckley, and J. Slingo, 1988: Tropical forecasting at the ECMWF: The influence of physical parameterization on the mean structure of forecasts and analyses. *Quart. J. Roy. Meteor. Soc.,* **114,** 639–664.

Tollerud, E. I., and S. K. Esbensen, 1983: An observational study of the upper-tropospheric vorticity fields in GATE cloud clusters. *Mon. Wea. Rev.,* **111,** 2161–2175.

Trier, S. B., D. B. Parsons, and J. H. E. Clark, 1991: Environment and evolution of a cold-frontal mesoscale convective system. *Mon. Wea. Rev.,* **119,** 2429–2455.

Tripoli, G. J., and W. R. Cotton, 1982: The Colorado State University three-dimensional cloud-mesoscale model—1982. Part I: General theoretical framework and sensitivity experiments. *J. Rech. Atmos.,* **16,** 273–304.

Vonnegut, B., 1987: Importance of including time in the specification of ice nucleus concentrations. *J. Climate Appl. Meteor.,* **26,** 322.

Wagner, P., and J. Telford, 1976: The measurement of air motion in and near clouds. Preprints, *Int. Conf. on Cloud Physics,* Boulder, CO, Amer. Meteor. Soc., 669–672.

Wang, T.-C. C., Y.-J. Lin, R. W. Pastken, and H. Shen, 1990: Characteristics of a subtropical squall line determined from TAMEX dual-Doppler data. Part I: Kinematic structure. *J. Atmos. Sci.,* **47,** 2357–2381.

Warner, C., 1981: Photogrammetry from aircraft side camera movies: Winter MONEX. *J. Appl. Meteor.,* **20,** 1516–1526.

Warner, J., 1955: The water content of cumuliform cloud. *Tellus,* **7,** 449–457.

——, 1969: The microstructure of cumulus cloud. Part I: General features of the droplet spectrum. *J. Atmos. Sci.,* **26,** 1049–1059.

——, 1970a: The microstructure of cumulus cloud. Part III: The nature of the updraft. *J. Atmos. Sci.,* **27,** 682–688.

——, 1970b: On steady state one-dimensional models of cumulus convection. *J. Atmos. Sci.,* **27,** 1035–1040.

——, 1971: Comments "On cumulus entrainment and one-dimensional models." *J. Atmos. Sci.,* **28,** 218–220.

——, 1973a: The microstructure of cumulus clouds: Part IV. The effect on the droplet spectrum of mixing between cloud and environment. *J. Atmos. Sci.,* **30,** 256–261.

——, 1973b: The microstructure of cumulus cloud. Part V. Changes in droplet size distribution with cloud age. *J. Atmos. Sci.,* **30,** 1724–1726.

——, 1977: Time variation of updraft and water content in small cumulus. *J. Atmos. Sci.,* **34,** 1306–1312.

Warner, T. T., and N. L. Seaman, 1990: A real-time mesoscale numerical weather prediction system used for research, teaching, and public service at the Pennsylvania State University. *Bull. Amer. Meteor. Soc.,* **71,** 792–805.

Warren, S., C. Hahn, J. London, R. Chervin, and R. Jennes, 1988: Global distribution of total cloud cover and cloud type amounts over the ocean. DOE/ER-0406, NCAR Tech. Notes NCAR/TN 317 + STR.

Webster, P. J., and G. L. Stephens, 1980: Tropical upper-tropospheric extended clouds: Inferences from Winter MONEX. *J. Atmos. Sci.,* **37,** 1521–1541.

Weisman, M. L., 1992: The role of convective generated rear-inflow jets in the evolution of long-lived mesoconvective systems. *J. Atmos. Sci.,* **49,** 1826–1847.

——, and J. B. Klemp, 1982: The dependence of numerically simulated convective storms on vertical wind shear and buoyancy. *Mon. Wea. Rev.,* **110,** 504–520.

——, ——, and R. Rotunno, 1988: Structure and evolution of numerically simulated squall lines. *J. Atmos. Sci.,* **45,** 1990–2013.

Wu, W. S., 1991: Helical buoyant convection. Ph.D. thesis, University of Oklahoma, 161 pp.

Wu, X., and M. Yanai, 1991a: Effects of vertical wind shear on cumulus parameterization. Preprints, *19th Conf. on Hurricanes and Tropical Meteorology,* Miami, Amer. Meteor. Soc., 318–323.

——, and ——, 1991b: Cumulus–environment interactions in midlatitude mesoscale convective systems. Preprints, *Int. Conf. on Mesoscale Meteorology and TAMEX,* Taipei, Taiwan, Meteor. Soc. Republic of China and Amer. Meteor. Soc., 471–480.

Xu, K.-M., 1991: The coupling of cumulus convection with large-scale processes. Ph.D. dissertation, University of California, Los Angeles, 250 pp.

——, and K. A. Emanuel, 1989: Is the tropical atmosphere conditionally unstable? *Mon. Wea. Rev.,* **117,** 1471–1479.

——, and S. K. Krueger, 1991: Evaluation of cloudiness parameterizations using a cumulus ensemble model. *Mon. Wea. Rev.,* **119,** 342–367.

——, and A. Arakawa, 1992: Semiprognostic tests of the Arakawa–Schubert cumulus parameterization using simulated data. *J. Atmos. Sci.,* **49,** 2421–2436.

——, ——, and S. K. Krueger, 1992: The macroscopic behavior of cumulus ensembles simulated by a cumulus ensemble model. *J. Atmos. Sci.,* **49,** 2402–2420.

Yamasaki, M., 1975: A numerical experiment of the interaction between cumulus convection and larger-scale motion. *Pap. Meteor. Geophys.,* **26,** 63–91.

Yanai, M., 1961: A detailed analysis of typhoon formation. *J. Meteor. Soc. Japan,* **39,** 187–214.

——, 1964: Formation of tropical cyclones. *Rev. Geophys.,* **2,** 367–414.

——, S. Esbensen, and J. Chu, 1973: Determination of bulk properties of tropical cloud clusters from large-scale heat and moisture budgets. *J. Atmos. Sci.,* **30,** 611–627.

——, J.-H. Chu, T. E. Stark, and T. Nitta, 1976: Response of deep and shallow tropical maritime cumuli to large-scale processes. *J. Atmos. Sci.,* **33,** 976–991.

——, C.-H. Sui, and J.-H. Chu, 1982: Effects of cumulus convection on the vorticity field in the tropics. Part II: Interpretation. *J. Meteor. Soc. Japan,* **60,** 411–424.

——, C. Li, and Z. Song, 1992: Seasonal heating of the Tibetan Plateau and its effects on the evolution of the Asian summer monsoon. *J. Meteor. Soc. Japan,* **70,** 189–221.

Yao, M.-S., and A. D. DelGenio, 1989: Effects of cumulus entrainment and multiple cloud types on a January global climate model simulation. *J. Climate,* **2,** 850–863.

Yau, M. K., and R. Michaud, 1982: Numerical simulation of a cumulus ensemble in three dimensions. *J. Atmos. Sci.,* **39,** 1062–1079.

Zipser, E. J., 1969: The role of organized unsaturated convective downdrafts in the structure and rapid decay of an equatorial disturbance. *J. Appl. Meteor.,* **8,** 799–814.

——, 1977: Mesoscale and convective-scale downdrafts as distinct components of squall-line structure. *Mon. Wea. Rev.,* **105,** 1568–1589.

——, and M. A. LeMone, 1980: Cumulonimbus vertical velocity events in GATE. Part II: Synthesis and model core structure. *J. Atmos. Sci.,* **37,** 2458–2469.

——, R. J. Meitin, and M. A. LeMone, 1981: Mesoscale motion fields associated with a slowly-moving GATE convective band. *J. Atmos. Sci.,* **38,** 1725–1750.

Zhang, D.-L., 1989: The effect of parameterized ice microphysics on the simulation of vortex circulation with a mesoscale hydrostatic model. *Tellus,* **41A,** 132–147.

——, and J. M. Fritsch, 1986: Numerical simulation of the meso-β scale structure and evolution of the 1977 Johnstown flood. Part I: Model description and verification. *J. Atmos. Sci.,* **43,** 1913–1943.

——, and ——, 1987: Numerical simulation of the meso-β scale structure and evolution of the 1977 Johnstown flood. Part II: Inertially stable warm-core vortex and the mesoscale convective complex. *J. Atmos. Sci.,* **44,** 2593–2612.

——, and ——, 1988a: Numerical sensitivity experiments on the structure, evolution and dynamics of two mesoscale convective systems. *J. Atmos. Sci.,* **45,** 261–293.

——, and ——, 1988b: A numerical investigation of a convectively-generated, inertially-stable, extra-tropical, warm-core mesovortex over land. Part I: Structure and evolution. *Mon. Wea. Rev.,* **116,** 2660–2687.

——, and K. Gao, 1989: Numerical simulation of an intense sqall line during 10–11 June 1985 PRE-STORM. Part II: Rear inflow, surface pressure perturbations, and stratiform precipitation. *Mon. Wea. Rev.,* **117,** 2067–2094.

——, and H. R. Cho, 1991: Parameterization of the vertical transport of momentum by cumulus clouds. Part I: Theory. *J. Atmos. Sci.,* **48,** 1483–1492.

——, and ——, 1992a: The development of negative moist potential vorticity in the stratiform region of a simulated squall line. *Mon. Wea. Rev.,* **120,** 1322–1341.

——, and ——, 1992b: Moist symmetric instability in the stratiform region of a simulated squall line. Preprints, *Fifth Conf. on Mesoscale Processes,* Atlanta, GA, Amer. Meteor. Soc., 276–281.

——, K. Gao, and D. B. Parsons, 1989: Numerical simulation of an intense squall line during 10–11 June 1985 PRE-STORM. Part I. Model verification. *Mon. Wea. Rev.,* **117,** 960–994.